Heroes in Space

To the men and women whose courage
enabled the human race to leave
planet Earth

Heroes in Space

From Gagarin to Challenger

Peter Bond

0177877

Basil Blackwell

Copyright © Peter Bond 1987

First published 1987
Reprinted 1988, 1989

Basil Blackwell Ltd
108 Cowley Road, Oxford OX4 1JF, UK

Basil Blackwell Inc.
432 Park Avenue South, Suite 1503
New York, NY 10016, USA

British Library Cataloguing in Publication Data

Bond, Peter R.
Heroes in space: from Gagarin to Challenger.
1. Manned space flight – History
I. Title
629.45′.009   TL873
ISBN 0–631–15349–7

Library of Congress Cataloging in Publication Data

Bond, Peter R., 1984–
Heroes in space.
Includes index.
1. Manned space flight – History.   I. Title.
TL373.B63   1987   .629.45′009   86–31004
ISBN 0–631–15349–7

Typeset in 10 on 12pt Plantin
by Columns of Reading
Printed in the USA

Contents

Preface

At the time of writing, the American space programme is in turmoil, with no more manned flights planned before 1988 at the earliest. While the Soviets press ahead with their plans for a permanently manned space station, NASA is beset by internal and external criticism. The astronauts are openly sniping at a management which allowed them to ride a potentially lethal spacecraft without fully informing them of the risks involved. Some are voting with their feet and leaving for more secure employment. The military is demanding greater funding and more control over the launch schedule of the remaining Shuttles. Lower down the list of priorities, the scientists are unhappily watching their multi-million-dollar satellites and experiments deteriorate in storage while they await their turn on the revised launch schedule. The President has approved the construction of a new orbiter but no-one seems sure where the funds will come from.

When I began writing *Heroes in Space* the atmosphere could not have been more different. The 25th anniversary of manned spaceflight was approaching; the 200th human space voyager was soon to leave this planet to live and work in the hostile environment of outer space; the Shuttle was scheduled for a record 15 flights during 1986; and space was to be opened to a new generation of astronauts, including a schoolteacher, a journalist, and a Briton. To mark these watersheds, it was my intention to write a history of manned spaceflight as a tribute to the courageous pioneers who volunteered to venture beyond Earth's atmosphere for the first time. The tragic events of 28 January 1986 simply reinforced my commitment to focus attention on the human aspects of space travel rather than the technology.

On the whole, the men and women who have risked their lives to explore this new frontier have been masters of the understatement, continually playing down the dangers of their occupation and reluctant to reveal their true inner feelings. Most of them are more at home piloting the latest flying machines or unscrambling complex engineering and scientific problems than playing the uncomfortable role of superhero. In this book I have attempted to show something of this secret world, the elation and frustration, the fear and the determination which have accompanied the incredible achievements of the past quarter century. Above all, however, the book is a tribute to the courage of the men and women who began the greatest exploration of all time. Most of us will never experience the rigours and delights of space travel, but perhaps this account will serve as a poor substitute for all gravity-bound spectators on spaceship Earth.

I would like to extend my appreciation to my wife, Edna, without whose encouragement this book would never have been written, and to Kim Pickin whose belief enabled it to get off the ground. My thanks also extend to the staff at Basil Blackwell, particularly Julia Mosse and Gillian Bromley; to Ralph Gibson at Novosti Press Agency; and to the staff at Johnson Space Center.

Finally, anyone wishing to learn more about the latest developments in space exploration would do well to join the British Interplanetary Society. The address is 27/29 South Lambeth Road, London SW8.

Peter Bond

Acknowledgements

The publishers wish to thank the following for permission to use their photographs: NASA for plates 1, 2, 7, 8, 11, 12, 13, 14, 17, 19, 20, 21, 22 and 23; Novosti Press Agency for plates 3, 4, 5, 9, 10, 16, 18, 24, 25 and 26; Tim Furniss and NASA for plates 6 and 15.

Figures 1 and 13 are based on illustrations from the HMSO publication *Exploring Man On The Moon*; figure 2 is based on a diagram from *The Observer's Book of Manned Spaceflight* by R. Turnill, 1972, by kind permission of Penguin Books Ltd; figures 3 and 9 are based on illustrations from *The Illustrated Encyclopedia of Space Technology*, 1981, by kind permission of Salamander Books Ltd; figure 6 is copyright © National Geographic Society; figure 8 is based on a painting by Alexei Leonov in *Space Age*, by R. Turnill, 1980, courtesy of Theo Pirard; figures 10, 17, 18, 19, 20 and 32 are based on diagrams from the *Sunday Times*, copyright © Times Newspapers Ltd, London; figure 11 is reproduced by courtesy of US Information Service; figures 12, 38 and 39 are reproduced by courtesy of NASA; figure 22 is based on a painting by Davis Meltzer, copyright © National Geographic Society; figure 24 is from *The Observer's Spaceflight Directory* by R. Turnill, published by Frederick Warne, 1978; figure 26 is based on a map by Raymond Turvey from *The Planets* by Peter Francis, 1981, copyright © Peter Francis, by kind permission of Penguin Books Ltd; figures 28, 29, 34, 35, 37 and 42 are copyright © Novosti Press Agency; figure 30 is based on a diagram from *Space History* by Tony Osman, 1983, by kind permission of Michael Joseph Ltd and A. P. Watt Ltd; figure 31 is based on an illustration from the *Telegraph Magazine*, by kind permission of The Daily Telegraph Ltd; figure 33 is based on a diagram from *Space Age* by R. Turnill, published by Frederick Warne, 1980; figure 36 is based on a diagram from *Spaceflight*, reproduced by courtesy of The British

Interplanetary Society; figure 40 is based on an illustration from *New Scientist*, copyright © IPC Magazines Ltd; figure 41 is based on an illustration in *The Times*, copyright © Times Newspapers Ltd, London.

1

Into the Unknown
The Challenge of Living in Space

In retrospect, it seems hard to believe that many people once doubted mankind's ability to adapt and live in space, so great have been the advances during the past 25 years. As the 1960s dawned, there was no doubt that a man could, and would, soon leave Earth for his first momentous step into the cosmos: Soviet satellites had already reached and photographed the far side of the moon and American robot craft had begun to discover the secrets of interplanetary space. But one lingering doubt remained; the one unknown which could only be solved by the lone heroism of the true pioneer. Could a human being, adapted so perfectly for life on a planet, survive in the totally alien, hostile environment of space, or would the barriers be so formidable that humans would be forever confined to their tiny outpost in the vast universe? Many experts feared the answer would be negative as the volunteers in the USA and the Soviet Union went through the final stages of their exhaustive preparation.

The dream of space travel had haunted scientific minds since the end of the last century, though to the general public it was simply the stuff of science fiction, rather than fact. Books such as Jules Verne's *From the Earth to the Moon* and H. G. Wells's *The First Man in the Moon* were seen as flights of literary imagination rather than predictions of future technological advances. In some ways this was certainly true. Jules Verne, for example, suggested a huge 'space gun' as the means of boosting his star voyager, or 'astronaut', towards the moon. The occupant of such a spaceship would in reality have been crushed to death by the tremendous G (gravity) forces if he was suddenly accelerated to the velocity necessary to escape the Earth's gravity.

The old ideas and misconceptions began to disappear around the turn of the century, largely due to the vision of one man, the Soviet scientist Konstantin Tsiolkovsky. Overcoming the handicap of deafness from the age of nine, he became fascinated with the idea of rockets being used for interplanetary travel. He realized the possibilities raised by Newton's law that for every action, there is an equal and opposite reaction. His first article on rocketry appeared in 1903, but was only partly published because the magazine's second issue was seized as politically subversive! The second part had to wait until 1911. Tsiolkovsky's breakthrough was to realize the superiority of liquid over solid fuels, such as gunpowder, which had propelled rockets since the Chinese invented fireworks in the first century AD. One of the fuel combinations he proposed became the basis of modern rocketry: liquid oxygen and liquid hydrogen. But his imagination did not stop there: he went on to envisage earth satellites and space stations, and voyages to other planets, all boosted on their way by multi-stage rockets. Such boosters would be necessary to overcome the weight problem: in order to reach escape velocity of 25,000 mph (40,000 kmh) and overcome the Earth's gravitational pull, very powerful and very greedy rocket motors would be required. This meant that between 90 and 95 per cent of the weight of a typical rocket would be taken up by the fuel, compared with perhaps 2 or 3 per cent for the weight of the spacecraft itself. One way partially to overcome these weight problems would be to build a rocket with several stages, each with its own motors and fuel. As each stage used up its fuel, it would be jettisoned and fall back to Earth, burning up in the atmosphere. The remainder of the rocket would continue on its way, relieved of the dead weight and with full fuel tanks.

Tsiolkovsky's work was largely ignored in the turbulent conditions which tore the Soviet Union apart during the First World War and the Communist Revolution. It was left to an American pioneer, pursuing his own independent line of inquiry and experiment, to take the first practical step towards the stars. Robert Goddard had become a professor of physics at the age of 31, obsessed with the search for a successful liquid-fuelled rocket. Having exhausted his own funds, he persuaded the Smithsonian Institution to give him a grant of $5,000. Although the popular press ridiculed his efforts, he continued his tests undaunted. The design of stabilizing and guidance systems took many years to perfect. When the historic moment came, in a field near Auburn, Massachusetts, on 16 March 1926, it was witnessed by only four people: Goddard, his wife and two assistants. The rocket soared to a height of 184 feet and reached a maximum speed of 60 mph (97 kmh). Further successes followed, bringing in more funds from the Guggenheim Foundation, but the military remained uninterested in the possibilities of long-range rocket research, unlike scientists in Nazi Germany.

Hitler's most devastating secret weapon, and the forerunner of all modern rockets, was a tribute to Goddard's pioneering work, and to the innovative minds of two Germans: Herman Oberth and Wernher von Braun. The theoretical background provided by Oberth, including designs for a two-stage unmanned rocket, inspired widespread interest in the new technology, but it was the backing of the Army, deprived of heavy artillery as well as fighter aircraft, which proved the decisive factor. By 1937, the brilliant young designer von Braun had been appointed technical director of a new top secret Army Experimental Station on the Baltic island of Peenemunde. Progress was slow, partly due to lack of interest from the Fuehrer, but in October 1942 came triumph. The V2, the most powerful rocket in the world, soared to an altitude of more than 50 miles (80 km), heralding the beginning of one of the most frightening periods of aerial bombardment ever endured. Fortunately, the V2's value as an offensive weapon was not immediately recognized. Of around 3,000 V2s built, just over half were fired in anger.

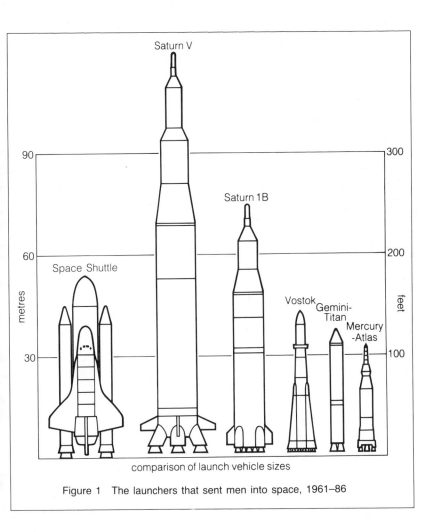

comparison of launch vehicle sizes

Figure 1   The launchers that sent men into space, 1961–86

In the hectic final months of hostilities, the German scientists were faced with a crucial decision: whether to await the arrival of advancing Soviet forces or head south towards the American army. Von Braun persuaded most of his colleagues to offer their services to the Americans, much to the delight of the American authorities. A hundred V2s were secretly transported back to the USA from the Nordhausen underground assembly plant before the Soviets arrived, a hoard which formed the backbone of post-war rocket development at White Sands, New Mexico. Von Braun also brought with him the blueprints for more advanced rockets.

The Soviets, however, had not been idle. Valuable documents missed by the Americans and some 3,500 Peenemunde personnel were sent back to complement Soviet research and development. While the American Army and Navy competed to produce a booster of sufficient power to carry a nuclear bomb or lift an artificial satellite into orbit, a Soviet team led by an anonymous 'Chief Designer' was preparing to steal the American thunder. Years later, the Soviet authorities identified the man who was the driving force behind their success as Sergei Korolev. He had been inspired by the visions of Tsiolkovsky, and, surviving a period of imprisonment under the Stalin regime, had developed a talented team of designers and engineers. As early as 1954, Korolev was informing the Soviet Union's Council of Ministers that an artificial Earth satellite could be launched in the next few years. He was as good as his word.

The astounding Soviet advances took place largely beyond the gaze of Western analysts. The announcement from Moscow that a satellite, named Sputnik, had been successfully launched on 4 October 1957 hit the West, and in particular America, like a bombshell. As the world watched American rockets explode with embarrassing regularity, further Soviet satellites followed: Sputnik 2 carried a dog named Laika, the first living creature to enter space and experience weightlessness. Laika survived for seven days before her oxygen supply ran out, and the Soviet intention was clear. A man from the Soviet Union would soon follow, once a way could be found of returning him safely to Earth. The USA finally succeeded in sending a satellite into orbit in January 1958, marking the start of the first Space Race, the race to get a man into orbit. It was a matter of international prestige and national security; a way of demonstrating to the world the advantages of a particular political ideology.

The search began early in 1958 for the men with the right qualifications and attitude who would risk a potentially suicidal debacle for the chance of writing a new page in history while carrying out their patriotic duty. The advantages in selecting experienced pilots, men who risked their lives every day in the most powerful jet fighters or the latest research aircraft and who relied on their own wits and speed of reaction to survive an emergency, soon became apparent to both sides. The fact that there would be hardly any actual 'piloting' to do – virtually every manoeuvre would be automatically controlled – made little difference to anyone in the selection process except the guinea pigs themselves. The suggestion that the first flights would simply be 'a man in a can', equally suitable for a chimpanzee or a human, brought sharp reactions from the men whose dignity and self-respect were at stake. Modifications were hurriedly made, providing a small window and manual controls, but the fact remained that no one knew exactly what the first man would have to endure or how the human body would react to the unfamiliar environment.

The approach in the two countries could hardly have been more different, though the end result was the same. To all the ballyhoo of an American press conference, the seven men chosen to break through the final frontier – the Select Seven – were presented to an adoring public in Washington, DC on 9 April 1959. They had beaten more than 500 other candidates and endured weeks of physical and psychological tests before attaining the prestigious title of astronaut. There were three navy men, three air force men and one marine. Apart from the possibilities of furthering their careers, the seven gained a substantial financial bonus: on top of their modest service salaries, each man received an equal share of $500,000 from *Life* magazine for his exclusive life story. The next two years would prove a frustrating wait for these men as they endured rigorous training interspersed with speech-making and publicity campaigns while their hardware struggled to a state of readiness. Meanwhile, their counterparts in the Soviet Union were undergoing a similar series of tortures, but in the secretive atmosphere peculiar to that country. Twenty men were finally selected towards the end of 1959 to represent the Soviet Union in its glorious quest, but there was no press conference or public celebration. The members of the group remained anonymous: six of them began immediate training in competition for the first mission, but their names were only released to the world when they actually achieved the distinction of a flight in space. Any cosmonaut who successfully accomplished his mission was assured of a place in the national hall of fame and an ecstatic hero's welcome, but eight of the 20 original cosmonauts were never to fly in space and their names have only recently become known in the West.

The selection procedure for these potential heroes had been rigorous in the extreme. A list of some criteria for an American Mercury astronaut shows the type of man envisaged by the space agency authorities as most suitable for the forthcoming challenge:

1 Less than 40 years of age.
2 Less than 5 ft 11 in in height.

3 Excellent physical condition.
4 Bachelor's degree (or equivalent).
5 Graduate of a test pilot school.
6 1,500 hours of flight time.
7 Qualified jet pilot.
8 Citizen of the USA.

These qualifications were later relaxed a little as requirements changed: for example, larger spacecraft meant that the maximum height was raised to 6 feet. Later still, scientists were encouraged to join both the American and Soviet space programmes and the pilot experience was relaxed. Physical fitness also became less important as these science specialists entered the lists, but the medical and psychological examination remained the most complete and exacting available

The extent of these examinations can be judged from the ordeal of the Mercury astronauts. The Lovelace Foundation in Albuquerque was the site for a week-long series of medical tests. Every aspect of the human body was focused upon after a detailed scrutiny of the applicant's past medical record. Doctors studied blood and tissue samples, ears, eyes, nose, throat, heart and circulatory system, digestive system, teeth, lungs, nerves and muscles. There seemed no end to the probing of the curious medicos. Candidates were asked to supply samples of urine, faeces and sperm, cold water was poured into the inner ear, electric shocks were sent into nerves and muscles causing uncontrolled spasms. The list seemed endless. The endurance test was followed by a radiation count measured by experts from Los Alamos Laboratory. No sooner was this over than they were whisked to Wright-Patterson Air Force Base for psychological and stress testing, as well as more fitness tests. Candidates were asked to step on and off a 19.5 inch (51 cm) high platform for five minutes in time with a metronome; motivation was tested by placing both feet in ice water for seven minutes; stress was generated by placing the men in a claustrophobic, blackened, soundproofed 'sensory deprivation chamber' for three hours; seated in front of a special simulator, they were told to press buttons and pull switches in response to an ever-increasing array of lights on the console; psychological inquiries followed more traditional lines – the men were asked for their interpretation of ink blot patterns and asked to write 20 answers to the question, 'Who am I?' Altogether, there were some 40 tests designed to determine the identity of the most adaptable volunteers.

Before men could be sent into space, they had to become familiar with every aspect of their mission. Astronaut training involved continual practice on a variety of simulators designed to duplicate as closely as possible the actual conditions that would be encountered. The extreme stresses or G forces caused by sudden acceleration or deceleration during ascent and re-entry were relatively straightforward to simulate. Most familiar is the accelerator, or centrifuge, a giant arm with a gondola at one end in which the spacemen sit as they are whirled around at ever-increasing velocities. The larger versions contained a capability to vary cabin pressure, temperature and humidity as well as the various types of G force likely to be experienced. There was much to worry about during such times: experiments showed that movement of the body was impeded as the men were pressed back on to their seats, blood circulation slowed down as the heart struggled to pump blood to the brain and blurred vision could interfere with gauge readings. The best answer was for the astronaut to lie on his back with his legs bent above him, like a chair tilted on its back. This enabled blood to be fed to the brain at up to 10G. Mercury astronauts were provided with individually moulded couches, but these were found to be too restricting and were dropped in later flights.

Weightlessness was one of the big unknowns. Humans have evolved over millions of years to walk upright in a one gravity environment, developing bones and muscles in such a way that gravity could be overcome. How would humans adapt to a continual floating sensation in an environment that no longer had an obvious up or down? Even more worrying were the unknown effects of re-entry stresses and a return to normal gravity on a body deprived of such forces for long periods. The only way to simulate zero gravity on Earth was to take a roller coast ride in an aircraft, but such effects could only be maintained at the top of the arc for about 60 seconds. During such brief episodes, spacemen had to practise such activities as feeding themselves with soups and paste squeezed from plastic tubes. Reactions varied considerably: some liked the strange sensation, others suffered a type of motion sickness, leading to the aircraft's nickname, 'The Vomit Comet'. For prolonged simulations of weightlessness, the only alternative was to enclose the men in their pressure suits and dangle them in a water tank. This underwater training was first introduced by the Americans in 1966 as they practised complex tasks to be performed outside their spacecraft. It is now an essential part of training in both America and the Soviet Union.

A number of side effects were discovered as spacemen were asked to survive for ever-longer periods of weightlessness. Space sickness was first reported by Soviet cosmonauts but not experienced by American astronauts in the early flights. It is now generally accepted that the Americans, tightly strapped to their couches within their smaller cabins, were less likely to float freely and thus less likely to be affected. The sensation was similar to seasickness on Earth, resulting in nausea and possibly vomiting, a potentially serious hazard should a man vomit in his pressure helmet. Other less obvious but potentially disastrous physiological effects were noted. Since the heart and other muscles had less work to do in space, they became lazy and weaker. Blood and other body fluids tended to move towards the chest and head, causing a facial puffiness and slit-eyed appearance. More significantly, the volume of blood in the body decreased and this, coupled with a less active heart, led to pooling of blood in the legs and partial blood starvation in the brain on returning to Earth. The answer was obviously exercise, but that was not easily done in a cramped spacecraft. Bungee cords were provided for early astronauts, and bicycle ergometers were carried in larger, modern craft. The loss of blood and body fluids mostly accounted for the decrease in body weight experienced by spacemen – Frank Borman, for example, lost more than 6 per cent of his body weight during the 14-day Gemini 7 mission – but there was also a loss of calcium and other minerals from the bones. The problem seemed to be similar to that facing people who are bedridden for long periods, and, once again, the only answer seemed to be more exercise combined with a modified diet.

Eating in space caused the experts many anxious moments, but turned out to be less difficult to solve than expected. Fears that food would be hard to swallow and would stick in the throat proved to be unfounded. The Mercury astronauts had to face the unappetizing prospect of cold food in bite-sized cubes, aluminium tubes or freeze-dried. The cubes left crumbs which floated into any available crevices, the freeze-dried food had to be rehydrated then squeezed from the tubes, and the aluminium tubes weighed more than the food they contained. On Gemini missions, the cubes were coated with a gelatin to reduce crumbling, and plastic containers replaced the aluminium. Menu selections were also enlarged, but it was not until the Apollo missions that hot water was provided to enable consumption of hot drinks as well as hot meals. The astronauts were eventually able to choose their own selections as long as they totalled 2,800 calories per day and balanced properly. Another concession was a spoon for eating; the moisture content in the food was found to be sufficient to adhere the food to the spoon and prevent its floating away.

Breathing was always going to be an obstacle to the conquest of the vacuum of space. The spacecraft had to carry their own atmosphere with them. The question concerning the type and pressure of this atmosphere was solved in different ways in the two competing countries. For the USA, NASA favoured a pure oxygen atmosphere in the cabin at 5 psi (lb/square inch), about one-third of sea level air pressure. This was the result of a need to reduce spacecraft weight while ensuring sufficient oxygen was entering the body. Clearly one of the main dangers resulting from this approach was the potential fire hazard, as was tragically witnessed by the fatal Apollo 1 launch pad fire in 1967. The Soviets favoured a mixture of oxygen and nitrogen at normal sea level pressure, since their more powerful booster rockets were able to overcome the weight constraints involved in carrying the gases aloft. The main problem which might arise from this alternative was the possibility of cosmonauts suffering from the bends in the event of a cabin leak or sudden decompression, just like deep sea divers. So confident were the Soviets in their craft, however, that many of the cosmonauts did not wear pressure suits, even during re-entry, a policy which led to the deaths of three cosmonauts in 1971. Removal of carbon dioxide from the atmosphere was carried out by renewable lithium hydroxide canisters, but water supply was a problem. Only with the development of fuel cells was a continuous supply of fresh water ensured as a by-product of electricity production.

While man was content to remain within the shelter of a spacecraft, pressure suits played a secondary role. They were more of a precaution than a necessity. But if men were to venture outside the craft to work in the vacuum of space or on the airless surface of the moon, a suitable protective suit and helmet were essential. The first spacesuits were based upon the suits worn by high altitude fliers. In a vacuum, the suits would be pressurized to prevent the spaceman's blood from boiling and to keep him from getting the bends. Oxygen would circulate through the suit to the helmet to enable the astronaut to breathe while keeping his body cool. Apollo astronauts wore a pair of longjohns with a network of tiny tubes which carried cool water to and from the backpack. Early spacewalkers had to rely on a long tether to provide their oxygen and communications, as well as to prevent them floating away to a lonely death. This was clearly unsuitable for lunar extra-vehicular activity (EVA), so backpacks were developed which eventually enabled the men to work on the moon for more than seven hours at a stretch. The main problem with the suits when pressurized was the lack of mobility they afforded their occupants. The first spacewalker, Alexei Leonov, had considerable difficulty in re-entering his craft when his suit ballooned; he was only able to redress the situation by lowering the pressure to about 4 psi. Multi-layered structures were provided to protect the men from all the hazards lingering in that hostile environment. Insulation kept out the intense cold of space and the excessive heat of the lunar day; other layers lessened the effects of cosmic rays and micro-meteorites.

Mercury spacesuits were silver-coated to help reflect heat, while Soviet cosmonauts sometimes wore orange coveralls to increase visibility on landing. Apollo suits were one-piece and specially made to measure; each one cost several million dollars. Today, the suits are a two-piece unisex design and come in standard sizes. Gloves, boots and helmet were an integral part of the suit, attached by pressure sealing rings. A special outer visor was provided for Apollo crewmen to give added protection against ultraviolet radiation. Their bubble helmets were also an advance on earlier versions in that they were fixed so that the man's head was free to move about; Mercury and Gemini helmets were close-fitting and moved with the man's head.

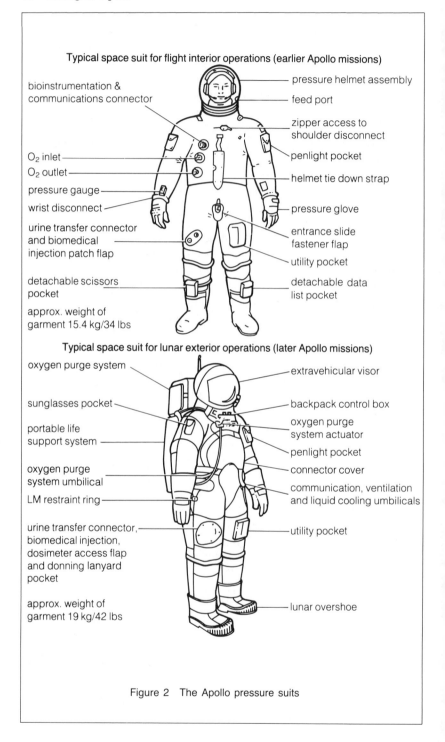

Typical space suit for flight interior operations (earlier Apollo missions)

bioinstrumentation & communications connector

pressure helmet assembly

feed port

zipper access to shoulder disconnect

O₂ inlet

O₂ outlet

penlight pocket

pressure gauge

wrist disconnect

helmet tie down strap

pressure glove

urine transfer connector and biomedical injection patch flap

entrance slide fastener flap

utility pocket

detachable scissors pocket

detachable data list pocket

approx. weight of garment 15.4 kg/34 lbs

Typical space suit for lunar exterior operations (later Apollo missions)

oxygen purge system

extravehicular visor

sunglasses pocket

backpack control box

portable life support system

oxygen purge system actuator

penlight pocket

oxygen purge system umbilical

connector cover

LM restraint ring

communication, ventilation and liquid cooling umbilicals

urine transfer connector, biomedical injection, dosimeter access flap and donning lanyard pocket

utility pocket

approx. weight of garment 19 kg/42 lbs

lunar overshoe

Figure 2   The Apollo pressure suits

Performing toilet functions was obviously a problem in weightless conditions, especially in a 17-layer spacesuit. Faeces were collected in special plastic bags, a packet of germicide added, toilet tissue also placed inside, and then the bag was sealed and stored for later scientific examination. Urine could be dumped directly overboard under normal bladder pressure, but if the suit was worn, it was collected in a neoprene-coated plastic bag strapped around the waist. This bag connected to a valve on the right thigh of the suit and could be emptied via a special fitting to the spacecraft waste disposal system.

It was the re-entry which it was feared could prove a fatal final leg to any space journey. This has been particularly true in the Soviet manned programme, when inaccurate and hurried returns to earth have sometimes left the crews stranded and alone in hostile terrain: the crew of Voskhod 2 landed some 600 miles off course in deep snow, and spent an uncomfortable night being stalked by hungry wolves. Survival training was, therefore, essential for all spacemen. The Gemini astronauts, for example, were taken to the jungles of Panama and the scorching deserts of Nevada, learning how to enjoy such delicacies as roasted boa constrictor, iguana or wild fruit, and how to find shelter and water in arid wastelands. As astronaut Alan Bean explained, when asked what the training had taught him: 'I learned that the best thing to do is to try very hard to keep from coming down in the jungle.' Astronauts were not expected to endure Arctic conditions since their craft only flew between latitudes 32 degrees, compared with 52 degrees for Soviet craft.

With so many potential sources of danger, it seems amazing that so few deaths have occurred in space during the past 25 years. The most dangerous periods have been during ground training, launch or re-entry: seven astronauts and four cosmonauts have died on flight duty. But people have survived the inhospitable environment with no major long-term effects and the efforts of those first pioneers have led mankind to the brink of establishing permanently occupied space stations. The pessimists of 25 years ago have been proved wrong.

2

'Have You Come from Outer Space?'

The Vostok Missions, 1961–3

At 10.55 am Moscow time on 12 April 1961, a man wearing an orange flight suit and a white pressure helmet landed in a field watched only by a cow and two bemused peasants. As he pulled himself together and staggered to his feet, he saw the woman and girl staring at him as he unhitched his parachute. Mrs Anya Takhtarova stepped towards him and doubtfully stammered, 'Have you come from outer space?' to which the young man triumphantly replied, 'Yes. Would you believe it? I certainly have.' The poor woman must have looked so frightened that he hastily added, 'Don't be alarmed . . . I'm Soviet!' He then walked over with them to inspect the charred space capsule which was still smouldering nearby. They were soon joined by excited tractor drivers who had been working near the landing site. Some of them had been listening to the radio and so knew the identity of the stranger. As they helped him remove the protective suit, they jubilantly shouted his name, 'Yuri Gagarin! Yuri Gagarin!' Someone gave him a cap to wear during the short interval before a helicopter brought members of the spacecraft landing support team. After a brief check-up, the exhilarated but weary cosmonaut was whisked away to the nearby city of Saratov for a period of rest and medical observation. Two days later, Gagarin was the centrepiece of a magnificent hero's welcome in Red Square, Moscow, which was televised live all over Europe, and later broadcast to the rest of the world.

Today, the landing site at Smelovaka, near Saratov, is marked by a 130 ft (40 m) high titanium obelisk, and thousands of visitors travel to the flat, fertile steppes to pay homage to the first human who travelled through space. His unique achievement makes Gagarin's place in history assured, despite the fact that he never flew in space again. No matter what miracles of human endurance or technological excellence the future may hold, Yuri Gagarin was the first; the pioneer who entered the unknown and opened the gate for his successors.

The man who made his mark on world history was born in a wooden house in the small village of Klushino, near Smolensk in the western Soviet Union. His father was a carpenter, and times were hard during the Nazi occupation. During the war, the young Yuri saw an aeroplane for the first time. The family moved in 1945 to the town of Gzhatsk (now named Gagarin) where Yuri completed school and went to a trade-training school, then worked in a foundry. He was sent to an industrial training college in Saratov and there he joined an aero club where he gained experience in solo flight and parachute jumping. Gagarin then progressed to flying jets at the air force training school in Orenburg, from which he graduated in 1957. The final step came in the spring of 1960 when his application was accepted to join a small group of airmen who were to be trained as future cosmonauts. By this time, he had clocked a mere 230 hours in the air.

There followed a year of intensive training at the specially constructed cosmonaut training centre, Zvezdniy Gorodok (Star City). Prophetically, perhaps, Gagarin was the first of the group of 20 to sit inside the Vostok ship on the day chief designer Korolev introduced it to them. However, it was not until 8 April 1961 that the 27-year-old air force lieutenant Gagarin was selected by the state commission for the momentous flight which was to take place only four days later. Apparently, Gagarin was nominated by General Nikolai Kamanin, head of the cosmonaut team, and seconded by Korolev with the words, 'I find in him an analytical mind and rare industriousness. We need profound information about outer space and I have no doubt that Gagarin will bring it.' The vote for Gagarin was unanimous.

The morning of 11 April saw the rocket assembled and transported in a horizontal position along the rail tracks from the flat-roofed assembly building to the launch pad. There it was raised to a vertical position, and held in the grip of four supporting arms. Maintenance technicians and service engineers busied themselves as the main rocket and the four strap-on boosters were loaded with liquid oxygen and kerosene, and all the mechanical, electrical, communications and life-support systems were checked and re-checked. Amidst this feverish activity, Yuri Gagarin was introduced to the service personnel and greeted by rapturous applause. Afterwards, Gagarin and Korolev spent over an hour alone at the top of the rocket beside the tiny Vostok capsule.

There were three identical wooden houses, surrounded by young poplar trees, only 15 minutes drive from the launch pad. It was there that the main participants in the historic events to follow spent a fairly restless night. Korolev's heart condition was causing him pain and as he opened a window he noticed that Konstantin Rudnev, head of the state commission, could not sleep either – the lights in both the neighbouring houses were shining brightly. The two men met in the garden and walked around discussing the launch scheduled for six hours time. Korolev even crept to the door of Gagarin's room and peered in, but both he and his back-up, Titov, seemed sound asleep, despite the biosensors attached by doctors to monitor their responses to stress. They apparently had little difficulty, however, in being roused by Yevgeni Karpov, Kamanin's lieutenant, at 5.30 on the morning of 12 April.

After breakfast, Gagarin and Titov were helped into the 'space wardrobe': sensors were attached to their bodies to monitor their physical condition while in flight, the woollen undergarment was put on, then the pressure suit to protect against radiation and failure of the heating system, then the orange coveralls to aid spotting on landing, and finally the gloves, boots and helmet. For the time being, the transparent visor was kept open as the two men walked awkwardly from the cottage door to the waiting bus. Within a few minutes they were at the Baikonur space centre, and beneath the towering rocket. Gagarin declared himself fit and ready to the state commission.

The final act was a moving farewell speech from the platform at the foot of the lift, addressed to 'people of all countries and continents'. Gagarin reflected on his role in the forthcoming mission:

At this instant, the whole of my life seems to be condensed into one wonderful moment. Everything that I have experienced, everything that I have done hitherto, was experienced and done for the sake of this moment . . . Of course I am happy. In all times and epochs the greatest happiness for man has been to take part in new discoveries. To be the first to enter the cosmos, to engage, single-handed, in an unprecedented duel with nature – could one dream of anything more!

Gagarin then dedicated the flight to 'the people of a communist society' and bade the emotional audience farewell.

Yuri Gagarin raised his hands aloft then turned and entered the lift. On leaving the lift at the top, he climbed a short metal ladder to the platform which led to the Vostok cabin. He was helped through the hatch and into the specially designed couch, then the hatch was closed and Gagarin was left to contemplate his fate. As the hour dragged by, some music was played over the intercom to help pass the time. After a short delay due to a faulty valve, the final countdown began. The last commands were issued: 'Switch to "go" position!' Gagarin settled back in the reclining seat. 'Air purging! Idle run!' The fuel tower slowly withdrew from alongside the rocket. 'Ignition!' The arm carrying the power cable swung away. 'Lift-off!' At 9.07 am Moscow time the supporting arms gently opened like the petals of a flower, and the huge white rocket rose, imperceptibly at first, into the clear blue sky. 'Off we go!' came the jubilant voice of Gagarin as his ship rapidly disappeared from view, leaving only a trail of smoke and flame.

The world's first cosmonaut was pressed harder and harder into his padded seat, unable to move, just as he had experienced in the centrifuge during his training. As the rocket broke the sound barrier, Gagarin lay back in an eerie quiet, broken only by the separation of the four strap-on boosters. Korolev listened worriedly to the buzzing on the intercom, then sighed with relief as Gagarin reported, 'The fairing has been discarded . . . I see the Earth. The loads are increasing. Feeling fine.' Pressure reached about 6G as the second stage boosted him into orbit before shutting off and throwing him suddenly forwards. This time, however, he did not rebound into the seat as the straps held him back, but, instead, felt himself suspended above the couch as everything not fastened down began to float past him. The second stage separated and Gagarin got down to work, noting instrument readings, checking equipment and recording the effects of weightlessness and the appearance of his surroundings. 'The sky looks very, very dark and the Earth is bluish.' Later he reported,

The sunlit side of the Earth is visible quite well, and one can easily distinguish the shores of continents, islands and great rivers, large areas of water and folds of the land. Over Russia I saw distinctly the big squares of collective-farm fields, and it was possible to distinguish which was ploughed land and which was meadow. During the flight I saw for the first time, with my own eyes, the Earth's spherical shape.

Soon after Vostok entered Earth orbit, Moscow Radio proudly announced to the world: 'On April 12 1961, in the Soviet Union, the Vostok, the world's first manned spacecraft, has been launched into orbit around the Earth. . .' Crowds began to gather in Red Square, cheering, embracing, holding placards. Meanwhile, Gagarin was watching the Earth flash by as he travelled at the unprecedented speed of nearly 18,000 mph (29,000 kph) over Siberia, Japan, Cape Horn and back towards Africa. The only moment of panic on the ground came while Gagarin was out of direct radio contact; the planners opted for brevity and clarity so a number code was used with 'five' indicating that all was well. Suddenly the machine began repeating 'threes' and everyone in the control room turned pale as they stared at the tape. Then the numbers changed again . . . 'fives' once more! Korolev dropped heavily into a chair and sighed, 'It's seconds like that which shorten a designer's life.'

Gagarin flew on, unaware of the emotional scenes on the ground. He happily practised eating and drinking from a supply provided in a small container by his right shoulder. Weightlessness he found very relaxing, giving a sense of increased room in the cramped cabin. He reported, 'Handwriting did not change, though the hand was weightless. But it was necessary to hold the writing block, as otherwise it would float from the hands.' He was not needed to pilot the craft, however, since it was designed to operate automatically from launch to touch-down. It seems that the original intention of the designers was to use the cosmonauts as passive passengers and so no manual control was provided. The cosmonauts, all experienced pilots, objected to this, and so a manual back-up was added. To ensure the system was not inadvertently activated, a combination lock was fitted with the numbers (1-4-5 in Gagarin's case) kept in an envelope attached to the cabin wall.

The automatic system operated on schedule, and Gagarin later admitted a sense of disappointment as the craft swivelled round ready for retro-fire. As Vostok flew over East Africa, only 1 hour 18 minutes after launch, the retro-rockets fired to brake the capsule. Had they failed, the orbit was such that friction with the upper atmosphere would have returned the craft to earth within ten days. The rocket section was safely jettisoned, and Gagarin began re-entry, facing backwards as so many other cosmonauts and astronauts were to do in the years to come. Through the portholes he was able to witness the terrifying firework display as the craft's exterior heated to thousands of degrees and the protective coating burned away. The acceleration forces built up to more than 8G before a small drogue parachute was deployed through a hatch at a height of 2½ ml (4 km) above the ground. This was followed at 1½ ml (2½ km) by the main parachute which slowed the steel capsule before its bumpy landing.

The official Soviet version of events was that Gagarin landed inside the charred capsule, but this has been questioned by a number of Western experts who believe that he ejected from the Vostok at about 20,000 feet (6 km) and landed using his own personal parachute. Certainly this was the method used by all subsequent Vostok crews. Was this because Gagarin's landing in the capsule was so rough, or was it a method common to all Vostoks, including Gagarin's? The Soviet motive for hiding the truth seems obscure, although it has been suggested that the flight would not have been registered as a record by the Federation Aeronautique Internationale, the organization which certifies all flights records, if Gagarin was known to have landed separately from his craft.

Soviet secrecy covered most aspects of the flight and only today is it possible to piece together the whole story. It was four years before any photographs of the Vostok were released, by which time the craft was obsolete, while the launch vehicle was not seen in the West until the Paris Air Show of May 1967. Inevitably this secretiveness by the Soviet authorities led to wild speculation and rumours in Western newspapers concerning the Soviet space programme. The *New York Herald Tribune* of 19 April 1961 even published a headline, 'Did Gagarin Do It?', alleging that a dummy may have been in the capsule. British newspapers quoted a Mr Bobrovsky who claimed 'reliable sources' for a story that a test pilot, Sergei Ilyushin, son of the famous Soviet aircraft designer, was launched into space a few days before Gagarin but returned to earth after three orbits suffering from loss of balance and was rushed unconscious to hospital. Other stories circulated concerning alleged cosmonaut deaths: some were said to have died in suborbital flights, others in rocket explosions and launch failures, yet others were supposedly trapped in Earth orbit. In 1986, the Soviet media specifically denied all of these allegations, although they did admit the death of one cosmonaut during a training accident on 23 March 1961. Valentin Bondarenko, the youngest member of the group, died as the result of a fire in an isolation chamber. The flames spread rapidly in the oxygen-rich atmosphere. Protected only by his woollen training suit, Bondarenko was so badly burned that the doctors were unable to save him. He was buried in Kharkov, his birthplace.

Yuri Gagarin became a legend in his own lifetime. The small, friendly, unassuming man with the boyish smile travelled the world as a roving ambassador, met the world's leaders and received the highest honours and decorations. He became a deputy of the Supreme Soviet, representing the Smolensk region, and always continued to help and advise his fellow cosmonauts. He graduated with honours from the Zhukovsky Air Force Academy and became Commander of the Cosmonauts' Detachment. Eventually he seems to have tired of the life of a celebrity. As he put it, 'Being a cosmonaut is my profession, and I did not choose it just to make the first flight and then give it up.' He went back into training and was appointed back-up pilot to Vladimir Komarov for the first flight of the new Soyuz craft in 1967. A year later, Gagarin too was dead. He was killed, along with his training instructor, in a plane crash on 27 March 1968. His ashes were buried in the Kremlin Wall, alongside other Soviet heroes. Some seven years earlier, he had stood in Red Square and spoken to the assembled multitudes: 'One can say with assurance that on Soviet spacecraft we will fly even over more distant routes. I am boundlessly happy that my beloved homeland was the first to accomplish this flight, was the first to reach outer space. . .'

Nearly four months after Yuri Gagarin had blazed a historic trail around the Earth his prophecy came true, when his back-up pilot, Gherman Titov, was blasted into orbit aboard Vostok 2. At a time when the USA was still experimenting with non-orbital 'lobs', the Soviet Union amazed the world with a 17 orbit, 25 hour space spectacular. Today, Titov tends to be regarded very much as second fiddle to Gagarin, and certainly he has totally faded from the limelight, but his flight caused a sensation at the time. The youngest person ever in space to this day, at 25, with clean-cut good looks, he was a hero in the classic Soviet mould. Titov was born in a village in Siberia, the son of peasant stock. His father became a teacher of languages, but Gherman did not want to follow in his footsteps and instead enrolled in an elementary aviation school at the age of 13. Nine years later, the young man fulfilled his main ambition and graduated as an air force pilot with top marks for pilot technique. In 1960 he joined the elite band of 20 cosmonauts who were to be trained for the first space flights. Chief designer Korolev described him: 'In general he has a good head on his shoulders and a sharp eye. He has the ability to pick out the main thing, but neither do details escape him. Some of his suggestions on the craft were truly useful.'

Just before the Vostok 2 was launched, Korolev explained the background to this new space milestone.

> Now we must accumulate experience in construction of space technology, elaborate the methodology of training man to live on board spacecraft. I won't conceal that the Soviet space programme envisaged follow-up space flights [after Gagarin]. The Vostok 2 was ready to go. We awaited the opinions of medical specialists and biologists who analysed the results of the first manned flight. We wanted to know, in particular, what impact space factors had on man's organism. Opinions on this differed. By no means everyone agreed, and certainly not at once, to the proposal of a 24 hour flight. It was suggested that we should limit ourselves to three or four, a maximum of six orbits. But why should we mark time?

This is not the whole story, however, for the choice of 17 orbits and 25 hours for the flight duration had a practical side: after that period, Vostok would once more be flying over Soviet territory, making a safe (and secret) recovery so much easier. As the spacecraft followed a fixed orbit in space, the Earth beneath it slowly rotated, so that Titov was able to observe a different part of the surface every time he flew around the world. After 24 hours or so, the Earth completed one revolution and Titov was more or less back where he started.

The events preceding the launch were a repeat of those involving Gagarin in April. At 9 am Moscow time the supports moved back and the ignition sequence began. Tass correspondent Alexander Romanov described the scene:

> The rocket, impelled by a powerful force, slowly, very slowly, rose over the Earth. Flames shot from the tail. Another few seconds and the rocket was up in the sky. It seemed as though there were two suns over the Earth. But one of them grew smaller and smaller. Gradually heading east, the ship set course for orbit. The radio transmitted Gherman Titov's first words from outer space addressed to Earth, 'Flying over the Earth, over our homeland.'

The first orbit went well and Titov reported that he was feeling 'magnificent' but from then on things went downhill. The unfortunate young man became the first human to experience that bane of all spacemen, space sickness. He felt nauseous and dizzy and his inner ear began to play tricks with his sense of balance and motion in the weightlessness of space. He resolved to fight the distress and continue the flight programme, even though the sensation worsened. Titov relayed the traditional propaganda messages to the peoples of the continents over which he flew, but he broke new ground by carrying out physical exercises and by twice manually testing the attitude control and orientation system. Titov must have been glad when his sleep period arrived at 6.30 pm Moscow time at the end of the sixth orbit. His tiredness overcame his feelings of sickness and he later stated, 'I slept the sleep of the just and spent 35 minutes longer than envisaged by the programme. My sleep was good, without dreams. In contrast to Earth conditions, I didn't feel the necessity of turning from side to side.'

After his eight hour sleep, the first ever in space, he had a meal using the special tubes and canisters provided so that nothing would be spilled or escape to float around the cabin and cause a potential hazard. At the beginning of the 17th orbit, the Vostok automatically reorientated prior to re-entry, and soon afterwards the rocket engines fired to brake the craft and then separated from the spherical capsule. Titov witnessed the brilliant firework display through the portholes as the acceleration forces built up. At an altitude of 22,000 ft (6½ km) he ejected from the capsule and landed safely using his personal parachute after separating from the ejection seat at 8,000 ft (2½ km). The landing site, more than 400 ml (700 km) south east of Moscow, was near Saratov and close to the place where Yuri Gagarin returned to Earth.

Titov was given a hero's welcome on his return, but he had inner ear trouble for some time after he landed. His problems with space sickness were not revealed for some time, but space doctors began to fear that humans might not be able to adjust to long periods of weightlessness. This may have been why there were no more Soviet manned space flights for a year. Another revelation after the flight was that Titov had injured a wrist many years earlier, unknown to the space authorities, and that the young major would not have flown in Vostok had the facts been known earlier. Certainly, Titov never flew again and he stepped back from the spotlight into obscurity.

By the summer of 1962 the American Mercury space programme was in full swing, although astronauts Glenn and Carpenter had both only logged three orbits each. The next Soviet space venture involved two separate manned craft in missions lasting several days and not surprisingly took the world by storm. Korolev was committed to longer flights which would extend Soviet knowledge about the effects of weightlessness, and Premier Khrushchev wanted more propaganda victories showing the superiority of the socialist system over the capitalist

Americans. Korolev explained the purpose of the missions to his fellow scientists in March 1962:

The proposal is to launch two craft – Vostok 3 and Vostok 4 – one after another. Moreover, they have to be orbited so accurately that they will find themselves in close proximity to one another. To accomplish this, a stay of three or four days in space is required. Why? Because the second spaceship must be launched the following day, when the first will be passing over the cosmodrome.

The gathering took some time to be convinced of the wisdom of this joint mission, although the readiness of the two pilots to take part may have helped to sway them. The men involved were both experienced air force pilots and significantly older than Gagarin and Titov. The man who went into space on 11 August 1962 in Vostok 3 was 32-year-old Major Andrian Nikolayev, a bachelor noted for his quiet, unflappable nature. He had studied as a physician and then at forestry school before finally entering air force college and graduating as a pilot in 1954. Nikolayev joined the team of cosmonauts in 1960 and acted as Titov's back-up for Vostok 2. An indication of his cool courage in a crisis is the story of how he landed his MiG fighter in a field after the engine cut out, despite orders from ground control to abandon the plane. 'The main thing is to keep calm,' he said.

Nikolayev settled into what seemed a routine extended flight of four or five days, though a first was chalked up when live TV pictures of the cosmonaut were shown to the Soviet people. Concern over the effects of weightlessness and the need for more detailed scientific data concerning the body's physical reactions led to far more intensive monitoring of the cosmonaut's responses. Two television cameras were focused on Nikolayev's face, one from the front and one from the side. Also for the first time, the cosmonaut was monitored by a whole series of electronic devices: electrodes were attached to his chest to measure his pulse, electrodes in the corner of each eye detected eye movements and blinking, and yet more electrodes attached to his foot and right shin recorded the electrical potential of the skin. Another electronic device measured his breathing rate. At one stage of the flight Nikolayev unstrapped himself and 'moved freely in the cabin' though there was no room for any acrobatics. A step towards civilizing space flight came with the provision of normal food, though specially prepared to reduce crumbs to a minimum; Nikolayev, and later Popovich, were able to tuck into chicken, veal, sandwiches, cake, fruit and sweets, to be washed down with water, coffee or fruit juices.

Nikolayev's sleep period began at 10 pm Moscow time after his third space meal and Tass news agency reported that his pulse dropped from the normal 80–90 beats per minute to between 60 and 65, and that his breathing was even and regular. The next morning Tass warned that an important announcement was imminent, a statement which led many Western observers to predict the return to Earth of Vostok 3. Instead, they were astounded to read on the teleprinters, 'The spaceship Vostok 4 piloted by pilot-cosmonaut Pavel Popovich was put into the orbit of an Earth satellite at 11.02 hours Moscow time on 12 August.' This timing was chosen, of course, so that Vostok 4 would be able to join Vostok 3 in a close encounter as it flew over the cosmodrome.

The fourth Soviet man in space was a 31-year-old air force lieutenant-colonel. He began life as a shepherd, then qualified as a carpenter before joining a local aeroclub and then the Soviet air force. He had the distinction of being the first cosmonaut to move into Star City, becoming the 'quartermaster' for the newly arriving recruits. Yet he was not the original choice for the mission: a man we knew only as Anatoli was the first choice, but he developed irregular heart rhythms during a centrifuge test which raised artificial gravity to 12G, a level that might occur during re-entry.

On attaining orbit, Popovich was able to see his companion in the distance. He later described the scene: 'Knowing where Vostok 3 would be in relation to me once I was in orbit, I looked for it immediately and saw it at once. It was something like a very small moon.' For a short time, the two craft were little more than 3 ml (5 km) apart, but the orbits were not exactly identical so that Vostok 4 gradually fell back and the two craft drifted out of view of each other. The media, both Soviet and Western, had a field day. 'Nik and Pop meet in space' was one banner headline as the press hailed the 'first rendezvous on the road to the moon'. With hindsight it is clear that this was a gross overstatement of the Soviet achievement: certainly the first dual flight had been carried out, and for the first time two men in separate craft had been able to speak to each other – Popovich was able to congratulate Nikolayev on breaking Titov's endurance record soon after Vostok 4 reached orbit – but neither craft had manoeuvring ability to enable them to adjust their orbits, and there was no question of a proper rendezvous with the two craft docking or joining with each other. Nevertheless, the joint mission did demonstrate the Soviet capability to track an orbiting spacecraft accurately, to launch a craft into a pre-determined orbit accurately, and to control simultaneously two craft in orbit.

For the next three days the two cosmonauts entertained TV viewers on Earth and carried out a number of experiments. Both craft carried plant seeds, bacteria and human cancer cells to enable studies on Earth of the effects of space travel; film and visual observations were made of the Earth, Sun and stars, and the way water behaved in zero gravity was studied. The subtle changes in the two orbits were useful evidence of the effects of atmospheric drag and of the variations in the Earth's mass and gravitational pull. Both men reported feeling well and neither of them apparently suffered from space sickness, though it was later revealed that Popovich suffered some disorientation.

By Wednesday 15 August the craft had opened up a gap of 1700 ml (2850 km) but they were still able to re-enter automatically almost simultaneously so that Vostok 4 landed only six minutes after Vostok 3, south of the town of Karaganda. Popovich parachuted to the ground 120 ml (193 km) away from his companion in space, both of them having ejected from the capsules. Nikolayev was the new space endurance record holder with 64 orbits and 94 hours 22 minutes to his credit, while Popovich marked up 48 orbits which lasted 70 hours 57 minutes. Ten months later, it was Nikolayev's girlfriend who stole the headlines in the final Vostok spectacular.

In June 1963 preparations for the second joint space mission were nearing completion, and it was generally assumed that the two pilots concerned would be the back-ups for Vostoks 3 and 4, Valeri Bykovsky and Boris Volynov. Tass correspondent Alexander Romanov described how 'in those days none of us had paid any attention to a slender attractive girl who now and then turned up in the company of Yuri Gagarin and Andrian Nikolayev. In answer to our question as to who she was, Yuri Gagarin told us to keep it quiet, but that she was Andrian Nikolayev's fiancee'. It was only just before the launch that the girl, Valentina Tereshkova, was revealed at a press conference as the pilot of Vostok 6. The earlier secrecy may have been because she was one of four female candidates, with the final selection being made only at a late stage – the names of the other three have never been revealed, and they returned to their normal lives after the flight since there was never any intention to send any more women into space.

This woman, who was to become the heroine of the Soviet people, the figurehead of women's movements all over the world and the diplomatic representative abroad of her nation, had to have an impeccable character as well as the correct social and political background. The 26-year-old Tereshkova was born in Yaroslavl to a peasant family. Her father died in the Second World War so her mother had to struggle to bring up Valentina and her brother and sister. Tereshkova went to work in a cotton textile factory after leaving school, but she continued her studies at evening school and completed a correspondence course at a technical school. She was also active in public life and was elected to the factory committee as well as the Communist Party. Significantly perhaps, she was also an amateur parachutist with 126 jumps to her credit. Tereshkova was enrolled in the cosmonaut group in March 1962 so that she had little more than one year's training. Korolev spoke about the significance of her selection: 'Until now, all spacemen have been jet pilots, trained men accustomed to G-loads and [high] speeds. Whereas the commander of Vostok 6 encountered G-loads only in the process of training for the flight.' He went on to add that the way she adapted to

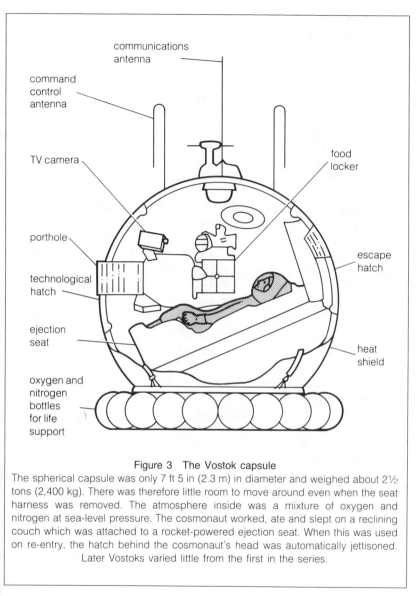

command
control
antenna

communications
antenna

TV camera

food
locker

porthole

escape
hatch

technological
hatch

ejection
seat

heat
shield

oxygen and
nitrogen
bottles
for life
support

Figure 3   The Vostok capsule
The spherical capsule was only 7 ft 5 in (2.3 m) in diameter and weighed about 2½ tons (2,400 kg). There was therefore little room to move around even when the seat harness was removed. The atmosphere inside was a mixture of oxygen and nitrogen at sea-level pressure. The cosmonaut worked, ate and slept on a reclining couch which was attached to a rocket-powered ejection seat. When this was used on re-entry, the hatch behind the cosmonaut's head was automatically jettisoned. Later Vostoks varied little from the first in the series.

weightlessness was of special interest to the space doctors.

Tereshkova was to join the already orbiting Vostok 5, piloted by 28-year-old air force lieutenant-colonel Valeri Bykovsky. He had always been interested in flying, and learned to fly when he was only 17. Selected as a cosmonaut in 1960, he was the first to try out the centrifuge and other special equipment at Star City. His air force superiors summed him up as follows: 'A bold, intelligent pilot, calm in flight, makes fast decisions in complex situations.'

After a day's delay due to bad weather, Bykovsky was launched aboard Vostok 5 at 2.59 pm Moscow time on 14 June 1963 amid broad official hints that he would soon be joined by a woman cosmonaut. Twenty-four hours later the expected companion did not materialize and rumours spread in the West of the chosen pilot being incapacitated. Whether this story has any truth is unknown, but on the afternoon of 16 June it was Valentina Tereshkova who went up to meet Bykovsky.

There was speculation in the Western press that the two craft might actually dock in the manoeuvre that would be vital for any future moon landing. It was even suggested that Bykovsky was piloting a much larger craft, weighing up to 10 tons (9,000 kg) and equipped with rocket thrusters. In fact, the mission was very similar to the previous dual flight. Valentina arrived in orbit only 3 ml (5 km) from her companion so that they were visible to each other. Neither Vostok was equipped to change orbit and so they gradually separated as they followed their slightly different orbits. Soon after reaching orbit, Tereshkova's excited voice was heard by millions of Russians on Moscow Radio: 'Here is Seagull [her code name]. I see a yellow strip. I see the Earth. Everything is in order. I'm feeling fine. The machine is working well.' Later, her compatriots were able to see TV pictures of their heroine laughing and smiling, and the former textile worker was able to speak to the most powerful man in the Soviet Union, Premier Khrushchev, as he radioed his congratulations.

Valentina's happy beginning to her three-day flight does not seem to have lasted, however. Unofficial reports stated that she succumbed to space sickness with a most unpleasant combination of nausea and disorientation, and that she pleaded to be allowed to return early. If that was true, it lends support to the idea that the effects of space sickness can be reduced by intensive high-G training, which her male colleagues had encountered as jet pilots. It might also explain why no other woman was sent into space for another 19 years. Whatever the facts, the flight continued for its full duration until Tereshkova parachuted to Earth 390 ml (630 km) north east of Karaganda after 48 orbits and 70 hours 50 minutes in space. To add insult to injury the unfortunate 'cosmonette' bruised her nose on landing, but she had more than doubled the American space flight record, held at that time by astronaut Gordon Cooper.

The forgotten partner in the 'Adam and Eve' mission, Valeri Bykovsky, completed two further orbits before ending his flight after a record 81 orbits and almost 5 days in space. Twenty years later, this remains the longest solo space flight; it was also the overall flight endurance record for more than two years until beaten by Gemini 5. Bykovsky landed nearly 2½ hours after Tereshkova, and about 500 ml (800 km) west of Vostok 6. Official sources quoted him as saying, 'I could have repeated the performance all over again.' While he was quietly carrying out his mission away from Tereshkova's spotlight, a series of manual control, medical and observational experiments were completed. Korolev described some of these:

The flight programme included a series of astronomical observations of constellations. The sun was photographed many times. At sunrises and sunsets pictures were taken of transitional spectrums. The value of these photographs is that they were made without atmospheric interference. The factual measuring of the radiation background and the ion fluxes along the course of the Vostok 5 and Vostok 6 is of exceptional significance. The observations of the Earth – both visual and with the aid of optical instruments – are also of great importance.

Within a week, the two heroes of the Soviet Union were given a rapturous reception by the Moscow population and the national leaders. In November there was once more cause for celebration as Tereshkova married Andrian Nikolayev. The following June there was a third happy event when the world's first 'space baby' was born to the proud couple – a girl named Yelena. There have been suggestions that Soviet scientists favoured Tereshkova's candidacy for the seat aboard Vostok 6 once they saw the close relationship between the two, taking the opportunity to carry out a unique experiment. This cannot be confirmed, but, whatever the truth, the doctors were able to confirm that Yelena was a normal, healthy baby, who was unaffected by her parents' exposure to cosmic radiation. Valentina became a leading spokeswoman for Soviet propaganda at home and abroad, and became the President of the Soviet Women's Committee. She never flew in space again, though her husband was commander of Soyuz 9 in June 1970, and he is now involved in cosmonaut training. No further children were born to the couple, and they divorced in 1982.

The 'Adam and Eve' mission ended the remarkably successful Vostok programme. The six flights had logged nearly 16 days in space compared with little more than two days for the entire American Mercury programme. Governments and media in Western countries were convinced of Soviet superiority in what was now becoming known as the 'Space Race', misled to a large extent by Khrushchev's space spectaculars and by ignorance of the true state of Soviet technology encouraged by the scarcity of reliable information. It was this which led President Kennedy to commit the USA to landing a man on the moon by the end of the decade. Unwittingly, the Soviet leadership had stung the pride of the richest and most technologically advanced nation in the world, and had started a race which they could not win, at least in the short term.

3

From Freedom to Faith
The Flights of the Mercury Seven, 1961–3

By the late 1950s, President Eisenhower was eager to set up a new civilian-controlled agency which would direct the thrust of the American effort to regain ground lost to the Soviet Union. Taking the advice of Senator Lyndon Johnson's sub-committee, the President decided to create the new agency around the long-established National Advisory Committee for Aeronautics (NACA). Within days of the creation of the National Aeronautics and Space Administration (NASA) on 1 October 1958, the Mercury Project to place a man in space and return him safely to Earth was initiated. No one was in any doubt concerning the difficulty of this project, involving as it did new technological developments and solutions to problems that had previously scarcely been studied. It was clear also that there would be a high cost to the taxpayer, and perhaps a price to pay in human life before the successful completion of the project. Despite the risks involved, more than 100 people answered NASA's advertisement for astronauts, though these were gradually whittled down to the 'Mercury Seven' or 'Select Seven' who were presented to the press in April 1959.

NASA had high hopes of sending a man in a Mercury capsule on a suborbital flight before the end of 1960, with the astronaut ensuring his place in history by becoming the first man ever in space. Shortly afterwards, it was proposed to launch a man into Earth orbit, less than two years after the project contracts were signed by the companies which were to provide the capsules and their booster rockets. Looking back, this time schedule seems impossibly optimistic, and so it proved. The inevitable faults, delays, errors and launch mishaps all put back the date of the first manned mission as public discontent increased. In the glare of worldwide publicity, the first test flights of unmanned Mercury capsules on both the Redstone and Atlas boosters were failures. By April 1961, the only living creatures to have flown aboard American rockets were two Rhesus monkeys and a chimpanzee.

Against this background, the Soviet Union was conducting its own series of tests using a satellite called Korabl Sputnik and crewed by dogs. The Soviet programme seemed to be going well and the American Space Task Group, happy at last that their Redstone rocket had proved itself, urged the go-ahead for a manned flight. Others were neither so happy nor so confident, and their opinion won the day: another Mercury–Redstone test was scheduled for 24 March 1961. As a result of this decision, which with hindsight can be seen as totally correct, it was the Soviet air force lieutenant Yuri Gagarin who became the first man in space. There was an immediate emotional outcry by Americans, whether politicians or the ordinary man in the street. One response was a sense of damaged pride and incredulity that a 'backward' peasant society had once again beaten the richest, most technologically advanced country in the world. This emotion was further fired by the Soviet boast that Gagarin's feat clearly indicated the superiority of the Soviet socialist way of life. The Cold War was at its height – it was the time of the abortive attempt organized by the American CIA to overthrow Fidel Castro in Cuba – and many Americans feared the consequences of a Soviet monopoly of space, with their country vulnerable to a strike by Soviet orbital weapons and exposed to the gaze of space spies, whether human or automatic.

On 22 February 1961 the shortlist for the first American in space had been cut to three: Alan Shepard, Virgil Grissom and John Glenn, although project manager Robert Gilruth had already informed Shepard that he was the first choice for the missions. Virgil 'Gus' Grissom was later named as Shepard's back-up. So it was US Navy commander Shepard who was awakened at 1.05 am in his quarters at the Cape Canaveral launch area, Florida, on 5 May 1961. It was the third launch attempt in four days. On 2 May Shepard spent three hours in his pressure suit waiting on the ground until bad weather forced postponement. Only then did the press and public discover the identity of the space pioneer. Another disappointment followed on 4 May, but the next day the countdown proceeded normally. Shepard ate a hearty, protein-rich breakfast of filet mignon wrapped in bacon, scrambled eggs and orange juice, and after a brief medical check, was fitted with the electronic sensors that would monitor his body temperature, respiration and heartbeat. Finally he was helped into his silver pressure suit and white helmet, and just before 4 am the astronaut boarded the transfer bus for the 40 minute ride to the launch pad.

On Pad 5 Shepard alighted near the 80 ft (25 m) high Redstone rocket glinting in the searchlights against the black sky. At 5 am America's first astronaut ascended in the lift to the air-conditioned 'greenhouse' at the top of the gantry, about 60 ft (18 m) above the ground. There he was greeted by Glenn, who had given the capsule a thorough check-over, and gave him an 'A-OK' hand signal. All possible safety precautions were carried out: Shepard wore plastic overshoes which were removed before he entered the capsule to prevent grit being carried in and floating around in weightless conditions, and a pad was placed over the lower edge of the hatch to prevent snags to the pressure suit. Grasping an overhead brace, Shepard was helped by the white-gowned attendants as he climbed feet first through the side hatch of the capsule he had named Freedom 7. He settled into the specially contoured couch and was carefully strapped in before his suit was connected to the craft's life support system. On the instrument panel in front of him was a typical piece of astronaut humour – a note from Glenn which read, 'No handball playing here'. The technicians wished him 'Happy Landings' as they pushed the hatch door into place and Shepard was enclosed in the tiny 9 ft × 6 ft (3 m × 2 m) capsule.

The countdown was held up for an hour and a half as low cloud closed in over the Cape, and later as a computer had to be checked. Shepard took the four hour wait calmly as he surveyed the launch area through his periscope and spoke to capsule communicator (Capcom) Gordon Cooper in the blockhouse. At 37 years of age, Shepard was well used to tense situations. After graduating from the US Naval Academy, he served in a destroyer in the Pacific at the end of the Second World War, then became a pilot and served on aircraft carriers before becoming a test pilot. Nevertheless, the excitement and nerves breached the calm exterior of even this experienced pilot as the countdown entered its final minutes. The 'cherry-picker', an access gantry on a hinged crane available in case of a launch pad emergency, was moved back a short distance as the countdown reached T minus two minutes. As Shepard's left hand grasped the abort handle, he confirmed that the periscope had retracted into the capsule's wall, and his pulse jumped to more than 100 per minute. The announcer's voice rang out over the public address system: 'Ignition. . . . Mainstage. . . . Lift-off!' At 9.34 am Cape time the Redstone rocket carrying the first American to venture into space climbed into a sky of billowing clouds and blue patchwork.

Shepard's first report came through loud and clear, 'Roger, lift-off and the clock is started.' His body's reaction was monitored as he left the Earth behind – pulse 124, respiration 30, deep body temperature 99. In the blockhouse Dr Carmault Jackson kept a Bible nearby: it had been sent to President Eisenhower by children in Toledo, Ohio, in the hope that it would one day be carried on the first manned American space flight, but unfortunately the tight weight limits on Freedom 7 had kept the book on the ground. Deke Slayton, Capcom in the Mission Control Centre, replied, 'OK José, you're on your way', a reference to TV spaceman José Jimenez.

As the rocket approached the speed of sound, Shepard reported

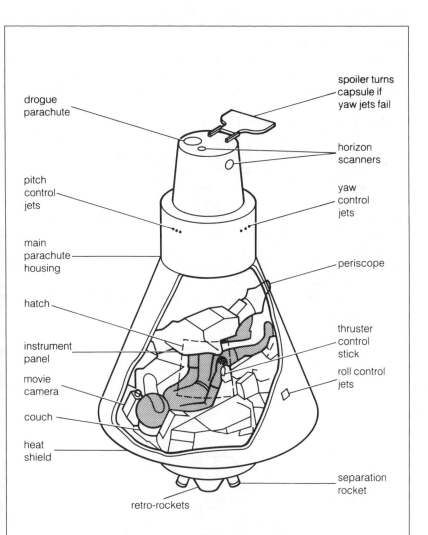

Figure 4 The Mercury spacecraft
There was little room for the pilot to move in this tiny capsule even if he had not been strapped to the special contoured couch. Its length was 9 ft 6 in (2.9 m) and maximum diameter − at the base, by the heatshield − was 6 ft 1 in (1.9 m), hardly sufficient for the pilot to stretch to his full length. All instrument displays had to be easily visible, and manual attitude and abort controls were within easy reach. There were 18 small thrusters for attitude control. Cabin atmosphere was pure oxygen at a pressure of 5 psi (0.36 kg per cm^2). There was no ejection seat: a rocket-powered escape tower was mounted on the nose instead.

increasing vibrations as his head shook and acceleration forces reached 6G, pressing him back into the couch. However, he was able to read the dials and confirm a steady cabin pressure; there was no need to turn the abort handle. At T plus 2 minutes 22 seconds the Redstone cut off and the rocket-powered escape tower, now redundant, automatically fired and jettisoned. Shepard, waiting anxiously for the explosive bolts to fire in order to separate his capsule from the Redstone, registered a pulse rate of 138, the highest of the flight. However, 10 seconds after shut-down and exactly on schedule, the separation occurred and Freedom 7 was flying on its own. A relieved Shepard reported, 'Cap sep. Periscope is coming out, and the turnaround has started.'

Shepard was now facing in the opposite direction to that in which he was moving, and it was time to take over manual control in order to practise attitude changes of the craft. He later explained the manoeuvre: 'I made this manipulation one axis at a time, switching to pitch, yaw and roll in that order until I had full control of the craft. I used the instruments first and then the periscope as reference controls. . . The spacecraft movement was smooth and could be controlled precisely.'

There followed five minutes of weightlessness as Freedom 7 headed towards its highest altitude (apogee) of 116.5 ml (187.5 km). Shepard

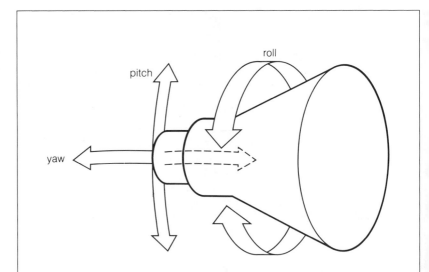

Figure 5   Attitude control in Mercury spacecraft

The attitude of the Mercury capsules was controlled either automatically or manually. Hydrogen peroxide gas was released from tiny thruster jets to manoeuvre the craft. Pitch is movement up or down from the craft's lateral axis; yaw is movement to either side of the craft's longitudinal axis; roll is the circular motion about the craft's longitudinal axis.

had time to exclaim, 'What a beautiful view!' before the three solid fuel retro-rockets fired, one after the other, in an essential test for a future orbital flight. Six minutes into the flight the exhausted retro-rockets were due to be jettisoned, but although the pilot sensed that the jettison had worked there was no indicator light on his panel. Once he switched to manual override the light appeared – it was the only cockpit malfunction in the whole flight.

The capsule was now beginning its long curving descent towards Earth. The automatic stabilization control system once again operated perfectly as the capsule swung into a nose-up position. With the periscope safely retracted, the 0.5G light came on indicating the build up of gravitational forces once more at a height of about 38 ml (61 km). Shepard tensed his muscles and pushed against the harness to combat the force which was trying to crush the breath out of him: in only 32 seconds the G force rose from zero to 11. Shepard was only able to grunt, 'OK. . . OK!' to reassure the ground controllers that all was well, as his weight increased to seven-eighths of a ton (844 kg). Nevertheless, the frictional heating experienced on Freedom 7's relatively slow, shallow re-entry was insufficient to produce the glowing fireball experienced by craft returning from Earth orbit. Temperature in the cabin rose to 44 °C, though his suit kept Shepard at a temperate 26 °C. The astronaut's pulse rate jumped to 132 as the G forces pressed him back on to the couch, but his pulse gradually dropped as the rate of descent slowed. Less than 10 minutes from launch, the drogue parachute deployed at an altitude of 21,000 ft (6.4 km), followed at 10,000 ft (3 km) by the red and white striped main parachute. Shepard was able to watch these successful operations through the periscope, and now felt able to relax.

Freedom 7 drifted slowly down with its triumphant passenger to splashdown in the Atlantic to the east of the Bahamas where the aircraft carrier *Lake Champlain* was waiting. At the Cape, 302 miles (480 km) away, they were no longer able to receive the craft's telemetry and could only listen to the progress of the recovery forces. As soon as the capsule hit the water, a helicopter was poised overhead ready to hoist up the astronaut and take him to the carrier. Shepard's matter-of-fact report gave no mention of the feeling of exhilaration and triumph which spread through all those who had witnessed the landing or participated in the first American manned spaceflight.

The landing did not seem any more severe than a catapult shot from an aircraft carrier. The spacecraft hit and then flopped on its side so that I was on my right side. I felt that I could immediately execute an underwater escape should it become necessary . . . I could see the water covering one porthole, I could see the yellow dye marker out the other porthole, and later on, I could see one of the helicopters through the periscope . . . The capsule righted itself slowly and I began to read the cockpit instruments . . . the helicopter was already calling me . . . I sat on the edge of the door sill until the helicopter sling came my way.

Shepard was greeted by the cheering crew of the *Lake Champlain*. There he was briefly examined and interviewed before the main two-day debriefing on Grand Bahama Island. Shepard passed all tests with flying colours: the way to the stars seemed open. Describing weightlessness, he simply said, 'Movements, speech and breathing are unimpaired, and the entire sensation is most analogous to floating.'

The flight had lasted a mere 15 minutes 22 seconds, but following its success a new expression, 'A-OK', entered the English language and America was at last challenging the Soviet Union in what had previously seemed a one-horse race. On 8 May, Shepard was awarded the Distinguished Service Medal by President Kennedy; at the ceremony the

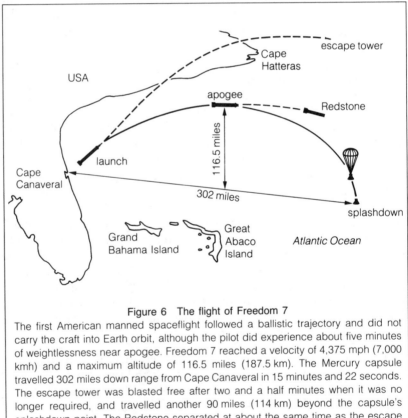

Figure 6   The flight of Freedom 7
The first American manned spaceflight followed a ballistic trajectory and did not carry the craft into Earth orbit, although the pilot did experience about five minutes of weightlessness near apogee. Freedom 7 reached a velocity of 4,375 mph (7,000 kmh) and a maximum altitude of 116.5 miles (187.5 km). The Mercury capsule travelled 302 miles down range from Cape Canaveral in 15 minutes and 22 seconds. The escape tower was blasted free after two and a half minutes when it was no longer required, and travelled another 90 miles (114 km) beyond the capsule's splashdown point. The Redstone separated at about the same time as the escape tower and flew on a further 20 miles beyond Freedom 7 (32 km).
(Map copyright © National Geographic Society.)

President dropped the medal, then quipped, 'This decoration has gone from the ground up.' Less than two years later it was Shepard's turn to be dropped – he developed an inner ear infection and was banned from flying. He was given a desk job as Chief of the Astronaut Office, but he never gave up hope of returning to active duty and in 1969 he achieved his ambition, eventually becoming the sole Mercury astronaut to step on the moon.

On 25 May, less than three weeks after Shepard's breakthrough, President Kennedy threw down the gauntlet to the Soviet Union, and thereby committed his country to a multi-billion dollar space programme for at least the next nine years. He asked Congress to provide the funds to achieve the goal, 'before this decade is out, of landing a man on the moon and returning him safely to Earth'. Kennedy had been informed by his advisers, including Vice President Johnson, that the feat was possible. Already there were plans on the drawing board for an Apollo circumlunar spacecraft, while a giant Saturn rocket was under development and an even larger Nova rocket was projected. Not all NASA scientists were pleased with the increased emphasis on manned spaceflight, which they saw as a frivolous propaganda exercise, to the detriment of serious scientific research. However, the way was now open for massive expansion of the Space Agency and its ground facilities, and the result would be an explosion in research and technology unprecedented in peacetime.

The second of the planned seven suborbital flights was already being prepared as Kennedy made his request to Congress. The USA's second man in space was to be Virgil 'Gus' Grissom, a 35-year-old Air Force test pilot who had flown 100 combat missions in the Korean War, though without claiming a 'kill'. Grissom had the misfortune to be sandwiched between the two American firsts – Shepard's historic lob into space and Glenn's orbital spectacular. Somehow, Grissom never gained recognition; he simply did not seem to fit the image of a popular idol. Gus was the shortest of the Select Seven, touching 5 ft 6 in in his bare feet, and the most introverted of the group – he himself admitted 'I tend to clam up at press conferences' – so that some of the press nicknamed him 'Gloomy Gus' or 'Great Stone Face'.

Grissom awoke to physician Bill Douglas's call at 1.10 am on Friday 21 July 1961. He had already gone through the mental trauma of a cancelled launch two days previously, but the technicians and weathermen were confident of a 6 am lift off this time. At the medical check-up after the usual low-residue breakfast, Grissom was passed as fit despite feeling a little tired and reporting a slightly sore throat. The biosensors and the suit were fitted and tested successfully, and at 4 am a tense Grissom climbed through the hatch of the capsule he had named Liberty Bell 7. His pulse rate climbed to 96 per minute as the technicians bolted down the hatch, but dropped to 80 as he settled down for the long wait.

This highly variable pulse rate was to be a feature of the entire flight, in contrast to the relatively stable pulse of 'ice man' Shepard; Grissom's pulse was above 150 throughout the five minute period of weightlessness, and peaked at 171 during the firing of the retro-rockets before re-entry into the atmosphere.

At least Grissom had the benefit of Shepard's previous experience. He had sat in on his predecessor's debriefing and had been able to contribute to a streamlining of the flight plan. His capsule was an improved version too: there was a central window instead of two circular portholes, additional foam padding was provided to support his head and vibration was also reduced by more streamlined wires linking the capsule to the booster rocket. The capsule's instrument panel was also rearranged, partly because an indicator had been added to show the capsule's position above the Earth (unnecessary in this flight but needed for future orbital missions). In answer to the astronauts' request for a more responsive craft which could be piloted, Liberty Bell received an improved attitude control system, while explosive bolts were fitted to blow off the hatch in order to enable rapid evacuation of the capsule in an emergency after splashdown. Finally, almost as an afterthought, Gus was provided with a makeshift urine receptacle beneath his suit. No one seems to have thought about the necessity for this piece of equipment since the early Mercury flights were only scheduled to last for 15 minutes. Alan Shepard had therefore been forced to relieve his aching bladder inside his suit during the prolonged four hour wait in Freedom 7, a fact not released to the millions who were admiring the efficiency and know-how of the manned spaceflight team!

The flight itself proved a straightforward affirmation of the progress the USA was making in the development of its manned spaceflight programme. Grissom obtained a much better view of Earth than Shepard, and was able to use his observations as a reference for the craft's attitude control. The modifications worked well, and the re-entry was virtually identical to that of Freedom 7. During the 15 minute 37 seconds flight, Grissom reached a maximum speed of 5,310 mph (8,500 kmh) and an apogee of 118 miles (190 km). Apart from the usual minor hitches during final countdown, the flight seemed a great success. Then disaster struck.

Grissom had removed his straps and 'unplugged' his suit in preparation for the helicopter pick-up. He had spoken to the pilot of the hovering helicopter to explain that he was completing the final checklist of switches and readings, and he had armed the explosive bolts ready to blow the hatch. Then as the helicopter crew moved in to hook on to the capsule, the hatch blew without warning and water started rushing through the gaping opening. The crew then witnessed Gus Grissom squeezing out through the hatch and swimming away from the capsule, supported by the buoyancy of his pressurized suit. He pulled away from the sinking Liberty Bell, shrugging off a snared line, and watched the helicopter struggling, with all three of its wheels in the sea, to save the capsule which was now weighed down with water. At first it seemed as if the crew might succeed, but then a red light came on to warn the pilot of impending engine failure.

By this time, Grissom was also struggling to stay afloat; he had forgotten to close the oxygen inlet valve on his suit and all the oxygen was bubbling into the sea. Instead of acting as a raft, the suit was now a dead weight which threatened to pull him down. A second helicopter came over, but the crew interpreted Grissom's waves for help as a sign that he was all right, and one of them began to take photographs of the floundering astronaut, who was now swallowing water. Eventually, about five minutes after he entered the water, Grissom was able to grab the 'horse-collar' harness and lie in an exhausted state with his arms dangling over the side as he was slowly winched aboard. His first, irrational, act was to attempt to don a life jacket in case the helicopter crashed and sank. He was still shaking when he reached the aircraft carrier *Randolph*, and kept repeating, 'I didn't do anything. I was just lying there and it just blew.' Meanwhile, the crew of the first helicopter decided to heed the warning light and gave up the struggle. The Liberty Bell disappeared from view as she headed towards her final resting place on the bed of the Atlantic.

Grissom always maintained that he never fired the explosive bolts, though he did explain that his mind was more on retrieving a knife from the survival kit as a souvenir than on anything else when the hatch blew. Exhaustive tests on the system were subsequently carried out, but a reason for the accident was never discovered. Grissom returned to Florida under a cloud, despite the official reports of a successful mission which suffered a minor setback at the end. There was no Washington presentation and no tickertape parade, only the award of the Distinguished Service Medal at Patrick Air Force Base by NASA administrator James Webb. Grissom later wrote:

> When my Mercury flight aboard the Liberty Bell capsule was completed, I felt reasonably certain, as the programme was planned, that I wouldn't have a second space flight. By then Gemini was in the works, and I realized that if I were going to fly in space again, this was my opportunity, so I sort of drifted unobtrusively into taking more and more part in Gemini. What it amounted to, in fact, was that they just couldn't get rid of me, so they finally gave up and programmed me into Project Gemini.

This dedication and determination to remain an astronaut would eventually lead to his death some five years later.

Only two weeks after the flight of Liberty Bell, the pre-eminence of the Soviet Union was emphasized by the 24 hour, 17 orbit marathon of Gherman Titov. To add insult to injury, his craft passed over the USA three times, a fact which was used by press and politicians alike to emphasize the nation's vulnerability and the need for urgent steps to be taken to redress the balance. In particular, NASA was under pressure to curtail the planned ballistic Mercury flights and to pull out all the stops for a manned orbital flight as soon as possible. This was not viable, however, until the Mercury–Atlas combination was flight-proved; the Atlas rocket had suffered a number of embarrassing failures, and some believed the booster would never be safe enough to carry a man into space. Only after a chimpanzee named Enos was safely recovered after a two orbit flight which was terminated early because of a control system malfunction, was NASA prepared to burn its bridges and go for an orbital manned Mercury mission. In this manner did John Glenn become the choice for the most prestigious and most coveted of all the Mercury flights, rather than just the third American to be lobbed a few hundred miles into the Atlantic.

Although Shepard had the distinction of being the USA's first spaceman, it was Glenn's first orbital flight which really caught the public's imagination and resulted in his remaining a national hero to this day. By 1961 he was already a well-known figure, partly because he was never afraid to pursue the limelight or to seek the attention of his superiors. At the age of 40, Glenn was the most experienced pilot among the Select Seven: he had flown 59 combat missions in the Pacific towards the end of the Second World War, followed by a further 90 in Korea, where he shot down three MiG fighters during the last nine days of combat. Already bedecked with medals including the Distinguished Flying Cross five times and the Air Medal with 18 clusters, Glenn enrolled as a naval test pilot. Within three years of becoming a test pilot, he was in the news again. This time he organized and flew the first transcontinental supersonic flight, completing the journey from Los Angeles to New York in 3 hours 23 minutes. He was subsequently invited on to a TV show called 'Name That Tune' where he appeared for several weeks.

Not surprisingly perhaps, Glenn became the most well-known of the Mercury astronauts. There were a number of factors in his favour, apart from his outstanding service record. From the time of the first press conference in 1959, Glenn established himself as the leading spokesman for the astronaut fraternity. He was a devout Presbyterian – the man who loved God, his country, his wife and his children. As the pilot who revelled in publicity, a characteristic not commonly found in his comrades, it was Glenn who gave the long, eloquent speeches, and Glenn who provided the quotable quotes. His solid good looks attracted admiration from women of all ages and made him the most photogenic of the seven. Glenn had worked hard for his fame and success, had lived cleanly and continually pushed himself both physically and mentally. He believed he had earned the right to make history by being the first American to go into space, and he had eagerly looked forward to the announcement of the prime Mercury pilot.

Imagine, then, the disbelief and dismay when NASA administrator Bob Gilruth told the assembled seven that Glenn was to play third string to Alan Shepard and Gus Grissom. In the opinion of Gilruth and his fellow astronauts, Glenn was considered less suitable to be the first American spaceman than Commander Shepard. Indeed, his straightlaced attitude, his disapproval of the ways his colleagues let off steam, and his obvious ambition and determination to succeed did not endear him to his fellows, so that he was far from the most popular of the Mercury team. His being the only Marine in the group also did little to help. Nevertheless, Glenn threw himself into his work as back-up for Shepard and Grissom, and did as much as anyone to ensure the readiness of them and their craft.

Then the clouds seemed to part and the Lord seemed to smile on astronaut number three, John Herschel Glenn Junior. Following cosmonaut Titov's flight and the successful test of the Atlas rocket, Glenn found himself the nation's choice to be its orbital spaceflight pioneer and only the third man ever to circle the Earth. It was as if he had fulfilled his destiny after all. Then frustration set in as the mission was put back from 20 December to 16 January 1962, then 23 January and 27 January. On the last date he lay, tense and hopeful, in his tiny capsule for nearly six hours as the countdown was held again and again, only to be postponed finally by bad weather. Was he to be denied his moment of triumph? Glenn's weary comment was, 'Oh well, there'll be another day.' Yet more delays followed as the launch date slipped from 1 February to 13 February, then the 16th and ultimately to the 20th. Many of the press gave up and packed their bags; President Kennedy called in Glenn for a personal briefing on the Mercury programme and its prolonged hold-ups, and Glenn became an almost permanent resident of Hangar S, the mission pilot's home on the Cape.

At last the sun broke through again, and there was a buzz of expectancy among the technicians and the multitudes of spectators who had gathered on the nearby beaches as the countdown proceeded. At 2.20 am on 20 February Glenn awoke and, following a small breakfast, his biological sensors – including a blood pressure monitoring system for the first time – and the pressure suit were fitted by Dr Bill Douglas and technician Joe Schmitt. By 5 am the astronaut was ready for the 4 mile (6½ km) drive to Pad 14 where the stately Atlas booster stood bathed in arc lights against a dark, overcast sky.

At 6.03 am Glenn was helped into Friendship 7, the capsule that was to be his home for the next nine hours. As he lay on his back listening to the creaking noises made by the Atlas and the progress in the countdown, Glenn stayed calm – he had been through so many simulations of a launch, both as back-up and as prime pilot for this mission, together with the recent postponed launches, that all the fear had drained away. His pulse rate of between 70 and 80 suggested a cool detachment bordering on boredom, and compared favourably with his two predecessors. Finally, at 9.47 am Cape time came the historic lift-off that all Americans had waited and prayed for. Seated at the top of the 95 ft (29 m) rocket, Glenn heard his back-up, Scott Carpenter, send him a final prayer, 'May the good Lord ride all the way', and then the end of the countdown recited by Alan Shepard in the Mercury Control Centre. Glenn's pulse jumped to a respectable 110 as he felt the three giant engines burst into life and struggle to lift the rocket free of the pad. A few seconds later he was able to report, 'The clock is operating. We're under way!' as the gleaming Atlas, bathed in liquid oxygen frost and belching orange flame, carried him into a clear blue sky.

Glenn looked in the small rear-view mirror as the Cape disappeared from view and the rocket rolled on to the correct path to orbit. Vibrations began to build up as Glenn reported, 'A little bumpy about here'. As the Atlas broke the sound barrier and approached the point of maximum air resistance these vibrations worsened and the noise almost drowned the roar of the engines. Just over one minute into the flight, although acceleration forces were building to 6G, he could report, 'We're smoothing out some now'. Glenn was not bothered by an increase in body weight to half a ton – he had experienced much worse in the centrifuge.

A little over two minutes into the flight the Atlas's two booster engines shut down, causing such a sudden deceleration that the astronaut shot forwards against the seat restraints. Then, as acceleration built up again, Glenn saw smoke through his window and believed his escape tower had fired earlier than scheduled. Half a minute later he realized his mistake as he saw the tower shoot clear and the green 'Jettison Tower' light came on. What he had seen earlier was merely a cloud of smoke from the redundant boosters as they separated and slid away. Then, according to plan, the craft pitched down briefly to give him his first glimpse of the Earth's horizon sharply delineated against the blackness of space. Friendship 7 was pitched up again as it was boosted by the main Atlas engine towards orbital velocity of 17,500 mph (28,000 kmh).

Glenn related after the flight how pleasant it had been to experience acceleration in a straight line rather than going in circles around a centrifuge: the return to more than 7G did not distress him. However, four minutes into the flight, he experienced something he was not familiar with: the Atlas fuel tanks were almost spent and filled only with gas, causing the rocket to sway from side to side as the guidance system struggled to keep it on course. Glenn felt as though he was sitting on the end of a springboard, a sensation which ended only when he was violently pitched forwards again at T plus 5 minutes. The sustainer engine cut off suddenly to leave Glenn with the feeling of falling head over heels, something he had also experienced on the centrifuge when the velocity was sharply reduced. Almost immediately there was another lurch forwards as the clamp ring which held the capsule to the Atlas was released by explosive bolts, and the small thrusters pushed Friendship 7 free. 'They really boot you off', commented the buffeted astronaut.

For the first time, Glenn could feel himself being lifted slightly out of his seat, and he realized he was weightless at last – an American was in orbit! The capsule yawed through 180 degrees and its nose pitched forward 34 degrees, exactly as planned, so that Glenn could look back the way he had come. The spent Atlas could be seen a short distance away, but it gradually dropped back as Friendship 7 sped on its way. At T plus 5½ minutes Alan Shepard passed on the message Glenn was awaiting eagerly, 'Seven, you have a go, at least seven orbits.' The happy astronaut replied, 'Probably the best words I have ever heard!'*

Soon afterwards he was handed over to the care of Capcom Gus Grissom in Bermuda. Now began the monotonous but essential switch checks, status checks and copying of the retro-sequence times which would be required at the end of the third orbit if all went to plan. To everyone's relief the attitude controls worked perfectly during a manual test. Contact with Bermuda faded out as the capsule continued on its nearly flawless journey, and Glenn drifted for nearly a minute before he entered the Canary Islands communication area. When he had time, he looked through the periscope at the surface drifting through his field of view, then observed the same area through the window 40 seconds later. Glenn reported, 'The horizon is a brilliant, brilliant blue. There, I have the [African] mainland in sight at the present time . . . and have the Canaries in sight out through the window.' To supplement the verbal description, Glenn used the special pistol-grip camera to photograph the view, though he was not always able to see much. 'I can see dust storms down there blowing across the desert. A lot of dust. It's difficult to see the ground in some areas.' Most of East Africa was obscured too, this time by high clouds: Glenn concluded that space was an ideal environment for a weatherman since he had no difficulty in distinguishing cloud types and heights.

The capsule and the astronaut continued to function well as Zanzibar centre took over control. Glenn had repeatedly to carry out a blood pressure check by pumping up the cuff around his upper left arm. He opened his visor and swallowed a white xylose pill which was slightly

---

* Shepard's message meant that the craft could complete seven orbits though three were planned.

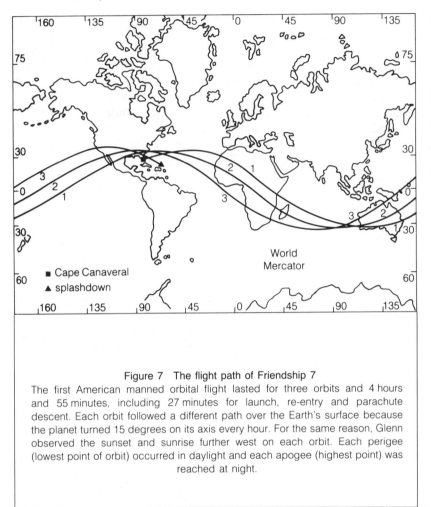

Figure 7   The flight path of Friendship 7
The first American manned orbital flight lasted for three orbits and 4 hours and 55 minutes, including 27 minutes for launch, re-entry and parachute descent. Each orbit followed a different path over the Earth's surface because the planet turned 15 degrees on its axis every hour. For the same reason, Glenn observed the sunset and sunrise further west on each orbit. Each perigee (lowest point of orbit) occurred in daylight and each apogee (highest point) was reached at night.

radioactive in order to enable the doctors to monitor the effects of weightlessness on his digestive system. Over Zanzibar he pulled hard on a handle located between his knees and extended its elastic cord to his chin once every second for 30 seconds: once again the doctors were happy with the results. He reacted normally when he tested his reactions – no nausea or dizziness yet.

At 38 minutes after launch the astronaut prepared for his first night passage. The main cabin lights were covered with red filters and the camera lights were switched off since no photographs of him were to be taken during this period. A patch refused to stick over his left eye so he decided simply to close one eye while observing the sunset. A small hand-held photometer equipped with a polarizing filter enabled him to look at the sun directly as the glowing ball sank rapidly towards the Indian Ocean. Glenn attempted to describe the vision which only two men had seen previously.

Orbital sunset is tremendous . . . a truly beautiful, beautiful sight. The speed at which the sun goes down is remarkable. The white line of the horizon, sandwiched between the black sky and dark Earth, is extremely bright as the sun sets. As the sun goes down a little bit more, the bottom layer becomes orange, and it fades into red and finally off into blues and black as you look farther up into space.

The bright strip along the horizon shrank to a small blue streak before Friendship 7 was plunged into total darkness.

This did not last long, however, as Glenn soon spotted a bright light through the window. 'I am having no trouble at all seeing the night horizon. I think the moon is probably coming up behind me. Yes, I can see it in the scope back here and it's making a very white light on the clouds below.' He could also see the stars clearly for the first time, enabling him to take navigational fixes. While verifying this as a valid method for future space travellers, he noted a strange phenomenon.

There seems to be a high layer way up above the horizon, much higher than anything I saw on the daylight side. The stars seem to go through it and then go down towards the real horizon. It would appear to be possibly some 7 or 8 degrees wide. I can see the clouds down below it, then a dark band, then a lighter band that the stars shine right through as they come down towards the horizon.

Fifty minutes and a half a world away from Cape Canaveral, John Glenn contacted a friendly spirit in Muchea, western Australia, as he was greeted by Capcom Gordon Cooper. Glenn confided, 'That was sure a short day. . .' to which Cooper lightheartedly replied, 'Kinda passes rapidly, huh?' in his Oklahoma drawl. He then told the man who was passing 162 miles (260 km) above him to look out for the lights of Perth down to his right since all the inhabitants were turning on their lights to give him a beacon in the night. Glenn spotted them soon afterwards, appearing like a small town as seen from a high-flying aircraft. 'The lights show up very well and thank everybody for turning them on will you?' said the lone night voyager.

He had a snack of apple sauce from a tube which he squeezed into his mouth, and put a brave face on it, though he couldn't resist a wishful comment. 'I wish now I had brought along that ham sandwich someone once put in the ditty bag as a joke. I'm sure I could eat it with no trouble. Only crumbly food would be a problem – no crackers in space.'

Only 35 minutes after watching the spectacular sunset, Glenn could look through the periscope and see the sun rising behind him. The blue band along the horizon grew wider and brightened until a brilliant red sun appeared. 'It's blinding through the scope on clear. I'm going to the dark filter to watch it come on up.' The Canton Island communicator sat up in astonishment as Glenn then described a swarm of brilliant specks which suddenly surrounded his capsule.

> I'll try to describe what I'm in here. I am in a big mass of some very small particles that are brilliantly lit up like they're luminescent. I never saw anything like it! They're round a little. They're coming by the capsule, and they look like little stars. A whole shower of them coming by . . . They swirl around the capsule and go in front of the window and then depart back the way I am looking. There are literally thousands of them!

Capcom was worried as well as bewildered: 'Can you hear any impact with the capsule?' Glenn's reply was reassuring: 'Negative. They're very slow. They're not going away from me more than three or four miles per hour (5–6 kmh).' He went on to explain that they were yellowish-green in colour but appeared white when in the shade of the capsule. 'It is as if I were walking backward through a field of fireflies.' Their slow relative motion suggested to Glenn that they were originating from Friendship 7, so he fired his hydrogen peroxide jets to see whether they were the source, but with no noticeable effect. However, as the sun rose and the Earth's surface brightened, they gradually disappeared from sight. Glenn remained fascinated by what he had seen, and repeated his observations to the control centre in Guaymas, Mexico, although they did not seem impressed.

Then, one and a half hours into the flight, Friendship 7 began to behave in a distinctly unfriendly manner. For most of the time Glenn had played the role of passenger, with the capsule flying on automatic control. Now it began to yaw about 20 degrees to the right, swing back to its proper position, and then repeat the manoeuvre. Apparently a small thruster had stuck and the main thruster was continually having to correct the attitude, so Glenn elected to 'fly-by-wire' and manually override the automatic system. Alan Shepard recommended that he should continue to do so for the remainder of the mission, even though this meant curtailment of in-flight experiments and keeping a close watch on control gas supplies to ensure that he could correctly orientate the craft for re-entry. A planned congratulatory call from President Kennedy was cancelled, and Glenn began his second orbit over a cloud-covered Cape with a small cloud hovering over his capsule too.

He continued to fight the attitude problem over the Atlantic, but was not helped by a reversal of the previous malfunction: now he began to drift to the left instead of the right. He deliberately turned the capsule through 180 degrees and found it a pleasant change. 'Just like sitting up front in a Greyhound bus. I like this attitude better than flying backwards', he thought to himself. Reluctantly he swung Friendship 7 back to her normal attitude, and observed his second sunset that day as he once more passed over Zanzibar. 'I can see a thunderstorm down below me somewhere.' The cloud tops glowed 'like a light bulb wrapped in cotton'.

Unknown to the astronaut, there was feverish activity on the ground as the flight controllers and technicians discussed the implications of a telemetry signal from the craft which suggested that the landing bag was deployed. This was a skirt of rubberized glass fibre which extended 4 ft from the base of the capsule and pushed the heatshield away from the inner bulkhead of the craft in order to cushion the impact during splashdown. At present the heatshield was held in place by the retro-rocket package, but once this was jettisoned the heatshield would apparently drift free from the bulkhead. As a result, it was highly likely that the shield would be torn away during re-entry, leaving no protection for the capsule or its pilot. Glenn would be incinerated like any meteor entering Earth's atmosphere. It was, of course, possible that the switch was malfunctioning or that Glenn had accidentally knocked it into the 'on' position. Accordingly, a puzzled astronaut received a message from the Indian Ocean communications ship 2 hours 19 minutes after launch: 'We have a message from Mercury Control Centre for you to keep your landing bag switch in off position.' Glenn glanced down and saw that it was off. 'Roger' he acknowledged.

Seven minutes later, Gordon Cooper in Muchea asked for confirmation that the switch was off. Glenn confirmed, so Cooper went on to ask, 'You haven't had any banging noises or anything of this type?' to which the reply was, 'Negative'. Glenn did not ask why Mission Control wanted this information, and continued with his checklist as he passed over Woomera in Australia. He was more interested in the 'fireflies' as he headed for his second sunrise of the flight.

The Capcom on Canton Island assumed that Glenn had been fully briefed on the problem with the landing bag, so the astronaut was surprised by the message which reached him, 'We also have no indication that your landing bag might be deployed.' He quickly asked, 'Did someone report the landing bag could be down?' 'Negative', came the reply. 'We had a request to monitor this and to ask you if you heard any flapping when you had high capsule rates [motions].' Glenn jumped to the wrong conclusion: 'Negative. . . I think they probably thought these particles I saw might have come from that.' Canton Island did not disillusion him.

Despite the attitude problems and the worries over the landing bag, Glenn was confident that he could complete three orbits, so ground control gave him the go-ahead. He had a fine view of the Cape as he set out over the Atlantic for the third time in three hours: 'I can see the whole State of Florida just laid out like a map. It's beautiful. I can see clear back along the Gulf coast.' The planned recovery area was clear and he noted that there should be no problems with the weather. Sailing backwards over Africa, the sun caught his window at an angle which showed all the dirt from the launch, and red stains which appeared to be blood – 'smashed bugs perhaps' he thought.

Into the third sunset he rode, noting a spectacular thunderstorm over southern Africa. He had time for a joke with Gordon Cooper during the passage over western Australia; then, over the Pacific, he turned the capsule around so that he could watch the sun rise ahead of him. His 'fireflies' appeared yet again, though not as noticeably as before in the glare of the sun. Then he returned to a backward-facing attitude, using the horizon as a guide since his gyroscopes were not functioning properly, and packed all loose equipment ready for re-entry. Only when he spoke to Hawaii for the last time did the possible danger of his position finally come home to him. Capcom tried to break the news gently: 'We have been reading an indication on the ground of segment 5–1, which is landing bag deploy. We suspect this is erroneous. However, Cape would like you to check this by putting the landing bag switch in auto position and see if you get a light. Do you concur with this?'

At last Glenn was fully in the picture. He could not believe that no one had told him of his peril until he was almost ready for re-entry. They seemed to be letting him complete his three orbits before telling him that he would not get back to Earth alive. So that was why there were all those 'dumb' questions about the position of the landing bag switch. Well, now came the moment of truth – if the green light came on when he threw the switch, it meant the landing bag was deployed. 'OK, if that's what they recommend, we'll go ahead and try it.' He reached out and flipped the switch. There was no light, so he switched it back to off. Hawaii seemed happy: 'Roger, that's fine. In that case re-entry sequence will be normal.'

Just as everything seemed back on schedule, Wally Schirra in California delivered another blow to his fellow astronaut in Friendship 7. A mere 30 seconds before Glenn was due to experience retro-fire, Schirra came through on the radio. 'John, leave your retro-pack on through your pass over Texas.' Glenn was so busy trying to keep the craft's attitude within the strict limits for safe re-entry that he had little opportunity to react to this unnerving request. The retro sequence began exactly on time, and Glenn felt a lurch as he returned to normal weight once more. 'It feels like I'm going back to Hawaii', to which Schirra replied lightheartedly, 'Don't do that, you want to go to the east coast!'

After 22 seconds the retro-rockets stopped firing: they had slowed the capsule just enough to send it on a long curving descent towards the Atlantic. Glenn asked impatiently for more information on when he should jettison the retro-package. Finally, Texas control gave him the confidence-sapping message, 'We are recommending that you leave the retro-package on through the entire re-entry. This means you will have to override the .05G switch [which automatically began re-entry operations]. This also means you will have to manually retract the scope.' Glenn's patience finally evaporated. 'What is the reason for this? Do you have any reason?' The Capcom tried to sound casual: 'Not at this time. This is the judgement of Cape Flight.'

As the astronaut wound in the periscope, Alan Shepard finally told him the whole story, though Glenn had more or less worked it out by then. 'While you're doing that, we are not sure whether your landing bag has deployed. We feel that it is possible to re-enter with the retro-package on. We see no difficulty at this time in that type of re-entry.' Were they still keeping back the bad news, in case he might panic? If they were, Glenn thought, it was a further insult to the test pilot and the man. He switched to a double manual system which enabled him to use the control jets used by the autopilot as well as the manual jets. He then pushed the .05G button at the exact time computed by ground control. Shepard's voice came over the radio, 'Seven, this is Cape . . . We recommend that you . . .' and then the voice faded as the ionization increased and the period of communications blackout began.

Tension built on the ground as the staff who knew of the dangers hoped and prayed for a safe return. Meanwhile, Glenn was too busy controlling the slowly revolving capsule as it plummeted towards the ocean, surrounded by a fiery orange glow. He heard a sudden crack and, looking up, saw a metal strap dangling in front of the window. The glow changed to a deeper shade of orange and pieces of debris began to sweep past the window: the heatshield was dissipating the heat as it burned away – ablated was the technical term. The outside temperature exceeded 1,600°C, and the interior of the capsule was noticeably warmer. For a moment, he felt that the capsule must be burning up and disintegrating, but his fears were soon allayed. This was a new experience for an American spaceman – neither Shepard nor Grissom had re-entered at sufficient velocity to experience this inferno. Glenn forced his mind to concentrate on the attitude control as he was pressed back into the contoured couch. Four and a half minutes after communications blackout began, he heard the welcome sound of the Cape calling, 'How are you doing?' He breathed more easily as he replied, 'Oh, pretty good.'

His problems were not quite over, however, as the capsule began to swing alarmingly from side to side. The manual fuel supply had run out, and the automatic supply soon ran out too. He decided to deploy the drogue parachute early, but as he reached for the switch he felt a sudden jerk and a dampening of the oscillations – the drogue had deployed early by itself. Although the velocity was reducing, the heat had at last penetrated the capsule walls and he began to perspire. Relief soon followed as fresh air began to enter the capsule through the snorkels, followed by the deployment of the main parachute at 10,800 ft (3,500 km). Even the landing bag light was on green; there was nothing wrong with it after all! Glenn kept repeating, 'The chute looks good . . . The chute looks very good.' The destroyer *Noa* radioed that she was on her way. The landing bag which caused so much worry and fear deployed perfectly as Friendship 7 splashed into the Atlantic 4 hours 55 minutes and 24 seconds after launch. Following instructions from the closing destroyer, Glenn stayed in the rocking capsule without blowing the hatch: there was to be no repeat of the fiasco which had marred Gus Grissom's recovery. The astronaut was not exactly comfortable, however, during the 15 minute wait for the ship to pull alongside. Apart from the erratic bobbing motion on the waves, the early afternoon heat, the enclosed cabin and the tight-fitting suit all conspired to make him perspire profusely. Glenn radioed the *Noa*, 'I'm very warm. I'm just remaining motionless here trying to keep as cool as possible.'

Friendship 7 was secured and hoisted on to the deck of the destroyer. Not surprisingly, Glenn was impatient to escape from the stifling cabin, so he requested that the area around the capsule be cleared so that he could blow the hatch. This impatience caused the only injury of the flight, for, as he turned his head away from the hatch and hit the firing pin with the back of his hand, the firing pin handle rebounded onto his knuckles. Despite the cut hand, Glenn's flight was an American triumph which was a much-needed boost to national pride and prestige. A few minutes after leaving the capsule, he was on the phone to President Kennedy, and three days later the President himself arrived at the Cape to present Glenn with the Distinguished Service Medal. On 26 February there was a triumphal procession through Washington and a visit to the White House; the ultimate American hero gave a speech before a special meeting of the Congress, an honour normally reserved for visiting presidents or prime ministers. From then on there was a national outburst of patriotism unprecedented in peacetime as John Glenn, space pioneer, became John Glenn, proud patriot and leader of the struggle against Soviet domination. The New York welcome given to Glenn and his six fellow astronauts reached new heights of emotion verging on hysteria as millions of people lined the streets to glimpse their champions and shower them with paper in a tribute reminiscent of that given to the aviation pioneer Charles Lindbergh, 35 years earlier.

Assessing the flight, perhaps the most significant comment came from Glenn himself: 'I think the fact that I could take over and show that a pilot can control the capsule manually . . . satisfied me most.' Unlike the Soviet cosmonauts who were virtually passengers, Glenn had 'flown' his Mercury capsule for two-thirds of the mission, including re-entry, and landed only 7 miles (11 km) from the *Noa*. He was, however, highly critical of the failure of ground control to inform him of the landing bag problem until 2 hours 40 minutes after it was first noticed. He observed, quite correctly, that he might have been able to contribute more to ground control's queries over the craft's status if they had confided in him. From that time on, ground controllers were instructed to be much more open over mission problems, thus paving the way for greater co-operation between the ground and the pilot. As for the technical difficulties, the thrusters were modified for the subsequent flights, while the erroneous landing bag signal was blamed on a loose switch.

'It is hard to beat a day in which you are permitted the luxury of seeing four sunsets' was the way Glenn summed up his historic flight. It was also hard to beat the wave of adulation which followed him wherever he went, something even Alan Shepard had not experienced despite his nominally being the first to fly in space. For a while Glenn was satisfied to continue his role as an astronaut, and he was assigned to help with the design and development of the Apollo spacecraft in January 1963. However, he soon became restless, and within two years he had resigned from both NASA and the Marine Corps after being promoted to the rank of colonel. He became a successful businessman and, later, a politician after being elected Senator for Ohio in 1974. Today he is a millionaire and in 1984 made an unsuccessful attempt to become the first astronaut to reside in the White House. His legend lingers on, as autograph hunters still ask him to sign his name, 'John Glenn, astronaut'. As a friend commented, 'He's part of everybody's history.'

The man scheduled to fly the next Mercury–Atlas mission was 38-year-old Donald K. (Deke) Slayton, one of the USA's leading test pilots. Slayton lived for flying and had little time or patience for the world of the socialite or the celebrity. He flew 56 bombing missions in Europe during the Second World War, with a further 7 over Japan, but he subsequently left the US Air Force to graduate as an aeronautical engineer, spending two years with the Boeing Aircraft Corporation. In 1951 he was recalled to active duty, this time with jet fighters, an occupation which he enjoyed so much that he attended the USAF Test Pilot School at Edwards Air Force Base in California in order to fly the newest fighter planes. Not surprisingly, once he was chosen as one of the Select Seven, Slayton became a leading opponent of the then prevailing concept of the astronaut as a 'man-in-a-barrel'. In September 1959, during an address to the Society of Experimental Test Pilots, he openly attacked those in the armed forces who argued that 'a college-trained chimpanzee or the village idiot might not do as well in space as an experienced test pilot'. Eventually his way of thinking won the day and John Glenn, for one, had good reason to be thankful that it did.

For four months, Slayton trained and worked for the Mercury–Atlas 7 mission with Wally Schirra as his back-up. The one cloud on the horizon, a slight heart murmur which had been detected in August 1959 during a centrifuge test, seemed to have been dismissed long ago since general medical opinion was that the problem was insignificant. Imagine Slayton's shock and disbelief when, only ten weeks before the scheduled May launch date, NASA administrator James Webb suddenly declared to a bewildered public that Slayton was unfit for operational duty. He had his flight status withdrawn, and, perhaps most hurtful of all, was banned from flying solo ever again. Apparently Webb had kept on file a letter from one of the air force doctors who had examined the astronaut soon after the problem was first noticed. The letter included a recommendation that Slayton not be assigned to a mission since, although little was known about the ailment, there was no doubt that it did reduce the heart's efficiency. Webb seems to have been stirred into action as he saw Slayton's Mercury mission approaching, so that the opinions of a new set of medical experts were sought. Since Slayton was an Air Force pilot, and only on loan to NASA, Webb referred him to the Air Force Surgeon General: an eight-man board concluded that he was fit and capable of withstanding the stresses of spaceflight. Webb then brought in three eminent cardiac specialists from Washington. They concluded that although there was no firm evidence that Slayton's heart could not stand up to orbital spaceflight, they were unable to state categorically that his performance would not be impaired during a mission. The inference was clear: if there was somebody else who did not suffer from heart murmurs in line for the mission, then that person should be given precedence. This was enough for Webb, and so Slayton was grounded.

It was then that the true character of the man shone forth. He managed to bite back the bitter words he must have felt like expressing and, at least in public, stuck to the official description of 'keenly disappointed'. Whereas many another man would have abandoned hope and drifted into a civilian job, Slayton accepted the post of co-ordinator of astronaut activities. In November 1963 he resigned from the Air Force only one year short of qualifying for his 20-year pension in order to become Director of Flight Crew Operations. As such, he became the immediate overlord of the growing astronaut corps, responsible for the astronaut office, crew training and simulation and crew procedures. He took these responsibilities very seriously, and made sure that others followed his example. Slayton became a hard task-master, a man with a decisive, sometimes aggressive, attitude who did not suffer fools gladly. Above all, however, he maintained his determination to regain the prize that had so callously been snatched from his grasp. He continued to fly high performance aircraft, accompanied by the obligatory co-pilot, in order to maintain his flight status, hoping against hope that one day the wheel would turn. Finally, in March 1972, ten years after his astronaut career had been shattered, apparently for ever, a NASA review board gave him a clean bill of health and restored his precious astronaut flight status. Deke Slayton was destined to fly in space after all, not now as a Mercury astronaut leading the American response to the Soviet space challenge, but as a member of the joint Soviet-American docking mission known as the Apollo–Soyuz Test Project.

On 15 March 1962, the same day that Slayton was officially removed from the flight programme, the new pilot of Mercury–Atlas 7 was announced – it was not Slayton's back-up, Wally Schirra, but the back-up to John Glenn, Malcolm Scott Carpenter. The official reasoning was that Carpenter had much more experience of pre-flight training, flight procedures and simulations, although, as a test pilot, he had less experience than any of the other astronauts. Carpenter himself was not overjoyed by NASA's springing the mission on him with only ten weeks in which to familiarize himself with the flight plan, including a variety of scientific experiments which were not carried on the first proving flight. Schirra remained as back-up for the approaching mission, with the promise that he would fly the third orbital mission later in the year. This practice of the back-up on one mission being designated the prime pilot for the subsequent mission became the preferred procedure from then onwards.

It is probably fair to say that more than one of his colleagues sometimes wondered how Scott (his preferred first name) Carpenter had become one of the Select Seven, chosen from more than 100 original volunteers. True, he had gained a degree in Aeronautical Engineering and had flown US Navy patrols between Korea and Formosa during the Korean War. He had also attended the Navy Test Pilot School in 1954 before test flying a wide variety of naval aircraft over the next few years. However, he had recently changed over to Air Intelligence Office on the aircraft carrier *Hornet*, so that his flying time in jet aircraft totalled only 200 hours by 1959. A devoted family man, and the father of four children, Carpenter seems to have deliberately avoided the elite but highly dangerous fighter squadrons which most pilots craved, yet felt unable to ignore the lure of the equally risky manned spaceflight pro-gramme. A superbly fit, athletic man, Carpenter excelled in the physical tests both during and after his selection: he broke five records, including holding his breath for an incredible 171 seconds. Undoubtedly it was this physical prowess, linked with his powers of perseverance and endurance, which enabled him to stand out from the crowd. Yet he also had imagination, a sense of curiosity and wonder, more of an artist's view of the world and the universe than the purely practical, matter-of-fact outlook of the majority of test pilots or astronauts. It was probably this different way of looking at things which distanced Carpenter from his colleagues, with the important exception of John Glenn. Unfortunately, it also led to a number of self-inflicted difficulties during the flight of Aurora 7.

The capsule was similar to Friendship 7 though the Earth-path indicator, the instrument panel camera and the red window filter were removed. More significantly, the manual attitude control system had been modified so that slight pressure on the handle gave a small spurt of gas from the thrusters, while a stronger push on the handle gave a much larger boost. Aurora 7 carried a number of scientific experiments which Carpenter was expected to carry out during the three orbits, in addition to the routine checks and reports previously undertaken. One experiment involved a multicoloured balloon, which was to be released outside the capsule in order to enable studies of atmospheric drag and reflection of sunlight. Carpenter was also expected to observe the behaviour of water during weightlessness, to photograph the weather formations and Earth's surface, and to study the atmospheric layers above the horizon.

Aurora 7's launch was put back several times, but on 24 May 1962 the countdown was almost perfect. Carpenter awoke at 1.15 am and, following the now normal pre-flight schedule, he climbed into the bus two and a half hours later for the short journey to pad 14. The man who some criticized for his lack of test pilot experience was almost unbelievably cool as he lay on the couch and chatted to the control centre – his pulse rate before, during and after launch was even lower than Glenn's. The morning dawned bright and clear after the thin mist was burned off by the sun, so that Aurora, named because she represented the 'dawn of a new age' achieved a perfect, smooth lift-off at 7.45.

Everything continued in the same vein as the Atlas booster operated exactly as planned, raising the capsule into an elliptical orbit which varied between 100 and 167 miles (160 × 270 km). The craft separated with hardly a jolt, then swung through 180 degrees to face backwards with its nose pitched down at an angle of 34 degrees so that Carpenter could see the jettisoned booster venting a 'steady stream of gas, white gas, out of the sustainer engine'. There soon proved to be a fault in the pitch horizon scanner, but Carpenter was too busy with photographing the Earth, manoeuvring the craft and conducting the experiments to worry about it. He found weightlessness a benefit rather than a handicap; he was amazed at the clarity of the features of the Earth's surface, spotting features such as roads and even dust tracks over Africa, though the weather over Australia was too bad to enable observation of a flare set off on the ground. The astronaut was spellbound by the magnificent coloured sunsets and sunrises, and able to confirm the existence of Glenn's 'fireflies' which Carpenter likened to snowflakes although he noted that they were not as numerous as reported by his predecessor. 'It could be frost from a thruster' was his considered opinion.

However, despite his confidence and the relatively problem-free nature of the flight so far, there were difficulties which had not been foreseen. The temperature control on the pressure suit was malfunctioning so that he was overheating and perspiring freely, while the highly active astronaut was using far too much control gas in swinging the craft around to make his observations. To some extent the fault lay with Carpenter, since he seemed oblivious of the amount of fuel he was using, and several times he inadvertently cut in the manual system while flying on automatic so that he was using fuel from both systems at once. By the end of the first orbit, Aurora 7 was down to 64 per cent of its manual fuel supply and 56 per cent of its automatic supply. Yet there were also mitigating circumstances: the astronaut proceeded through the flight plan with remarkable speed, yet the activity list was so crowded that he continually struggled to keep up to schedule. The flight of Aurora 7 illustrated a problem that irritated many astronauts on subsequent missions, namely an unrealistic work load combined with a poorly planned order of activities. The man in space was expected properly to carry out the routine checks and reports while conducting numerous experiments and manoeuvres dictated by the requirements of ground-based scientists.

The balloon experiment began towards the end of the first orbit, but the balloon did not inflate properly. Nevertheless, Carpenter attempted to make some meaningful observations of its irregular motion: 'The balloon not only oscillates in cones in pitch and yaw, it also seems to oscillate in and out toward the capsule; and sometimes the line will be taut, other times it's quite loose.' He also noted that the orange and silver segments were brighter than the others when the sun illuminated the balloon. The balloon subsequently failed to release, so that it completed all three orbits and eventually burnt up during re-entry.

During the second orbit Carpenter tried to revise the flight plan, so that capsule manoeuvres were cut down with most of the flight conducted under automatic control. Then, as reserves for the automatic system became further depleted, he switched to manual control. The overheating problem refused to go away too – at one stage the bio-telemetry registered a body temperature of 38.8°C. The hot and bothered astronaut decided to cut off all attitude control systems as the third and final orbit began. He quite enjoyed the sensation of tumbling un-controlled, carrying on with his experiments while gradually cooling down.

Then, as he observed his fourth sunrise of the day, Carpenter found himself surrounded by 'beautiful lighted fireflies'. As he reached for the densitometer in order to register their brilliance, he accidentally hit the capsule's hatch, causing a swarm of 'fireflies' to swim past his window. 'They are capsule-emanating,' the elated astronaut announced to Hawaii ground control. 'I can rap the hatch and stir off hundreds of them.' The fuel shortage was immediately pushed to the back of his mind as emotion replaced logic. 'Let me yaw around the other way,' he said, announcing his intention to re-activate the control system. Apart from using valuable fuel, this chase after the elusive 'fireflies' consumed time which had been put aside for the final checklist before retro-fire.

The harassed astronaut consequently dropped behind re-entry schedule. As so often happens at such times, fate was cruel, since he now found that the automatic system would no longer properly stabilize the craft. Forced to switch to override of 'fly-by-wire', he neglected to cut off the manual system and thereby wasted more of his dwindling reserves. In a matter of minutes, just as the craft was re-entering the atmosphere over California, Carpenter spoke to Alan Shepard: 'I am out of manual fuel, Al.' Shepard knew that the craft was not yet at the stage of re-entry when it became dynamically stable, so he advised his colleague to use the fly-by-wire system sparingly while attempting to achieve the correct re-entry attitude.

Meanwhile, Carpenter was trying to reconcile conflicting attitude indicators: the periscope and the window did not coincide with the pitch attitude indicator. At retro-fire the capsule was slewed 25 degrees to the right, unknown to its occupant. He was several seconds late in hitting the manual retro-fire button as well, and this, combined with the incorrect attitude and a slight shortfall in thrust from the rockets, meant that Aurora 7 was considerably off course as it blazed its fiery path towards the Atlantic Ocean. Carpenter was so busy concentrating on his attitude problems, which were at least partially self-inflicted, that he had to be reminded by Shepard to withdraw the periscope, and then by Gus Grissom at the Cape to secure the faceplate on his helmet in case the air drained away during re-entry.

The craft began to oscillate as re-entry began, but the automatic damp mode was able to compensate. Carpenter continued to record his observations on his voice recorder as acceleration forces built up to 7.5G while an orange glow spread around the capsule. The oscillations returned when the attitude control gas was exhausted, causing the capsule to swing wildly with the threat of overturning completely. Like Glenn before him, Carpenter decided to hit the switch which deployed the drogue parachute as a method of stabilizing the descent. The main parachute followed within a minute, but another problem had arisen. Carpenter was unable to raise Gus Grissom on his radio, so he knew that he must have drifted well off course and beyond Cape communication range. In fact, there was equal concern at Mission Control over the lack of communication: no one knew whether Carpenter had survived re-entry or where the craft had landed if re-entry had been successful.

Aurora 7 splashed down some 250 miles (420 km) down range from the planned target area where the carrier *Intrepid* waited. The carrier's commander put into action a contingency rescue plan, sending recovery ships towards the landing area as soon as the position was known. Meanwhile, Carpenter was concerned about the buoyancy of his capsule and so decided to abandon ship. After the ordeal of his 4 hour 56 minute spaceflight, the astronaut now had to compose himself for a long wait in the tiny liferaft. One hour later, he was joined by a pararescue team dropped from an Air Rescue Service plane. An amphibious aircraft was soon circling overhead, but was ordered not to attempt a landing unless the Navy helicopters, which had only just left the *Intrepid*, were unable to effect the retrieval. Carpenter, resigned to his watery resting place, opened his survival rations – he deserved something to eat apart from paste after the experiences he had been through. It was three hours after splashdown when he was finally winched aboard a helicopter: his first act was to cut a hole in the toe of his suit to drain the sea water which he had inadvertently let in by leaving the suit inlet hose open on leaving the capsule. Yet another hour passed before the exhausted astronaut set foot on the deck of the *Intrepid*. Aurora 7 was eventually retrieved by the USS *Pierce* at around 7 pm Cape time.

Carpenter was elated by his successfully completed mission and by the sights and sensations which so few people had yet experienced. He wanted to tell the crew of the *Intrepid*, the doctors and scientists, the President, indeed the whole world about his journey into the unknown. There was so much to relate: the fuel shortages, the overcrowded work schedule, the experiments and the 'fireflies'. He seemed blissfully unaware of the scare stories which circulated for more than an hour after he splashed into the ocean, the tears and the prayers which greeted the news that an astronaut might have burned up or suffocated during re-entry. The returning hero subsequently received the DSM from the President, with parades and accolades all over the country, including Boulder, Colorado, his birthplace.

Yet behind the scenes NASA staff were very unhappy about the mission. Flight Director Chris Kraft apparently said very loudly to his colleagues, 'That sonofabitch will never fly for me again.' He and a number of other engineers and ground staff viewed the flight of Aurora 7 as a near disaster, largely due to the arrogant independence of an astronaut who deliberately ignored repeated warnings of overconsumption of fuel, finally running out of time with the result that he became rattled, forgot certain vital checks and misorientated the craft so that it nearly skipped off the atmosphere to be lost in space forever. There was even a suggestion, never publicly aired, that Carpenter might have panicked during the final stages of the flight, despite his remarkably low pulse rate – it had never risen above 105 during any part of the mission, including the hair-raising re-entry.

Carpenter continued to work with the Mercury programme, only resigning from NASA in 1967. However, long before this he realized that he had lost the internal political battle with the result that he would never fly in space again. Undaunted, he applied for and was granted a special leave of absence to participate in the Navy's 'Man-In-The-Sea' programme as an aquanaut. Helped by his astronaut training, as well as his unique physical abilities, he became the leader for two out of the three teams of aquanauts who lived 240 feet (75 metres) down on the bottom of the Pacific for periods of 15 days each. On 29 August 1965 he took part in a unique experiment when he spoke to Gordon Cooper, his old Mercury colleague, and Pete Conrad as they orbited the earth in Gemini 5 while he rested on the sea bed. Carpenter retained his interest in 'inner space' after he resigned from NASA, returning to help with the Navy's Sealab 3 experiment. Even after leaving the Navy in 1969, he continued to work in the fields of aerospace and ocean engineering technology as President of the Sea Sciences Corporation in Los Angeles.

By August 1962 NASA had further reason for displeasure. With the successful dual flight of Vostoks 3 and 4 the Soviet Union had once again demonstrated its lead in the space race, while at the same time pulling off a brilliant propaganda coup. Between them the Vostok crews logged more than six days in space, with a combined total of 112 orbits: the American total of six orbits by Mercury craft seemed positively paltry in comparison. The Western politicians, press and public were unaware of the basic difference between the two programmes: the Americans had a main objective which had been laid down by President Kennedy the previous year, whereas the Soviet Union's space strategy, though in general the same as that of the USA, was subject to the short term objectives of Mr Khrushchev – to beat the USA and to produce propaganda extravaganzas on demand. Although there was undoubtedly pressure on NASA, especially from the military, to compete with the Soviets on their own terms, the Space Agency continued steadily along its planned route towards Gemini, Apollo and eventually, the moon.

The Mercury spacecraft was originally designed for a maximum of three orbits, but from February 1962 the Manned Spacecraft Centre engineers had been working on possible ways of extending the mission duration to seven orbits, with an eventual aim of achieving 18 orbits. They succeeded in cutting flight consumption of electricity and fuel, while reducing the oxygen leak rate from the craft and utilizing the maximum potential to absorb cabin carbon dioxide of the lithium hydroxide cleanser. By the end of July, the pilot of Mercury–Atlas 8, Commander Walter Schirra, had been given a flight plan for six orbits (or more accurately five and a half). The re-entry over the Pacific and the final recovery north-east of Midway Island was easier for a six orbit mission than seven orbits, which meant a landing too far from mainland USA. For the first time the splashdown had to be in the Pacific, since the Earth would rotate 135 degrees during the nine hour mission.

Wally Schirra had been the back-up for Deke Slayton and then Scott Carpenter on Mercury–Atlas 7. Although both he and Carpenter were navy men, Schirra had more in common with test pilot Slayton, even though the latter belonged to the rival air force. Certainly their views on astronauts as pilots rather than mere passengers were very similar, and both of them hated the leg pulling by other test pilots, especially the X15 rocket plane pilots who risked their lives as they soared above altitudes of 50 miles (80 km) to the edge of space. The suggestion by these colleagues was that monkeys, the first 'pilots' of the Mercury series, could just as easily take over the seats occupied by the Select Seven astronauts. Schirra told a story to emphasize his point:

A marvellous new passenger plane takes off from Los Angeles. As it reaches flight altitude, a voice comes on the loud speaker: 'Ladies and gentlemen. Welcome aboard. You are now cruising at an altitude of 35,000 feet. My voice is coming to you by recording, for I am back on the ground. I am not needed in this machine because it is fully automatic, and with a machine nothing can go wrong . . . can go wrong . . . can go wrong . . .

Schirra was a hard-headed, independent pilot with ice cool nerve and a record to prove it. He had flown 90 combat missions in Korea as one of the top pilots, flying on an exchange basis with the USAF jet fighter squadrons, downing two MiG-15s and earning the DFC as well as two Air Medals. He went on to test the Sidewinder missile, then test flew jet fighters for a while at Edwards Air Force Base where Slayton also worked. Calm in tough situations, he could be charm personified yet stubborn and outspoken when necessary. Schirra drove a sports car, loved flying and carried out practical jokes as his way of easing the strain. At 5 ft 10 in and 185 lb, he was the largest of the seven, a stocky, jovial professional.

The nature of his mission as he saw it was exemplified by the name he gave his Mercury capsule – Sigma 7 after the engineering symbol used to denote summation. This was to be a practical test flight with precedence given to the successful launch, flight and return of the capsule rather than a series of weird experiments dreamed up by scientists who had little or no idea about the problems of flying a Mercury spacecraft. He already had to contend with a problem which had not applied to his predecessors, namely a period of nearly 30 minutes on later orbits when he would be out of contact with ground control since he would be passing over territory never overflown by Americans before. Schirra had learned from the mistakes of the previous missions, so he made sure the flight plan minimized attitude manoeuvres with plenty of time allocated for each activity. One way in which fuel and power would be conserved was a planned drift with no attitude control for one whole orbit, something which Carpenter had been forced to adopt through necessity on his flight.

The Atlas booster sprang a leak in its fuel tank, causing a postponement of the planned September launch, but on the morning of 3 October 1962 it was all systems go as Schirra settled into the couch in Sigma 7. He found the little mementoes left by his team-mates: an ignition key attached to his right-hand attitude controller and a steak sandwich in a plastic bag secreted in his personal compartent. At 8.15 am the gleaming rocket rose into the sky above Pad 14 with a suddenness which surprised the man perched in its nose. Engineers at the Cape were shocked a few seconds later as the rocket developed such an alarming roll rate that an abort was almost triggered. Happily, the remainder of the ascent was problem-free, apart from a minor communication difficulty, so that the booster cut off 5 minutes 16 seconds after launch to put Sigma 7 into a 100 × 176 mile (161 × 283 km) high orbit, marginally the highest altitude reached by any Mercury astronaut.

Schirra was at his determined best from the start: no fuel shortages and pilot errors for this flight. Using the manually operated fly-by-wire system, he turned the capsule round through 180 degrees with a minimum use of fuel, far below that used by his predecessors. Deke Slayton at the Cape heard him introduce a phrase which both of them understood only too well, 'I'm in chimp mode right now and she's flying beautifully.' Schirra was very pleased with the modified attitude control thrusters as he later recalled: 'It was a real thrill to realize the delicate touch that it is possible to have with fly-by-wire, low . . . it just amounted to a light touch and maybe a few pulses in either axis to get the response I wanted.'

The suit's temperature control system gave Schirra a less pleasant experience, one which worried the flight director, Chris Kraft, and the flight surgeon enough for them to consider terminating the mission after one orbit. By the time Sigma 7 was overflying Muchea in Australia, Schirra's suit temperature had risen from 23°C at launch to over 32°C despite adjustments to the regulator by the perspiring astronaut. Fortunately, Schirra's speciality in the Mercury hardware was the pressure suit, and he had closely studied the temperature control system, including the overheating suffered by Carpenter. The surgeon had no way of gauging Schirra's body temperature since the sensor was malfunctioning, but as Schirra gradually advanced the regulator the suit temperature seemed to stabilize, encouraging Kraft to give the go-ahead for one more orbit. He had read the signs correctly, for by the time Schirra reached Australia again his suit temperature had dropped to a comfortable 22°C.

During the second orbit the astronaut successfully orientated the capsule using only visual references without using the attitude indicator on the control panel. Schirra's conclusions backed up those of Carpenter: retro-fire attitude could be successfully attained even if the instruments failed, while the periscope was a useful but non-essential back-up. For part of the third orbit the capsule flew in the 'drifting mode' to conserve fuel, a practice which was repeated for the whole of the fourth orbit after Gus Grissom in Hawaii passed on the good news that he had 'a go for six orbits'. Schirra was thoroughly enjoying himself, hardly pausing for breath as he cruised around the world. 'I'm having a ball up here drifting. Enjoying it so much I haven't eaten yet.' The VOX (voice actuated) tape recorder was continually in use when he was out of contact with the ground, so that the Capcoms often heard him talking to himself as he came over their radio horizon.

Schirra found no difficulty carrying out his flight plan: the experiments were not allowed to intrude in the smooth running of the flight. The experiment with the flares over Australia failed once again because of cloud cover; other experiments did not involve the astronaut at all.* The astronaut passed disorientation tests with flying colours, experiencing no discomfort or physical problems in the weightless environment. He took numerous photographs of the Earth's surface and clouds, and also duplicated Carpenter's creation of a snowstorm by thumping on the side of his capsule. At the end of the fourth orbit he was persuaded by John Glenn to say a few words to the millions listening to the live radio coverage of the flight both in the USA and Europe. Wally's response was typical: 'I suppose an old song, "Drifting and Dreaming" would be appropros at this point, but at this point I don't have a chance to dream, 'I'm enjoying it too much.' Even with the craft powered up again during the fifth and sixth orbits, fuel consumption was unbelievably low – Schirra could have spent several days going round and round the world if ground control would have let him.

Schirra took plenty of time during the final orbit to complete his checklist and orientate the craft into retro-fire attitude – no last minute rush for him. The retro-rockets fired soon after Sigma 7 entered the control zone covered by Alan Shepard on the Pacific Command Ship. As the capsule began re-entry, Shepard was able to report the startling news that the automatic fuel supply still read 68 per cent while the manual supply was 78 per cent. Although the retro-rockets fired two seconds late, Schirra knew that he was going to land right on target, or, as he told Shepard, 'I think they're gonna put me on the number three elevator'. Schirra followed the safety procedure by dumping the remaining fuel before he splashed into the ocean, and, sure enough, he came down only 5 miles (8 km) from the carrier *Kearsarge* in full view of its crew. Schirra remained cool to the last, preferring like any good captain to remain with his ship to the end. A five-man whaleboat towed the capsule, which now had a flotation collar as a safety precaution, to the carrier where it was hoisted aboard with its pilot still inside. Just over 40 minutes after splashdown, the triumphant hero hit the hatch jettison knob, cutting his knuckles in the process, and stepped on to the deck bathed in sunshine. Schirra's comment on the mission summed it up nicely: 'It was a routine,

* Test samples of heat-resistant material were attached to the outside of the capsule, while radiation was detected by special films and emulsion packs.

textbook flight. I was able to accomplish everything I wanted during the flight. It was a honey of a machine.'

Schirra's flight had restored some pride and prestige to the USA, although as a TV spectacle it has been rather dull. The NASA engineers had seen their modifications to Mercury tested and come through with flying colours, so much so that they were more than ever convinced that the capsule could be extended to a mission lasting more than a day. Schirra received the now traditional parades and congratulatory speeches which greeted the conquering hero, including the special 'Wally Schirra day' in his home town of Oradell, New Jersey. There was also the visit to the White House where President Kennedy presented him with the DSM, though this time it seemed a rather hurried, peremptory affair. Kennedy was involved in a deepening crisis over the shipment of Soviet nuclear missles to the island of Cuba, a threat which promised to bring the world to the brink of a nuclear war, so the reception for the returning hero had to be short and sweet. Schirra, the archetype of the test pilot-cum-astronaut, continued to work with NASA as the development of the two-man Gemini programme progressed towards a start date of 1964–5 and promised him further challenges.

The next, and last, flight in the Mercury programme was also a major challenge for the engineers and the pilot. Originally planned as an 18 orbit mission, this was altered to 22 orbits when the constraints of re-entry and rapid recovery were considered. A great deal had been learned from the 'textbook' mission of Wally Schirra; nevertheless, Mercury–Atlas 9 would last more than 34 hours, nearly four times the duration of Schirra's previous record-breaking flight. Clearly, the Mercury capsule would have to be considerably modified to enable the astronaut to survive such a marathon: more oxygen, more water in the cooling system, more lithium hydroxide to absorb carbon dioxide in the atmosphere, more food and more electricity had to be provided for the tiny capsule without breaking the launch weight barrier. Any equipment considered superfluous, such as the periscope carried on all previous missions, was removed from the capsule in order to save weight and space. One major piece of surgery involved the removal of the rate stabilization and control system, one of the four attitude control systems which were originally built into the Mercury capsule. In all, 183 modifications were made to the last capsule of the series compared with Sigma 7, yet the difference in weight added up to a mere 4 lb (1.8 kg)!

The man entrusted with this new challenge was the youngest of the Select Seven, Leroy Gordon Cooper Junior, a major in the USAF, had packed a lot into his 36 years though many might query his selection above other astronaut candidates back in 1959. Cooper was the only southerner in the group, his Oklahoma drawl making him instantly recognizable. He began his military career in the Marines, but flying was in his blood – his father had been a colonel in the Army Air Force – so he enrolled at the University of Hawaii to obtain his commission. There he met his wife, who was also a pilot, and, following pilot training for the USAF in 1949, he was assigned to a fighter-bomber unit in Germany. Returning to the States, he obtained a degree in aeronautical engineering, qualified as a test pilot and became Experimental Flight Test Engineer at Edwards Air Force Base. Nevertheless, among the test pilot fraternity, Cooper was a mere novice: he had done some test flying as part of his job, but to a red-hot fighter pilot who risked his life daily, the phrase 'Cooper was in engineering' summed it all up.

The fact remained that 'Gordo' was a natural pilot, as cool as any of his colleagues when the chips were down or when they pulled his leg about being 'in engineering', and determined to prove that he was as good as the next man. The fact that he had been left until the final Mercury mission did not dishearten him; neither did the possible medical problems associated with prolonged spaceflight. The Soviets were giving little away, but there were stories of Titov suffering from nausea and disorientation during his 17 orbit flight aboard Vostok 2, while Wally Schirra had experienced pooling of blood in his legs for a short time after returning from only six orbits. The only thing Cooper hated about being an astronaut was the public appearances. Cooper was a man who liked action – fast planes, fast boats and fast cars – but hated receptions and social functions.

The Mercury–Atlas 9 mission was originally scheduled for April 1963, but a series of delays put back the launch date to 14 May. Early on that day, Cooper awoke, breakfasted and donned his pressure suit before being driven to Pad 14 at about 6.30 am, but after four hours in the capsule the launch was postponed until the next day due to technical problems with the launch tower and with the tracking station in Bermuda. The morning of 15 May saw the same procedure repeated, but this time the countdown was almost perfect. Cooper was clearly not upset by the events of the previous day or the tensions of the final launch preparations: the biomedical readings coming through to the doctors in mission control showed that he had gone to sleep! The astronaut had aptly named his capsule Faith 7.

At 9.04 am the Atlas booster took off for the last time with a human passenger aboard. The trajectory was perfect with a minimum of oscillations or vibrations, although Cooper's pulse jumped noticeably when the booster separated from the capsule. Five minutes after launch Cooper was weightless in an orbit that varied between 100 and 166 miles (160 × 266 km). Wally Schirra, Capcom at the Cape, told him, 'You look real good here, Gordo' to which Cooper replied in confirmation, 'Real good, buddy.' The elated astronaut kept repeating, 'What a thrill! What a thrill!' There was no question of his being carried away, however, for his careful turnaround of the capsule was the most frugal of all the Mercury missions, beating even ultra-professional Wally Schirra.

The first orbit was mainly taken up with routine checks of the spacecraft and communications systems. Satisfactorily completely, he was given the go-ahead for seven orbits. At this stage, he was following the well-trodden path of his three predecessors as he headed over Africa and the Indian Ocean, viewed the first of many sunsets that day, and observed the lights of Perth far below. Back into daylight he rode, surrounded as usual by the 'fireflies'*, then across the Pacific where he would splash down a day and a half later, and back to the voice of Alan Shepard at the Cape. America's first astronaut, as well as the public, was able to watch TV pictures of Cooper relayed directly from the craft for the first time. The slow-scan camera sent back TV pictures every two seconds: for Gordon Cooper there was nowhere to hide. 'You look pretty casual on TV', commented Shepard. 'So I am', came back the southern drawl of Cooper.

As the second orbit began, the astronaut powered down some of the craft's systems as part of the planned conservation of electricity and fuel. Although he was suffering from the problem encountered by all the Mercury astronauts – a space suit temperature irregularity – he carried out the programmed blood pressure test and exercises with the elastic cord, then promptly dozed off for 10 minutes or so. This was to be a feature of the entire mission, one which few of the experts on the ground would have dared predict, and after the mission Cooper described his experience.

> One indication of my adjustment to the surroundings was that I encountered no difficulty in being able to sleep. When you are completely powered down and drifting, it is a relaxed, calm, floating feeling. In fact, you have difficulty not sleeping. I found that I was catnapping and dozing off frequently. Sleep seems to be very sound. I woke up one time from about an hour's nap with no idea where I was and it took me several seconds to orient myself to where I was and what I was doing.

Not that Cooper spent almost the entire flight asleep. He had a number of experiments built into the flight plan, both medical and scientific. On the third orbit he released a small sphere with two flashing xenon gas lights: at first he could not see anything, but on the night passes of the fourth and fifth orbits he could clearly see the lights despite a gradual separation to a distance of more than 13 miles (20 km). This was an observation of major significance for future rendezvous and docking missions in Earth orbit. On the sixth orbit the balloon experiment attempted by Scott Carpenter should have been repeated, but

---

* The fireflies were derived from frozen water vapour released into space by the capsule's cooling system and smaller dust particles.

this time the balloon refused to release so the experiment had to be written off.

One of the medical experiments involved the collection of the astronaut's urine for subsequent analysis in order to determine his body's reaction to weightlessness. The astronaut was provided with a syringe to transfer the liquid from the receptacle to a special container, but unfortunately the syringe leaked, so that the unfortunate Cooper spent the rest of the mission surrounded by a drifting swarm of amber droplets. At least it was diluted somewhat when he had a mishap while rehydrating some of the food:

> I'm eating a pot roast of beef. I've had considerable difficulty getting the water in it from this water device on the McDonnell water tank. I spilled water all over my hands and all over the cockpit here trying to get some in it. I have succeeded in getting about half of it dampened and am proceeding to eat . . . I am washing my face with a damp cloth now . . . Certainly feels good.

For this mission Cooper had the benefit of a menu specially devised for him, including such delicacies as spaghetti with meat sauce, and roast beef with gravy, cold of course. No wonder he remarked, only half jokingly, to the Capcom in Muchea that, 'hot black tea would go very well right now!' as he neared the end of the fourteenth orbit.

From the sixth orbit onwards Cooper was breaking new ground. His craft began to drift over territory never seen from orbit by an American before, but also away from the network of communications relay stations which had served his colleagues. NASA had deployed three ships in the Pacific in order to fill some of the gaps, but there were still alarming omissions, especially on orbits 10 to 14, when the astronaut was out of contact for periods of up to half an hour. This was the part of the mission during which Cooper's eight hour rest period was scheduled, but, for once, he found difficulty in sleeping for long spells, spending some of his wakeful time in photographing the magnificent panoramas of South America, the Himalayas and south-east Asia which his colleagues had missed. Indeed, photography of the Earth's surface, of the Earth's horizon, and, in infrared, of the cloud structures, was one of the main tasks and major successes of the mission.

Even more significant, and unexpected, were the observations which Cooper reported without any visual aids. Earlier astronauts had reported a wide variety of surface features from their orbital vantage point, but Cooper's observations were so detailed that they were greeted with incredulity by the experts on the ground. Only later were the scientific and military connotations of such high resolution observations from space realized, with confirmation by the Gemini astronauts of Cooper's reports.

I could detect individual houses and streets in the low humidity and cloudless areas such as the Himalayan mountain area, the Tibetan plain, and the south-western desert area of the US. I saw several individual houses with smoke coming from the chimneys in the high country around the Himalayas. The wind was apparently quite brisk and out of the south. I could see fields, roads, streams, lakes. I saw what I took to be a vehicle along a road in the Himalaya area and in the Arizona–west Texas area. I could see the dust blowing off the road, then could see the road clearly, and when the light was right, an object that was probably a vehicle. I saw a steam locomotive by seeing the smoke first; then I noted the object moving along what was apparently a track. This was in northern India. I also saw the wake of a boat in a large river in the Burma–India area . . . and a bright orange light from the British oil refinery to the south of the city [of Perth].

Following the rest period, Cooper continued his tasks of photographing and observing the Earth, taking blood pressure readings, carrying out exercises, status checks and all of the routine jobs expected of an astronaut. All seemed to be plain sailing, despite a gradual fall in the oxygen pressure in the cabin and a higher than predicted carbon dioxide reading. As yet, there was no need to worry about the astronaut's having breathing difficulties or suffering from slowed reactions. Then, on the nineteenth orbit, a warning light indicated that the capsule was slowing down, the first part of the re-entry sequence. Cooper reported the problem to Hawaii, and a series of tests carried out as Faith 7 flew over friendly home territory confirmed his belief that the automatic control system was shorting out. From the quiet calm of the preceding hours there was a sudden change to frenzied activity as the flight controllers wrestled with this unprecedented problem. With the gyroscopes and the horizon scanners out of action, it looked as if Cooper would have to conduct a manual re-entry, although the automatic system could take over as normal once the descent passed the 0.05G point.

On the twenty-first orbit even this degree of automatic control was lost when a fuse blew in the automatic control system. There was no question that from now on Cooper would have to use manual control for the complete re-entry, from firing the retro-rockets to deployment of the parachutes. Apart from aligning the capsule's angle of re-entry by visual reference to the horizon, he would have to hold the capsule in position by manual control of the yaw, pitch and roll, and fire the retro-rockets at the precise time for an accurate splashdown. Cooper's response indicated a feeling of nonchalance rather than fear or excitement: 'Things are beginning to stack up a little. ASCS inverter [a device to convert alternating current to direct current] is acting up, and my CO_2 is building up in the suit. Partial pressure of oxygen is decreasing in the cabin. Standby inverter won't come on the line. Other than that, things are fine.'

The flight surgeon obviously did not think things were fine, for Cooper was told to take a dexedrine tablet to keep his mind at maximum alertness. Over Zanzibar on the final orbit he was told to use his emergency oxygen supply if the breathing problem worsened, but Cooper was unwilling to do this unless absolutely necessary. Approaching the western Pacific for the last time, Faith 7 entered the communication zone of Capcom John Glenn on the ship Coastal Sentry Quebec. Final instructions were relayed to the astronaut in good time for the re-entry to begin. The intensive training now paid dividends as Cooper held the capsule in perfect alignment while firing the retro-rockets exactly on schedule. Mercury pilot Cooper showed that he could beat any automatically controlled re-entry when it came to an accurate splashdown: the capsule came down in full view of the carrier *Kearsarge*, about 4 miles (7 km) distant and only 1 mile (1.8 km) off target. Even Wally Schirra had not hit the bullseye with such accuracy. Within 40 minutes, Faith 7 had been fitted with her flotation collar and carried by helicopter to the carrier where Cooper blew the hatch. Following a brief medical inspection, the new American space endurance record holder was helped out of the capsule, staggering a little as he overcame a dizzy sensation while he adjusted once again to normal gravity.

Gordon Cooper was rightly hailed by the nation and the world as a hero, having overcome all the technical difficulties and proved once again that, when it came to the moment of truth, there was no subsitute for a well-trained, dedicated pilot. In the age of computers and machines it was reassuring to have faith in human ability and character reaffirmed. It was also reassuring to the American public to know that the time and money invested in the space programme was beginning to pay off, so that a serious challenge could be mounted to the Soviet supremacy in outer space. The astronauts argued that a three day Mercury mission was necessary as a final test for long endurance space flights that could match those of their Soviet rivals, but the NASA administrators remained firm, so Alan Shepard never got his orbital mission in a Mercury. Only three of the Select Seven went on to fly in the Gemini programme, though Shepard and Slayton did make a comeback when Apollo got off the ground. Meanwhile, the expansion of manned spaceflight activities required an increase in the size of the astronaut corps. On 17 September 1962 a further nine recruits had been added to the handful who had so far monopolized the limelight. Fourteen more were chosen in October 1963, bringing the total number of astronauts to 30. As the honour was spread more widely, the status accorded to spacemen would never be quite the same again.

4

Sunrise, Sunset

The Voskhod Missions, 1964–5

More than a year passed without a manned spaceflight by either the USA or the Soviet Union. The Gemini programme to launch two-man crews into Earth orbit was in its early test stage, but was suffering from teething problems which continually put back the date of the first manned mission. From behind the Iron Curtain came little news concerning the next step in Soviet space exploration, though the one-man (or woman) Vostok craft had clearly outlived its usefulness by reaching the limits of its design capability. There was no longer any scientific or technological advantage in simply sending one astronaut or cosmonaut into space and then safely returning him or her to Earth; neither was there any propaganda value in repeating such events now that they had become almost common place. The question was, had the Soviets reached a technological barrier beyond which they could not venture, or were they planning some new spectacular advance which would stun the Western world?

If there were doubts about the latter among the Soviet spacecraft designers towards the end of 1963, the Soviet leadership had no such qualms. Khrushchev had made considerable political capital out of the Soviet space triumphs which, since the launch of Sputnik in 1957, had proclaimed to the world the superiority of Soviet technology and of the communist system in general. Apparently, it was Khrushchev who, more than anyone else, was responsible for the birth of the Voskhod (Sunrise) project. Certainly, the programme did not long survive his removal from the political arena.

If Western reports are to be believed, it seems that Khrushchev summoned Chief Designer Korolev to the Kremlin in 1963 and presented him with an ultimatum. The Americans were forging ahead with their Gemini programme, and had publicly announced their aim of launching a three-man Apollo craft by 1966 as a preparation for the manned landing on the moon by the end of the decade. The Soviet Union's new generation of spacecraft, the Soyuz, was still on the drawing board and unlikely to materialize before 1965 at the earliest. This would mean a time lag of at least two years before the next cosmonauts would be blasted into space, a delay which was unacceptable to Premier Khrushchev. He therefore demanded an interim programme which would maintain Soviet prestige and steal much of the thunder from the American competition. The answer was to modify the existing Vostok capsule and its launch rocket so that a Soviet multi-manned mission could be sent up ahead of the first Gemini launch – however, since Gemini held two men, it was clearly preferable for Voskhod to contain not two, but three, cosmonauts. Furthermore, a subtle variation on the crew's membership would create an even more favourable impression both at home and abroad: instead of the usual highly trained pilots, a scientist and a doctor would be carried aloft, so that Voskhod I could be called 'the world's first space laboratory'.

The official Soviet version does not mention any of these behind-the-scenes machinations. Voskhod is described as the logical successor to Vostok, a project envisaged by Korolev even before Gagarin's first orbital flight. The Chief Designer's guiding principle since the days when he designed gliders was said to be: 'Build yourself, fly yourself, test yourself.' When this came up in a discussion with other experts in 1963, Korolev is said to have accepted that he was too old to fly in space, adding, however: 'But the time has come when the specialist must fly. Konstantin Feoktistov is in my place.'

Feoktistov was on Korolev's design team, and a graduate of the same Moscow Higher Technical School which his superior had attended as a pupil and lecturer. He was born in 1926 in Voronezh, served and was wounded in the Second World War, and eventually gained an influential position in the space programme. He explained: 'I was one of those who lectured to the future cosmonauts. Later, they took their exams, and I was a member of the examination board.' The story was later told of how he checked and double-checked Gagarin's capsule before the historic launch in 1961. Descending in the lift, he bumped into Korolev and two companions, and was unable to hold back the comment: 'It wouldn't be a bad idea for us engineers to go up in space.' Korolev apparently frowned, and, as Tass correspondent Alexander Romanov reported, 'glancing at the spare figure of the engineer, at his pale face and prematurely greying temples, he thought doubtfully: will he be able to stand the stress of launching and descent? Besides, he needed him so much on earth.' So Korolev made light of the matter: 'When we create a multi-seater craft and develop a soft-landing system, then we'll go up together for three days. Agreed?' Feoktistov did not smile, 'but replied in such a way as though the question of his flight had long been decided and it was only a matter of settling the exact date: "It is essential for me to fly. Essential." ' Whatever the truth concerning that encounter, it is a fact that Feoktistov and his doctor companion, Boris Yegorov, only had four months in which to cram all of their flight training; allowing for their knowledge of the space programme, this was still barely adequate, and would certainly not have been possible in the USA.

Yegorov, a lieutenant in the Soviet air force, was only 27 years old. He was born in Moscow, the son of a famous neurosurgeon and was himself a graduate of the First Medical Institute in Moscow. He specialized in aviation medicine, particularly the working of the inner ear, and was involved in studies on the effects of weightlessness on the human balance mechanism. He was also one of the doctors who examined Gagarin on his return.

The lives of these two 'rookies' were in the hands of air force colonel Vladimir Komarov, one of the original cosmonaut detachment. The 38-year-old commander was born in Moscow and entered pilot training when he was only 15. A graduate of the Air Force Engineering Academy and three other air force colleges, he was one of the senior cosmonauts, and would probably have flown earlier but for the discovery of a minor heart irregularity similar to that which grounded astronaut Deke Slayton. He had previously been Popovich's back-up on Vostok 4 and, as an engineer, had been encouraged to participate in the design programme by Korolev.

Discussing the Voskhod 1 mission, the Chief Designer rationalized the choice of crew for the Soviet press:

The flight of the multi-seater spacecraft Voskhod in October 1964 was a qualitatively new step. That was the world's first space laboratory, the first time a crew of three went up. It was composed of a specialist in the field of physics and technology, a doctor and a commander who was an engineer. It is hardly necessary to explain how important it is for a scientist to see and perceive for himself what happens in the state of weightlessness, in the world of unknown radiation, the hitherto unseen brilliance of the sun and distant stars, amidst the blackness of the space night, on the border of light and dark which replace each other. In other words, from now on the scientist has access to more than just dry figures, the recordings of instruments, photo and telemetric films, the readings of sensors. Now he has access to a live perception of events, the sensations of what he has experienced and seen, he has the enticing possibility of conducting research as he wishes, of analysing the results at once and proceeding further.

There is no doubting that Feoktistov and Yegorov were the first specialists in their fields to go into space, but Korolev's eulogy to

Voskhod hardly related to the true significance of the flight, and modern readers could be forgiven for thinking they were reading about Spacelab or Salyut rather than the cramped, primitive Voskhod capsule which only spent one day in Earth orbit.

There are other respects in which the quotes attributed to Korolev at the time were misleading: 'Space technology has taken a stride forward. The crew of the Voskhod made the flight in a comfortable module equipped with everything needed for work and relaxation.' Certainly the launch weight of the capsule was, at 5.2 tons (5,320 kg), more than half a ton heavier than the Vostok craft, but part of this increase was due to the two extra crew members. There were some improvements on the Vostok, notably an extra retro-rocket to ensure a safe return from orbit and a further retro-rocket system attached to the top of the spherical capsule which would brake the descending Voskhod just before it crashed to Earth with a bone-shattering thud. As Korolev put it: 'On the Vostok, at a certain height the cosmonauts catapulted out and came down to earth by parachute. This time the crew descended to Earth whilst sitting in their seats. They could have come down on water equally successfully. The Voskhod has the buoyancy to float. Soviet designers have taken care to minimize the loads on the human organism and the landing system ensures that when the craft touches Earth its speed is down to zero, and it lands softly.' The craft was also fitted with improved TV, radio and navigational equipment, and a more refined attitude control system.

In order to cram three men and all this extra equipment into such a tiny capsule, certain safety measures employed on Vostok had to be discarded, making Voskhod 1 possibly the most dangerous mission ever. The men were seated in three lightweight couches, instead of the normal ejection seats, with that of Yegorov slightly above and in front of the other two. This meant that there would be no escape should a malfunction occur during the launch. Furthermore, there was insufficient room in the capsule for the usual pressure suits, though the Soviets insisted that they were omitted since they simply were not needed. The cosmonauts boarded Voskhod wearing 'sky-blue jackets over lightweight, dark grey suits and white helmets'. Until recently, no photographs of the exterior of Voskhod have been released, but from the available evidence there seems no doubt that it was simply a Vostok capsule modified for multi-manned missions – a stop-gap compromise which met premier Khrushchev's demands for more headline-grabbing feats.

Alexander Romanov described the scene on the morning of 12 October 1964:

Three men march down the concrete strip of the cosmodrome's launching pad, lit by the pale autumn sun . . . They walk in silence, unhurriedly. Their bearing is impressive. We admire them.

Even old hands at the cosmodrome stand rooted to the spot, watching them. Having reported to the Chairman of the State Commission on their readiness, the crew go up to Korolev. The Academician embraces each one warmly, then says, 'It's time!' The cosmonauts climb up the light metal ladder to the platform in front of the lift which will take them to the ship.

The countdown proceeded without major delay, so that Voskhod blasted off from Baikonur at 10.30 am Moscow time, flame belching from the twenty main rocket nozzles. The four strap-on boosters were soon spent and fell away, followed by the first stage, leaving the new second stage to boost them into orbit with a faster rate of acceleration than any previous Soviet manned mission. This second stage had been developed to carry robot probes to the moon and planets, but had never before been used to carry men aloft. Utilizing the extra power of this stage, Voskhod was carried into the highest orbit yet reached by man with a maximum altitude (apogee) of 254 miles (409 km). The Vostok orbits had been lower, enabling the use of a relatively low-powered rocket and with the added advantage that the capsule would be slowed by atmospheric resistance which would ensure re-entry after 10 days should the retro-rocket fail; the latter possibility seemed to have been excluded by the back-up retro-rocket on Voskhod.

Each crew member had his specific task to perform. Pilot Komarov later described his duties:

As commander I was responsible for the work schedule on board the craft. Using the new attitude control system, I orientated the Voskhod in various directions. This enabled Feoktistov and Yegorov to carry out the planned scientific observations. I conducted practically all the radio contacts with Earth, the reports to the State Commission as to how the flight was proceeding, and made entries in the ship's log. Moreover, my personal observations, naturally, will also form part of the broad scientific research that was conducted. The crew knew exactly what their duties were and worked in harmony. Before drawing any conclusions, we usually consulted each other on the wording, trying to arrive at the most accurate description of the results of the investigation.

Feoktistov, meanwhile, carried out experiments on the effects of weightlessness on liquids, together with various geophysical and astronomical observations. He was shown during a live TV broadcast rolling his head around and, later, lying back with his eyes closed while Yegorov grinned at him, turned his watch to the camera and laughed. Dr Yegorov was involved with medical tests to investigate physical and mental adaptation to weightlessness. Each crew member at all times wore three different sensors to record heart beat and respiration rate, but Yegorov was able to apply a variety of other sensors to monitor the responses of the brain and eyes, as well as motor coordination, at specific times during the mission. Altogether, the Voskhod flight returned a mass of medical information by taking advantage of the doctor's expertise. Yegorov described his work later:

> During the orbital flight I collected much data on the functioning order of the central nervous and cardiovascular systems and blood circulation. I was interested in man's performance in a state of weightlessness, in an evaluation of the hygienic conditions, and, of course, in the functioning of the vestibulary (inner ear) apparatus.

One additional observation was the description of Glenn's 'fireflies' outside the portholes. Feoktistov said they were tiny, like terrestrial dust particles, which 'shone in the sunlight and were visible for a few dozen seconds'. They did not quite keep pace with the capsule, but showed 'a slow relative movement' against the antennae, enabling them to calculate that they were drifting within 3 ft (1 m) of the craft.

The usual propaganda messages were broadcast as they passed over six of the seven continents, including greetings to the Soviet competitors in the Tokyo Olympic Games. Both Khrushchev and leading Politburo member Mikoyan spoke to the cosmonauts, giving them greetings and best wishes for a safe landing. At the end of his speech, Khrushchev complained that Mikoyan was 'pulling the receiver out of my hand', a comment whose full significance was not immediately apparent.

Prior to the launch the mission was described as an 'extended' one, which Western observers took to mean a duration of at least a week. There was, therefore, considerable surprise when Voskhod returned to Earth after a mere 16 orbits. The cosmonauts pleaded to continue, and Feoktistov informed Korolev: 'We have come across many interesting things.' However, the decision was final, with the reply in the form of a quotation from Hamlet: 'There are more things in Heaven and Earth, Horatio, than are dreamt of in your philosophy.'

Alexander Romanov described the final stages:

> The spacecraft landing support teams aboard planes and helicopters were already in the designated touchdown area. Korolev and other members of the State Commission came to the command post. The spacecraft was due shortly. Sergei Korolev, the technical director of the flight, sat down at a small table with a microphone on it. His face was strained, dark eyes narrowed, two deep furrows ran across the bridge of his nose. By habit, his left hand propped up his chin. One after another, comments were relayed aboard the craft prior to its entry into a descent trajectory, and immediate monitoring of their execution was carried out. The Academician was all attention, prepared to interfere in the work of the monitor service at any moment should things go awry.
>
> The tension in the room was as great as that one minute before

launching. Absolute quiet. Understandably, everyone was holding his breath. Even though the system had gone through a full cycle of trials both on the ground and in space, a soft-landing system was being tried for the first time in manned flight conditions. Through the loudspeaker standing on the operator's table, a voice was heard: 'The retro-engine's on . . . Separation has been effected . . . The parachute has opened . . . The craft is descending smoothly.' It was the voice, taut with anxiety, of the pilot Mikhailov, who was over the landing site.

Minutes . . . minutes . . . minutes . . . A few hundred metres separated the Voskhod from Earth. The joyous voice of another pilot came: 'I see the craft clearly. It's coming down smoothly, smoothly . . .' A tense hush enveloped the room. Everyone was mentally calculating the distance between craft and Earth, which was slowly decreasing: 15, 14, 13, 10, 8, 4 metres. I noticed that a blue vein in the temple of an engineer I knew was throbbing more obviously than usual. A man standing next to him was unsuccessfully trying to shove a packet of cigarettes into the breast-pocket of his coat. These were the creators of the soft-landing system. How would it function? 'Down! They're down! Everything's fine!' the pilots observing the descent shouted all together. It is hard to describe the next few minutes. People were applauding, embracing, kissing each other.

The capsule returned to Earth on the steppes 194 miles (312 km) north-east of Kustenai in Kazakhstan, 24 hours 17 minutes after launch. The crew were met and congratulated first by local farmworkers who invited them into their homes, but they were quickly whisked away to Kustenai where they were examined and debriefed by the doctors and scientists. Within hours they were back at Baikonur cosmodrome. Photographs released all over the world showed a happy, smiling crew, and at the post-flight press conference, Yegorov confirmed the impression these gave: 'Our commander had a longer period of training. However, we, too, withstood the G-loads during launching and descent well.' He went on: 'A thorough post-flight examination of the Voskhod crew did not reveal any kind of disturbances in the organism. The only thing that was found was a temporary upset in the morphological make-up of the blood, the water-saline balance. This testifies to general fatigue and a reaction to the tension caused by the physical and psychological factors of the flight.' Feoktistov was not quite as buoyant: 'The impression could be created that it is just as easy to live and work in the cabin as on Earth. But a spaceflight is not an excursion. It demands considerable effort, a mustering of physical and moral powers, special concentration.'

Despite strenuous denials from the Soviet officials, and the cosmonauts, there was a lingering impression among Western observers that the flight had been curtailed for some reason. It was later revealed that, although Komarov felt no ill effects, both Yegorov and Feoktistov suffered from disorientation throughout most of the mission. According to Tass: 'One of them imagined he was half crouched and facing downwards, while the other thought he was hanging upside down. These illusions remained both when their eyes were open and shut.' They also experienced brief dizzy spells during rotation of the head, and Yegorov increasingly lost his appetite due to queasiness in his stomach, reaching a peak about seven hours after launch. Neither man was said to be incapacitated by these unpleasant sensations, however, and this was supported by evidence of normal eye movements and near-normal physical movements as measured by Dr Yegorov. Soviet scientists concluded that the extra training carried out by Komarov had enabled him to adapt more successfully to the weightless conditions.

The triumphal return to Red Square was delayed for several days until the Soviet leadership struggle was resolved. It turned out that premier Khrushchev's talk with the cosmonauts had been his last public utterance – indeed, there have been suggestions that he was ousted from office during the final hours of the flight, and that this may have been the reason for the hasty return to Earth. Certainly, Khrushchev never returned to Moscow from his holiday home beside the Black Sea, and it was the new power in the land, Leonid Brezhnev, who eventually greeted the three heroes in the traditional Moscow celebration. As for the crew, Feoktistov and Yegorov both returned to their normal duties and never flew in space again. Feoktistov was able to use his personal experiences to advantage in helping to design the next generation of Soviet spacecraft, becoming a leading commentator on the space programme. Yegorov returned to the field of space medicine, also armed with his first-hand knowledge of the problem of weightlessness. He now heads the Soviet space medicine laboratory. Komarov was regarded as one of the leading cosmonauts of his day, accordingly being granted the privilege of flying the prototype Soyuz craft in 1967 and becoming the first Russian to make two space flights.

Although the instigator of the Voskhod programme had now been supplanted, the new leadership was prepared to allow the second flight to continue. Thus Voskhod 2 went into orbit on the morning of 18 March 1965, five days before the scheduled launch date of the first manned American Gemini mission. Once again, the Soviet Union had stolen the limelight from the Americans, and in the most spectacular manner possible, for this time one of the cosmonauts left his capsule to take the first ever walk in space. The exploit was hailed as yet another demonstration of Soviet superiority over the USA as the latter struggled to get the Gemini programme off the ground. More recently, it has tended to be dismissed by Western analysts as a mere publicity stunt, a propaganda coup which had little practical value. So what was the truth?

Certainly, as a piece of propaganda it had a touch of genius, causing the press at that time to speculate over the future Soviet plans. One expert analyst wrote: 'Lieutenant Colonel Alexei Leonov's extra-vehicular space excursion leaves only a space link-up between two vehicles as the last remaining demonstration before Russia attempts a launch of men to the moon or on an interplanetary mission.' Another commented: 'The Russians showed a considerable lead with Voskhod 2 – particularly in components and techniques for rendezvous, docking, space assembly and life support.' It is interesting to compare these remarks with those of Chief Designer Korolev at a pre-flight press conference:

> Why do we need to go out in space, why do we attach so much significance to this experiment? We must be able to do it if we are going to have spaceflights, just as on a ship at sea one cannot be afraid of falling overboard and must be able to swim. It is all connected with a whole number of operations which might be required when two spacecraft rendezvous. Exit into space greatly simplifies the conduct of certain space observations. And the situation could arise when something has to be repaired on the outside of the craft. For example, we are seriously thinking that a spaceman should be able to carry out all essential repair work, including welding. Finally, a situation could well arise when one ship has to extend assistance to another. But how? The craft are well heat-insulated, which means they are firmly sealed. One could approach a craft and be helpless to do anything, since if it is unsealed by opening the entrance hatch, the people inside will perish. Therefore, a system of air-locks has to be devised, one that would ensure life support and transfer to and from the craft, which would make it possible to extend help.

Korolev went on to discuss prolonged space flights which would need air-lock systems to permit docking and transfer in orbit of crews and equipment. He also envisaged a time when crews on long-duration flights would need periodic visits from other cosmonauts to provide 'assistance, insurance . . . even the simplest human intercourse and aid'. If these quotes are to be believed, then Korolev was imagining a time more than ten years ahead, with a lunar landing by a manned craft having no place in his future plans. As for the relevance of the space walk, it is, perhaps, significant that the first Soviet transfer of crews in space was accomplished in 1969 by a two-man space walk rather than the more sophisticated American method of floating through a docking tunnel between the two craft.

The man chosen to carry out the first ever space walk was 30-year-old air force officer Alexei Leonov. A member of the original group of cosmonauts, he came from the Altai region of Siberia, entered flying school at the age of 19 and later qualified in aeronautical engineering at the Air Force Academy. Korolev described his qualities:

Alexei Leonov's main traits are quick-wittedness, liveliness, cleverness. That's first. Secondly, he has a good grasp of technical knowledge. Thirdly, he's easy to get along with. He is an artist, likes drawing, is very sociable – in my opinion, a kind, attractive personality. He is a daring pilot, and can handle any modern jet fighter plane. I think this man deserves the greatest trust.

Leonov had worked hard to enjoy this trust from the leader of the Soviet space programme. He later told of the day Korolev first showed him the modified Voskhod:

I examined the [air-] lock and halted, mentally calculating what was what. 'If you like, you can put on the spacesuit and try making your first exit,' I heard the Academician say. Only then did I notice the spacesuit, which, it seemed to me, was in no way different from those in which the spacemen had already flown. I got into it with the assistance of staff workers and, on Korolev's instructions, began to work. The rehearsal took nearly two hours. To be honest, I was quite tired out: I was unaccustomed to the business. The Chief wanted to know right away: 'Well, what do you say, what are your comments?' If I remember right, I said something to the effect that, before embarking on 'exits' and 'entrances' one should have a very clear plan of action. 'Very good,' the Academician replied, 'so start working on it.'

For nearly a year, Leonov familiarized himself with the spacecraft and the suit, taking part in vacuum chamber tests and practice 'exits' from a mock-up craft during the brief moments of weightlessness experienced in Tu-104 aircraft as they flew in parabolic curves. He was even confined to a soundproof room for a month, then, without warning, taken to a MiG-15 aircraft and subjected to a session of aerobatics in order to test the effects of the prolonged solitude.

The suit itself was not discussed in any detail by the Soviet press. One medical expert described it as 'a miniature hermetic cabin which consists of a metal helmet with a transparent visor, a multi-layer hermetic suit, gloves and specially designed footwear'. The cosmonaut's body was cooled by a special air-conditioning system through which air circulated at room temperature, while the oxygen for breathing came from cylinders carried on his back. The suit was inflated at a pressure of about 6 lb/sq in (0.42 kg/cm^2) to protect him from the vacuum of space, though this could be adjusted if necessary. Over the pressure suit was a white covering which acted as protection against extreme temperatures and micro-meteorites. There were also voice communication and a variety of physiological sensors provided so that his progress could be continually monitored. A similar suit was worn inside the capsule by the mission commander, Pavel Belyayev, in case some emergency might arise which would necessitate a rescue attempt.

Belyayev was a 39-year-old air force colonel who came from Chelishchevo, north-east of Moscow. A former metalworker, he learned to fly and joined the air force during the Second World War when he saw active service against Japan. He was recruited to the original cosmonaut group in 1960, but lost a year's training through a leg injury in 1961. Korolev said of him:

> He has the same qualities as Leonov. He was a squadron commander, which means he has command experience. He is a graduate of the Air Force Academy and his knowledge is broad. He is calm, unruffled, I would even say a trifle slow, but very thorough. He is not given to speechifying, but everything he does is done thoroughly. The two are just the combination we need.

Korolev was aware, perhaps more than anyone else, of the dangers that lay ahead. Leonov told of a conversation they had with the Chief Designer the night before the flight as a blizzard raged over the cosmodrome:

> The Chief came to see us in the evening, to the little cottage where we, like all the fellows before us, were staying the night before the flight. He came in, slowly removed his dark blue overcoat in which he always travelled to the cosmodrome, hung his fur cap on a peg and sat down heavily in a chair standing beside the round table. It was visible that Sergei Korolev was very tired. And no wonder – he was on his feet from eight in the morning until midnight, always on the go. Only his brown eyes, as always, were bright.
>
> 'Well, how are you feeling?' he came out with his perennial question. 'Fine', Pavel Belyayev replied. 'Everything's normal.' I backed up my commander. 'I've sharpened my coloured pencils. Getting ready to draw a little . . . I want to draw spacescapes.' In reply the Chief smiled, and cocking his head to one side studied first me, then Pavel, and we realized a serious talk was in the offing.
>
> 'The preparations for the launching are proceeding normally. There were a few kinks but they have been eliminated.' He paused for a second and embarked on the main topic of his talk: 'The flight and the experiment of walking out in space are complicated. From you we want a precise execution of the programme.' Another pause. 'You yourselves will have to take into consideration all

circumstances and make decisions. It is impossible to foresee everything on the ground. You will have to act in keeping with the situation. Naturally, ground control will be there to advise you, but on board, your own lives and the outcome of the experiment are in your hands. If you spot some trouble – anything might happen – don't take unnecessary risks. Understood?' the Chief warned sternly. 'We don't need records, we need a carefully executed scientific experiment. You know how much depends on it. What we accomplish tomorrow will open a whole new avenue in space research.' He got up, looked at the books lying on the table and said sternly, like a father to his sons: 'Go to bed. Tomorrow you have a hard day ahead of you.'

As the storm abated and the sun rose to heat the cosmodrome and disperse the haze, the rocket was moved along the rails from the assembly building to the launch pad. Tass correspondent Alexander Romanov described the scene:

The platform with the rocket stopped at the launching pad. In the same second, as though with the wave of a magic wand, the silvery arrow began to rise slowly above the launch installation, growing before our eyes. Next to the enormous rocket, service points went up. On all sides it was surrounded by assembly platforms which formed multi-tiered structures beside the rocket. It took only minutes to install the rocket in its place.

At 10 am Moscow time on 18 March, Voskhod 2 was blasted into a more elliptical orbit than its predecessor, with a record apogee of 308 miles (495 km) despite an extra launch weight of about 800 lb (360 kg). This additional weight, despite the reduction in crew members, seems to be due to the presence of the two pressure suits and the air-lock.

Once the initial orbital checks were completed, the cosmonauts began to breathe pure oxygen instead of the normal oxygen–nitrogen mixture used on Soviet craft. This was necessary to ensure that all nitrogen was removed from Leonov's respiratory system, thereby allowing the reduction of pressure from normal to 0.4 Earth atmospheres without endangering his health – deep sea divers experience this problem, commonly known as 'the bends'. Towards the end of the first orbit, the cosmonauts gave a final check-over to the air-lock, now extended fully outside the craft, and the suits. The telescopic airlock was cylindrical in shape, but, as Leonov found, a length of 6 ft (2 m) and a diameter of 3 ft (1 m) did not allow much room for manoeuvre. The cosmonaut was able to slide from his couch through the adjacent inner hatch without disturbing his companion, then, with the hatch closed, Belyayev watched on a TV screen as Leonov began to depressurize the air-lock.

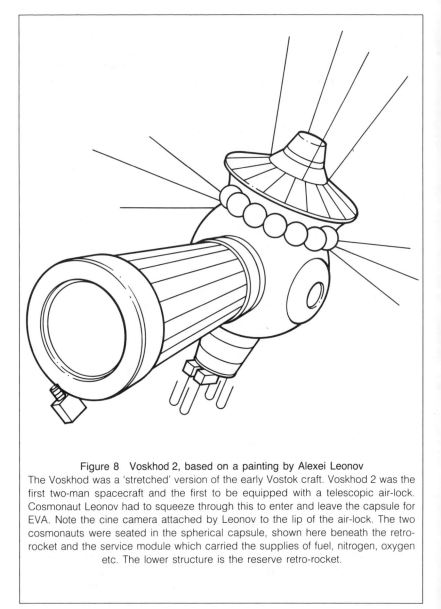

Figure 8   Voskhod 2, based on a painting by Alexei Leonov
The Voskhod was a 'stretched' version of the early Vostok craft. Voskhod 2 was the
first two-man spacecraft and the first to be equipped with a telescopic air-lock.
Cosmonaut Leonov had to squeeze through this to enter and leave the capsule for
EVA. Note the cine camera attached by Leonov to the lip of the air-lock. The two
cosmonauts were seated in the spherical capsule, shown here beneath the retro-
rocket and the service module which carried the supplies of fuel, nitrogen, oxygen
etc. The lower structure is the reserve retro-rocket.

With the operation completed, Leonov slowly opened the outer hatch
and gradually floated out into the vacuum of space, grasping a handrail
as he adjusted to the strange environment. Pictures of the historic event
were relayed to the TV screens of the watching millions back home by
two cameras, one of which Leonov mounted himself on a special bracket.

As a precaution against the cosmonaut losing his hold and drifting away from the craft to a lingering death, he was attached to Voskhod by a 16 ft (5 m) tether which included a telephone cable and wires carrying a telemetry from his sensors, as well as an emergency oxygen supply which was never used. As he looked around, he had a fine view of his homeland:

I looked through the light filter and through the glass of the spacesuit. The stars were bright and unblinking. I could distinguish clearly the Black Sea with its very black water and the Caucasian coastline. One could even see what the weather was like there – I saw the mountains with their snow tops looking through the cloud blanket covering the Caucasian range. The Volga appeared and disappeared; the Urals floated under us; I saw the Ob and Yenisei.

Everyone watched with bated breath as Leonov cast himself free with the words: 'I'm pushing off.' Belyayev gave him cautious encouragement: 'Don't be in a hurry, Lyosha, do it as you were taught. Go, but don't rush it!' As Leonov floated towards the rear of the craft, he exclaimed: 'Man has walked out into space!' He stretched out his arms like wings, experiencing complete freedom from gravity and restraint such as no one had ever experienced before. He later described what happened during the next seven minutes or so:

The lifeline that connected me to the ship stretched to its full length and then my movement away from the ship ceased. The slight effort I had made in detaching myself from the ship had caused it to move slightly, and I saw our wonderful spacecraft turning slowly before my eyes. I expected to see sharp contrasts of light and shadow, but there was nothing of this kind. The parts of the ship in the shadow were illuminated well enough by the sun's rays reflected from the Earth. I pulled the lifeline slightly and started slowly moving away again, turning about my transverse axis. I knew that it was impossible to stop my rotation by any movements, so I merely waited for the rotation to slow down when the lifeline became taut by twisting. And soon the speed of my movement gradually decreased. . .

Some time later, I made a pretty strong pull at the lifeline and was forced to protect myself from the spaceship which started moving swiftly toward me. My first thought was not to strike the spaceship with the visor of my helmet, so, as it flew towards me I softened the blow with my hands. This proved very easy to do . . . I felt fine, was in excellent spirits, and did not want to leave free space even after I had received the order to return to the spaceship. I pushed myself away from the hatch once more to check the origin of the angular velocity in the first moment after the push. I realized that the slightest divergence of the direction of the push produced rotation in a corresponding plane . . . As to the so-called psychological barrier which was supposed to seize a man about to meet the space void face to face, I must say that I did not feel any barrier at all and even forgot that such a barrier could exist.

After about ten minutes cavorting and somersaulting in space, Belyayev gave him the command to terminate the space walk. Leonov complied, though with some reluctance. However, his problems were just beginning, as he later explained:

Don't imagine that things went as comparatively easily in space as on Earth. No, I got pretty tired. Don't forget about the pressurized space suit. Even though it ensured safety, it was awkward to work in. Your movements are restricted. The gauntlets are not quite as elegant and convenient as the gloves we wear normally. For example, it required quite some effort to remove the cine camera from its bracket and send it into the air-lock.

The camera continually floated out from the air-lock as he tried to push it inside and manoeuvre his feet through the entrance. Eventually he kept the camera in place by placing his foot on it and succeeded in squeezing inside the cylinder, though by now he was perspiring profusely. The major difficulty was the inflation of his pressure suit like a balloon, which made the suit almost rigid and any movement extremely tiring. By the end of the eight minute struggle to re-enter the airlock, his pulse had reached 168 and he was breathing heavily. Despite the risk of contracting 'the bends', he had reduced the air pressure in his suit to 0.25 atmospheres in order to reduce the ballooning effect and increase its flexibility. It was a relieved cosmonaut who finally closed the outer hatch, repressurized the air-lock, and slid back into the safe confines of Voskhod's cabin. Leonov's official time outside the cabin was 23 minutes 41 seconds, of which about 12 minutes were taken up by the space walk itself.

With the main objective of the flight achieved, there was time for congratulations from the ground controllers and the politicians. The two cosmonauts were elated with their success, particularly Leonov, who could hardly wait to relate his experiences. When asked later how it felt to float in space he replied:

It's not like floating in water. In water you feel support, the slipping through the medium. In space you don't have that sensation. You're simply flying beside the craft (at 18,000 mph!), but at the same time it was nice to know that you were securely tied to the ship.

He was able to relax by carrying out his favourite pastime, using coloured pencils to sketch the beauties of the heavens in the ship's log. He described the stars as 'the colour of red gold', admitting to astonishment at the wealth of colours, particularly at each orbital sunrise and sunset. He added: 'It's not at all hard to write and draw in the state of weightlessness. One had only to press the pencil firmly down and to practise more often.'

One other incident of note occurred later in the flight:

> Our attention was attracted by an object bathed in sunrays. Examining it closely, we whooped with surprise and delight. To the side of the ship, floating about a kilometre (½ mile) away, was an artificial Earth satellite. The encounter rather excited us. We thought that the time will come when encounters in space with envoys of earth will become commonplace.

Having jettisoned the inflatable air-lock, Belyayev prepared the craft for an automatic re-entry on the sixteenth orbit in a carbon copy of Voskhod 1's flight. However, the crew were not destined to return on schedule; a solar sensor was not functioning properly, preventing correct alignment of the spacecraft for re-entry. For the first time in the Soviet space programme, a spacecraft would return to Earth under manual control. Capcom Yuri Gagarin passed up the bad news to Belyayev, but Korolev added his reassurance as the time neared on the seventeenth orbit when the commander had to fire the retro-rocket: 'Everything will be fine. We await you with impatience.' The extra orbit meant that the Earth moved 22 degrees to the east beneath the craft's trajectory, causing the landing zone to be far from the usual grassy steppes.

Voskhod 2 landed in a densely forested region of the Urals mountains near the town of Perm. The cosmonauts were some 1,200 miles (2,000 km) from the planned landing area, and, to add to their problems, an aerial had been incinerated during re-entry, causing a breakdown in communications and preventing the rescue teams from homing in on their distress beacon. When the cosmonauts opened the hatch and peered out at the wintry wilderness that surrounded them, they must have realized that they were in for a long, cold wait. Forced to keep their suits on for warmth, their hygienic facilities soon became overloaded. It was 2½ hours before they were spotted by a helicopter, but there was nowhere within 12 miles (20 km) for the would-be rescuers to land so the cosmonauts had to be content with a drop of food and supplies.

Towards sunset, with wolves prowling in the growing shadows, the men lit a fire, then decided that prudence was the order of the night and hurriedly retreated to the cold but secure shelter of their capsule. They slept fitfully, taking turns to keep watch, and, no doubt, cursing the ill fortune which had caused such an ignominious end to their heroic flight.

The next morning a ski patrol from the base which had been set up in the nearest clearing arrived with more food and warm clothing. With their spirits revived and their clothes changed, the cosmonauts donned their skis for the long trek to the waiting helicopters. Once there, they were whisked away to Perm before flying back to Baikonur cosmodrome on the second day after their return to Earth. They were none the worse for wear, though their dignity had suffered a little. Nevertheless, they had achieved another space first for the Soviet space programme. As usual, Tass gave an edited version of the final stages of the mission: 'On 19 March, at 2 minutes past 12 noon, Moscow time, the spaceship Voskhod 2, manned by commander of the craft, colonel Pavel Belyayev, and co-pilot, lieutenant colonel Alexei Leonov, landed successfully in the vicinity of the city of Perm. For the first time in the history of manned flights, the craft's commander landed it by manual control.' No explanation was offered for the four and a half hour delay in announcing the crew's safe return.

The two men were given the usual ecstatic welcome by the Soviet people, but no one realized they were witnessing the end of an era. Leonov was quoted as saying: 'Man will visit [the Moon] in the near future. I dream of this being accomplished by men of our detachment. If I am very lucky, I will get the assignment.' It was not to be – indeed, Leonov did not fly again in space until the Apollo–Soyuz Test Project of 1975, having suffered the frustration of training for the abortive Soviet moon flights and early Salyut space stations. As for Belyayev, he was destined never to fly again in space. In a private conversation at the 1967 Paris Air Show with astronauts Michael Collins and David Scott, he confided that he expected to make a circumlunar flight 'in the not-too-distant future'. That flight never materialized, and within two years the solid, trustworthy astronaut was dead: he contracted peritonitis and died despite emergency surgery. He had achieved one more, sad, first – the first spaceman to die of natural causes. He was not buried in the Kremlin Wall but in Novodevichy cemetery.

Many Western observers expected further Soviet advances after Voskhod 2, but Voskhod 3 never appeared. Indeed, there were no more Soviet manned missions for more than two years, and then the flight of the new Soyuz craft proved a fatal disaster. The attitude of the Kremlin leadership was no longer the same as when Khrushchev demanded space spectaculars to order as the price of his support for Korolev's scientific programme. However, the wind of change also blew from another direction which boded ill for the future. Korolev's deputy, Voskresensky, had died between the two Voskhod flights. Now the Chief Designer himself, weakened by years of imprisonment under the Stalin regime died on 14 January 1966 during what should have been a straightforward heart operation. Korolev had involved himself completely in his work, inspiring Soviet science and technology to heights previously unsuspected by the Western powers. Only after his death did the Soviet press acknowledge the anonymous 'Chief Designer', with his real name alongside the long list of his achievements. His ashes were buried next to other Soviet heroes in a solemn Kremlin ceremony, and with them went the Soviet hopes of winning the race to the moon. It took their space programme seven long, difficult years to recover from his death, by which time the USA was firmly in the driving seat. Perhaps the most apt epitaph came from ex-premier Khrushchev's memoirs: 'I'm only sorry that we didn't manage to send a man to the moon during Korolev's lifetime.'

5

The Heavenly Twins

The Gemini Missions, 1965–6

The American Mercury manned spaceflight programme was hardly off the ground when NASA announced in December 1961 its plans for a two-man spacecraft, soon to be known as Gemini, after the mythological Heavenly Twins. It was originally conceived as a way of gaining greater experience through long-duration flights in Earth orbit – the entire Mercury programme had provided less than 54 man-hours in space, hardly a sufficient basis for a week-long moon mission before the decade was out. However, once the space agency had decided on the method by which it intended to send men to the moon, the Gemini programme became the indispensable foundation for the Apollo venture of sending men to another world.

The option for achieving a moon landing finally chosen by the engineers was called Lunar Orbital Rendezvous. Many thought it was unnecessarily complex, but it had the advantage that only one Saturn V rocket was required. A lunar landing craft would need to be built, but a small purpose-built lunar module should be much easier to handle during a moon landing than a giant rocket. The main controversy was over the requirement for transposition and docking between the command and lunar modules after the craft had been fired out of Earth orbit, with additional separation and docking manoeuvres involving both craft when they were out of radio contact on the far side of the moon. Such additions to a lunar flight plan were considered risky by engineers who were concerned with ensuring the least hazardous method was chosen. Nevertheless, the savings on payload weight and cost proved decisive, so that Lunar Orbital Rendezvous was given the go-ahead in October 1962. Now it would be up to Gemini to prove that orbital rendezvous and docking were possible.

The new craft which emerged in 1964 looked much like an enlarged version of Mercury, but it was, in fact, a much more advanced and sophisticated piece of equipment. The manned section had the same conical shape as Mercury, with the blunt nose, and though it was only one foot longer and a foot wider than Mercury, it provided 50 per cent more room inside for the crew. The large nose on the front of this section contained the radar for tracking rendezvous targets, eight thrusters for attitude and re-entry control, and the parachute recovery system. The capsule was originally intended to parachute on to dry land attached to a paraglider, but this was never made operational, so the splashdowns continued. The equipment bays were placed between the inner and outer walls of the craft, with access doors through the outer skin which made maintenance quick and easy compared with Mercury. Attached to the manned section, beneath the heat shield, was the adapter section. This innovation was built to house the rockets for orbital manoeuvring, the life support system and the power supplies. Even the electricity was supplied in a new way, for fuel cells which made power from a chemical reaction between hydrogen and oxygen were installed in addition to the usual batteries. These fuel cells weighed about half that of batteries with comparable power and life, with the extra bonus that they supplied sufficient water as a by-product to meet the needs of the crew for even the longest missions. One Gemini engineer commented, 'It's like carrying your own well with you on a long desert journey.' The adapter section was jettisoned before re-entry.

Since Gemini's weight was around 8,000 lb (3,500 kg), double that of Mercury, the Atlas launch vehicle was no longer adequate to lift the craft into orbit. Instead, the Titan intercontinental ballistic missile was modified to create a powerful two-stage booster. It used a less explosive combination of propellants than the Atlas, which proved a highly reliable mixture, capable of long-term stability inside the rocket. This in turn meant a shorter countdown, and since the fuel and oxidizer burned instantly on contact, there was no need for an elaborate ignition system. The less explosive nature of the propellants meant that the emergency escape system for the astronauts could also be changed. On Mercury–Atlas a rocket-powered escape tower was available to boost the occupant clear in the event of a malfunction during the countdown or the journey into space. On Titan 2, this was considered unnecessary; the two astronauts were provided with special ejection seats. During an abort on the ground, a rocket would fire the seat from the capsule at an angle of 15 degrees to the horizon so that it reached a height of 500 feet (150 m) before setting the astronaut down by parachute some 1,000 feet from the Titan. For ejection up to 70,000 feet during the launch, the astronaut was given an emergency oxygen supply; a drogue stabilization device prevented the man blacking out from an uncontrolled spin before he could deploy the main chute. Above that altitude, the men had to return in their capsule: they shut down the Titan's engines, separated the craft from the booster, and used its parachute system, though ejection was possible at the normal height should the crew so desire.

The first Gemini–Titan (GT) launch took place in April 1964 as a test of the compatibility of the two components. NASA needed a success – the project was behind schedule and well over budget, causing both the press and the politicians to start asking awkward questions. The Soviet Union had larger, more powerful boosters, and seemed to enjoy an embarrassingly large lead in the space race. To NASA's delight, the GT1 flight was almost perfect, giving the green light for a full test of the entire Gemini system, including separation, retro-fire and re-entry for a mid-Atlantic splashdown. This GT2 flight was scheduled for August, to be followed by the first manned flight in November if all went well.

The crew allocated to GT3 were Mercury veteran Gus Grissom and newcomer John Young. After his sub-orbital flight in Liberty Bell 7, Grissom had nearly drowned and his capsule turned turtle and sank; many people had blamed the astronaut for the mishap, and there was some doubt about his future in the space programme. Nevertheless, he had continued to put his heart and soul into the Gemini project, finally receiving his reward as command pilot of Gemini 3. The Liberty Bell incident was obviously still on his mind, however, for he decided to name his Gemini craft Molly Brown after the successful musical comedy 'The Unsinkable Molly Brown': when his superiors questioned his choice, gutsy Gus simply asked, 'Well, what about the "Titanic"?' He got his way, but Gemini 3 was the only craft in the programme to be christened by her crew.

John Young had become an astronaut in 1962 when NASA began recruiting new blood in anticipation of the increased activity and demand for crewmen during the build-up to Apollo. To him went the honour of being the first group 2 astronaut to fly in space. Young followed in his father's footsteps by joining the US Navy after graduating in aeronautical engineering with highest honours in 1952. After serving on a destroyer for a year, he began flight training, eventually being assigned to a jet fighter squadron. In 1959 he attended the Navy Test Pilot School and during the next three years he evaluated fighter weapons systems. In his last months before he joined NASA, Young set new world time-to-climb records for 3,000 metres (9,600 ft) and 25,000 metres (80,000 ft) in the Phantom fighter. Neither he nor Grissom were renowned for their volubility, but Young had a dry sense of humour which helped when the chips were down. One of his favourite pastimes was drawing cartoons.

The launch of Gemini 2 was hit by delay after delay when the Titan 2 rocket had arrived at the Cape, the spacecraft was still not ready. Then, on 17 August the launch pad was struck by lightning, causing some damage to ground installations and a delay of at least two weeks. Only ten days later, Hurricane Cleo threatened the Cape, so the second stage on the rocket was hastily removed to prevent damage. The next few

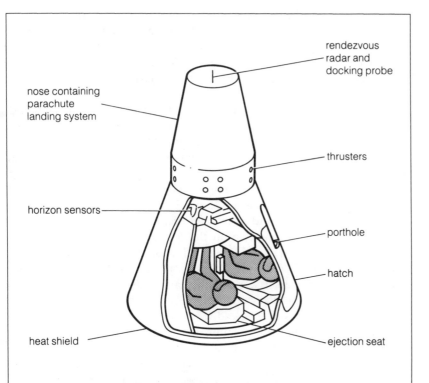

rendezvous radar and docking probe

nose containing parachute landing system

thrusters

horizon sensors

porthole

hatch

heat shield

ejection seat

Figure 9   The Gemini spacecraft
Gemini was the direct descendant of Mercury, a fact immediately apparent from their similar shapes. However, Gemini was much more sophisticated, with the capability of orbital rendezvous and docking. To achieve this, it was provided with the first space-borne computer, and a radar tracking system. Thrusters on the adapter section (not shown here) enabled the craft to change orbit as well as attitude. Cabin atmosphere was pure oxygen. Ejection seats replaced the Mercury escape tower.

weeks saw further delays as two more hurricanes swept by, with the Titan dismantled and sheltering in a hangar. The flight was rescheduled for November, then 9 December. At Mission Control, Grissom, Young and their back-up crew (Schirra and Stafford) watched and waited as the countdown progressed. Inside the capsule were two astro-robots, designed to record everything that happened and even carry out some of the tasks that would be done by men on Gemini 3. The countdown finally reached its conclusion. 'T minus zero. Stage one ignition,' came the voice of the announcer, then one second later the engine cut off – the flight was scrubbed. A drop in hydraulic pressure had caused number two engine's thrust chamber to be pushed sideways at an acute angle.

Although this had been corrected by a back-up system, the rocket was designed to shut down automatically if a secondary system was called on during the final three seconds of launch. Frustrated as they were, the astronauts could see that the malfunction detection system, designed with their safety in mind, really did work. That at least helped to put some of their doubts at rest.

At long last, on 19 January 1965, Gemini 2 was successfully launched. The rocket reached an altitude of 106 miles (170 km) before plunging back into the atmosphere for a splashdown 1 hour and 48 minutes after lift-off. The system passed the test with flying colours; Gemini 3 was 'go' for 23 March. Yet one more indignity was heaped upon the long-suffering American space programme before that flight took place. Just five days prior to the launch of Gemini 3, the Soviet Union sent up its own two-man spacecraft and cosmonaut Alexei Leonov stunned the world by stepping out of the capsule on the first-ever space walk. The USA seemed as far behind the Communist superpower as ever. No one realized that Voskhod 2 was the last of the line of Soviet space spectaculars and that for the next four years the Soviet space programme would languish in the doldrums while Gemini, and then Apollo, pulled the USA into an unassailable lead.

Grissom and Young moved into the astronaut quarters on Merritt Island on 15 March. They spent the next week familiarizing themselves with the Pad 19 launch complex, which was being used to send up a manned craft for the first time. Grissom likened their quarters to a 'brand new motel', though they made plenty of use of the gymnasium. On the launch date Deke Slayton woke them at 4.40 am. They had a brief medical check over, then settled down to a breakfast which included a two-pound porterhouse steak for each man. The normal low-residue meal was waived on this occasion because they were scheduled to be in space for less than five hours.

Before they left the building, John Young was presented with a 60 foot (16 m) telegram from 2,400 residents of his former home town, Orlando in Florida. Then out to the medical trailer, about a quarter of a mile from the launch pad, where the sensors that would send back readings on heartbeat, pulse and so on were attached. First the 'long john' undergarments, then the white pressure suits were donned. Wally Schirra brought a smile to everyone's faces when he appeared wearing his version of the suit – an old, threadbare Mercury suit which had definitely seen better days. Schirra explained: 'I've suited up just in case you two chicken out and turn the mission over to the back-up team.' In fact, he and Stafford had been up most of the night checking out the spacecraft systems for their colleagues.

At 7.06 am Grissom and Young left the trailer for the three-minute ride to Pad 19. Wasting no time, they were soon strapped on to their couches in the capsule. There was one hold during the countdown due to a leak in a fuel line, delaying the launch by 24 minutes, but little more than two hours after stepping into Molly Brown, the first manned Gemini flight blasted off from the Cape, now renamed after the assassinated President Kennedy. Gus Grissom gripped the ejection ring between his knees, but there was no last-second abort. 'The clock has started,' he calmly informed Capcom Gordon Cooper, to which the other Mercury veteran replied, 'You're on your way, Molly Brown!' Grissom noticed the much more rapid acceleration on this ride compared with his Mercury experience. He also noticed that the ice-cool John Young was not even holding his ejection ring.

The Gemini went into an elliptical orbit of 100 miles by 140 miles (160 km × 225 km) some five and a half minutes after launch. Separation from the second stage went perfectly, but a potential threat arose as they passed over the Atlantic. John Young noticed that the oxygen pressure readings for their suits and the cabin had suddenly dropped, causing an automatic, but useless, reaction from Gus Grissom who closed his helmet visor. Young quickly realized that the fault lay in the instruments and switched to a back-up system, while a rather sheepish Grissom raised his visor with a sigh of relief.

Gemini 3 carried a limited number of experiments, one of which was described as 'taking sex into space for the first time'. It involved Grissom turning a knob to artificially fertilize some sea urchin eggs in synchronization with a similar experiment on Earth. However, all did not go to plan, for Grissom twisted the handle so hard that it broke off – as he later admitted, 'Maybe after our oxygen scare, I had too much adrenalin pumping.'

The two men were much more at home test flying their new craft with the aid of the first computer ever used in space. As they passed over Texas at the end of the first orbit, they fired the large translation thrusters for 74 seconds in order to change Gemini's orbit. The firing slowed the craft so that it dropped into a roughly circular orbit at 100 miles altitude, the first-ever such manoeuvre. During the second orbit Grissom fired forward and aft thrusters over the Indian Ocean for 15 seconds, successfully accomplishing minute changes in the orbital inclination. A third and final lowering of the orbit was completed on the third orbit prior to re-entry, a safety precaution in case the retro-rockets failed.

NASA was delighted with the performance of its 'Heavenly Twins', though the agency officials were not amused by an unscheduled meal provided courtesy of Wally Schirra. John Young amazed his companion when he suddenly produced a corned beef sandwich from a pocket in his spacesuit. Neither man was enthusiastic about the dehydrated food they had carried aloft, so Schirra had asked a Cocoa Beach restaurant to make them a sandwich. Unfortunately, when Grissom took a bite, crumbs of rye bread began floating all over the cabin, so the meal went unfinished.

The men received a severe dressing down after the flight, with officials making it clear to future crews that such non-regulation meals were banned. However, Gus Grissom could not help admitting that the incident was one of the mission highlights.

The adapter section separated on schedule with a sudden jolt, the retro-rockets fired to slow down the craft, then the retro section fell away. Molly Brown was on her way home. The aerodynamic character-istics of the craft were not quite as predicted by ground tests, causing it to head away from the planned recovery zone, so the astronauts performed two banking manoeuvres to reduce the error shown by their computer. Molly Brown splashed into the Atlantic 58 miles (93 km) short of the target near Grand Turk Island. The final stages of re-entry were not without incident: the sudden lurch when the main parachute deployed caught both men off guard. They were thrown against the windows, with Gus Grissom's faceplate being smashed and Young's badly scratched.

When they hit the water Grissom's window lay below water level. He quickly realized that he had not released the parachute, so that the wind was tugging it along with a half-submerged spacecraft in tow. The situation was remedied by triggering the parachute release mechanism, but more agonies were to follow. The cabin soon became like an oven, and this, combined with the rocking of the capsule on the waves, soon made the men feel distinctly queasy. Poor Gus Grissom, never at home on water, succumbed to sea sickness while his Navy colleague managed to keep his dignity. As Grissom later commented, 'Gemini may be a good spacecraft, but she's a lousy ship.'

Although they were soon joined by a pararescue team and a group of Navy swimmers, the main recovery craft, the aircraft carrier *Intrepid*, was still more than an hour's steaming time away, so they requested a helicopter pick-up. Removing their spacesuits in order to cool off, they clambered out of the craft, Grissom first since his hatch was above the waterline – this caused John Young to comment jokingly that it was the first time he had ever seen a captain leaving his ship first. On the recovery helicopter they were given blue bathrobes to wear over their long johns, enabling them to maintain some composure when they met the huge reception committee on *Intrepid*'s flight deck. The greetings over, they were whisked below deck for their physical examinations and first debriefings. As soon as possible they were flown back to the Cape where a special reception and parade greeted them. The next day they flew to meet President Johnson at the White House, where both men received the Exceptional Service Award. A tickertape parade through New York was repeated in other cities as the public were given the opportunity to acclaim the men who had given the USA a much-needed boost on its way to the moon. For little Gus Grissom the celebrations must have tasted all the sweeter since he, alone, of all the Mercury astronauts had been denied the hero's welcome the first time around.

For the next Gemini flight, a four-day mission which would break new ground for the Americans, Gus Grissom was appointed chief Capcom. He would be based in the new Manned Spacecraft Centre in Houston rather than at the Cape, which had been the control centre for all previous manned flights. The flight plan for Gemini 4 was subjected to much greater scrutiny than most flights as the NASA administrators tried to make up their minds about a possible space walk. At the beginning of 1965 it seemed unlikely that any astronaut would even pop his head out of the hatch until the Gemini 5 mission. Then came the challenge thrown down by Leonov's walk in space; NASA's response showed uncertainty about picking up the gauntlet. At first it looked as though one of the men on Gemini 4 would be assigned to admire the view from his open hatch, but the engineers who had been working for two years on the development of a pressure suit and other equipment for extra-vehicular activity were certain they could complete their task in time for Gemini 4. Not until 19 May did everything pass the strict qualification tests, leaving the way open for the final agreement to go ahead with the full EVA. On 25 May, NASA announced that astronaut Edward White would leave his cabin for a walk in space early in the flight of Gemini 4, scheduled for launch on 3 June. In fact, White was not really thrown into a completely new project only nine days before the launch date: he had been practising EVA for some months in the high altitude pressure chamber at McDonnell in St Louis.

Ed White was no newcomer to weightlessness either, even though he had never flown in space before. Prior to joining NASA he had flown a total of five weightless hours while piloting Air Force C-135 jet transports during astronaut training. Graduating from West Point in 1952, he had gained considerable experience before becoming a test pilot. In 1959 he gained a master's degree from the University of Michigan where he first met the Gemini 4 command pilot, James McDivitt. That same year, he narrowly missed selection as a Mercury astronaut, but, not to be deterred, he applied again in 1962, this time with success.

The other crew member was also a group 2 astronaut, White's friend from student days, James McDivitt. Although White was a Texan and McDivitt came from Chicago the chatty, friendly McDivitt perfectly complemented his colleague. Both men were Christians; both had risen to the rank of air force major, and both had become first-rate test pilots, though McDivitt had become a flyer in time to complete 145 combat missions during the Korean War. While at Michigan University he had studied aeronautical engineering, finishing top of a class of 607. By 1962, his flying had so impressed the authorities that he had been chosen to fly the X-15 rocket plane; then came the call from NASA.

The launch of Gemini 4 went ahead as planned on 3 June 1965, though there was a 76 minute hold during the countdown due to difficulties in lowering the erector tower at the launch pad. Eventually, at 10.16 am Cape time, the Titan 2 rocket ignited with a burst of orange flame before lifting off into a clear blue sky. By the time the crew had acquired orbit, a little more than five minutes after launch, mission control had been switched to Houston. One of the first tasks was to turn the craft around in order to rendezvous with the spent second stage of the Titan which was following some 650 feet (200 m) distant. The dead rocket, however, was tumbling out of control as well as flying in a different orbit to Gemini. McDivitt tried to carry out the planned station-keeping, but despite his considerable efforts, the hulk continued to drift away. Finally, mission control decided that a fuel expenditure of more than 40 per cent was enough, instructing the pilot to save his manoeuvring fuel for the rest of the four days.

The original schedule called for Ed White to exit from the cabin during the second orbit, but the preparations took much longer than expected, so it was postponed until the third orbit. The delay made little difference to the flight plan; it was still early enough in the mission for White to be fresh and alert, while the activity could still take place largely over the USA and in daylight. Partial depressurization of the cabin proved successful, so the remainder of the cabin oxygen was vented as they headed towards Hawaii on the third orbit. The Hawaii Capcom informed the waiting crew, 'We just had word from Houston, we're ready to have you get out whenever you're ready.'

Ed White stood on his couch with his head poking out through the open hatch. The door had given Jim McDivitt quite a shock when it had popped open, causing him to hold grimly on to the lanyard so that it would not be damaged – a stuck hatch would mean certain death for both astronauts during re-entry. Now White was attaching the 16 mm sequence camera which would record the next 20 minutes for posterity; he checked it three times before he was satisfied. 'I wanted to make sure I didn't leave the lens cap on,' he said later. 'I knew I might as well not come back if I did.' Then he assembled a unique piece of apparatus which was being tried for the first time – a hand-held manoeuvring unit, a kind of gas gun which propelled the astronaut in whichever direction he wished by spurts of oxygen emitting from small thrusters as he pulled the trigger.

Twelve minutes after opening the hatch, the first American space walk began. As White floated out through the rectangular opening, he fired the gun to distance himself from the craft. A controlled roll to the left took him out beyond the Gemini's nose, followed by a white overglove which he had not worn in order to get a better feel for the gun. A pull on the tether caused him to loop over the craft to the adapter section at the rear, then a prolonged gas burst sent him floating back between the two hatches to the nose section until he reversed the thrust to stabilize himself. Soon afterwards, he pressed the trigger but failed to get a response; the gas supply had run out after about four minutes. No one was more disappointed than White: 'OK I think I've exhausted my air now. I have very good control with it, I just needed more air.'

The space walker now had to resort to tugging on the 27 foot (7 m) long tether, just as Leonov had done before him. The Russian, however, had carried his own oxygen supply while White's oxygen was fed from the craft through the gold-coated umbilical, along with his power supply and telemetry lines. Only in a sudden emergency would the astronaut have to rely on the nine-minute oxygen supply in his chest unit. White later commented, 'Changing my position by pulling on the tether was easy, like pulling a trout, say a two- or three-pounder, out of a stream on a light line.'

The 17,500 mph (28,000 kmh) walk continued as White turned somersaults on the end of the line: 'Right now I'm standing on my head and I'm looking right down and looks like we're coming up on the coast of California, and I'm going into a slow rotation to the right. There is absolutely no disorientation associated with it.' He soon added, 'I can sit out here and see the whole California coast.' Meanwhile, James McDivitt was having to concentrate hard on holding Gemini steady as his companion cavorted outside: 'When Ed gets out there and starts whipping around, it sure makes the spacecraft hard to control.'

Ed White was finding it much more difficult to hold his position with only the tether as an aid. He floated all around the Gemini, even scraping his helmet on the craft's outer skin on one occasion. While White attempted to use a small camera attached to the now-redundant gun, McDivitt also tried to focus a camera through his window. 'Let me take a close-up picture of you,' he requested. White tugged on the tether to move in closer, enabling his colleague to obtain some of the most spectacular shots ever taken. Unable to stop his progress, however, White's left shoulder and elbow touched the window as he collided with the craft. 'You smeared up my window shield, you dirty dog,' exclaimed the command pilot.

White later described the final stages of his adventure: 'I was taking some big steps, the first on Hawaii, then California, Texas – lightly, in deference to the President – Florida, and the last on the Bahamas and Bermuda.' He went on:

There was absolutely no sensation of falling and very little sensation of speed, other than the same kind of sensation we had in the spacecraft, very similar to flying over the earth at about

20,000 feet. The views continued to be spectacular. The one I remember best is as we came over Florida. Looking down, I could see all the lower part of the state, the island chain of Cuba and Puerto Rico. And that was about my last look, since that was when Chris Kraft [Mission Director] gave me the word to come back in.

In fact, Gus Grissom had been trying to contact Gemini for some time, since White had been outside longer than expected and Gemini would soon be out of radio contact over the Atlantic Ocean. This was partly due to the astronauts' preoccupation with their own activities, and partly due to McDivitt leaving a switch in the position which cut off ground communication for much of the EVA. The failure of White to come back in when instructed was interpreted by some reporters as a case of 'space euphoria', but White simply commented:

I can say in all sincerity and honesty that I enjoyed EVA very much, and I was sorry to see it draw to a close, and I was indeed reluctant to come in. But when the word came that the EVA phase was over, I knew it was time to come in and I did. There was no euphoria, but getting back into the cabin took just as much time as getting out; I had to do the same things, only in reverse order, handing my gear in to Jim, and so on.

The total time between the opening of the hatch and its closure was 36 minutes, of which 21 were spent by White in the space walk – he had doubled the time set up by Leonov.

During the walk, White had been quite happy and comfortable in the special suit, and the helmet with its three visors. His pulse was fairly steady at around 150, so that he was able to tell Dr Berry, the NASA physician, that he felt fine. It was later revealed that his pulse had risen to 178 in trying to close the hatch once he had re-entered the capsule. By the time the two men succeeded in locking the hatch, White was perspiring profusely. Small wonder that Dr Berry prescribed a four-hour rest period. White made one further revelation after his return to Earth: he had carried three religious symbols during his EVA, a St Christopher's medal, a gold cross and a Star of David. He explained:

I had great faith in myself and especially in Jim, and also I think I had a great faith in my God. So the reason I took those three symbols was that I think this was the most important thing I had going for me, and I felt that while I couldn't take one for every religion in the country, I could take the three I was most familiar with.

The remainder of the flight proved to be a routine, unspectacular three days, apart from a sensationalized episode reported during the second day. McDivitt grabbed the headlines as he told ground control, 'I just saw something else up here with me. It had big arms sticking out of it.' His partner confirmed the sighting before it disappeared in the glare of the sun. Controllers were surprised since no Earth satellites should have been visible from Gemini. The astronauts' capacity to see surface details on Earth also surprised many experts, who had disbelieved Gordon Cooper's reports on the final Mercury flight. It was Cooper's endurance record which they overtook during the second day: the astronaut's response to the news was simple and to the point: 'Roger, we've got quite a few more to go.'

Much of the crew's time was spent carrying out photography of the Earth's surface and weather systems. Ed White told of seeing roads, street lights, airfield runways, ship wakes and smoke from trains and chimneys. Another experiment involved continual monitoring of radiation levels in the spacecraft. Since Gemini 4 was breaking new ground for America, the space agency was anxious to get as much information as possible on the effects of prolonged exposure to cosmic radiation and weightlessness; information released by the Soviet Union indicated some cause for concern. Medical experts suggested a programme of exercise might help to mitigate the weakening of the muscles, including the heart, though this inevitably caused problems for engineers when faced with a spacecraft in which there was hardly room for a man to stretch. In the end, they came up with a bungee cord and an elastic foot strap for the astronauts' use while seated in their cramped cabin, with maximum exercise scheduled prior to re-entry when stresses on the body reached a peak.

The supply of freeze-dried and dehydrated food was stored in a compartment above McDivitt's left shoulder. The meals were marked and stacked in order of use, four meals for each man per day to be reconstituted by means of a special water gun. This provided a problem which NASA had more or less ignored on previous short-duration missions; the digestive system functioned normally in weightless conditions, leading to a problem of waste disposal. Urine was simply collected, then flushed into space, but solid waste had to be collected in special bags which adhered to the astronauts' buttocks, then a disinfectant added before the contents were kneaded and finally stored. Inevitably, no washing or shaving facilities were available, though the cabin still became crowded with bags of rubbish as the days went by. The sleep schedule did not improve the astronauts' discomfort, for they were supposed to alternate their rest and work periods so that one man slept while the other carried out experiments, photography and manoeuvres. Predictably, the movements of the active crewman, together with the humming of electrical circuits and banging of activated thrusters, led to many sleepless hours for the resting partner. Their experiences caused NASA officials to think again, eventually resulting in a common sleep period for both astronauts on future missions.

The only major difficulty on the entire mission arose on the 48th orbit when the small computer broke down. The main significance of the failure was that the automatic re-entry had to be replaced by a Mercury-type ballistic re-entry. As on Gemini 3, the orbit was lowered prior to firing the retro-rockets as an insurance policy in case of failure. However, all went to plan, with McDivitt initiating a slow roll to counteract the aerodynamic lift built into the craft's design. The more direct re-entry method created acceleration forces up to 8G, twice those experienced in automatic re-entry, but the splashdown was rather more accurate than on the first Gemini mission. Gemini 4 came down in the Atlantic swell 50 miles (80 km) from the target area; they also managed to endure the craft's motion better than their predecessors before being picked up by helicopter nearly an hour after splashdown. They were given a joyous reception on the deck of the carrier *Wasp* as the band played and sailors cheered. As they put their arms around each other, smiled and gave a thumbs-up to the massed cameramen, the unshaven heroes gave no real sign of ill-effects. They did admit, however, 'We are feeling a bit tired and hungry . . . Now we'd like something to eat.'

The medical verdict on the flight verified that neither man had suffered any major physical problems; there had been some loss of calcium in their bones, and their blood pressure had fallen, but there was no reason to call off the planned eight-day mission of Gemini 5. Despite their fatigue, the two men were in high spirits. As he lay in *Wasp*'s sick bay, McDivitt cracked, 'I always knew I'd end up in hospital after four days in space.' White admitted, 'I felt so good I didn't know whether to hop, skip, jump or walk on my hands.' In fact, he did a jig step as he walked to the captain's quarters where they spoke to President Johnson. One outcome of the flight was their promotions to lieutenant-colonel, though as the President commented, 'If I had seen your space films before, I might have promoted you to full colonels.' A present of a different kind came from the news reporters at Houston – an abacus in case their computer broke down on a future flight.

Ed White had secured a place in history with his record-breaking spacewalk, but his moment of triumph was not to be repeated. After serving on the back-up crew for Gemini 7, he was assigned to Apollo 1, the first scheduled manned flight of the new moon craft, but he was not destined to reach the moon. For Jim McDivitt, the Apollo programme would bring a leading role in the race to land men on the moon.

The Gemini programme was now in full swing, with the next mission scheduled only two months after Gemini 4. Confidence was high, so the NASA administration saw the next logical step as an extended, eight-day mission which would prove once and for all whether men could survive a flight to the moon and back. Gemini 5 was designed to carry the new fuel cells which would make such long-duration flights possible. The craft was ready but Deke Slayton, Director of Flight Crew Operations, was not satisfied that the same could be said of the crew: they had been in training less than six months for this mission, half the time given to Grissom and Young for their flight, which lasted only three orbits. His will prevailed to the extent that Gemini 5's launch date was put back by a couple of weeks.

The crew for Gemini 5 was made up of old-timer Leroy Gordon Cooper and a group 2 newcomer, Charles 'Pete' Conrad. Cooper, now 35, had flown the last Mercury mission, the longest of the series. During the flight, he had demonstrated his amazingly acute vision as well as great composure in carrying out an entirely manual re-entry. Conrad, three years younger than his command pilot, had narrowly missed joining Cooper on the Mercury team, but had succeeded second time around. A mere 5 ft 6 in tall, Conrad was one of the smallest astronauts ever selected, but his lack of height was, if anything, a positive advantage in the cramped Gemini cabin. He had attended the Navy's Test Pilot School at Patuxent River, Maryland, at the same time as Wally Schirra, having graduated in aeronautical engineering from Princeton University in 1953. A lively extrovert, he loved living up to the devil-may-care image of the test pilot and astronaut, yet beneath this noisy, humourous exterior lay a first-rate pilot, a man who had seen many of his best friends lose their lives in flight test training and who remembered the lessons he had learned.

Cooper's motto for the forthcoming flight, 'Eight days or bust', and a mission badge depicting a pioneer's covered wagon, demonstrated the crew's challenge to the hostile environment of outer space. However, on 19 August, with the two men strapped on to their couches, the Titan booster and the weather combined to defeat their ambitions. A thunderstorm moved in over the Cape, causing impossible launch conditions as well as power surges which led to dubious readings on some of the instruments. Cooper proclaimed his frustration to the Cape weathermen: 'Oh gee, you promised us a launch today and not a wet mock-up.' A few hours later, a fire broke out in an underground cable shaft. Following a complete check-out of the whole system, the launch date was reset for Saturday 21 August.

The astronauts were roused at 4.30 am that day with the news that all was going well. After the usual breakfast the men donned their suits and were driven to the launch pad. At 9 am Cape time precisely, exactly on schedule, the Titan lifted them towards orbital altitude at the beginning of their record-breaking attempt. However, the unusually rough ride proved a foretaste of what was to come. On separation from the second stage, Gemini moved into an elliptical orbit which would carry it to a record altitude for an American flight of 217 miles (350 km). Meanwhile, down below, the Titan's first stage was recovered from the ocean for the first time.

As they passed over Australia on the first orbit, the first sign of possible trouble was detected by Conrad; the pressure in the oxygen tank which fed the fuel cells began to fall, even after he switched on the tanks' heating system. Although the decline continued towards a critical level, Cooper was given the go-ahead on the second orbit to release a 76 lb (34 kg) radar evaluation pod (REP) from the adapter module. One of the main tasks on the road to the moon was rendezvous between two craft, so the REP experiment was a vital forerunner designed to test the radar system fitted in Gemini's nose. The pod carried a radar transponder similar to one planned for an Agena target rocket on future missions, enabling it to drift more than 50 miles (80 km) from the Gemini before the astronauts, using data from the radar and computer on board, manoeuvred their craft into a rendezvous position. The crew had time to begin evaluation of the radar navigation system before ground control advised them to abandon the experiment; oxygen pressure had fallen from 800 to 120 psi in little more than two hours, leading to dismay both in the spacecraft and on the ground. Pete Conrad's 'Little Rascal' – the name he gave to the REP – was allowed to escape as flight director Chris Kraft told the crew to power down their Gemini as a prerequisite for a return to Earth during the sixth orbit. The men were in no danger since the Gemini carried the traditional batteries for use during re-entry, with sufficient power to last for up to 13 hours. Meanwhile, aircraft from Hawaii were ordered to the contingency splashdown zone in the Pacific: it looked as if Gemini would struggle to complete six orbits, let alone eight days in space. Cooper showed his annoyance by doodling a drawing of a wagon teetering on the edge of a cliff. As oxygen pressure fell by 65 psi towards the end of the third orbit, there seemed no doubt that his pioneer vehicle would have to give way.

Then the miracle happened: the pressure began to stabilize, though at a very low level. Tests during the fourth orbit suggested that the fuel cells were still operating correctly, the problem being the heater system which had ceased to function. Pressure dropped to a low of 60 psi then began to rise a little as some heat infiltrated the tanks from outside. By the end of the fifth orbit, confidence among the flight controllers was high enough for a unanimous decision to continue the flight for at least a few more orbits. A limited increase in power consumption was allowed, something for which the astronauts were very grateful since the cabin temperature had dropped low enough for breath to freeze on the windows. Every so often they could glimpse the REP through those windows, though it was gradually drifting further away, beyond their reach. However, on the ground, astronaut and rendezvous specialist Buzz Aldrin was dreaming up a substitute exercise which would relieve the monotony for the disappointed astronauts while enabling at least some evaluation of the radar navigation to take place. It was hardly surprising, though, when the two men reported only about two hours sleep each during the first day of the mission. In the Soviet Union, the official reaction was to criticize the Americans for taking risks with untested equipment in order to win the space race at all costs. Flight controller John Hodge gave NASA's reply: 'This is a simple failure which can happen to any piece of equipment, no matter how much you test it.'

By the third day, the craft seemed in good enough shape to attempt the rendezvous with a 'phantom' target. The radar had already been tested on its range-finding capability when it measured the distance to a transponder attached to a tower at Merritt Island next to the Cape: the range reading was almost identical to that given by ground-based radar for the altitude of the Gemini. Now there was an opportunity for the crew actively to manoeuvre the craft in search of an imaginary target which was fed into their computer by mission control. The task began on the 32nd revolution, and for the next two revolutions or 40,000 miles (64,000 km), the astronauts completed four orbital changes which brought Gemini within ½ mile (1 km) of the imaginary target's position. The way seemed clear for future crews to carry out the real thing.

As they approached the record set by Gemini 4, the crew were beginning to lose their customary good humour. They were still having trouble getting a good sleep: Cooper told ground control he had hardly slept at all, while Conrad had managed to drop off for some six hours. The temperature was still too low for comfort: 'We've been sitting here shivering for the last few hours,' was Cooper's comment. A further source of ill temper was the work schedule, since Gemini 5 carried 17 experiments, more than any previous flight. Most of these were concerned with medical evaluation of the effects of long-term weightlessness, or various photographic and observational assessments. The cabin was beginning to resemble a rubbish tip as the men struggled to stow their litter while packing and unpacking various cameras, lenses, rolls of film and so on; they had discarded their helmets and gloves long before. Not for the first time, and certainly not the last, the astronauts complained about the number of tasks which the flight programmers had crammed into a limited amount of time. When asked if he had any questions concerning the experiments, Conrad gave the reply of a harassed man: 'No, I would say we've got a full day. I hope we can get them all done.' Was it all the fault of the programmers? One additional factor was the relatively short period of time given to the men for training in carrying out all of their tasks. A third point was also worth considering: perhaps test pilots were no longer the ideal raw material for flights which were becoming increasingly scientifically orientated. Significantly, perhaps, NASA had selected its first six scientist-astronauts in June of that year, an indication of the agency's changing attitudes and priorities.

Nevertheless, the crew of Gemini 5 returned a wealth of data at the end of their mission, thereby verifying the numerous civilian and military uses of spaceborn observation platforms. Of particular interest were their observations of two Minuteman missiles as they were launched from Vandenberg Air Force Base in California. Cooper reported picking up the trail of smoke and flame as the missiles broke through the cloud layer, then watching until they burnt out. They even picked out the smoke from a rocket-sled which was specially fired as they sped overhead, though they failed to spot a number of huge checkerboard markers which were laid out on the ground. Just as on his first flight, Cooper reported seeing ships' wakes, including that of their recovery carrier on station in the Atlantic. 'This time he had a witness,' said his companion. Their photographs picked out very fine detail on the surface: in Florida, for example, they showed the launch pads, including 'good old pad 19', at the Cape, airfields and even highways.

The record of Gemini 4 was smashed, to be followed less than 24 hours later by a new world record as the time of Vostok 5 was passed. Cooper's only comment was, 'At last, huh?' The last three days were spent powered down once more, the result of another fuel cell problem. This time they were producing too much water as a byproduct of the chemical reaction which created the electricity. Since there was no way of dumping the excess, the only solution was to reduce power once again, thereby subjecting the astronauts to near-freezing temperatures. Furthermore, the excess hydrogen from the fuel cell supply was venting overboard, causing a slow roll by the craft, and control of the Gemini's attitude was made even more complicated by suspected valve-clogging which put two thrusters out of action. Mission control tried to make the men's lives less burdensome by broadcasting a selection of dixieland jazz, including 'Muskrat Ramble' and, appropriately perhaps, 'Birth of the Blues'. Conrad's reply was a poetic rendition: 'We were drifting along by the CSQ, when the radio said, "Here's word for you, your controls are dead, but you're not through." So here we are for three days more, with the end quite far.'

Ordered now to save the previous fuel for the thrusters until re-entry, the astronauts continued to drift and tumble, passing time by cracking jokes and making up verses. In reply to a query over his exercises by Dr Berry, Cooper replied, 'I hold Pete's hand once in a while, and once in a while I use a skin-cleaning towel, and then a couple of days we chewed gum.' Later, he came up with another idea: 'We thought we'd start taking long walks.' There was a suggestion that the flight be curtailed after seven days, but Conrad registered their pleasure when informed that they would go for the full eight days after all: 'Over the ocean, over the blue, from Gemini 5 we thank you.'

After what must have seemed endless drifting through space, the final day dawned at last. One interesting diversion came when they spoke to former Mercury astronaut Scott Carpenter as he worked in a submerged sea laboratory 200 feet (62 m) down on the bed of the Pacific off the

Californian coast. Then back to the realities of space travel as weathermen informed the crew that Hurricane Betsy was heading for their planned recovery zone. The decision was taken to end the mission one orbit early. The precautionary lowering of the orbit before retro-fire was ignored for the first time, but there was no need to worry as the rockets fired perfectly. Unfortunately for the crew, the splashdown turned out to be the least accurate of the Gemini series at a distance of 103 miles from the carrier, *Lake Champlain*. The reason was the failure by ground control to send up the correct computer co-ordinates for the re-entry. Before the crew realized the error, the craft had travelled far off course, too far to correct the path sufficiently. Gemini 5 splashed down at 7.56 am Cape time after a record-breaking 7 days 22 hours and 56 minutes in space. During that time, the two astronauts had circled the earth 120 times: on this one flight, Cooper and Conrad had spent longer in space than all previous astronauts combined.

The astronauts had to wait about 45 minutes before the swimmers dropped from a helicopter to attach the flotation collar to the Gemini, with another three-quarters of an hour elapsing before they set foot on the carrier's flight deck. There were some worries over their physical condition – some experts expected they would have to be carried because their legs would be unable to support them – but the men literally jumped for joy as they clambered on to the deck, then walked arm in arm along the red carpet. The medical verdict was: 'Absolutely nothing wrong with them. It is just as though they had taken a short plane ride.' Meanwhile, President Johnson phoned from his ranch in Texas to congratulate them on their 'great achievement', followed by their wives – it was Cooper's wedding anniversary – who had not been allowed to speak directly to the spacemen during the mission due to a tightening up of procedure by Deke Slayton. The next eleven days were taken up with exhaustive medical examinations, debriefings and scientific evaluations. Then came the medal presentations at the White House and the appearance before Congress, a proud moment for Cooper since he was the first military man to give two addresses to joint sessions of Congress. In mid-September came a new departure at the express wish of the President, a six-nation propaganda tour which took them to the Mediterranean and Africa. The Manned Spacecraft Center staff tended not to approve of this new role as roving ambassadors with the Gemini programme building to a climax and important work on Apollo development well under way; there were enough demands on the astronauts' time already.

As if to emphasize the point, Pete Conrad was selected for the Gemini 8 back-up crew while he was still out of the country. On his return, he would be plunged straight back into the training treadmill, first for Gemini 8, then as the prime command pilot of Gemini 11. The latter mission was scheduled to take place only a year after the Gemini 5 exploits. Small wonder that the increasing pressure hit hard at the astronauts both physically and mentally. An important factor here was the amount of time spent way from home, leading to often unbearable stress on marital and family relationships. As for Cooper, he stayed with NASA for another five years, serving on the back-up crews for Gemini 12 and Apollo 10 before deciding that enough was enough. He took with him the distinction of being the first man to orbit the Earth on two occasions. Unable to settle in any field for some time, he eventually joined Walt Disney Enterprises as vice president for research and development.

One interesting side issue arose after the record-breaking flight of Gemini 5. On board, amongst all the equipment and rubbish, were two dollar bills provided by the Federation Aeronautique Internationale. Pete Conrad told the story:

> The serial numbers of the bills were carefully recorded by the Federation, and a representative was aboard our recovery carrier, the *Lake Champlain*, to retrieve them. Then he solemnly compared the bills with the serial numbers he had on record. If they agreed, the Federation would, again with all due solemnity, certify that Gemini 5 had indeed surpassed the Russian record. What I can't for the life of me figure out is how you'd go about faking a space flight with the whole world watching, and I sometimes wonder what the consequences would have been if Gordo and I had substituted a couple of Confederate bucks for the others.

Clearly, it was the physical achievement which held more significance for the astronauts than any world record, important though that may have been to American prestige and pride. Besides, Cooper and Conrad realized that their record was unlikely to survive more than a few months, for NASA had planned a 14-day endurance test with Gemini 7 early in 1966. First, however, the agency intended a demonstration of the orbital rendezvous necessary for any flight to the moon.

Gemini 6 was scheduled to catch up and dock with an Atlas–Agena rocket specially launched 90 minutes earlier. The operation would be repeated several times in a thorough test of the Gemini radar, rendezvous and docking hardware. The Agena target was a modified version of the normal second stage booster, the most reliable available at that time. It was provided with a docking cone which would mate with the probe on the approaching Gemini, together with the radar and tracking aids fitted earlier to Pete Conrad's 'Little Rascal'. On the other end of the metal cylinder was the single rocket motor, capable of being restarted by the Gemini astronauts so that its orbit could be changed, though this was considered an unnecessary risk for the initial brief Gemini 6 test flight.

The men who had rigorously trained for the complex orbital manoeuvring were the veteran Mercury practical joker, 42-year-old

Wally Schirra, and group 2 newcomer, 35-year-old Thomas Stafford. Both had graduated from the US Navy Academy at Annapolis, but Stafford had taken a commission in the air force rather than the senior service. In 1959, when he came top of his class in the Experimental Flight Test Pilot School at Edwards Air Force Base, the older, more experienced Schirra was moving on to his new career as an astronaut. Stafford eventually wrote two flight test manuals while in command of the Performance Branch, before his selection by NASA in 1962. The call from Houston prevented him taking his place at Harvard Business School, though this never affected his prospects in post-astronaut business activities. The two had worked together since their assignment as Gemini 3 back-up crew; all they wanted was the opportunity to prove themselves.

On the morning of 25 October 1965 the two men lay on their couches on top of the Titan booster, awaiting completion of countdown. All seemed to be going to plan as the Atlas rocket carrying the Agena target took off from Pad 14 at 10 am Cape time; the astronauts expected to follow 100 minutes later. As they listened to the Agena's progress, elation turned to despair. The Atlas launch seemed perfect, shutting down right on schedule and separating from the Agena. Then, a little over six minutes into the flight, ground control lost track of the rocket just as it was due to fire its main engine for the final boost into orbit. A hold was ordered on the Gemini 6 countdown as engineers feverishly tried to re-establish contact. The phantom rocket failed to materialize, leading to the reluctant conclusion that the ignition system had malfunctioned, causing the Agena to blow apart and disintegrate over the Atlantic. 'Old Reliable' had let everyone down at the worst possible moment: there would be no launch of Gemini 6 that day, or for many days to come. At 10.54, flight director Chris Kraft called 'Scrub', and two desperately unhappy astronauts returned to Earth with a bump.

NASA officials were now faced with a dilemma: should they postpone Gemini 6 until the Gemini 7 flight had been completed and another Agena prepared, or should they delay the Gemini 7 mission until the Gemini 6 rendezvous had been successfully accomplished? In fact, they elected for neither of these, but for a third, more risky, alternative. In order to ensure at least one more flight before the end of the year, and finally to pull off the elusive rendezvous between two separate pieces of orbiting hardware, it was proposed to launch Gemini 7 on its marathon 14-day mission as originally planned, then send up Gemini 6 to effect a rendezvous only nine days later. To many this plan seemed incredibly optimistic, possibly even downright dangerous. Since Pad 19 was the only site suitable for a Gemini-Titan launch, the new plan meant readying the scorched pad, erecting the second Titan, mating it with the spacecraft and completing all of the checks within the unprecedented period of nine days. The previous record was two months. Another problem which had never before been faced would be the dual control of two manned craft in orbit at the same time, but engineers simply saw this as a useful pre-trial of the techniques which would be used during the Apollo programme. The Gemini craft would not be able to dock or even touch since Gemini 7 was not equipped for this purpose and there was a danger of an electrostatic discharge between the two craft. However, this did not nullify the value of a full trial of rendezvous equipment and techniques. The only questionmark lay over the space agency's ability to deliver the goods: the astronauts were ready to take the risk of sitting on a combination which had only partially been checked out, and the engineers and technicians were prepared to attempt the impossible. In the full spotlight of world publicity, NASA put its reputation on the line, not to mention the astronauts' lives and the nation's international standing. It was unprecedented in the history of the American manned space programme, and at any other time would probably have been dismissed as a crackpot idea suitable only for the waste bin, but the Agency was riding on the crest of a wave of confidence from which they did not plunge until the completion of the Gemini flights.

The Titan carrying Gemini 7 was duly launched as planned. It was 2.30 pm local time on 4 December 1965. A few hours earlier, the two crewmen had walked from the transfer van to Pad 19 looking for all the world like two characters from some Dan Dare comic story. Astronauts Frank Borman and James Lovell were wearing lightweight pressure suits specially designed for this mission topped by a hood which was zipped on to the suit and looked like a giant face mask. The normal suit and helmet had been replaced for this mission on the grounds of practicality; they were very bulky and cumbersome, so with cabin space at a premium and permission for the first time on an American flight to remove their pressure suits while in orbit, the astronauts were happy to wear the 16 lb (7.3 kg) suit and aviator's helmet which could be easily taken off and stowed when not required.

The first crewman to benefit from the change in regulations was Jim Lovell, the larger of the two men. The 37-year-old Navy lieutenant commander was free of his suit while floating out of his couch in his longjohns, to the considerable envy of the command pilot, 37-year-old air force lieutenant-colonel Frank Borman. Officials had decided it was too risky to allow both men to slip out of their suits at the same time so Borman had to suffer in the stifling suit and try to snatch some sleep while Lovell slept in comfort. The men were the same age, all but 11 days. Their careers had some parallels too, with Borman serving as a test pilot at Edwards Air Force Base at the same time that Lovell was test flying Phantoms at Patuxent River. It was while he was in Maryland, serving with Pete Conrad and Wally Schirra, that Lovell had been inflicted with the nickname 'Shaky' by Conrad, supposedly because of his intense nervous energy and the fact that it was the name his friends knew would most annoy him. Borman was the more aggressive of the two, a man with ambition, a decisive character whom it was comforting to have with you in a crisis. Finally, in 1962, like so many of their companions at Edwards and 'Pax' River, they had been drawn like moths to a lamp by the lure of a NASA recruitment campaign. Following their teaming as the Gemini 4 back-up crew, they had spent months training for the most rigorous flight yet attempted, 14 days orbiting the Earth in a capsule which had no more room than 'the front seat of a Volkswagen', as Borman put it.

Only one day out of the two-week mission was to be taken up with the Gemini 6A rendezvous (Gemini 6 had been redesignated). The remainder of the time was given over to a record 20 experiments, mostly medical studies of the effects of prolonged weightlessness though studies of radiation, the Earth's magnetic field and spectral analysis were also included. It was noticeable that photography was strictly limited due to the problems Cooper and Conrad had in stowing and changing film, lenses and cameras. Gemini 5 had splashed down looking like a miniature rubbish tip, and the crew of Gemini 7 were determined to do their housekeeping in a more orderly fashion. Apart from a fuel cell warning light which raised visions of a Gemini 5 repeat, but later proved to be faulty, the first day went according to plan. Borman was able to station-keep successfully with the nearby second stage, unlike Jim McDivitt on Gemini 4, then fire the thrusters to raise the low point in the orbit. Later that day they saw some unidentified satellites in the distance, but there was no need for alarm. As Borman said after hearing reports from Sunday newspapers of a mid-air crash near New York, 'I think we're safer up here'.

Meanwhile, back on the ground, the launch pad had scarely cooled before the army of engineers, welders, plumbers and so on descended like a swarm of locusts to begin the repair work needed before the next Titan could be installed. The race was on. As the Sunday dawned, the first stage was installed on Pad 19, to be followed during the day by the second stage and then the spacecraft itself. Preparations continued to go well, so that by the middle of the week confidence was high that Gemini 6A could be launched on Sunday the 12th, a day earlier than planned. The Gemini 7 crew continued their undramatic routine as the days ticked by, though Borman put in a request to Deke Slayton concerning removal of his spacesuit to match his undressed companion. In ground visibility tests they failed to sight markings laid out on the ground at Laredo, Texas, but succeeded in viewing the launch of a missile fired by a Polaris submarine. A planned laser experiment was called off due to cloud cover over New Mexico. On Thursday, they were given the go-ahead to circularize their orbit at an altitude of 185 miles (300 km) in readiness for the rendezvous attempt. They had already passed another historic landmark; their flight meant that the USA had overtaken the total space hours clocked up by the Soviet Union.

On Friday the 10th, Borman was relieved to get permission at last to remove his suit, particularly since the cabin temperature had risen to a distinctly warm 29°C. The medical team, however, wanted to compare the telemetry readings from both men under similar conditions, so Jim Lovell had to relinquish his freedom and don his pressure suit. Their first week in space ended with another laser experiment failure, but the signs were good for the launch of Gemini 6A as the crew of Gemini 7 began their second Sunday in space. The gamble seemed to have paid off.

Astronauts Schirra and Stafford were roused early for their second attempt to ride Gemini 6 into orbit. While they breakfasted with Gordon Cooper, the back-up crew (Gus Grissom and John Young) checked over the capsule. All seemed perfect as the countdown proceeded on schedule. The crew, wearing the standard Gemini suits, rode the elevator to their capsule perched on top of the Titan, clambered into their couches and settled down for the roar and vibration which would tell them they were on their way. Unfortunately, things did not work out as they expected. The Capcom informed them, 'You are cleared for take-off . . . adios'. The countdown came to an end as the seconds were calmly ticked off, then came the announcement 'ignition' to coincide with the smoke and flame issuing from the rocket motors. It was just after 9.54 am Cape time. Little more than a second later, the engines cut out and silence reigned – the Titan and its human cargo were still grounded on Pad 19! Confirmation from the Capcom swiftly followed: 'We have shutdown, Gemini 6.' Their colleagues on Gemini 7 watched the incident as they passed overhead. 'We saw it ignite. We saw it shut down,' they reported.

For Wally Schirra, seated on top of a potential fireball, the situation was nowhere near as clearcut. All the safety rules dictated that he pull the abort ring immediately in order to trigger the escape mechanism, jettisoning the hatches and ejecting the men for a safe parachute landing. As he watched the fuel pressure rise and the clock in the capsule begin to run, the temptation must have been immense. Technicians monitoring their heartbeats saw Schirra's rise to 98 and Stafford's to 120, both remarkably low considering the stress they were under. The command pilot rapidly weighed up the situation. The rocket seemed securely attached to the pad – the explosive bolts had not had time to free the rocket from the launch pad – and as long as the fuel and the oxidizer in the Titan's tanks did not come into contact, the booster was safe. It took ground control several minutes to confirm that Schirra had made the right decision; he was later awarded the DSM for 'his courage and judgement in the face of great personal danger, his calm, precise and immediate perception of the situation that confronted him, and his accurate and critical decisions [which] made possible the successful execution of the Gemini 6 mission'. Had he pulled the ring, no one would have blamed him, but instead he had kept the capsule intact and available for another attempt in three or four days time. Both the astronauts and their mission had lived to fly another day. As they waited to be released from the capsule, the men must have felt mixed emotions: fear, relief, frustration, annoyance, disappointment. Schirra simply commented, 'These things happen. It could happen to anyone. No one was hurt.'

Once again, the race was on. This time the fault had to be traced before any possible launch date could be contemplated, and there were only six days remaining before Gemini 7 was due to return to Earth. Preliminary investigations blamed a small electrical plug in the rocket's tail which had prematurely come free due to vibration during the brief ignition. However, detailed analysis of telemetry records showed that this could not be the whole story. Then, on the Monday morning, an engineer discovered a plastic dust cover which had inadvertently been left obstructing an oxidizer inlet. This cover had been inside the rocket since its original assembly, and was sufficient on its own to have prevented lift-off on either 25 October or 12 December. With the problem solved, the Cape staff once again swung into a prolonged period of non-stop activity in order to prepare the unlucky ship for her new date with destiny – Wednesday, 15 December.

By this time the crew of Gemini 7 were beginning to suffer some difficulties too as they started to break new ground, having passed the old endurance record set by Gemini 5. On the sixth day, one stack of fuel cells began to cause trouble, a problem that lasted for the remainder of the mission although it did not pose a major threat. Potentially more serious was the tumbling of the spacecraft which was reported when the astronauts awoke on their seventh day; it was caused by excess water being ejected into space. Although partly overcome by ground engineers, it still caused control problems for the crew, a crew which was becoming increasingly tired, dirty and disenchanted with the reality of long-duration flight in a cabin that was too small to swing a cat. Borman, with the extra responsibility of command pilot resting on his shoulders, was sleeping more fitfully than his companion, and his body was adapting less rapidly to a combination of space food and weightlessness. He reported: 'Jim and I are beginning to notice the days seem to be lengthening a little. We're getting a little crumby!' The cabin was inevitably beginning to resemble a refuse tip, complete with the associated odours, despite all the astronauts' efforts. Its condition was not improved when a urine bag split open in Borman's hands. 'Before or after?' came the Capcom's voice. 'After', came the rueful reply. Appropriately, his choice of reading material for the flight had been Mark Twain's *Roughing It*; Lovell chose *Drums Along the Mohawk* by James Fenimore Cooper.

One successful flight which the crew were able to watch came on 14 December when they observed a Minuteman missile re-enter the atmosphere over Eniwetok Atoll in the Pacific. All fingers were crossed that Gemini 6A would be able to reverse the manoeuvre the next day. The weathermen predicted a threat of dense fog, but when the morning came visibility was perfect. For the third time in three months Grissom and Young gave the capsule a final check-out while Schirra and Stafford consumed the pre-flight breakfast. Three hours after they were awakened, the crew once more climbed into the cabin. Right on schedule, at 8.37 am Cape time, the countdown reached zero, the mighty engines roared into life and, almost as if it were being lifted by concentrated human willpower, Gemini 6A rode a white smoke trail across the clear blue sky on its long-awaited journey. The spirit at the Cape was exemplified by Wally Schirra's prayer from the heart: 'Go! Do you hear the man, go!' The successful launch gave a much needed boost to the watching Gemini 7 crew as they passed over the Cape, as well as to the men on the ground who had worked so hard to turn despair into triumph.

Some six minutes after launch, Gemini 6A entered Earth orbit. For the first time there were two American manned craft in space, with four men on board. Now came the real test as Schirra and Stafford attempted to utilize their faster, lower orbit to make up the gap of 1,380 miles (1,990 km) which separated the craft. Borman and Lovell could only put on their suits and wait for their pursuer to appear. A preliminary height adjustment was required on the first pass over New Orleans since Gemini 6A's orbit was just a little lower than expected. Soon after, the orbit was made more circular, thereby reducing the rate by which Gemini 6A was closing the gap. Nearly three hours into the joint mission, the orbital planes of the two craft were exactly matched by Schirra's 40 second thruster firing; with the gap down to 439 miles (700 km) and closing, the way was clear for the straightforward raising of the orbit until they could fly along identical paths.

In reality, the apparently simple manoeuvres involved a mass of detailed instructions fed to Schirra by the ground and by Stafford who was operating the onboard computer as well as completing navigational plots. Small wonder that Stafford at one stage described himself as 'busier than a one-armed paperhanger'. The distance closed during Gemini 6A's third orbit until radar contact between the craft was established at a distance of 288 miles (460 km). With the exact position of the target now fixed, the computer now had solid data with which to advise Schirra and Stafford. Three hours and 45 minutes into the mission, another thruster burn virtually circularized Gemini 6A's orbit, though it was still about 17 miles (27 km) below Gemini 7 and 200 miles (320 km) behind.

Using angular and distance measurements, Schirra and Stafford slowly edged towards their target during the next two hours. Soon after firing the thrusters to raise their orbit for the last time, the crew of Gemini 6 were able to obtain visual contact at a distance of about 30 miles (48 km). The steady docking light and the two flashing acquisition lights looked like tiny stars against the black sky as the craft crept towards each other above the dark Indian Ocean. Jim Lovell, calmly awaiting the visitors, later recounted the view from Gemini 7:

> It was night time, just becoming light. We were face down, and coming out of the murky blackness below was this little pinpoint of light. The sun was just coming up and was not illuminating the ground yet, but on the adapter of Gemini 6 we could see sunlight glinting, and as it came closer and closer, just like it was on rails, it became a half moon. At about half a mile we could see the thrusters firing, like water from a hose. And just in front of us it stopped. Fantastic!

Five hours and 56 minutes into the mission, Wally Schirra brought Gemini 6 to a halt a mere 120 feet (36 m) from their companions. As the men on the ground brought out their miniature American flags and cheered themselves hoarse, the crew of Gemini 6 were able to relax at last: Stafford reckoned he had spent no more than 15 minutes looking out of the window during the whole six-hour period since take-off. Schirra and he had used only half of their allocated fuel, a remarkable achievement.

The delighted astronauts lapsed into lighthearted banter as they swept around the world in tight formation at a speed of 17,500 mph (28,000 kmh). 'Hello there. What kept you?' came Borman's Arizona drawl. Station-keeping with masterly precision, Schirra brought his craft to within a foot of Gemini 7, though he was very careful not to let them touch. At such close quarters, the four men could peer at each other through the small, slightly smoke-smudged windows. 'You guys sure have big beards,' commented Schirra. 'For once we're in style!' replied Borman. And so it continued as they waltzed around each other, then rode together nose to nose across the heavens, for the next 5 hours 15 minutes and nearly four revolutions. Schirra was almost blinded by the sunlight reflecting from the Gemini's white skirt: 'Hey, Frank, I see your hatch is on fire!' Passing over Hawaii, he joked: 'There seems to be a lot of traffic up here!' to which Borman replied: 'Call a policeman.' Schirra later taunted West Point graduate Borman with a sign in his window, 'Beat Army'.

As Gemini 6 drifted around behind the adapter section of its companion, Schirra gave Borman a surprise: 'You've got a lot of stuff all around the back end of you. You guys are really a shaggy-looking group with all those wires hanging out.' Soaring over the blue Pacific west of Chile, Stafford was taking some breathtaking photographs of Gemini 7's gold coloured rear insulation with its streamers flying out like tail feathers. There was nothing to worry about; they were simply remnants of the insulating tape used to minimize damage during separation from the rocket. The Gemini 7 crew reported to Schirra: 'You have some too.' Each astronaut took turns in taking over the controls in order to gain experience in formation flying. On the night passes they could watch each other through the dimly lit windows, drifting silently over a sleeping world. All too soon, it was time for the astronauts themselves to grab some sleep. Schirra fired the thrusters to move his craft into a safe orbit, drifting slowly apart as much as 40 miles (64 km) while they waved farewell. As the crews settled down for a well-earned rest, ground control picked up a touching comment from Gemini 7: 'We have company tonight.'

The next morning the Gemini 6 crew prepared for their return to Earth: all the mission objectives had been achieved, and their craft, with its conventional batteries, was not equipped for a long stay in space. Suddenly, tingles of alarm began to run up and down the hitherto relaxed lines of ground controllers as Schirra reported: 'This is Gemini 6. We have an object, looks like a satellite, going from north to south, up in a polar orbit. He's in a very low trajectory . . . looks like he may be going to re-enter pretty soon. Stand by . . . it looks like he's trying to signal us.' The warning transmission was immediately followed by the sound of Wally Schirra on harmonica and Tom Stafford on bells giving their rendition of 'Jingle Bells'. There were just nine days to Christmas and the irrepressible practical joker Schirra had struck again, though as he later explained, he did have another motive: 'It was to relieve the tension . . . I think we convinced Chris [Kraft] and many of the people on the flight control team that we did, in fact, have an unidentified flying object there. And I think the children of this country are happier for the fact that we might have seen something there.'

A few hours later, Schirra radioed a farewell to their Gemini 7 colleagues: 'Really a good job, Frank and Jim. We'll see you on the beach.' Retro-fire sequence began over Canton Island on their sixteenth revolution, slowing the craft for its fiery descent towards the Atlantic Ocean. Using the computer guidance system, Schirra and Stafford became the first crew fully to control their re-entry. They splashed down 630 miles (1,000 km) south west of Bermuda, only 13 miles (21 km) from the recovery carrier *Wasp*. In true naval style, the astronauts opened the hatches and casually sat chatting to accompanying frogmen while awaiting pick up by the carrier – no helicopter rides or abandoning ship for those two sailors. Watched live on TV by millions of Americans, the two men stepped from their craft on to the red carpeted deck to the sound of a band playing 'Anchors Aweigh'. Schirra's delight with the mission's success was tempered only by a slight criticism of their press reports: 'I'm real sorry they made the rendezvous sound so easy. It may have looked easy, but it was only because we had practised so much.' Stafford backed him up: 'I figure we must have worked out at least 80 times in the simulators, an hour and a half each time.' What about the Soviet rendezvous with Vostoks 3 and 4 in 1962? Schirra scornfully commented:

> If anybody thinks they've pulled a rendezvous off at three miles, have fun! This is when we started doing our work. I don't think rendezvous is over until you've stopped, completely stopped with no relative motion between the two vehicles at a range of approximately 120 feet. From there on it's station-keeping. That's when you can go back and play the game of driving a car or driving an airplane or pushing a skateboard, it's about that simple.

Back on board Gemini 7, life wasn't quite as straightforward, and there were fears that the mission might have to be curtailed by one day. The crew, both clad in their longjohns since the rendezvous, were feeling the strain, resulting in more irritability and less tolerance. Their spacecraft was also feeling the strain of the record-breaking endurance test: the fuel cells continued to cause problems as another stack failed in the same section as had previously caused trouble, and two attitude thrusters jammed, an irritating malfunction that had to be corrected by using the aft firing thrusters. As the time dragged by, the men took the opportunity to catch up on their reading. Their sleep periods were adjusted to give them the maximum rest before re-entry, and when Saturday morning arrived, they were ready and eager to return to Mother Earth. The astronauts, looking for all the world like two tramps, wriggled back into their pressure suits, cleared the cabin of all equipment and stashed the rubbish as best they could. As Lovell later related:

> We were worried that we'd sort of get pushed out of the spacecraft with all the debris that would accumulate. So we spent many hours prior to the flight finding little spots and crevices in the spacecraft where we could pack things. We would eat three meals a day, and Frank would very nicely pack the containers in a small bag, and at the end of the day he would throw it behind the seat. We managed to get nine days' debris behind those seats.

As they passed over the dark Pacific for the 206th time, the re-entry sequence began. The kick from the engines gave them their first experience of gravity for two weeks: 'We felt we were going backwards when the retro-rockets fired – I think John Glenn said back to Hawaii – I felt we were going back to Japan,' said Borman later. Like their Gemini 6 friends, Borman and Lovell successfully piloted their craft to an accurate splashdown only 8 miles (13 km) from the *Wasp*. Little more than half an hour later, TV viewers on both sides of the Atlantic watched as they staggered from the recovery helicopter on to the sunlit deck. The beaming, bewhiskered Borman commented: 'It seemed more like six weeks than a fortnight.'

The men were in much better condition than the physicians had dared hope. Borman explained:

When we got back on the carrier, if we had any deterioration at all, it was that our legs were heavy because they hadn't been used. We were able to run a mile the day we got back to the Cape. In my opinion, with proper crew-comfort provision, people will have no difficulty going a month, two months, or as long as they want to in space.

The scheduling of two exercise periods per day for each astronaut probably helped to maintain their physical condition: they had a pair of elastic cords attached to a handle at one end and a nylon foot strap at the other which required 70 lb of pressure to pull the exerciser to its maximum extension. Both men also emphasized the value of flying suitless in their cramped, uncomfortable environment: 'If we would have had the suits that had been used prior to our flight and not been able to remove them, I think our physiological effects would have been tremendous when we got out of the spacecraft,' commented Borman. 'I am not certain that we would have been able to bounce back the way we did and it would have been a matter of survival rather than a matter of operating efficiently in space. Our firm recommendation is that on any long duration flight, the crew go suitless.'

As for Borman's difficulty in sleeping, the official explanation was given as 'command pilot syndrome' – the nagging fears and doubts which caused him to awake in order to check the instruments. Borman was reluctant to accept that reasoning: 'Jim Lovell slept very well. I was envious many times when I'd look over and see him snoring away. I think that perhaps the lack of space and the fact that we weren't able to stretch out at all led to it.' He also had a few complaints about their provisions: 'I guess the people didn't realize we were going to operate on a regular day, and try to eat breakfast, lunch and dinner. The meals weren't prepared that way, and several times we had shrimp cocktail and peas for breakfast.' Their daily routine was considered one of the chief contributory factors to their success:

We worked on a Houston day. Our watches were set on Houston time. We had a regular work day, had three meals a day, and then at night we went to bed. We put up light filters in the windows and didn't look out, and to us it was night time. We had absolutely no sensation of movement. Our world was inside the spacecraft.

The historic rendezvous and long-duration missions of Geminis 6 and 7 brought to a highly satisfactory conclusion the most successful year yet in the American space programme. Any lingering doubts over man's ability to survive a trip to the moon and back had been finally dismissed by the crew of Gemini 7, and, with the rendezvous techniques thoroughly tried and tested, all that remained was the operational proof of the docking between two separate craft. There was to be no let up, however, in the hectic pace being set by the flight programmers. Borman, now promoted to colonel, and Lovell, promoted to captain, acted as special presidential ambassadors on a foreign tour before plunging back into the humdrum world of flight training and preparation. They were to fly again on the Apollo 8 mission three years after their Gemini 7 epic. Wally Schirra was the back-up command pilot on the ill-fated first Apollo, later taking over the improved version for its first operational flight, and Tom Stafford was quickly thrown back into the fray, commanding Gemini 9 only seven months after his debut in Gemini 6A.

Three months after Gemini 7's return, the next mission was ready to launch. Due to the failure of the Atlas–Agena the previous October, Gemini 6 had not been able to fulfill its original task of docking, so it was now left to the inexperienced crew of Gemini 8 to carry out the one remaining objective laid down for the Gemini programme. Minor problems arose during final preparations with both the spacecraft and the Atlas booster, so the mission was delayed for one day until 16 March 1966. The men on whose shoulders lay the responsibility for the first ever docking between two spacecraft were 35-year-old Neil Armstrong, an astronaut since 1962 but making his first spaceflight, and 33-year-old David Scott, the first group 3 astronaut selected in 1963 to go into space. Armstrong had already led a full, adventurous life, having served in the Korean War, survived being shot down, flight tested high performance jet aircraft at the Lewis Flight Propulsion Laboratory and at Edwards Air Force Base, and flown the X 15 rocket plane to the edge of space at speeds of 4,000 mph (6,400 kmh). He had the reputation of being a cool, careful aviator, and now had the distinction of becoming the first civilian astronaut to fly in space. Scott, a major in the USAF, came from San Antonio in Texas, the son of an air force brigadier general. He was one of the most highly qualified astronauts, with a bachelor's degree in science from the US Military Academy, a masters degree in aeronautics and astronautics, an an engineering degree from Massachusetts Institute of Technology. He had also graduated from the Air Force Experimental Test Pilot School and the Aerospace Research Pilot School at Edwards. He survived a serious jet crash in 1963 to become regarded by his peers as the leading member of the 14 group 3 astronauts.

The men were awakened at 7 am Cape time, had the usual physical check followed by the traditional breakfast, and moved from the crew quarters on Merritt Island to the suit trailer at launch complex 16. They wore the heavy EVA pressure suit, for, apart from the rendezvous and docking with the Agena, the three-day mission would include an ambitious two-hour EVA by Dave Scott. To aid his space walk, Scott was provided with a specially designed hand-held manoeuvring unit and a backpack which included a portable life support system stored in the craft's adapter section and intended to be strapped on by the astronaut during the space walk. Back-up crew Pete Conrad and Richard Gordon checked out the spacecraft, then, shortly before the Atlas was due to boost the Agena target into orbit, the astronauts settled into their couches. Right on schedule the Atlas lifted off from pad 14, and this time both stages worked perfectly. The relieved astronauts continued the final checks, and at 11.41 am their Titan lifted them on the first leg of their Agena pursuit.

Six minutes after launch, Gemini 8 entered an elliptical orbit 1,200 miles behind its target. Just as in the case of Gemini 6, Armstrong and Scott spent the next four orbits circularizing and raising their orbit, slowly but surely catching up on the Agena which was cruising in its 185 mile high orbit. Aided by the onboard computer, they established firm radar contact with the target at a distance of 200 miles (320 km), with visual contact reported at 87 miles (140 km) 4 hours and 40 minutes into the mission. Just over an hour later, Armstrong began the final manoeuvre which brought Gemini 8 to within 150 feet of the cylindrical Agena. He described the target as 'looking fine' as he carefully examined the target docking adapter. Creeping towards the Agena while they passed over the eastern Pacific, the craft was steadied in a docking position only two feet (50 cm) from the docking adapter while they awaited contact with the tracking ship, *Rose Knot Victor*, which was stationed in the South Atlantic. With all systems 'go', Armstrong was permitted to close that tantalizingly small gap and thus conclude the first ever space docking 6 hours and 34 minutes into the mission. The mooring latches engaged and the Gemini was pulled into a 'hard dock' position with the craft's cone firmly inside the Agena's docking collar. Armstrong reported: 'Flight, we are docked. It was a real smoothie.' The Agena was said to be very stable, with no noticeable oscillations. There had been no sparks or discharge of static electricity on contact.

The two craft were not just physically linked, they now shared the same control systems. Soon after docking, the crew were able to utilize the Agena's attitude control system to yaw the whole combination through 90 degrees. Everything seemed to be working perfectly, but as the first half hour of docked operations approached, Dave Scott noticed that the craft had drifted about 30 degrees out of alignment with their original stationary position; something seemed to be pushing them around. They were unable to see or hear any thruster activity, so they assumed the problem lay in the Agena control system. They spent about three minutes steadying the assembly by using Gemini's thrusters, then suddenly the rates of yaw and roll accelerated. By now they were passing over the Pacific out of radio contact with the ground network, so they were unable to get advice from that direction. Still assuming the Agena to be the source of the trouble, even though its systems should have been switched off, Armstrong managed to slow down the motion sufficiently to undock from the Agena. There seemed a real danger that the docking assembly would give under the strain of the wild bucking motion, but if the Agena was the cause of the problem, the Gemini should stabilize while the rocket should start to gyrate out of control once separation was complete.

Instead, to the astronauts' great concern and discomfort, the Gemini began to cavort more wildly than ever – it had been their craft, not the Agena, which had been at fault. They rapidly built up to the terrifying state of tumbling once every second, sixty revolutions a minute. There was now a threat to the integrity of the spacecraft as well as a danger of the crew becoming nauseous, even blacking out. The control panel began to blur before their eyes as their pulse rates shot up to 150. Armstrong's rapid, accurate assessment was picked up by the puzzled Capcom on the tracking ship *Coast Sentry Quebec*: 'We're in violent left roll . . . We seem to have a stuck thruster.' Unable to bring the craft under control in the normal way, the first priority seemed to be to disable the malfunctioning thruster: without knowing which one was the culprit, the only way was to shut down the entire orbital attitude manoeuvring system. The next question was, how to damp down the tumbling and restore control? The only way was to activate the re-entry control system, and gradually, by firing periodic bursts from the small thrusters, the craft was once more stabilized. It had been a damn close run thing, the closest yet to a disaster in space. Nearly 75 per cent of the re-entry fuel supply had been used up, and under mission rules the flight had to be terminated as soon as possible. For about ten minutes, the whole Gemini programme had been in jeopardy: the one positive result was the proof of the effectiveness of the astronauts' training.

Flight Director John Hodge, in the chief controller's seat for the first time, had the unfortunate task of ordering a return to Earth on the seventh orbit, the first available opportunity to splashdown in one of the contingency zones specially planned for just such an emergency. The destroyer *Leonard F. Mason* was despatched immediately to the recovery zone, about 500 miles (800 km) east of Okinawa in the Pacific. Meanwhile the crew were testing the main thrusters one by one in order to analyse the cause of their problem. They concluded that a short circuit in number 8 yaw thruster had caused it to fire continuously without command, an interesting discovery since a short circuit in the thruster heating system had been found prior to launch.

Most of the necessary information was passed up by the two tracking ships, since Gemini 8 was far from all land stations apart from Hawaii, an effort which was gratefully acknowledged by the astronauts. They had one orbit in which to stow all loose equipment, with retro-fire taking place over central Africa, only 10 hours and 4 minutes into the mission. The eventful mission came to a successful conclusion a little over half an hour later as Gemini 8 parachuted into the ocean within three miles (5 km) of the target point, watched by a circling aircraft. The crew were soon joined by three pararescue men, who attached a flotation collar to the craft. During the three hour wait for the destroyer to arrive, the unfortunate astronauts were violently seasick. They then spent another 18 hours at sea as the destroyer steamed to Okinawa where they were greeted by Wally Schirra and a group of NASA officials: there was no red carpet welcome this time.

Not surprisingly, neither astronaut was very happy with the outcome of the mission, particularly Scott, who had lost the opportunity to test the new EVA equipment – there were those who said the EVA had been too dangerous anyway, and that the cancellation was a blessing in disguise. NASA officials made the best of things: the Director of the Manned Spacecraft Center, Dr Robert Gilruth, said,

> The flight crew and the ground crew, I feel, reacted extremely well and ably to an inflight emergency and we feel very fortunate to have experienced a problem like this and to have been able to overcome it and bring the craft back successfully. We missed the spacewalk, of course, and we missed doing some experiments, but by and large we feel that we got in a very important day's work. We have learned a lot . . . perhaps we have learned more than we set out to learn.

The astronauts were awarded the Exceptional Service Medal for 'the remarkable job they did in carrying out the mission . . . and of bringing back much useful data'. They were praised especially for their cool command of the situation when all hell was breaking loose around them: 'During the emergency period they had the presence of mind as they were undocking to leave the Agena response to ground command and with the tape data intact so that it could be read out to the ground.' The astronauts reported last seeing the Agena in a stable condition, and so it proved. Thanks to their efforts, ground control was able to carry out ten orbital manoeuvres with the rocket, thoroughly testing its systems in order to improve its future performance. The Agena was parked in a high orbit awaiting a possible visitor from a later mission. Armstrong was assigned to the back-up crew for Gemini 11, while Scott became involved in the development of the Apollo spacecraft, both fully aware of how close they had come to death on their first venture into space.

Tragedy had already struck the next mission, Gemini 9, when the prime crew, Elliott See and Charlie Bassett, died while coming in to land in poor visibility at the McDonnell plant in St Louis. Their T 38 aircraft struck the roof of the building which housed their Gemini capsule and exploded in a giant ball of flame as it struck the courtyard beyond. It was 26 February 1966. For the first time in the Gemini space programme, the back-up crew had to be called upon to fill the gap: Tom Stafford, fresh from the Gemini 6 success, was partnered by a group 3 newcomer, Eugene Cernan. The replacement crew had until 17 May to complete their training for the complex mission, a three-day flight which would include rendezvous and docking with an Agena target, and a two hour space walk by Cernan. It was a lot to expect from the 32-year-old Cernan, one of the quiet men in the astronaut corps. Unlike many of his colleagues, he had not trained as a test pilot, serving instead on Naval Attack squadrons in California after receiving a commission through the Naval Reserve Officer training programme at Purdue University. He later gained a master's degree in aeronautical engineering from the US Navy Postgraduate School at Monterey, and had been closely involved with spacecraft propulsion systems, particularly the Agena, since joining NASA in 1963.

Ironically, it was the Agena which let them down on 17 May – it was the second time Stafford had lain in his couch while his target rocket malfunctioned and plunged into oblivion a few minutes after launch. Fortunately, or so it seemed at the time, NASA had covered itself in case of such an eventuality, preparing an Augmented Target Docking Adapter to fill in for the lost Agena. Although not provided with a propulsion system, the adapter was made from Gemini spare parts to replace the Agena in all other respects. Gemini 9 should be able to rendezvous and dock with the 12 foot (3½ m) long adapter once it was lifted into orbit on board an Atlas booster. The new launch date was set for 1 June, and the mission was redesignated Gemini 9A.

Came the morning of 1 June, and all seemed to be going well. Once again the two astronauts rode up in the lift at Pad 19, clambered into their spacecraft and awaited events at Pad 14. Right on schedule, at 10 am Cape time, the Atlas lifted off perfectly, sending the adapter into a circular orbit 185 miles (295 km) high. A few seconds after the engine shut down, ground control commanded the shroud which was protecting the adapter to separate from it, but incoming telemetry did not confirm separation. No one was sure whether the shroud had jettisoned, thereby uncovering the adapter's docking cone; the only way to confirm was for the astronauts to see for themselves. They waited in their craft for an hour and a half as the countdown proceeded to T minus 1 minute and 40 seconds. Then came the news that the countdown had been recycled to T minus three minutes and holding: a problem arose with updating the spacecraft's computer through the ground equipment. Twice more, the ground controllers tried to pass the essential guidance commands that would enable Gemini to located the adapter, but each time they failed. The launch window closed as they ran out of time, leaving Mission Director Bill Schneider with no option but to scrub the launch. Gene Cernan could not believe what had happened, and was visibly upset. From Stafford, who had been left on the launch pad four times out of five, the comment was a philosophical 'Oh shucks!' However, he was concerned over his younger partner's attitude:

> You know, I think Gene was starting to think that I was jinxed. A long time ago, Wally Schirra and I went up and down the elevator a few times and then after I had been up and down the elevator with Gene, I finally decided what it was – it was Wally that was jinxed and Gene that was jinxed, and I wasn't jinxed at all!

His supposed jinx went on trial again two days later. The astronauts were awakened at 3.11 am Cape time, went through the routine medical check for the third time in three weeks, ate breakfast and then went over to the suiting up room at Pad 16, where Cernan was fitted with a new suit specially developed for use with the astronaut manoeuvring unit during EVA. The leg covering was made of a woven stainless steel cloth designed to withstand temperatures of 700 degrees Centigrade, a necessity to protect Cernan's legs from the hot gases emitted by the AMU's thrusters. The protection was attained at the cost of extra weight and lack of flexibility, a fact which was not lost on the man who would make America's second space walk. Finally, the crew rode up in the familiar lift to the white room where they were briefed by the back-ups, Jim Lovell and Buzz Aldrin. At T minus 105 minutes the hatches were closed, and the waiting began. The computer update failed again at T minus three minutes, but it had been programmed direct from Mission Control in Houston at T minus 15 minutes to allow for such a problem repeating itself, and at 8.39 am, right on schedule, the Titan rocket roared into life. Stafford, who had earlier presented pad technicians with a 3 ft long match to ensure ignition, was heard to utter a quiet exhortation: 'For the third time, go.'

The initial manoeuvres went according to plan as the Gemini gradually caught up with the target which had been awaiting its visitors for the past two days. On the third revolution, 4 hours 11 minutes into the mission, Cernan reported that they were about 1 mile (1.6 km) from the adapter. Passing over a darkened New Guinea, the crew were able to see the adapter's docking light, suggesting that the nose shroud had jettisoned after all. However, as they closed in, all their hopes were dashed as Stafford reported:

> We have a weird looking machine here . . . both the clam shells of

the nose cone are still on but they are open wide. The front release has let go and the back explosive bolts attached to the ATDA (Augmented Target Docking Adapter) have both fired . . . The jaws are like an alligator's jaw that's open at about 25 to 30 degrees and both the piston springs look like they are fully extended . . . It looks like an angry alligator out here rotating around.

The frustrated command pilot put forward a suggestion to ground control: 'We might put out our docking bar and go up and tap it.' The response was not favourable; the steel band holding the two 'jaws' together would be difficult to sever, and could damage the craft or an astronaut trying to cut it during a space walk. Stafford was told to back away while ground signals were sent to the adapter in order to move the jaws and hopefully free them. Unfortunately, the only result was to close them slightly. There was no choice but to abandon the attempt and revert to a contingency plan which had been worked out in case of a problem with the shroud.

Stafford fired the thrusters to raise Gemini's orbit and carry it in a great loop up to 12 miles (20 km) behind the adapter while Cernan used the hand-held sextant to track the target and eventually re-rendezvous an hour and a half later. Soon afterwards, the two craft separated once again as Gemini's orbit was slightly altered in order to allow it to drift 70 miles (110 km) ahead while the crew attempted to get some sleep, without much success. Nearly 18½ hours into the mission, the crew began their third rendezvous, this time in a simulation of a lunar module approach from above. The crew found difficulty in visually picking up the target against the bright, ever-changing background of the Earth's surface, but the radar contact was firm, so that the manoeuvre was successfully completed after more than three hours of dropping back towards the disabled adapter. Stafford edged to within 3 in (7½ cm) of the shroud to obtain detailed close-ups of the shroud and its steel binding 'making sure the alligator wouldn't bite us'. The hectic revised schedule had put considerable strain on the two men, leading Stafford to request a postponement of Cernan's space walk which had been planned for that day: 'We're pretty well bushed.' Permission was passed on by Capcom Neil Armstrong, so they backed away from the potentially dangerous 'alligator' for a well-earned rest.

The third day saw the men preparing for the longest space walk yet, scheduled for two and a half hours. Half-way through the complicated procedure, Stafford told ground control, 'We've got the big snake out of the black box', indicating that they had managed to retrieve the 25 ft (8 m) umbilical from its stowage space. Minutes later, he told ground control of a worrying thruster malfunction and a faulty reading on the attitude indicator. The panic was soon over, however, when one of the designers realized that they had inadvertently knocked a switch into the wrong position as they were moving around in their cramped cabin. Finally, after nearly four and a half hours getting ready, the moment of truth arrived. The Carnarvon station passed up permission to begin cabin depressurization.

As Gemini 9 passed into the sunlight above Hawaii, 49 hours 22 minutes into the mission, the crew started to open the hatch, and shortly afterwards, a delighted Gene Cernan poked his head into the void as he stood on his seat. Before launching himself into space, he retrieved a micro-meteorite package attached to the craft's skin, deployed the handrails which would help him to manoeuvre, installed the docking bar mirror and set up the 16 mm EVA camera. He found the tasks hard going as everything tried to float away from him, but he was compensated by the magnificent view beneath him. Twenty minutes after opening the hatch, Cernan moved outside the Gemini. Passing over the Californian coast, America's second space walker reported that he could see Los Angeles and Edwards Air Force Base, the former haunt of so many astronauts, including Tom Stafford. One of the early tasks was to test the umbilical as a means of controlling his movements, together with evaluation of the handholds and Velcro pads as aids to an astronaut working outside his craft. He used the new extra-vehicular life support system which had been prepared for the Gemini 8 mission, deciding that it operated satisfactorily on its medium flow in maintaining temperature control and oxygen supply. He did admit, however, that his movements were rather erratic and out of control the umbilical tended to snake around him, and there was no way of bracing himself so that he could free his hands without tending to drift out of position. More reassuring was the report that he experienced no disorientation or vertigo.

While Cernan moved above, installing cameras, attaching rearview mirrors for Stafford to use, and changing film packs with the command pilot's help, the Gemini swung around all over the sky – at one stage the craft was pulled around through 150 degrees, pitched down 40 degrees and rolled over into an 'upside down' position. The space walker's pulse was generally between 140 and 160, rising at times to a peak of 180. The two men partially closed the hatch, leaving it open about 2 in (5 cm): this was to protect the hatch seals from temperature extremes, to prevent overheating of the spacecraft interior, and to give more protection to Tom Stafford, who was only wearing the standard pilot's pressure suit. With the external lights on for the first night pass, Cernan began to move back towards the rear adapter section for the main part of the EVA, the testing of the astronaut manoeuvring unit (AMU).

In the sunlight once more as Gemini flew over Africa, Cernan began to plug into the AMU circuits, but the work seemed much harder than in any of the simulations. In particular, the attitude control arms 'presented far more difficulty in zero G than they did in the simulation'. Special stirrups were provided in the adapter section to hold his feet in place in

The Select Seven The seven Mercury astronauts selected in April 1959, here wearing their silver-coloured pressure suits, composed the first group of American astronauts. Front row, left to right: Walter M. Schirra Jr, Donald K. Slayton, John H. Glenn Jr and M. Scott Carpenter. Back row, left to right: Alan S. Shepard Jr, Virgil I. Grissom and L. Gordon Cooper Jr.

The first man in space Yuri Gagarin, who on 12 April 1961 made history by becoming the first human to travel in space. The 27-year-old Soviet Air Force lieutenant orbited Earth once in his Vostok capsule. On his return, he became a roving ambassador for his country. He died in a plane crash on 27 March 1968 while training for his second spaceflight.

(Left) **Mercury–Redstone 3 Liftoff** The first American manned spacecraft was launched by Redstone rocket on 5 May 1961. On board was US Navy lieutenant-commander Alan Shepard. The Redstone did not have the power to boost Shepard into orbit, so he had to settle for a 15½ minute sub-orbital lob into the Atlantic Ocean. Note the escape rocket system above the Mercury capsule.

(Right) Valentina **Tereshkova**   The first woman in space was part of a major propaganda spectacle in June 1963 when Vostoks 5 and 6 flew in tandem. The 26-year-old former factory worker and amateur parachutist spent nearly three days in orbit. Here she is seen during wireless communication drill, part of her 15 months of training for the flight. She married another cosmonaut, Andrian Nikolayev, and the couple produced a daughter named Yelena.

(Below) **Vostok 6 after landing**   Valentina Tereshkova seated beside the charred Vostok capsule. Note the open circular hatch and the parachute. The Vostok cosmonauts ejected during the final descent and parachuted to Earth close by the empty capsule. Vostok 6 landed in one of the designated zones on the flat, open Soviet steppes, north-east of Karaganda.

Astronaut survival training Gemini astronauts endured survival training in hostile jungle and desert environments to prepare them in the event of some future crash landing. Here Walter Schirra samples roast wild pig during a three-day course in the Panama Canal Zone in 1963. To his left are Gus Grissom, Frank Borman, Charles Conrad and Alan Shepard. Seated in front are (left to right) John Glenn, Scott Carpenter, Donald Slayton and Elliott See.

Edward White's spacewalk The first American spacewalk took place on 3 June 1965 when Edward White spent 21 minutes outside the Gemini 4 spacecraft. Secured to the craft by a 25 ft (7½ m) umbilical line which carried his oxygen supply, White held a Self-Manoeuvring Unit in his right hand to control his motion. Note also the emergency oxygen chest pack.

(Above) Gemini 7 photographed from Gemini 6
The first dual US mission took place when
Gemini 6 was launched on 15 December 1965
to rendezvous with Gemini 7. Walter Schirra
and Thomas Stafford on board Gemini 6 spent
the best part of a day carrying out rendezvous
and station-keeping manoeuvres vital to the
plans for a moon landing. Frank Borman and
James Lovell remained aloft in Gemini 7 for two
weeks, an endurance record which lasted until
1970.

(Right) Saturn V aims at the moon   The Apollo
moon missions were launched by the three-
stage Saturn V booster. Developed by Wernher
von Braun's team, America's most powerful
rocket stood 363 ft (111 m) tall and generated
7.5 million pounds of thrust. The Saturn V was
flown on 13 occasions between its first flight
test in November 1967 and the launch of
Skylab on 14 May 1973.

(Above) **Vladimir Komarov** The first Soviet cosmonaut to fly twice in space was 40-year-old Komarov, veteran of the three-man Voskhod 1. Selected as a cosmonaut in 1960, his career was threatened by a minor heart irregularity, but he was passed fit for the prestigious first flight of the Soyuz spacecraft.

(Left) **Soyuz launch** The Soyuz launch vehicle (also known as the A-2) has changed little in the past 20 years. It is about 160 ft (49 m) long and is divided into three stages. The first stage consists of four strap-on boosters attached to the core second stage which continues to burn after the boosters separate. The Soyuz spacecraft is perched on the third stage beneath the escape rocket system. Launches take place from Baikonur Cosmodrome at Tyuratum.

(Overleaf) **Man on the moon** Apollo 11 astronaut Edwin 'Buzz' Aldrin photographed by Neil Armstrong on the Sea of Tranquillity, 20 July 1969. The black sky demonstrates the lack of atmosphere on the moon. Protection against the lunar vacuum, the high temperatures and the glare of the sun is provided by the multi-layered pressure suit and gold-plated visor. Note the probe and part of the footpad belonging to the lunar module Eagle in the bottom right corner.

order to allow use of both hands, but, in practice, his feet tended to float out of the stirrups, causing great difficulty in maintaining the proper position. His predicament was not aided by the partial separation of some suit insulation, which resulted in solar radiation heating up the small of his back. This overheating and the unexpected exertion required to back on to the AMU, connect to its life support systems and deploy the attitude control arms was too much for the chest pack to cope with. The visor began to fog over, and there was little change even after Cernan increased the oxygen flow rate to maximum. The astronauts decided to wait for sunset and hope that things would improve.

As the sun disappeared below the horizon, Cernan was relieved to see that his suit temperature was falling, but when he resumed his attempts to fit the AMU, the same problems of stability arose once more. The fogging refused to go away, so he was virtually working blindfolded with the EVA lights taking on the appearance of streetlights struggling to shine through a dense fog. To make matters worse, when he connected the AMU communications link to his suit, the quality of voice contact between the astronauts noticeably deteriorated to what Stafford called 'a log of garble'. Cernan groped his way through the well-rehearsed procedure until he had only to connect the restraint harness, the equivalent of the AMU safety belt, and the oxygen hose before Stafford could throw the switch to set him off on a Buck Rogers style adventure. However, the two men agreed that this was not practicable with a half blind astronaut at the controls, and they agreed to wait once more for the clearing of the visor. Shortly after sunrise, 1 hour 34 minutes into the EVA, they reluctantly came to the conclusion that no improvement was likely in the near future; consequently, the AMU would remain untried and a major mission objective remained unaccomplished. A dejected Cernan reported that the AMU was 'in top notch shape' and he was 'convinced it was a flyable machine'. He reattached the umbilical communications lead, enabling Stafford to hear clearly his facetious comment: 'Hey, Tom, what's that guy doing with the Texas driver's licence out there on the California highway?' Moving slowly back to the hatch, the visor almost half cleared, though the fogging returned as soon as he exerted himself to retrieve the docking bar mirror, and the temperature problem arose once more. A plan to photograph the approaching sunset was cancelled as Cernan was ordered back inside the Gemini. Helped by Stafford, the tired and frustrated Cernan floated carefully back on to his couch, the umbilical was gradually stowed away and the hatch was closed 2 hours and 8 minutes after it was first opened. As the cabin repressurized, the astronauts were left to reflect on a record-breaking space walk which had been a virtual flop as a result of inexperience of the problems of weightlessness and of inadequate equipment.

Cernan and Stafford settled down to a well-deserved rest and meal.

The respite from onerous duties was short-lived, however, for ground control told Stafford to fire the thrusters to lower Gemini's apogee in preparation for retro-fire the next day. He found he was still able to get a radar contact with the distant docking adapter. Cernan fed the computer with the necessary re-entry information, and updated information was sent up from the *Coastal Sentry* stationed in the western Pacific. After their evening meal and sleep period, the crew were aroused by a call from Carnarvon early on the morning of 6 June. Final preparations went to plan, with Cernan confirming 'four good retros' to the Canton station. The splashdown turned out to be the most accurate to date. Stafford inquired, 'Have you got us in sight?' only to be told by the carrier *Wasp*, 'The whole world has you in sight.' TV cameras on board the carrier relayed the final moment of the mission via the Early Bird satellite to the watching millions in the USA, as the blackened craft plopped into the Atlantic only 1½ miles (2 km) from the carrier and a mere 0.4 miles (0.7 km) off target. Within four minutes the swimmers attached a flotation collar to Gemini, enabling the crew to throw open the hatches and survey the scene in a cheerful, relaxed mood while recovery operations went on around them. Following the example of Gemini 6, Stafford and his companion remained in the capsule while it was swung on to *Wasp*'s deck to the sounds of 'Anchors Aweigh' less than one hour after splashdown. They had spent three days orbiting the Earth, completing 45 orbits. There followed the usual congratulatory telephone call from President Johnson, the post-flight medical and two-week debriefing, climaxed by the awards of NASA's Exceptional Service Medal. Cernan told reporters that his space walk 'opened up a whole new dimension in space flight' though it was 'no cake walk'. Stafford summed up the mission: 'It was a two-man job all the way.' NASA officials fended off media criticism by reminding reporters that 'the Gemini programme is experimental', therefore, by definition, there was still a lot to learn.

Even as Stafford and Cernan told the press corps of their adventures, the engineers were analysing the reasons for the problems with the Agena and the AMU while the next two astronauts in line were only a month away from what was described as 'the most ambitious manned spaceflight ever attempted by the United States'. The command pilot of Gemini 10 was the unflappable John Young, the Mr Cool of the Gemini programme. Already the holder of two world records from his Navy test pilot days, he was set to break the world altitude record with the aid of an Agena target rocket on the forthcoming mission. His partner was a former air force test pilot, Michael Collins, also 35 years of age, who was venturing into space for the first time. Their flight plan involved rendezvous and docking with two different Agena rockets, utilization of the power from an Agena to raise their altitude beyond anything achieved previously, a space walk to an Agena by Collins with the aid of a hand-held manoeuvring unit, and the execution of 14 experiments. There would be little time for relaxation during the three-day mission!

The final countdown was, however, more civilized than normal, so that the crew were able to sleep until noon due to the unusually late launch. The suspense of their two hour wait was lessened by the successful launch of their Agena target and news of its perfect orbital insertion. At 5.20 pm Cape time, it was their turn to head for orbit and begin the chase. Only seven minutes after entering orbit, Gemini 10 crossed the day-night boundary or terminator with Collins hastily preparing to take star sightings for orbital calculations. The harassed pilot found using the sextant much more difficult than in simulations, so that his predicted manoeuvres did not coincide with the ground results. Young tried to cheer him up: 'What the heck! If you can't see the stars, you can't see them. I've been telling you this for six months.' They decided to use the ground's computations, and, beginning at 2 hours 10 minutes into the mission, Young edged Gemini towards its distant quarry. All went well until the terminal phase approach when they suddenly discovered an error in their alignment; the slip was not fatal, but the cost had to be paid in extra fuel consumption and probable curtailment of future EVA and rendezvous experiments. Eventually, the 'go' for docking was passed up by the *Coastal Sentry Quebec* in the western Pacific during the fourth revolution, and 5 hours 53 minutes into the mission the second mating in orbit of two spacecraft was completed.

Unlike the previous occasion, Gemini 10 did not have to beat a hasty retreat, though the planned manoeuvres around the Agena had to be cancelled due to the fuel shortage. The lack of fuel had no influence on the next experiment, however, for the docked Agena's engine was fired to raise the two craft to a new world altitude record of 474 miles (763 km). Young later described the view of the Agena as the unprecedented manoeuvre began: 'At first, the sensation I got was that there was a pop, then there was a big explosion and a clang. We were thrown forward in the seats. Fire and sparks started coming out of the back of that rascal. The light was something fierce.' The 'glorious Fourth of July spectacle', as Collins described it, lasted only 14 seconds, then the 'golden halo encircling the entire Agena' faded as the astronauts once more returned to weightlessness. The elation of the wild ride evaporated as they looked back on the day; perfectionist John Young was particularly annoyed over the wasted fuel, but he slept well compared with his less experienced companion. Collins managed a mere two hours sleep as he swallowed aspirins for an aching knee and fretted over his sleeping posture, especially his hands which floated perilously close to the array of switches.

The next day saw the Agena engine fired twice within about two hours as the Gemini's orbit was lowered and circularized to enable a close approach to the Agena left behind by Gemini 8 in March of that year. Young, a highly experienced test pilot familiar with high performance aircraft remained impressed with the sudden jolt from the engine: 'It may be only one G, but it's the biggest one G we ever saw. That thing really lights into you!' Meanwhile, Collins prepared for his stand-in-the-hatch EVA, but his keenness nearly backfired for the kick from the Agena engine threw the ultraviolet astronomical camera against the cabin wall, breaking the automatic timer. He would now have to operate the shutter manually while John Young ticked off the necessary twenty seconds. Deke Slayton was worried by another aspect of the mission: 'You guys are doing a commendable job of maintaining radio silence . . . Why don't you do a little more talking from here on?' The overworked astronauts did not feel much like putting on a public relations exercise for the press, but they did their best.

Nearly 23½ hours into the mission, just as darkness enveloped the craft, Collins emerged from the hatch and installed the camera in its bracket. Finding he had not released the left shoulder harness, he struggled to free himself. Once successful, he spent the rest of the night pass photographing the stars, punctuated by periodic glances at the world below.

> Down below the Earth is barely discernible, as the moon is not up and the only identifiable light comes from an occasional lightning flash along a row of thunderheads. There is just enough of an eerie bluish-grey glow to allow my eye to differentiate between clouds and water and land, and this in turn allows motion to be measured. We are gliding across the world in total silence, with absolute smoothness; a motion of stately grace which makes me feel god-like as I stand erect in my sideways chariot, cruising the night sky.

With sunrise came the next photographic experiment to take pictures of a four-colour plate which could lead to accurate reproduction of the colours in space. Collins soon found his efforts hampered by watering eyes, despite lowering his visor and avoiding the sun; John Young also began to suffer, with the result that both men were virtually blind. They had no choice but rapidly to terminate the EVA and attempt to retreat to the safety of the cabin. They managed to grope their way back inside, close the hatch, and repressurize the cabin, relieved that Collins had not been swimming at the end of a tether in the dark void. Collins thought he could smell lithium hydroxide in their oxygen supply; whatever the cause, it was a distinct threat to the major EVA scheduled for the next day. The first excursion had lasted about 50 minutes.

The rest of the day went smoothly as the eye problem cleared. The men slept well in readiness for the next exacting session. On the morning of 20 July, the astronauts made final adjustments of their orbit before separating from their Agena and commencing rendezvous manoeuvres with Agena 8. After more than 38½ hours locked together, the two craft gently pulled apart, but it was a further three hours before the crew reported they were closing in on their target. Unable to use radar since Agena 8 was without a transponder, they relied on ground instructions until visual contact was established. To their relief, they arrived alongside the Agena with 15 per cent of fuel still remaining, and while Young practised station keeping with the floodlit booster, Collins unstowed the chest pack and 50 ft umbilical required for his space walk.

As Gemini 10 began its third day in orbit, Collins swam out of the hatch while Young remained at the driving seat. Using the handrail, Collins moved along the craft behind the cockpit to remove a micro-meteorite detection plate. His first attempt to fit the nitrogen line that would power his hand-held manoeuvring unit resulted in Collins crashing into the craft and shaking it with the impact, but he succeeded at the second try. All the time the two men kept each other informed, especially when Young was about to fire the potentially dangerous thrusters. The command pilot then edged closer to the Agena until he dared approach no nearer: 'If I get in closer to it, I won't be able to see you or it.' Collins weighed up the distance. 'OK, I can almost leap right now, but I'd rather not if you can get a little bit closer. I'll give you directions.' Young reluctantly agreed, so Collins guided the Gemini to within 6 ft (2 m) of the threatening booster. 'OK, you're in a good position. I'm going to leap for her, John.' Young issued a fatherly warning: 'Take it easy, babe.'

Collins gently pushed himself across the yawning gulf, floating for several seconds before colliding with the Agena's docking adapter. Seizing the lip of the smooth cone with both hands, he began to work around towards the main object of the excursion, the micro-meteorite package. Reaching the correct place, he tried and failed to stop, lost his grip and slowly cartwheeled away from the Agena; he had built up too much momentum to stop still suddenly. A worried John Young hoarsely inquired: 'Where are you, Mike?' Back came the reassuring reply: 'I'm up above. You don't want to sweat it. Only don't go any closer if you can help it.' Collins brought his hand gun into operation, squirting nitrogen into space in order to slow his motion and return to the Agena. Slowly following a curved path, he swung behind the Gemini. 'I'm back behind the cockpit, John, so don't fire any thrusters.' As he passed the open hatch, Collins grabbed the craft and came to an abrupt halt.

The second attempt took place with Young holding the Gemini back about 15 feet from the Agena to give him a wider field of view. Collins decided to use the hand gun for propulsion this time, but his aim was slightly off. Again he managed to cling to the docking adapter and move, hand over hand, around the lip. Young was now concerned about a stray metal loop: 'See that you don't get tangled up in that fouled thing.' At last, he reached the package and successfully removed it, but Young cautioned against a prolonged stay as he watched the Agena slowly tumbling due to the rough treatment meted out by its visitor: 'Come back . . . get out of all that garbage . . . just come on back, babe.' The command pilot was also concerned about the depleting fuel reserves, so Collins was unable to give the hand gun the exhaustive tests the engineers had wished. He returned to the open hatch, pulling himself in along the umbilical. Only when safely installed did he realize that he had lost the 70 mm still camera which he had wedged into his chest pack. There would be no close-ups from that space walk. One of the most difficult parts of the EVA followed as Collins and Young struggled to get the astronaut and his umbilical back inside the cabin. The humour of the situation struck even these two close-lipped professionals: 'He's down in the seat because there is about 30 feet of hose wrapped around him. We may have difficulty getting him out.' Collins added: 'This place makes the snake house at the zoo look like a Sunday-school picnic.' About an hour later, they lowered the cabin pressure for the third and last time, opened the hatch and dumped overboard the EVA equipment, empty food bags and other rubbish.

With fuel down to seven per cent, the priority was now to lower the orbit as a preliminary to retro-fire. Most of the fuel was consumed in the thruster burn, so the craft was left to tumble freely on the final stages of the mission. The crew had no difficulty in sleeping, much happier than earlier in the mission. Equipment was stowed and final housekeeping completed before a successful retro-fire over the Pacific 70 hours 10 minutes into the mission. They splashed down in the Atlantic only 3.4 miles (5.4 km) off target after 44 revolutions and nearly three days in orbit. The helicopters from the carrier *Guadalcanal* were overhead before they touched the water, dropping swimmers with the flotation collar. Although the day was hot, the sea was quite smooth and the astronauts felt relaxed: 'Hey, boys, take your time! We're not in any hurry. We don't want anybody getting hurt out there.' Some 27 minutes after splashdown, the two men stepped from the helicopter on to the carrier's deck. 'We had a lot of fun,' was Collins' comment.

Collins was able to escape some of the ballyhoo which traditionally greeted the returning hero; he had been born in Rome, so had no home town to put on the rousing welcome. He was soon assigned to the back-up crew for the second manned Apollo flight, tentatively scheduled for 1967. John Young continued for a while with the Gemini programme, but it was already closing down as the Apollo steamroller began to dominate NASA's resources and training. Eventually assigned to the back-up crew of Apollo 7, he flew in space again before Michael Collins when the latter was grounded instead of participating in Apollo 8. Both men had even greater fame ahead.

The penultimate Gemini mission was an all naval affair with Charles

'Pete' Conrad as command pilot accompanied by 37-year-old lieutenant commander Richard Gordon, a group 3 astronaut making his first trip into space. Both men had served at the Navy Test Pilot School at Patuxent River, though Gordon left in 1960 to assist in the training and introduction of the F4H Phantom in the Navy's carriers. In 1961 he set a new record of 2 hours 47 minutes in the trans-USA Bendix Trophy Race with an average speed of 869.7 mph, thus following in the footsteps of John Glenn. By the time he was accepted by NASA, Gordon had enrolled on a naval postgraduate course. The cool, competent Gordon and the flamboyant extrovert Conrad would make a good combination.

The mission was slow to get off the ground, however, as the launch was postponed twice due to a leak in the first stage oxidizer tank and a suspected autopilot malfunction. It was a case of third time lucky on the morning of 12 September 1966. The crew ate breakfast with Alan Shepard, breaking the astronaut tradition by substituting sirloin strip steaks for the usual filets mignons. They arrived at the top of the gantry at 7.25 am to be briefed by the back-up crew of Neil Armstrong and William Anders. Apart from a short hold due to a suspected hatch leak, the countdown went smoothly, so that the Atlas–Agena was launched into orbit at 8.05 am. The Gemini launch constraints were much tighter than on any previous mission, for the main objective was to practise an emergency ascent from the lunar surface; Gemini 11 had a two-second launch window if it was to rendezvous and dock with the Agena during the first revolution. Right on schedule, the Titan ignited to lift Gemini in pursuit of its target at 9.42 am Cape time. Using only their onboard computer and radar equipment, the astronauts successfully closed on the Agena and locked into position 94 minutes after launch. It was the fastest rendezvous and docking in space history, and, significantly, had consumed less fuel than the similar manoeuvre on the previous flight. Conrad later explained the only problem which arose:

> Due to our late lift-off time we came upon the Agena in daylight a little bit sooner than we expected. I think I shot two or three percent of the fuel fussing around getting used to seeing the bright Agena when I couldn't see the instruments. I fumbled for my sunglasses and didn't get them.

Before the first rest period began about eight hours after lift-off, the crew carried out a number of experiments, including each man undocking and redocking with the Agena. The booster's engine checked out satisfactorily, so the next day Dick Gordon went ahead with preparations for his main EVA. For once, the flight plan allowed too much time rather than too little; the crew completed their EVA preliminaries two hours earlier than expected, leaving Gordon sitting around in his bulky suit wasting oxygen. His environmental control system was not designed for long periods in the pressurized cabin and was unable to cope, forcing Gordon to return to the craft's system in order to prevent overheating. About half an hour before the EVA was due to commence, Gordon repressurized his suit, but suddenly found himself struggling to fit the gold-plated sun visor to his helmet. In the confined space Conrad was unable to help his comrade, and it was a hot and bothered Gordon who finally opened the hatch 24 hours and 2 minutes into the mission.

Gordon explained what happened next:

I stood up in the seat, or rather flew out of the spacecraft, because as soon as we opened the hatch all the debris and junk that we found in the spacecraft went floating out the window – or out the door really. And I was right along with the rest of the debris. The only thing I could say to Pete was 'Hey, grab me, I'm leaving you.' So Pete actually had to hold me in the craft.

In this position, Gordon set up the EVA camera, deployed a handrail and retrieved an experiment package, tasks which sapped his strength much more than expected. Equipped only with an umbilical to facilitate his movements, Gordon then moved out towards the Agena in an attempt to attach a tether linking the two craft. He missed and had to be reeled back in by Conrad. The second try was more successful, Gordon holding on to the nose of the Gemini by straddling it with his legs, thereby leaving his hands free to attach the tether. In simulations he had found this task very simple, but in reality it became a nightmare. Conrad yelled encouragement: 'Ride 'em, cowboy!' Eventually he succeeded, but his suit was unable to cope with the demands made upon it, leaving Gordon bathed in perspiration and partially blinded from the sweat running into his right eye. After a short rest, there seemed little improvement, causing Conrad to consider abandoning the remainder of the EVA. Gordon told him: 'I'm pooped, Pete . . . The sweat still won't evaporate.' The command pilot decided discretion was the better part of valour and ordered his companion back inside. He reported to ground control: 'I just brought Dick back in. He got so hot and sweaty he couldn't see.' The hatch was open for just 33 minutes; evaluation of the hand gun and a special power tool would have to wait, perhaps indefinitely. Once again an EVA had to be curtailed. An hour later, the hatch was opened once more to eject the equipment which was no longer needed. The next six hours leading up to the sleep period were more leisurely as they reported to the medical experts and took photographs of airglow above the horizon.

It was around midnight at the Cape on the second day of the mission when the crew of Gemini 11 repeated the Agena burn carried out by Gemini 10, boosting the combined craft to a new record altitude of 850 miles (1,373 km). Conrad whooped with delight as the brief firework display ended with their craft soaring higher than any man had ever flown before. For the first time, men could see for themselves that the world was indeed round. Even Conrad was almost stumped for words: 'It's utterly fantastic . . . the water really stands out and everything looks blue . . . the curvature of the earth stands out a lot . . . there's no loss of colour and details are extremely good.' The next two revolutions were taken up with photographic experiments, resulting in some of the finest Earth views ever snapped.

The main Agena engine was fired once more to lower the high point of the elliptical orbit. More photographs followed during a stand-up EVA which lasted more than two hours. Restrained by a short tether, Gordon and Conrad were able to proceed at a leisurely pace which proved much more relaxing than the previous EVA – almost too relaxing. Conrad later explained:

> Dick was going to hang out of the hatch and grab some pictures of Houston with the Hasselblad . . . It only took us four minutes to go from Houston to Florida and we had the whole rest of the Atlantic to go with nothing to do. So, lo and behold, I fell sound asleep in my hard suit with my arms extended and all of a sudden I woke up and realized that not only was I asleep on the job but I was asleep while we were depressurized. I said, 'Hey, Dick, would you believe I fell asleep?' And all I got out of him was, 'Huh, what?' So there we were out over the Atlantic. He was asleep hanging out of the hatch on his tether and I was sitting asleep inside the spacecraft.

The crew then prepared for the novel experience of rotating two spacecraft linked by the tether fitted earlier by Dick Gordon. At 49 hours 55 minutes into the mission, Gemini undocked and slowly backed away from the Agena. There was some difficulty keeping the two craft properly aligned after the tether proved reluctant to deploy smoothly, so that the craft were moving independently at the ends of the looping 100 foot (30 m) tether. 'Boy, this is wild,' shouted Conrad. 'It's like a skipping rope.' Advised by the ground to wait in a passive mode, Conrad watched as the line went taut of its own accord. He increased the rate of rotation from 38 to 55 degrees per minute, causing the whiplash motions to build up again, but was able to dampen them down by using the thrusters. After about 20 minutes, the rotating combination became very stable, and though the astronauts could feel no gravitational influence, their instruments registered a non-weightless condition. For the first time, artificial gravity had been created by rotating a spacecraft under experimental conditions. The link-up continued for two revolutions, a dramatic demonstration that such tethered station keeping was an economical and feasible method of orbiting without wasting precious fuel. The Agena was released and allowed to drift away as Gemini's orbit was raised then lowered again to place the craft at a safe distance behind the booster.

The next day, the astronauts manoeuvred their Gemini so that they could make up the 30 mile (46 km) leeway and re-rendezvous with the Agena. The entire operation took just under 1 hour 15 minutes, less than one Earth orbit. Soon afterwards they pulled away from 'the best friend we ever had' for the last time in order to prepare for re-entry. Retro-fire took place over the Canton Island tracking station at 70 hours 41 minutes into the mission. The crew placed the craft in the correct attitude for re-entry, checked the computer, found all was operating properly, and sat back to enjoy the first fully automatic re-entry in the Gemini programme. Everything went perfectly as Gemini 11 splashed down within 1½ miles (2½ km) of the carrier *Guam*. They decided on a quick helicopter pick-up, setting foot on the carrier's deck in a record time of 24 minutes. Conrad told the assembled newsmen: 'This old world looks good from the deck of the carrier. But boy, it looks really great from 850 miles out. We had a very good flight. A good night's sleep and we'll be ready to go up again.' Dr Berry later confirmed that Dick Gordon was none the worse for his exertions during EVA. The astronauts maintained their working relationship into the Apollo programme where they were assigned as back-up crew to Apollo 9, though they had to wait more than three years for their next joint spaceflight.

Gemini 12 was to be the grand finale marking the curtain fall on the two-man American space programme. Most eyes were turned on the exciting prospects offered by the Apollo moon landing project as the last quarter of 1966 arrived. But all was not well at the North American plant in Downey, California where spacecraft 012, the first Apollo command module, was falling further and further behind the scheduled completion date. Earlier in the year its crew of Grissom, Chaffee and White had been optimistic about a joint Apollo 1–Gemini 12 mission as a spectacular climax to Gemini and propaganda boost to Apollo, but, logistic problems apart, the idea never stood a chance: Apollo would not be ready for launch until early 1967 at the earliest. There was also a need for a rethink concerning the last Gemini flight following the experiences of Dick Gordon on Gemini 11. Clearly the problems associated with EVA had not been solved, and, although space walks were not planned for the Apollo missions, the astronauts would be expected to complete work programmes on the lunar surface wearing bulky pressure suits. NASA did not want the astronauts to become overheated or collapse from exhaustion on the hostile surface of a distant world, so EVA became the priority for the final Gemini mission.

The man most involved in solving these problems was a 36-year-old Air Force major, Edwin 'Buzz' Aldrin, a combat veteran from Korea who had also flown jet flighters in the USA and Europe. Even before his selection as an astronaut in 1963 he had been directing his thinking to spaceflight, having written a Doctor of Science thesis on 'Guidance for Manned Orbital Rendezvous'. He now had the opportunity to put his theories into practice, but his major task on the forthcoming mission was to evaluate the new training programmes and equipment produced by the NASA engineers and designers. In late September and early October he practised extended EVA with a Gemini mock-up in a giant water tank which gave him valuable experience unavailable to his predecessors and also gave the engineers and doctors a much more accurate idea of the demands that would be put upon the pressure suit's environmental control system. To his disappointment, the proposal to introduce the astronaut manoeuvring unit (AMU), carried on Gemini 9 but never used, was vetoed; the EVA time would be spent on 'fundamentals' rather than trials of exotic flying machines from science fiction. A whole series of aids were provided to overcome the main complaint of all the earlier space walkers, the difficulty in maintaining position without adequate hand and foot restraints. There would be three EVA periods altogether, totalling five and a half hours, outside the craft, but, for the first time, a dozen two minute rest sessions were allotted for the main two hour excursion. NASA was determined to solve the only significant problem which had so far eluded it during the Gemini list of successes.

The launch date slipped by two days as the secondary autopilot on the Atlas booster was checked, so Aldrin and his command pilot, Jim Lovell, a veteran of the record-breaking Gemini 7 duration mission, spent the time working in the simulator at the Cape and reviewing the flight plan. The new launch date was 11 November 1966, Veterans' Day, the first time a launch had taken place on a public holiday. The time set for launch was later than usual, so the crew were not aroused until 10.30 am. The next two hours were filled with the normal preliminary physical examination and breakfast. A minor problem arose with the air tubes in Aldrin's suit which meant that he had to unsuit and start again, but the schedule was unaffected. Their arrival at the launch pad was marked by typical NASA humour: the astronauts wore a placard around their necks each bearing one word, 'The' and 'End'. They were greeted by a poster which read: 'Last chance. No relaunch. Show will close after this performance.' Installed in the capsule shortly before 2 pm, there was only a brief interlude before their Atlas–Agena successfully blasted off from Pad 14. As on the previous eleven Gemini missions, the trusty Titan lifted off right on schedule to place them in orbit six minutes later.

The rendezvous with the Agena was not planned until the third revolution, a more relaxed timetable than on Gemini 11. Early manoeuvres suggested a straightforward approach with the added bonus of a radar lock-on at the tremendous range of 235 miles. The distance steadily closed to 75 miles (120 km) over the next couple of hours until the crew were ready for the last major orbital change. Jim Lovell later explained what happened next:

Buzz noticed that the computer wasn't giving any change of range. I looked down at the little green light that tells us we had a radar lock-on, and it was off. We just looked at each other. We said, 'Oh no, it can't happen to us. Anybody else or any other time but not this time.' Then it suddenly dawned on us that our radar had indeed failed. We went to the radar back-up procedures which we had practised quite a bit in pre-flight training but never really expected to use.

It was the first time the radar had malfunctioned in the entire Gemini programme.

Buzz Aldrin resorted to his sextant and successfully picked up the flashing lights on the Agena. Using only a set of bearings and some navigation charts, Aldrin fed the necessary information to Lovell. His calculations were so accurate that they accomplished the rendezvous on schedule and without undue fuel consumption 3 hours 46 minutes after launch. The craft successfully docked, 28 minutes later though a re-docking attempt ran into difficulties when they approached in the wrong alignment and had to back out again hastily. The second try proved more accurate, but then ground control informed them that the next manoeuvre would have to be cancelled.

As on the two previous missions, Gemini 12 was due to utilize the Agena booster to fire it into a high orbit, but ground controllers who had been monitoring the performance of the Agena during its initial ascent were concerned about a small drop in pressure in the rocket's thrust chamber. They decided to play safe rather than risk any danger to the astronauts, though the change of schedule gave a bonus in that a photographic session during a solar eclipse could be reintroduced to the flight programme. Two docked manoeuvres were made using the Agena's secondary propulsion system in order to place the craft in the correct position for eclipse observation on the tenth revolution. Between these burns the crew were able to get some rest, though the warning lights flashed on at regular intervals to suggest that the fuel cells were producing more water than intended – each time the crew drained their drinking water the lights were extinguished. For the brief period during which they were in the moon's shadow, a mere seven seconds, the crew managed to snap two photographs of the eclipse though they were unable to swing the docked Gemini round in time to try for photographs of the shadow on the Earth's surface.

The first session of EVA began 19 hours 29 minutes into the flight. For the next 2 hours 29 minutes Aldrin was able to become accustomed to working in the vacuum of space while standing on his couch. He installed a telescoping handrail and a motion picture camera which would be used on his main excursion the following day, retrieved a micro-meteorite package and some glass strips carried to pick up contamination originating during launch, and carried out photographic assignments. He also carried out some exercises, moving his arms to touch his helmet once a second, but his heart rate hardly rose at all. The only problem reported was some difficulty in adapting to the darkness once the sun had set. 'It's like coming in out of a snowstorm,' commented Aldrin. He was full of wonder as the blue Earth flew by beneath, causing Lovell to comment: 'What did I tell you? Four days' vacation and see the world.'

The lead up to the second EVA was marked by some minor troubles reminiscent of the early Gemini days. One of the fuel cell stacks failed and had to be disconnected, then two thrusters seemed to malfunction, slightly complicating the task of manoeuvring the craft. The hatch was opened at 42 hours 48 minutes into the mission, and Aldrin utilized the handrail he had installed the previous day as he slid along the nose of the Gemini. One of his main tasks was to evaluate the various aids that had been provided in an attempt to overcome the difficulties of his predecessors: adjustable nylon tethers were attached to his waist and could be hooked on to the handrail and on to rings on the Agena; portable Velcro handholds could be attached to matching areas on the two craft or used to fasten the tethers and a special foot restraint like a pair of large overshoes was located at the rear of the Gemini. After a short rest, Aldrin completed the tether link between the two craft and set to work on a work station equipped with various connectors and bolts designed to test man's ability to perform useful tasks in space. He then moved back to the cockpit, picked up another camera from Lovell, then pulled his way along the built-in handrails to the rear of the adapter section where he fitted his feet into the restraints. Lovell radioed a complaint to his partner: 'Hey, easy, easy, you're shaking up the whole spacecraft.' Aldrin, however, was full of praise for the new foot restraints: 'There really is nothing better than this type of restraint system. The situation is very similar to being in a one-G environment.'

The adapter work station contained 17 different tasks, including further trials with Velcro pads, rings and hooks, connecting and disconnecting various plugs, wire-cutting and use of a torque wrench on different sized bolts. Lovell, concerned over his companion's work rate, asked about his condition, but Aldrin reported that he felt no perspiration and had no trouble with fogging of the visor. The ground medical technicians watched their instruments with delight as the space walker's pulse reached an unexceptional 120. All the time he had been working in the artificially illuminated adapter section since flight planners thought the absence of sunlight would reduce the overheating problem. However, in order to compare the performance of the suit and the astronaut in daylight conditions, Aldrin was now required to slide to the nose section once more for a final session on the Agena work station. Again, he had no difficulty with the tasks, whether tethered or untethered, and it was a happy astronaut who moved back towards the cockpit for the last time. On his way he wiped Lovell's window, not merely out of consideration for his partner, but also to collect a sample of the contamination which had smeared the windows on all of the Gemini flights during launch.

Firmly attached once more inside the open hatch, he looked back along the craft to observe a troublesome pitch thruster, and was able to confirm that it was not functioning properly. He observed vapour coming from the thruster when Lovell fired the control, but there was nothing to be done about it. The only difficulty encountered during the 2 hour 8 minute walk was in installing a camera in the adapter section; after a brief struggle he gave up and brought it back to be examined by ground engineers. In a last exuberant gesture he dismantled the temporary handrail and flung it into space with the cry: 'There she goes. I'm the world's highest javelin thrower.' The redundant handrail joined a small pennant which Aldrin had launched into orbit earlier in the space walk. The small blue-bordered flag had been carried to commemorate Remembrance Day and 'to include all the people of the world who stand for now and continue to strive for peace and freedom in our world'.

A few hours later, the crew began the tether experiment first tried by Gemini 11. The Agena was commanded to pitch the docked combination to a vertical position so that both craft pointed directly down towards the Earth. The Gemini then undocked and backed away in order to extend the tether fitted by Aldrin during his EVA. The early stages proved frustrating as Lovell tried to control his craft:

> Both the two and four thrusters were out and every time I wanted to pitch up or yaw, I would roll. I got mad at it occasionally because I couldn't do anything. But we finally, through a learning curve, determined how to handle the situation by using a manoeuvre thruster – actually bipping it a little bit to bring it around and counteract this roll.

The two men worked through an entire night pass and into the next day pass to steady the craft so that it would remain in a stable position in relation to the Earth. Lovell told of their eventual success as their patience paid off:

> Buzz got the slide rule out and made a few fast calculations, and we got above the Agena again, maintained this position, and it appeared to us then that our rates had indeed dampened. We let it go for the next two revolutions and finally we let the Agena go too, and there we were – two dead vehicles captured by gravity in a vertical position going around the Earth.

As the two craft eventually separated after a flight time of 52 hours 14 minutes, the crew could be excused for believing that the rest of the mission was downhill all the way.

First, however, there was a final stand-up EVA to negotiate. Firmly harnessed to the couch for 55 minutes, Aldrin used the occasion to clear out the cluttered cabin by dumping overboard a pile of waste and unwanted EVA equipment, as well as continuing the programme of star and sunrise photography. Lovell commented: 'It looks after this one you can call us the litter-bug flight.' As Aldrin closed the hatch he established a new record of 5 hours 32 minutes for EVA during one mission. There was no time for self-congratulation, however, for the crew soon reported that two more thrusters were malfunctioning. The programme of experiments continued, including an attempt to observe a sodium cloud released from a French rocket launched from Algeria.

Minor problems continued to dog the Gemini during the fourth and final day in orbit. The crew found the water supply in the adapter section had run out and were forced to draw on supplies from the cabin container. Tests of the attitude control system confirmed four thrusters were malfunctioning, and output from the two remaining stacks of fuel cells in section number two dropped so low that the 4 main batteries had to be brought on line to compensate. After careful consideration, mission control decided to allow Gemini 12 to end the programme without the ignominy of an early return. Retro-fire came over Canton Island at 93 hours 59 minutes into the mission, and, following a comfortable re-entry, the last American two-man craft splashed down into the Atlantic within 3 miles (5 km) of the recovery carrier *Wasp*.

Millions of TV watchers heard the Capcom tell the astronauts: 'Smile – you're on the tube.' They were picked up by helicopter and whisked to the deck-top celebrations within 30 minutes of splashdown. Both men had reason to be proud and elated: they had brought the Gemini programme to a successful conclusion, despite niggling operational difficulties, and proved once and for all that men can carry out meaningful work in the weightless environment outside a spacecraft. In addition Aldrin had broken the record for total time in EVA, and Lovell had become the world record holder for time spent in space – 425 hours 9 minutes. Both men went on to make their mark in the Apollo moon programme, Lovell as the senior pilot on the first mission to the moon, and Aldrin as the second man to walk on that forbidding world.

The Director of the Manned Spacecraft Centre, Robert Gilruth, summarized the feelings of the NASA hierarchy at a champagne celebration in Houston shortly after the Gemini 12 splashdown:

In order to go to the moon, we had to learn how to operate in space. We had to learn how to manoeuvre with precision, to rendezvous, to dock, to work outside the spacecraft in the hard vacuum of outer space, to learn how man could endure long duration in the weightless environment, and to learn how to make precise landings from orbital flight. This is where the Gemini programme came in.

The list of achievements was, indeed, impressive, and there is no doubt that the experience gained during the ten manned Gemini flights had left the USA far ahead of the Soviet Union, which had been unable to launch one manned mission during the 20 months of concentrated American activity. The Americans looked optimistically ahead to a moon landing within the next couple of years, relegating any minor doubts to the back of their minds in their confidence that the most technologically advanced nation on Earth could safely overcome any obstacle. Had they known the true state of the first Apollo capsule scheduled to carry men into orbit, they might have had second thoughts.

6

Too Far, Too Fast

The Apollo 1 and Soyuz 1 Disasters, 1967

The new year of 1967 began with the American and Soviet superpowers in optimistic mood. Both countries were in the final stages of preparing their next leaps towards the conquest of space and the journey to another world. In the USA the two-man Gemini series had been successfully completed, opening the way to the three-man Apollo missions and the ultimate confirmation of American supremacy, the lunar landing before the decade was out. In the Soviet Union there had been a lull lasting nearly two years as the next generation of manned spacecraft, Soyuz, was developed with the similar aim of sending men around the moon. Progress had been slowed by the death of Sergei Korolev, the 'Chief Designer' of Soviet spacecraft, so that the previous 21 months had seen the complete dominance of Gemini and the position of forerunner in the 'space race' firmly grasped by the Soviet Union's capitalist rivals. Now 1967 was to be the year in which the Soviet Union once more took up the challenge; the race was on in earnest.

There had been a remarkable safety record during the six years of manned spaceflight. Admittedly there had been embarrassing failures with rockets going off course or exploding on the launch pad, while spacecraft malfunctions had threatened the lives of spacemen on more than one occasion. Yet there had been no fatalities during the 16 American missions or the eight Soviet missions to date. It almost seemed safer to fly in space than to travel by more conventional means. However, three astronauts had died in accidents before they ever had a chance to prove themselves at their chosen profession. The first was a 34-year-old US Air Force captain, Theodore 'Ted' Freeman: he had been an astronaut almost exactly one year when he took off for a routine training flight on a clear, sunny afternoon at the end of October 1964. On completion of the session, he turned his T 38 jet towards Ellington Air

Force Base, but on his landing approach he ran into a flock of wild snow geese. One bird smashed into the windscreen, while another was sucked into the air intake, causing the engines to cut out. At an altitude of 1,500 feet (460 km) he tried to glide the rest of the way home. Too late he realized he would not make it, banked sharply away from the built-up area and ejected, despite the wrong attitude of the plane and the low altitude. The parachute never opened properly; his body was found one mile from the base, not far from the wreckage of the plane.

The T 38 was an essential part of the astronaut's life. Apart from the psychological benefits of enabling these highly skilled pilots to maintain their flight status (with the resultant increase in salary), to keep in training and simply to get away from all the pressures down below, the jets on loan from the Air Force were needed to enable the astronauts to fulfil their crammed work schedule. They were expected to traverse the country, day after day, as they visited the different NASA training centres and the various companies contracted to build the hardware. Each astronaut was assigned to a specific aspect of the engineering development programme so that he could keep himself and his colleagues abreast of the latest developments, as well as give the point of view of the man whose life might depend on the performance of the new equipment. Inevitably, such a commitment increased the amount of travelling in the already hectic lives of the astronauts.

Thus it was that astronauts Elliott See and Charles 'Chuck' Bassett were flying their T 38 to Lambert Field, St Louis on the last day of February 1966. Both were due to fly later that year on Gemini 9; close behind in another plane was their back-up crew, Tom Stafford and Eugene Cernan. The four men were booked for a two week stint on the docking simulator at the St Louis plant of the McDonnell Aircraft Corporation. They left Houston on a fine morning, but ran into low cloud, rain and fog as they flew north. Both See and Bassett were experienced pilots: See was a 39-year-old civilian engineering specialist who had become an astronaut in September 1962, while Bassett was a 34-year-old Air Force major who had been selected in group 3 during October 1963. However, conditions were bad enough to concern everyone as they approached the runway with a cloud base at only 750 feet (240 m) and visibility little more than one mile (2½ km).

The first approach through the swirling rain and sleet was on line with the runway but flight control advised them they were too far along the approach path and travelling too fast to land this time around. The two men decided to swing low across the airfield in a wide left turn instead of boosting the power to gain altitude before returning for a second attempt. They must have suddenly realized they were dangerously low for the roar of the jet's afterburner reverberated around the airfield as they tried desperately to climb above the buildings looming out of the mist. The T 38's undercarriage struck the roof of Building 101, the home of the Gemini 9 spacecraft the astronauts were due to fly. Part of the roof caved in, though the spacecraft was undamaged. The plane ploughed its way across the roof before it plunged on to the courtyard beyond and exploded in a giant fireball. The astronauts must have died instantly. Their grieving companions witnessed the ceremonial burial the following day at Arlington National Cemetery. But these setbacks made little impact on the momentum of the space programme. See and Bassett were replaced by their back-ups and the Gemini missions were completed on schedule.

Even before the moon landing programme had been initiated by President Kennedy in 1961, the need for a giant rocket capable of lifting large spacecraft into Earth orbit or even sending them to the moon had been recognized. As early as 1959, Dr Wernher von Braun's Saturn project was under way, with the first successful test flight of the Saturn I rocket following only two years later. The succeeding tests were just as successful, with an unprecedented 100 per cent perfect record. While the world's largest rocket, the Saturn V, was being developed, von Braun modified the original Saturn to create the Saturn IB, a mid-range booster capable of lifting the Apollo moonship into Earth orbit. This passed all three unmanned flight tests with flying colours during early 1966, leading to optimism among NASA officials that the first manned Apollo mission would take place later that year. The spacecraft development programme, however, was running into difficulties; the rocket may have been ready for use, but the payload it would carry aloft most certainly was not.

In August 1966 astronauts Grissom, White and Chaffee were told that the first Apollo flight would be theirs – a 'shakedown' flight in Earth orbit with a duration of up to two weeks, probably in December. The command and service modules were accepted from North American Aviation, though further tests were planned, both in the Kennedy Space Centre altitude chamber and on top of the Saturn IB during launch simulations. Numerous minor ailments were discovered and had to be rectified, so that the engineers were soon behind schedule. The first manned simulation in the altitude chamber did not take place until 18 October, and then it was soon terminated by a transistor failure. A few days later another simulation involving the back-up crew of Scott, Schweickart and McDivitt had to be discontinued when the oxygen regulator in the environmental control system failed. The unit had to be taken out and replaced, causing another delay of nearly two weeks. Meanwhile, a rupture in a fuel tank of another service module at North American's California plant led to further time-consuming checks on the craft at Kennedy. The replacement environmental control unit soon developed a series of leaks, including, on one occasion, a spillage of coolant fluid across a number of electrical wires. By this time the December launch date had been scrapped, with a launch in February

1967 now seen as the earliest possibility. The command and service modules (CSM) were finally mated to their Saturn launch vehicle on 6 January, enabling the final ground tests to begin.

A few minor problems arose while the craft was linked to external power supplies, but by 27 January all was ready for the final check-out using internal spacecraft power. CSM 12 had been scrutinized, tested, modified and re-tested almost continually for the past year, far longer than any previous craft. Since its delivery to NASA in August 1966, there had been a remarkable 623 changes. Astronauts Grissom, White and Chaffee, together with the new back-up crew of Schirra, Eisele and Cunningham, must have often contemplated their role as guinea-pigs whose lives depended on an untried spacecraft that was virtually a prototype. On one occasion, Grissom hung a lemon on the craft to demonstrate their discontent. As he put it, their mission would be 'primarily concerned with checking out the spacecraft's systems and seeing whether it is both flyable and livable'.

On the morning of 27 January the three men walked in their white suits to launch complex 34, ascended in the lift to the white room at the top of the rocket gantry and prepared to climb into the cramped Apollo capsule for what was labelled a routine ground test. No danger was anticipated, for the rocket was empty of fuel, so the fire crews were on standby rather than maximum alert. In all other respects, the simulation was as close to reality as possible: the capsule would contain a pure oxygen atmosphere at 16 psi, higher than normal atmospheric pressure in order to prevent nitrogen leaking into the cabin, and the access hatches would be sealed and locked. An extra item on the itinerary was a simulated emergency at the end of the test, added at Grissom's suggestion in order to check that a rapid exit from the craft could be made if necessary.

Grissom was the command pilot for the Apollo 1 mission, so he was seated on the left of the capsule. In the right couch lay rookie Roger Chaffee, a lieutenant commander in the Navy who became an astronaut in October 1963. His role was communications officer on the forth-coming flight. He must have been quite awed by his illustrious companions, for the third member of the crew was, like Grissom, an Air Force lieutenant colonel and a veteran of Gemini. Ed White had already earned his place in history with the first American space walk. He was the senior pilot of Apollo 1, the last to clamber awkwardly into his seat, the middle couch of the three. Onlookers remembered how none of the men seemed to be looking forward to their afternoon's work on this occasion.

Their mood did not improve as the test proceeded. Gus Grissom noticed a strong odour from the suit oxygen supply, so the simulation was halted at 1.20 pm for gas samples to be taken and analysed. All seemed well, so the countdown resumed with the astronauts secured in the spacecraft. The next three hours proved frustrating for all concerned, with numerous hold-ups and problems with communications in particular. At one stage an irritated Grissom's voice crackled over the intercom, 'How do you expect to get us to the moon if you people can't even hook us up with a ground station? Get on with it out there.' He had been so concerned with the communication problem that he had discussed changing seats with Deke Slayton, Director of Flight Crew Operations, so that he could witness the gremlins for himself, but Slayton eventually decided against it – 'I'd be better off in the blockhouse,' he told the long-suffering astronaut.

After five and a half hours in their couches, the simulation was approaching its culmination as the countdown reached T minus 10 minutes. At 6.20 a hold was called while the communications links were checked once more, prior to the switch to fuel cell power. It was now dark, and the astronauts were looking forward to a respite from their trials and tribulations.

At 6.31, the calm at the Space Center was shattered by a completely devastating and unexpected event. The telemetry from the craft showed a momentary power surge somewhere in the 12 miles (20 km) of wiring surrounding the astronauts. It went unnoticed at the time, but the significance of the electrical short soon became apparent. Less than ten seconds later, there was an almost casual report from Roger Chaffee over the intercom: 'Fire, I smell fire.' Immediately there was a surge of activity in the spacecraft as the men initiated the emergency escape procedure. A more urgent appeal came across to the blockhouse: 'Fire! We've got a fire in the cockpit!' Ed White was struggling to find the hatch handle and undo the six securing bolts, but the temperature began to rise rapidly as the flames quickly spread in the pure oxygen atmosphere. The smoke was by now so dense that no one could see the men on the TV screens: Deke Slayton later related how he was more worried at that stage about the smoke than the actual fire. Then came the final report from the three astronauts, a heart-rending cry from Roger Chaffee: 'We've got a bad fire . . . We're burning up here!' There followed only the sounds of frantic movement, pounding on the wall of the cabin and unintelligible shouting.

Outside the craft a camera located in the white room showed the rapid spread of the fiery glow: the cameraman did not realize at first that he was looking at a fire as he reached to adjust for brightness. Once the full horror of the situation struck the pad personnel, they tried all they knew to release the trapped men until the capsule became too hot to handle. As the fire spread, the possibility of the pressure hull rupturing became apparent, persuading some staff to retreat from the immediate area. A mere 16 seconds after the first report from Chaffee, the cabin split apart in a violent explosion. Pad leader Don Babbit was standing by his desk when the blast hit him. 'The force of it slammed me against the desk,' he later stated. His white smock was burned through by fragments of molten fibre glass, the papers on his desk were charred by the intense heat and a small fire started away to his left. He staggered outside through the dense cloud of smoke, then he and technician Jim Gleaves forced their way back inside in an attempt to remove the outer hatch. Gleaves said, 'You couldn't see six inches from your face. I had to run my hands around the capsule to locate the hole the size of a dime into which the tool had to be inserted.' Eventually they succeeded, then had to retreat once more to avoid asphyxiation. Another technician staggered out carrying the middle hatch, a metal door which normally needed two men to carry it.

The first fireman arrived on the scene some four and a half minutes after the fire began, but he did not have a smoke mask – he explained that they did not think it was that kind of fire, and they simply wanted to reach the blaze as quickly as possible. Fireman Jim Burch later admitted: 'I realized I'd never seen a hatch before, and wasn't sure what it looked like.' He went on,

I found it, and there was somebody else grabbing it. It was loose, but it wouldn't come off. I fought at it for 20 or 30 seconds, but I had to come out. I got a good breath of air and ran back. This time I was by myself. I grabbed that hatch. I was frantic. I was grabbing it and shaking it and somehow it just fell over. When it did, someone was standing right behind me and helped me lift it away.

Burch was one of the first to peer inside the hot, charred interior of spacecraft 12:

I laid over inside the hatch and shone my light around, but I couldn't see anything I recognized. The test conductor was standing outside, and I told him there was no one in there . . . People were hollering for me to get the men out and I was confused. I couldn't see anybody. It didn't seem real. Where were they? I backed off with my light. Then I could see the bodies.

All three suits had burned through, though Grissom's had suffered the most and Chaffee's the least. Grissom was stretched out with his feet on his couch and his body seeking shelter beneath the centre couch. Chaffee was still strapped into his couch, while White's body lay below the hatch sill, a testimony to his heroic efforts to release the hatch before the interior pressure build up became so great that removal was impossible. All three had been incinerated by the rush of flame across the cabin as its hull violently ruptured, though the official cause of death was asphyxiation caused by inhalation of smoke. It was not until seven hours later that the bodies could be removed; their suits had melted along with the nylon netting used to prevent loose objects floating into dangerous places in zero gravity, the Velcro fastening material on the cabin interior wall and the wire insulation, causing a synthetic liquid which solidified as it cooled, welding the bodies to the capsule. Gus Grissom and Roger Chaffee were buried a few days later in the presence of President Johnson at Arlington National Cemetery in Washington; Edward White's funeral, attended by the President's wife and Vice President Humphrey, was held the same day at West Point Military Academy.

The Apollo moon programme was not buried with the unfortunate victims, but the multi-billion dollar space effort virtually ground to a halt as a special board of inquiry sought the explanation of the disaster. On the very day the three astronauts were buried, another flash fire in a pure oxygen atmosphere killed two airmen who were in a steel pressure chamber at Brooks Air Force Base in Texas; they had been taking samples from rabbits being used to test the effects of long-term exposure to pure oxygen. Nor was this the only such case: the previous April an environmental control unit had burst into flames in a pure oxygen cabin as a result of an electrical arc, though no one had been hurt. Rocket designer Wernher von Braun commented:

The pure oxygen in connection with inflammable material can cause a fire in the spacecraft, and that's apparently what happens. What is not understood yet is that all surfaces of the spacecraft were supposed to be oxygen compatible, and there shouldn't have been a fire.

Three months later the official report of some 3,000 pages was issued. There was no attempt at a cover-up; on the contrary, it was viewed by many as 'almost an orgy of self-criticism'. The cause of the fire was uncertain, though an electrical arc in the vicinity of the environmental control equipment was considered the most likely candidate, especially since inspection of the wiring which survived the fire showed 'numerous examples of poor installation, design, and workmanship' including damaged insulation and incorrect colour coding. The report concluded that a small fire broke out between the inner and outer hulls, spreading only slowly at first due to the lack of combustible materials in the immediate vicinity. Some ten seconds after it began, Gus Grissom spotted the flames beginning to enter the cabin. According to the report, 'The original flames rose vertically and then spread out across the cabin ceiling. The debris traps [nets] not only provided combustible material and a path for the spread of the flames but also firebrands of burning molten nylon.' After the cabin wall ruptured, the inferno reached maximum intensity, fed by the outrush of gases – holes burned in aluminium tubing showed temperatures of at least 760°C. The rapid spread of deadly carbon monoxide gas followed in a matter of seconds, with billowing black smoke and soot blanketing the whole cabin as the supply of oxygen ran out. The complete cycle lasted perhaps 35 seconds, though the three men lost consciousness only 20 seconds or so after their first fire report. The hatches were designed so that it was impossible to open them in less than 90 seconds. The official inquiry also found 'many deficiencies in design and engineering, manufacture and quality control'. As if to emphasize this, the investigators discovered a wrench socket which had been accidentally dropped between two bundles of wiring, despite apparently foolproof check-out procedures for all such equipment. Clearly the Americans were trying to run the moon race at too fast a pace.

A few weeks before the disaster Gus Grissom had completed the first draft of a book about the Gemini programme. In it he summarized the astronaut philosophy:

> There will be risks, as there are in any experimental programme, and sooner or later, inevitably, we're going to run head-on into the law of averages and lose somebody. I hope this never happens, and with NASA's abiding insistence on safety, perhaps it never will, but if it does I hope the American people won't feel it's too high a price to pay for our space programme. None of us was ordered into manned spaceflight. We flew with the knowledge that if something really went wrong up there, there wasn't the slightest hope of rescue.

The sacrifice was not in vain. The outcome of the tragedy was a much modified, but much safer Block 2 Apollo spacecraft. The fire had caused a delay in the moon programme of nearly two years, but it had given NASA and its contractors the breathing space necessary to create a viable spacecraft, and reduced some of the political pressure generated by the competition between the two superpowers.

Once Apollo 7 took to the skies, the direct successor to Apollo 1, America was well on the way to fulfilling the commitment inherited from President Kennedy. The recommendations of the board of inquiry were closely heeded: the craft was made as fireproof as possible with flameproof coatings on all wire connections, plastic switches replaced by metal ones, and even the spacesuits made of a fire-resistant material known as Beta cloth. There was considerable debate over the continued use of pure oxygen, finally resulting in a compromise which gave a much higher degree of safety; the cabin atmosphere in all future ground tests and during launch countdowns would comprise 40 per cent nitrogen and 60 per cent oxygen at normal atmospheric pressure. The astronauts would be able to breathe pure oxygen through their suit systems so that there would be no fear of contracting 'the bends', and the cabin atmosphere would be altered to one-third normal pressure in pure oxygen once the flight was under way. The escape valve was designed to vent all cabin oxygen into space within one minute if necessary, while the new aluminium and glass fibre escape hatch could be opened from the inside in less than ten seconds with the aid of a pressurized nitrogen cylinder. In the remote possibility that a fire did break out, causing the release of toxic fumes into the cabin, an emergency oxygen supply and a specially designed extinguisher were provided.

The year 1967 left more bad memories for the American astronaut corps as two more of the brotherhood lost their lives. Edward Givens had become an astronaut only one year earlier, having previously served in the Air Force as a major, when he was in charge of development of the astronaut manoeuvring unit. Although it was never used on Gemini because of astronaut exhaustion, it was the forerunner of the unit introduced on the Shuttle nearly 20 years later. On 6 June Givens was returning home from a flyers' reunion in Houston when his car went out of control on a sharp bend and crashed in a ditch. Givens died from chest injuries. A memorial service was held in Houston, followed by the funeral in his home town of Quanah, Texas.

Four months later, untimely death halted the career of Marine Corps major Clifton Curtis (C.C.) Williams. Aged 35, he had been selected as an astronaut in 1963 and had been on the Gemini 10 back-up crew with Alan Bean in 1966, with a more recent assignment as back-up to a future Apollo flight. On 5 October he was flying home to Houston from Cape Kennedy in order to celebrate his wife's second pregnancy. He was scheduled to stop for refuelling at Brookley Air Force Base near Mobile, Alabama, but at 1.24 pm as he approached Tallahassee, something went wrong with his T 38 jet. Air traffic controllers heard a faint report 'Mayday, Mayday, am ejecting' followed by silence. He never succeeded in ejecting, possibly because he lost consciousness during the headlong plunge towards the earth. The plane smashed into a wooded area, exploding on impact; the astronaut's body was found nearby. Williams was buried with full military honours at Arlington National Cemetery, the fourth group 3 astronaut to have died since they were selected, and the fifth American spaceman to lose his life that year. None had died actually carrying out a mission in outer space.

In the Soviet Union, the reaction to these setbacks was as expected; the authorities refrained from gloating over the disasters, concentrating instead on the propaganda points to be gained, and blaming the Americans for their own misfortunes. The astronauts who died in Apollo 1 were described as the 'victims of the space race created by American space programme chiefs'. In March 1967 General Kamanin, the man in command of cosmonaut training, was interviewed on Warsaw Radio about the possibility of a Soviet space spectacular to mark the 50th anniversary of the Communist Revolution. He said:

When the crews are ready for the flight, then we shall give the orders to take off. This might take place in the spring or the summer. We must be fully convinced, however, that the flight will be a success. The flights we are preparing will be more complicated than the previous ones, and thus the preparation for them will have to be appropriately longer . . . We do not intend to speed up our programme. Excessive haste leads to fatal accidents, as in the case of the tragic deaths of the three American astronauts last January.

The latter comment would prove particularly ironic when the Soviet Union attempted its next giant step in space, the long-delayed flight of its new spacecraft, named Soyuz (Union).

Western observers had been expecting a Soviet manned flight for some time, with some experts predicting the launch of a giant craft of 12–15 tons and capable of holding six or even nine men. Certainly, the launch of such craft was within the capability of the existing Soviet rockets as had been demonstrated by the unmanned Proton satellites of 1965 and 1966. Reports of a new craft under development had been circulating since 1965, yet no manned launch took place, despite a number of short duration missions by Cosmos satellites which seemed to resemble the low orbital paths flown by cosmonauts. Rumours circulated that the Soviets were planning an orbiting space platform from which they would launch a manned rocket to the moon; that they were about to send men around the moon; that they were building a huge rocket capable of matching the Americans' Saturn V; that the reason for the delay in launching the new craft was the shortage of money, with only half the amount being spent on the Soviet space programme as on the American competition. Officials and reporters in the West waited eagerly as reports from Moscow indicated a build-up in preparations during April 1967; a reliable source was quoted as saying that two ships would be launched a day apart before achieving a spectacular space docking and transfer of crews. This seemed a logical next step: the Soviets had achieved close rendezvous by two craft several years earlier, but had been overtaken by the Gemini flights which had successfully carried out docking manoeuvres, extra-vehicular activity and missions lasting up to two weeks. Now it seemed they were about to catch up.

At 3.35 am Moscow time on 23 April Soyuz 1 was launched with veteran cosmonaut Vladimir Komarov, aged 40, on board. As Soviet custom dictated, the father of two children had left his wife without telling her of the forthcoming mission. For him the flight must have been a personal triumph: at a relatively ripe age for a cosmonaut he had beaten all the opposition to be selected for the prestigious first flight of the Soyuz, despite the doubts over his health early in his career. The purpose of the flight was announced by Tass soon after Komarov reached an orbit varying between 125 and 139 miles (201–224 km): he would 'test the new piloted spaceship; check the ship's systems and elements in conditions of spaceflight; hold extended scientific and physical–technical experiments and studies in conditions of spaceflight; and continue medical and biological studies including the influence of various factors of spaceflight on the human organism'. There was no mention of any space rendezvous, but the Soviets were notoriously close-chested about their future plans, so a second launch was confidently expected. One curious aspect of the mission was the lack of information being released by official sources; for once, the Soviet propaganda machine did not seem to be functioning well. Apart from a radio interview with Komarov soon after launch, there was little hard news to grasp.

The official reports told of 'the flight programme being fulfilled' as Soyuz began its fifth orbit, six and a half hours after launch. The cosmonaut was comfortable in his 'shirt-sleeve environment' of oxygen–nitrogen atmosphere at near normal pressure, and cabin temperature of 16°C. The Soviet network of tracking stations was nowhere near as widespread as that of the Americans, so from 1.30 pm until 9.30 pm Komarov was mostly out of range of ground stations. This time was designated his rest period. Shortly after midnight, Tass announced that he had completed thirteen orbits, though the craft had, in fact, passed that mark at about 10.40 pm. On the morning of 24 April, the day of the predicted rendezvous, Komarov was said to be in good condition and carrying out experiments according to the flight programme. Little more than one hour later, at 6 am Moscow time, Komarov was ordered to return to Earth. As he flew over Africa, Pravda reported, 'a successful ignition of the retro-rockets followed by separation of the spaceship's instrument module'. Everything was said to be under control. There followed an ominous silence lasting nearly 12 hours until Soviet radio and television simultaneously announced the grievous news: Colonel Komarov 'was killed during the landing of the spaceship. According to preliminary data, on the opening of the main cupola of the parachute at an altitude of 4.3 miles (7 km), the spaceship descended at a great speed because the parachute strings had become twisted, and this was the cause of the cosmonaut's death'. The official sources maintained that the Soyuz mission was a straightforward test flight which had been successfully carried out until the fatal 'Roman candle' during re-entry. 'During the test flight, which had lasted more than 24 hours, Colonel Komarov had completed the testing of the systems of the new spaceship and had carried out the planned scientific experiments. He had manoeuvred the ship and tested its main systems under different regimes.' There had been no escape route for the cosmonaut during the final spiralling plunge to Earth; the Soyuz was not equipped with an ejector seat, nor did it have a back-up parachute.

The Soviet people were stunned by the first real disaster to strike the Soviet space programme – the press had not reported a terrible launch pad explosion which had taken place during October 1960, killing Field Marshal Nedelin and numerous other personnel. The Soviet leadership signed the cosmonaut's obituary: 'We shall for ever cherish the memory of the loyal son of our Motherland, wonderful Communist, courageous explorer of space, comrade in arms, and friend. He will be an example for ever of heroism, courage, and valour. His name will summon the glory of our great Socialist country to new feats.' Two days later, his ashes were buried in a moving ceremony watched by millions on television; they saw the mourning widow kneeling beside the niche in the Kremlin Wall where her husband's remains were interred. As she said, 'Pilots' wives always worry about their husbands.' But had she been given cause to worry during the flight, or was the official version that all had been well until the parachute became tangled the truth?

The mission was surrounded by mystery and doubt. Few details were released about the new craft, though reporters at the time described it as being 'as big as a double-decker bus'. However, later study of launch pictures showed that the rocket used was not a new booster but the same vehicle as had launched Voskhod. Launch weight was about 14,200 lb (6,000 kg), nearly double the weight of Gemini but far less than the Apollo craft which the Americans were now redesigning. Indeed, Soyuz was found to be comparable in performance and capability to Gemini, with the ability to adjust its orbit as well as its attitude, and to dock with other craft. Its main role seemed to be related to developments in Earth orbit, carrying up to three cosmonauts on quite lengthy missions which could involve a wide variety of scientific and medical experiments as well as practice in rendezvous and docking. If a space platform or space station had been available, it could be used as a ferry craft, a role which it later assumed quite successfully. A moon landing was impossible, though Western experts pointed out that it could be adapted quite easily to carry two men around the moon and back to Earth, a point which was verified the following year when Zond 5, a Soyuz-type unmanned craft, flew on a similar mission. Although the Soviet authorities never admitted their intentions, the cosmonauts were more open about their hopes and ambitions: at the 1967 Paris Air Show, for example, Pavel Belyayev told astronauts Collins and Scott that he expected to make a circumlunar flight in the not-too-distant future.

Belyayev's name was one of those mentioned in unofficial reports concerning a second Soyuz launch on 24 April; another was Valeri Bykovsky, who was supposed to have been the command pilot. The truth concerning this mysterious second Soyuz may never be known, though there is circumstantial evidence in its favour. The designation of Komarov's flight as Soyuz 1 seems significant, since neither the first Vostok nor the first Voskhod was given a number. A rendezvous and docking would have restored Soviet prestige much more than an 18-orbit routine test flight, even though it would be risky attempting such a manoeuvre with two untried craft. A photograph released in 1978 showed Komarov with Yuri Gagarin, the Soyuz 1 back-up pilot, Valeri

Bykovsky, and two space-suited companions, Yevgeni Khrunov and Alexei Yeliseyev, who transferred from Soyuz 5 to Soyuz 4 by space walking in January 1969. This photograph gives what seems to be conclusive evidence that such a crew transfer was intended for Soyuz 1 and 2 in April 1967. So why didn't it take place?

One possibility is that something went wrong with the launch of Soyuz 2, so that the mission had to be held or postponed altogether. This may well have been the case, but it seems more likely that the source of the trouble was Komarov's craft; a faulty Soyuz 1 would have immediately led to the cancellation of the proposed link-up. There is reason to believe that one of the two solar panels on the Soyuz did not deploy properly on attaining orbit, thereby reducing the craft's electricity supply. However, this fault cannot explain the fatal ending of the flight. One clue lies in the site of the crash; Komarov had missed two perfect opportunities to land on his seventeenth and eighteenth orbits, finally re-entering on orbit 19. The exact position of the crash was not immediately revealed, but it was later explained that Soyuz fell to Earth on the steppes to the north of the Caspian Sea, not far from the city of Orenburg. This position was some 600 miles west of the planned landing zone, a shortfall which may be accounted for by the capsule spinning during re-entry. Yet such spin would not normally be employed as a method of controlling the flight path, suggesting that the craft's guidance system was not functioning properly.

American tracking stations announced soon after the crash that the Soyuz had begun tumbling out of control during the fifteenth orbit. Despite valiant efforts to correct the spin, Komarov was unable to stabilize the craft, merely wasting his precious fuel. A night landing was decided upon during the seventeenth orbit, but the cosmonaut was unable to orientate the capsule with its heat shield facing forwards and at the same time hold it steady on the correct re-entry trajectory while he fired the retro-rockets. He probably tried once more on the next orbit, again with no success. By this stage he must have been getting desperate; years later, a National Security Agency technician who had been manning an American listening post in Turkey told his version of the last hours of the doomed cosmonaut. Between instructions from Soviet ground controllers, Komarov was able to speak to his wife until eventually he could stand the heartache no longer and told her to go home. Komarov tried all the suggestions offered: 'I'm doing it, I'm doing it, but it doesn't work,' came the man's frustrated voice. As the tumbling continued, he told the ground he was feeling sick from the perpetual motion. He kept asking, 'How far from re-entry?' Prime Minister Kosygin came on the radio link in an emotional tribute to the Hero of the Soviet Union, but it must have seemed more like an obituary. Komarov began to shout, 'You've got to do something. I don't want to die.'

On the nineteenth orbit ground control decided on a last desperate gamble. Komarov was told to put the craft into a stabilizing roll instead of attempting a controlled re-entry. He succeeded in turning the craft around and firing the retro-rockets, but the angle of entry was wrong. Amateur enthusiasts listening in Turin reported Komarov's final cry 'You are guiding me wrongly, you are guiding me wrongly, can't you understand?' With G forces building up to perhaps 10G, it is possible that Komarov blacked out during the wild, fiery plummet into the atmosphere. At any rate, the capsule was still rotating as the parachute deployed at an altitude of 4 miles (7 km), causing the canopy to twist around and fail to arrest the rate of descent. The Soyuz must have been travelling at around 500 mph (800 kmh) when it smashed into the ground, with the occupant killed instantly. It is possible, though not likely, that the body was partially consumed by a cabin fire. Komarov's couch was probably torn free, tossing its occupant, still strapped down, across the cabin like a rag doll.

The Soviet space programme went into a long decline, as if its designers and technicians were unable to function properly without Korolev's guiding light. The next proving flight of the Soyuz craft did not take place until 18 months after Komarov's death, with the first crew transfer in January 1969. By that time, the Soviet authorities were forced to admit defeat in the moon race, though their public stance revealed surprise that anyone should have thought they were interested in the moon.

Meanwhile, in March 1968, the world's first spaceman, Yuri Gagarin, was laid to rest with his comrade in the Kremlin Wall. He was back in training for spaceflight at his own request when he took off on a routine training flight in a two-seater jet aircraft. The plane crashed in a birch forest and disintegrated in a ball of fire. The previous year, Gagarin had written about his friend Komarov's death:

> Nothing will stop us. The road to the stars is steep and dangerous. But we're not afraid . . . Space flights can't be stopped. This isn't the work of any one man or even a group of men. It is a historical process which mankind is carrying out in accordance with the natural laws of human development.

The American astronauts understood what he meant; they, too, were part of that development. Both groups had suffered grievous losses, experienced the same fears and dangers, and felt a spirit of brotherhood which crossed all political boundaries. There would be an occasion, years later, when they would briefly share their spacecraft in that same spirit of friendship and comradeship. Meanwhile, they could only commiserate at a distance over the deaths of their brethren, the victims of the politicians' demands which caused the designers, engineers and technicians to move too far, too fast.

7

Fly Me to the Moon
Apollo 7 to Apollo 10, 1968–9

In October 1968, 20 months after they had witnessed the deaths of three of their colleagues in the Apollo 1 inferno, the astronaut team of Schirra, Eisele and Cunningham set foot on Pad 34 at Cape Kennedy in an effort to lay the ghost of gloom and despair while at the same time putting America back on course for a moon landing before the decade was out. The intervening period had been a trial for both men and machine as the engineers and designers from North American Aviation and NASA rebuilt the Apollo command module to transform it from a death trap to a viable spacecraft. The astronauts too worked hard and long in an effort to iron out any problems, living alongside their craft as it slowly matured from a shell into a sophisticated craft. To aid them in their task a three-man support crew was appointed in addition to the usual back-up: from now on there would be nine astronauts working on each Apollo mission.

The mission commander, Wally Schirra, was one of only three Mercury astronauts still on the active list, but at the age of 45 he had decided that his third flight – the first manned trial of the new Block 2 Apollo command module, designated Apollo 7 – would be his last. In contrast to Schirra's experience, the other two crew members had no space time under their belts. Air Force major Donn Eisele, 38, had become an astronaut in 1963, having previously graduated from the Naval Academy, obtained a master's degree in Astronautics and joined the Air Force as a test pilot. The junior partner, civilian Walter Cunningham, was also a group 3 recruit. He joined the Navy at the age of 19, qualifying as a pilot the following year. Following active service with the Marines until 1956, he then concentrated on a scientific career, gaining two physics degrees and working on classified defence studies for the Rand Corporation before astronaut selection. The crew chosen for the Apollo shakedown cruise was a strange combination: the Korean War veteran, nicknamed 'Mr Cool' by NASA staff; the quiet, unambitious Air Force pilot whose life revolved around flying; the outspoken, blunt Marine fighter pilot turned civilian scientist.

The Saturn IB booster was installed at Pad 34 six weeks before the command and service modules (CSM) arrived. The new craft, with more than 1,800 modifications to the original design, had been tested more than any of its predecessors. Now came the ultimate trial, an 11-day work-out in Earth orbit which would be the first major step in giving Apollo the certificate of moonworthiness. The astronauts were pleased to see the familiar, cheerful face of Guenter Vendt, the man who was responsible for pad supervision of the earlier Mercury and Gemini flights: significantly, Vendt had not been present during the Apollo 1 preparations, since he was employed by McDonnell. Deke Slayton arranged for his transfer to North American, but it was Wally Schirra who insisted that he was switched from the midnight shift to the morning shift for 11 October 1968. Wally wanted 'the Fuehrer of launch pad' on the job when they climbed into the command module. The lighthearted banter which came across the airwaves during the countdown showed the pleasure of the three astronauts at Wendt's presence; even as Apollo 7 lifted into a clear blue sky, Donn Eisele could not resist a final crack: 'I vunder vere Guenter Vendt?' The first manned flight aboard a mighty Saturn booster got away to a perfect start, cheered on by the watching crowds and the millions of TV viewers worldwide.

Apollo 7 went into an elliptical orbit ranging from 173 to 140 miles (285–228 km) above the Earth, drawing fulsome praise from the old hand, Schirra: 'We're having a ball . . . she's riding like a dream.' The second stage of the Saturn rocket remained attached until near the end of the second orbit. Once they had tested the manual attitude controls, the crew released the explosive bolts which linked the command and service modules (CSM) to the second stage, enabling Schirra to begin the vital separation, turn and approach manoeuvre which a moon mission must complete in order to withdraw the lunar module (LM) from its housing in the stage. This time there was no LM, only a white target between the open panels of the S-IVB stage. Schirra practised station-keeping about 100 feet from the target as they sailed over the Cape and headed out across the Atlantic to begin their third revolution. Soon after, the thrusters were fired to lower the CSM so that it would gradually pull ahead of the stage. For the first time, the crew were able to relax and remove their bulky pressure suits to reveal the latest in space fashion, Teflon two-piece coveralls worn on top of cotton one-piece 'long johns'. Compared with earlier spacecraft, the Apollo CSM was a palace: with their pressure suits removed and stowed, the astronauts actually had room to float around the cabin and stand upright without banging their heads on the wall. With a cabin area three times the usable space inside Gemini, each man had twice the room available to his predecessors.

'Bedrooms' were provided in the form of hammocks beneath the left and right couches with straps to prevent the men from floating in their sleep. Cabin temperature was maintained at an equable 24°C (75°F) and the menu consisted of more than 60 different items of 'freeze-dried bite-size rehydratable foods'. Water could be obtained from a portable drinking water dispenser, and from two spigots at the food preparation station; for the first time on a space flight hot water was available so that hot meals could be prepared.

Unfortunately, these 'mod cons' were not enough to keep Wally Schirra happy. Fifteen hours into the mission he reported that he had developed a bad cold and had been forced to resort to using paper handkerchiefs and aspirins. His mood had not improved 26 hours after launch when he refused to carry out a scheduled TV transmission, the first time this had ever been included on a flight. 'The show is off. The television is delayed without further discussion. We've not eaten, I've got a cold and I refuse to foul up our time this way.' Ground control tried to persuade the irate commander to change his mind, but Schirra was adamant: 'You have added two burns to this flight schedule, you have added a urine water dump and we have a new vehicle up here, and I tell you this flight TV will be delayed without further discussion until after the rendezvous.' Schirra prevailed, setting a precedent for the remainder of the flight. He was shaken out of his misery, however, when the main engine fired for ten seconds, giving them a sudden boost which jerked them back in their seats. Over the next three and a half hours the craft looped above the S-IVB then behind and below it before finally approaching once more using computer calculations based on sextant readings provided by the crew. The brief return visit was terminated by a five second burn of the thrusters, causing the final separation of the two craft. Over the next 24 hours the crew were able to watch the second stage slowly disappear into the void.

For a first trial run Apollo 7 was proving remarkably trouble-free, despite the crew health problems. Mission control watched with bated breath for the inevitable malfunction which could endanger the mission or even the lives of the astronauts, but it never came. The worst the engineers had to deal with were a tendency for part of the cooling system, which radiated surplus heat into space, to freeze up, a problem with a navigational telescope and trouble with an accelerometer which caused a small thruster to fire unexpectedly. The first TV special from space also went ahead with Schirra's permission on the third day and was rated a great success. Pictures sent back by the tiny 4½ lb black-and-white camera showed the men floating around their cabin in a lighthearted eight minute broadcast. Schirra held up a card to the camera declaring, 'Hello from the lovely Apollo room high above everything.' A second card followed: 'Keep those cards and letters coming in folks.' The three astronauts were described as 'grinning like exuberant schoolboys' during the broadcast, which they closed by pointing the camera out of the window to show the curved surface of the Earth, and a panoramic view of New Orleans, the Mississippi delta and Florida.

The systems tests continued to go well, but the mood of the astronauts deteriorated once more, possibly because all three of them had plenty of time to contemplate their head colds and feel sorry for themselves. Schirra complained to ground control at being woken too early on the fourth day: 'I asked for an hour and a half extra sleep for each last night and that was apparently ignored.' Capcom Bill Pogue replied: 'We acknowledge the error here.' Schirra was not mollified: 'Let's have the ground get to work on these sleep–rest cycles. We had only five hours per shift sleep scheduled this last night. I want the rest of these work periods worked out – and give us a chance to get some sleep.' A few minutes later, Pogue informed him that a ten hour sleep cycle was being set up for that night, but Schirra, as usual, came straight out with his opinion of the change: 'We can't do that! Let's get nearer an average of eight. That'll be plenty.' Once again the astronauts' mood changed, however, when the TV camera brought them before a wider audience. Schirra, at his extrovert best, introduced the broadcast as 'The one and only original Apollo roadshow starring the great acrobats of outer space.'

The next few days continued in much the same vein. The astronauts complained about the continually breaking wires on their bio-medical harnesses, culminating in Schirra's refusal to patch up any more of the wires which led to their body sensors to enable continuous monitoring of their physical condition. There was a minimal danger of a spark igniting the pure oxygen atmosphere, but Schirra admitted that their minds kept harking back to the Apollo 1 fire which had incinerated three of their colleagues. The craft's orbit was lowered to a safe altitude which would ensure a return to Earth even if the main engine malfunctioned, but an alteration to a final test burn caused renewed hostilities between the crew and the ground. Schirra voiced their feelings: 'We have a feeling you are believing that some of these experimenters are holier than God down there. We are a heck of a lot closer to Him right now.' A little later he added: 'I've had it up here today. We are not going to accept any new games like adding 50 feet to the Delta V for a burn, or doing some crazy tests we never heard of before. From now on I am going to be onboard flight director for the updates.'

The discontent had by now spread to all three crewmen. Walter Cunningham complained about the time scheduled for a TV broadcast: 'I think what they're trying to do is set this up so it will tie into somebody's TV show.' Ground control reassured him this was not the case, and brought the telecast forward one orbit. Neither was Cunningham enamoured with the 'Mickey Mouse' operations with which he was saddled by the engineers and planners on the ground, usually at short notice. 'I would like to go on record here saying that people who dream up procedures like this after you lift off have somehow or other been dropping the ball the last three years.' Even the normally amiable Don Eisele was unable to restrain himself after an additional test: 'I want to talk to the man, or whoever it was, that thought up that little gem. That one really got to us.'

Despite the strained relations between the astronauts and the ground – Schirra was being nicknamed 'The biggest Bolshie in space' – the TV shows continued to show the more amiable side of the crew and proved to be a big hit with viewers. The Saturday special showed the crew whirling weightless in the cabin to Schirra's 'close order drill'. The commander explained, only half jokingly, 'It is known in spacecraft talk that we have a crew commander, but what is not known is that we run a taut ship, and to maintain moral discipline we carry on a close order drill instruction period.' The final transmission on Monday was closed by Schirra: 'As the sun sinks slowly in the west, this is Apollo 7 cutting out now.' So ended the first ever series of live space broadcasts, one of the great technical and propaganda successes of the Apollo 7 mission.

The final confrontation between Schirra and ground control was the direct result of their medical condition. The cabin pressure would rise during re-entry from 5 psi to the normal sea level pressure of about 14½ psi over a time period of little more than a quarter of an hour. The best way to ease the pressure build-up in the eardrums was to pinch the nose and blow, but mission control was unhappy with this idea because the astronauts would have to leave off their pressure helmets, thus increasing the risks of suffocation due to a cabin leak and of banging their heads during splashdown. Schirra insisted that this was the only way to solve the problem without a risk of ruptured eardrums caused by blocked tubes in the ears, and Houston eventually gave way. The press enjoyed asking their readers to picture the three intrepid voyagers returning to Earth holding their noses.

The main engine retro-fire took place right on schedule, followed four minutes later by the bumpy jettison of the Service Module. The Command Module and its heat shield, designed to withstand the much greater velocities and temperatures generated during a return from lunar trajectory, functioned perfectly, so that the craft's three parachutes billowed open at an altitude of 27,000 feet to lower the astronauts gently into a choppy Atlantic Ocean, 330 miles (530 km) south east of Bermuda and 7½ miles (12 km) from the recovery carrier *Essex*. The capsule landed correctly, but tipped over almost immediately, resulting in a loss of contact and fears for the crew's safety. A few minutes elapsed before the automatically inflating floats restored the craft to an upright position, enabling Schirra to assure the waiting millions that all was well. Helicopters homed in on the radio signals from the capsule, then dropped frogmen and a dinghy alongside. To compound his misery with his health, Schirra had the misfortune to become seasick, but by the time the astronauts walked stiffly from the helicopter into the drizzle of the carrier's flight deck, they were in good humour and Schirra was able to smile at a crack about his 'first submarine service'. He was less forgiving, however, towards 'the idiot' who thought up some of the tests they had to endure. President Johnson telephoned his congratulations to the men on the *Essex*, but perhaps the most apt comment came from Mrs Walter Cunningham: 'It has been a very long 11 days. I am very glad it is all over.'

The verdict on the 163 revolution, 4.5 million mile flight was pronounced by Apollo Programme Director Sam Phillips: '101 per cent successful'. Apollo 7 had undergone more scrutiny than any previous spacecraft before it left the launch pad, and the hard work had paid off. The CSM could now be rated moonworthy. As for the problems of crew morale and strained relations with ground staff, these could be partly explained by the unfortunate head colds of the crew, the prolonged drift in space which left the three men with time to spare, and the keenness of the boffins on the ground to utilize this free time with extra tests dreamed up once they realized there were not going to be any major systems breakdowns on the new craft. Wally Schirra, who believed in doing things by the book, saw this as unwarranted meddling with the flight plan and did not hold back from saying so, especially since he had already announced his imminent departure from NASA and had nothing to lose. Schirra left in July 1969, immediately after the first landing on the moon, to become the President of Regency Investors Inc., a major financial company based in Colorado. He later went on to head a variety of major companies, including the ECCO Corporation which specialized in solving environmental problems. He was the only astronaut to fly on Mercury, Gemini and Apollo. Neither of his Apollo 7 colleagues flew again in space: Walter Cunningham stayed with NASA through the Apollo years, becoming Chief of the Skylab Branch of the Flight Crew Directorate when he was responsible for much of the manned space hardware. He eventually went into banking in Houston. Donn Eisele served on the Apollo 10 back-up crew, but retired from the Air Force and NASA in 1972 to become Director of the US Peace Corps in Thailand. Returning to the USA, he turned to business, eventually joining the investment firm of Oppenheimer & Co.

The Apollo programme was back on course for a moon landing during 1969, and was in better shape than anyone could have dared hope. However, the pressures were still on, especially since the Soviet Union sent two unmanned craft, Zonds 5 and 6, around the moon during September and November of 1968. Both craft carried animal passengers, including flies, meal worms and turtles, and returned safely to Earth, thereby fuelling speculation that a man would soon be sent around the moon to steal some of the American thunder. The official Soviet attitude had now changed in order to allow for the imminent success of the American moon programme: there was no moon race, there never had been such a race and the USSR was not concerned with landing a man on the moon. Strangely, despite all the previous statements to the contrary, the Western media tended to believe the new policy statement. NASA was more sceptical, since the possibility of a Soviet manned circumlunar flight was certainly one factor in the decision to send Apollo 8 around the moon during the Christmas holiday.

The pros and cons of swapping the Apollo 8 and 9 mission profiles had been the subject of an internal NASA debate for some time. The Saturn V booster, the most powerful rocket the world had ever seen, had passed its first unmanned test flight on 9 November 1967. A second test flight in April 1968 was less successful: there were vertical oscillations caused by the first stage, and fractured propellant lines in the second and third stages caused partial engine failure. Such problems could lead to potentially dangerous, even fatal, mishaps on a manned mission. Meanwhile, an unmanned lunar module had been tested in Earth orbit, but the first production line module was far behind schedule and unlikely to fly until well into the New Year. Rather than wait several months, it seemed to make sense to postpone the first lunar module (LM) mission and go for an extended Earth orbit mission or a circumlunar flight on the next Apollo. The success of Apollo 7 made the second option an attractive proposition, especially since it gave an excellent opportunity to test the CSM and the entire navigation and communication network which had been built up for just such a monumental journey. The decision was announced on 12 November, coinciding with the Soviet Zond 6 moon flight. Immediately there was widespread condemnation of what was seen as a gigantic and dangerous propaganda stunt. More surprisingly, perhaps, there was criticism within the scientific field, notably from Sir Bernard Lovell who declared his preference for automatic moon landing craft which were cheaper and posed no threat to human life. It was a controversy that continued as long as the Apollo programme and eventually helped to curtail the number of missions to the moon.

Behind the scenes, the astronauts had to consider the effects of the proposed changes. The crew originally designated for Apollo 8 while it was still envisaged as a test of the lunar module was McDivitt, Scott and Schweickart. NASA offered them the circumlunar flight, but, despite the prospect of a great adventure and a place in history, the crew turned it down in order to remain with the mission they had trained and lived for during the past months, now to be known as Apollo 9. So it was that Borman, Lovell and Anders moved up to take the limelight and also to take over the CSM which Scott had nursed through countless tests.

The Borman–Lovell team had worked together on Gemini 7 two years earlier, notching up the world space endurance record of just under 14 days. Jim Lovell had added another four days in Gemini 12 to give him an unprecedented 425 hours in space. Owing to the incapacity of Michael Collins, prevented from occupying the command module pilot's seat because of a spinal injury, Lovell was about to clock up another six days on a momentous voyage to another world. The two veterans worked hard to familiarize themselves with the new craft and new mission profile. They had no doubts about the correctness of the decision to risk their lives on top of a Saturn V rocket which had flown only twice before, in a craft which had never been beyond Earth orbit yet was now expected to carry them safely into lunar orbit then carry them home again. The third crewman, an Air Force major, William Anders, age 35, had joined NASA in 1963 having served as a fighter pilot in Air Defence Command before gaining a MSc in Nuclear Engineering and going on to become a technical manager at the Air Force Weapons Laboratory. Although one of the last group 3 astronauts to fly in space, Anders was a highly intelligent, energetic and dedicated pilot and engineer, though rather a different kettle of fish from his two senior test pilot colleagues. It was he who suffered most from the mission swap, for he had the misfortune to lose his craft – he became the lunar module pilot without a lunar module. Even a chance to swoop above the desolate lunar surface was insufficient compensation for such a loss. However, professional as he was, he just had to grin and bear it.

The morning of Saturday 21 December 1968 dawned early for the three astronauts. They were roused shortly after 2.30 am Cape time with the cheerful cry, 'It's go, go, go!' Breakfast was followed by a lighthearted session in the suiting room. Once the goldfish bowl helmets were donned, the white-suited pioneers walked to the special air-conditioned van which would take them the five miles (8 km) to Pad 39 where the towering Saturn V rocket awaited its first human passengers. The elevator climbed 320 feet (100 m) up to the command module in which they were finally closeted at 5.34 am, T minus 2 hours and 17 minutes. As they checked the spacecraft's systems, they had little time to ponder what lay ahead, though perhaps they did consider the good luck message sent by President Johnson: 'I am confident that the world's finest equipment will strive to match the courage of our astronauts. If it does that, a successful mission is assured.'

The President's confidence proved well-founded. The Saturn V, a white needle as high as a 36-storey building, burst into life precisely on schedule at 7.51 am Cape time. At first there was an eerie silence as the orange flames belched from the booster in an apparently vain attempt to lift the monster from the pad. Then, as the rocket slowly accelerated into the clear blue sky, the roar from the mighty engines burst upon the watching multitude, only to be reinforced by the accompaniment of thousands of car horns.

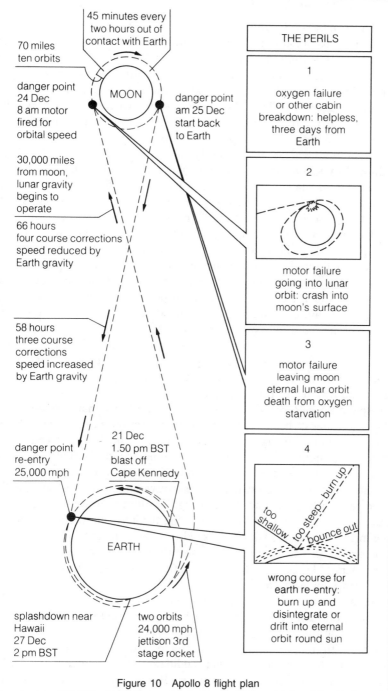

45 minutes every
two hours out of
contact with Earth

70 miles
ten orbits

danger point
24 Dec
8 am motor
fired for
orbital speed

MOON

danger point
am 25 Dec
start back
to Earth

30,000 miles
from moon,
lunar gravity
begins to
operate

66 hours
four course corrections
speed reduced by
Earth gravity

58 hours
three course
corrections
speed increased
by Earth gravity

danger point
re-entry
25,000 mph

21 Dec
1.50 pm BST
blast off
Cape Kennedy

EARTH

splashdown near
Hawaii
27 Dec
2 pm BST

two orbits
24,000 mph
jettison 3rd
stage rocket

THE PERILS

1

oxygen failure
or other cabin
breakdown: helpless,
three days from
Earth

2

motor failure
going into lunar
orbit: crash into
moon's surface

3

motor failure
leaving moon
eternal lunar orbit
death from oxygen
starvation

4

too shallow   too steep – burn up
bounce out

wrong course for
earth re-entry:
burn up and
disintegrate or
drift into eternal
orbit round sun

Figure 10   Apollo 8 flight plan
The six-day, 600,000-mile flight was the first to leave the vicinity of the Earth and
carry men to an area of space dominated by another world.

Thirteen seconds after lift-off, Launch Operations Manager Paul Donnelly was heard to announce, 'Tower cleared', the signal for mission control in Houston to take over from the staff at the Cape. Generating an incredible 7.5 million lb of thrust, the Saturn's first stage engines gradually accelerated the rocket and its three passengers on the first leg of their adventure. The Houston Capcom, Michael Collins, watched anxiously as the 'big bird' carried its 3,100 ton burden to an altitude of 38 miles (60 km) while consuming 15 tons of fuel a second. He watched as separation and second stage ignition were successfully completed, followed by second stage separation and the final boost into Earth orbit by the third stage. Nearly 11½ minutes into the mission, Borman announced third stage cut-off; Apollo 8 was cruising at around 17,400 mph a little more than 100 miles above the Earth.

The next two and a half hours kept the crew and ground staff fully occupied as they checked every system on the third stage, the spacecraft and on the ground. Jim Lovell unstrapped himself from the centre couch to use the scanning telescope and sextant fitted in the side of the spacecraft. Such freedom had not been possible in the cramped Gemini cabin, so it came as a surprise when he began to float 'all over the place' and experienced a queasy feeling in his stomach. Once he slowed down the feeling receded and doctors monitoring his physical condition found no cause for concern. The crew appeared very relaxed as the next critical event approached; a nonchalant Borman told Collins, 'It looks just about the same way it did three years ago.'

Apollo 8 was given the green light for the translunar injection burn, the crucial boost that would kick the craft from the Earth on course for a lunar encounter. Everyone at Mission Control held their breath as the third stage engine re-ignited 2 hours 50 minutes into the flight. For more than five minutes the burn continued as it accelerated Apollo to 24,200 mph: the three astronauts were travelling faster than anyone had ever previously experienced. More significantly, they were on their way to a strange, hostile world and venturing further from home than any of the intrepid explorers who had risked their lives in centuries past. As the friendly Earth began to shrink in size through the spacecraft windows, Chris Kraft, Director of Flight Operations, reassured the crew: 'You're really on your way now.'

In a dummy run of the manoeuvre required for a lunar landing mission, Borman fired the explosive bolts to separate the third stage from the rear of the service module then turned the craft to station-keep briefly with the redundant booster. On later flights the commander would have to withdraw the lunar module from its protective cocoon, but the main priority now was safety, so Borman soon fired the reaction control jets to pull away from the stage – an accidental collision was the last thing to be desired. The booster seemed reluctant to leave them, however, as Borman reported: 'It looks like I might have to do a couple more small manoeuvres to stay away from the front of this S-IVB . . . it sure is staying close. It's spewing out from all sides like a huge water sprinkler.' The stage was venting its unused fuel as it followed less than 1,000 feet (300 m) behind Apollo, a process which did not help Lovell's efforts to make navigational sightings: 'I am looking through the scanning telescope now and I see millions of stars.' A second short burn with the reaction control engines was finally agreed. Borman helped to cheer the ground controllers who hated any departures from their carefully worked out flight plan: 'OK., as soon as we find the Earth we will do it.' Another pre-flight worry was also quickly dispersed when Anders reported negligible radiation readings from their passage through the Van Allen radiation belts which surround the Earth.

The astronauts at last had some leisure time with which to examine the unique view from a distance of 21,000 miles. Lovell enthusiastically described the blue planet he was leaving rapidly behind: 'I'm looking out my centre window, the round window, and the window is bigger than the Earth is right now. I can clearly see the terminator. I can see most of South America all the way up to Central America, Yucatan, and the peninsula of Florida.' Bill Anders, busily photographing the remarkable panorama for posterity, chimed in: 'Tell the people in Tierra del Fuego to put on their raincoats. Looks like a storm is out there . . . You might be interested to know the centre window is pretty well fogged up.'

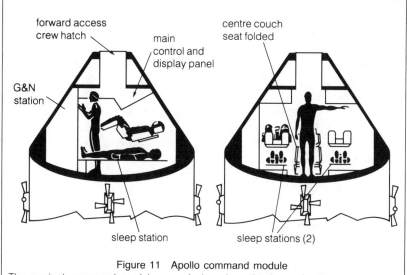

Figure 11 Apollo command module

The conical command module was designed as the home for three astronauts during a week-long lunar mission. It measured 10½ ft in height and 12 ft 10 in in diameter at the base, sufficient to allow men to stand upright and float freely. Cabin pressure in the pure oxygen atmosphere was 5 psi.

As the Earth pulled on Apollo in a vain attempt to prevent its escape, the craft noticeably slowed and the crew prepared for the long cruise through interplanetary space. They set up the 'barbecue roll', officially known as the passive thermal control, which ensured even heat exposure during a once-an-hour roll, then removed their pressure suits to enjoy the relative freedom of the large command module cabin. Following a minor course correction, Capcom Ken Mattingly radioed: 'OK, number one on the list of things is that the flight plan shows commander should hit the sack.' After a 17 hour period of almost continual work, stress and responsibility, Borman should have had little trouble cooperating, but he eventually requested permission to take a sleeping pill. This seemed to have little effect, so Lovell informed Houston, 'We're going to try to keep the conversation down here for a while so the commander can go to sleep.' At the end of his seven hour rest period, it was the turn of Lovell and Anders to get some sleep. They were awakened by a call from Michael Collins: 'The "Interstellar Times" latest edition says the flight to the moon is occupying prime space on both paper and television . . . We understand that Bill Anders will be in private communication today with an old man who wears a red suit and lives at the North Pole.' To the delight of his son, Greg, the astronaut replied: 'Roger, we saw him earlier this morning and he was heading your way.'

As Apollo 8 neared the halfway mark to the moon, Mission Control played back the tape which recorded spaceship telemetry and crew comments while the craft was out of touch with ground stations. It provided a nasty shock as they learned that, some ten hours earlier, Borman had suffered from a bout of nausea, vomiting and diarrhoea. Medical staff were immediately aware of the probability of any infection being passed on to his companions, as well as the hygiene problems in such a confined space. Collins contacted Apollo for an update on his health: 'We are on a private loop and would like amplifying details on your medical problems.' Borman replied: 'Mike, this is Frank. I'm feeling a lot better now. I think I had a case of the 24-hour 'flu.'

Dr Charles Berry, Apollo flight surgeon, took the unusual step of speaking directly to Borman instead of using the Capcom: 'Frank, this is Chuck. The story we got from the tape and from Jim a while ago went like this. At some 10 to 11 hours ago, you had a loose BM, you vomited twice, you had a headache, you've had some chills, and they thought you had a fever. Is that affirm?' Borman reassured the doctor: 'Everything is true, but I don't have a fever now. I slept for a couple of hours and the nausea is gone . . . I think everything is in good shape right now.' Berry persisted: 'As we understand it, Frank, neither Bill nor Jim have anything at the present time except some nausea. Is that right?' Borman replied: 'No, none of us are nauseated now. We are all fine now.' The ground controllers adjourned to an empty conference room for an emergency debate. Was the illness a threat to the mission and the lives of the astronauts? If so, how could the threat be minimized for men who were a matter of days rather than hours from Earth? There wasn't much to go on, but Dr Berry recommended continuation of the flight – he prescribed rest and plenty of fluids. As for Hong Kong flu, the astronauts had been immunized before take-off, so all should be well. Curiously, Borman later changed his mind and blamed the sleeping pill he took.

The doctor's optimism paid off. A few hours later a cheerful, healthy crew presented their first TV broadcast about halfway to the moon. The tiny 4½ lb camera showed Lovell injecting water into a bag of freeze-dried chocolate pudding, followed by Anders demonstrating how a toothbrush floats in zero gravity. Borman made fun of the bewhiskered Lovell: 'Let everyone see he has outdistanced us in the beard race.' He went on to point the camera at the Earth, but it simply appeared as a bright blur on the screens:

> I certainly wish that we could show you the Earth. It is a beautiful, beautiful view with predominantly blue background and just huge covers of white clouds – particularly one very strong vortex up near the terminator. Maybe we can get some help from the ground on this telephoto lens and show you next time.'

Twenty-four hours later, after a day filled with routine star sightings, shifting antennae as the craft turned, updating the computer and trying to sleep, the second broadcast came through. This time millions of TV viewers were treated to a near perfect view of the world they inhabited, a small half-illuminated ball hanging against an empty black backcloth. Lovell described the scene:

> We are manoeuvring now for the TV. The Earth is now passing through my window. It's about as big as the end of my thumb. Waters are all sort of a royal blue; clouds, of course are bright white. The reflection off the Earth is much greater than the moon. The land areas are generally sort of dark brownish to light brown. What I keep imagining is, if I were a traveller from another planet, what would I think about the Earth at this altitude? Whether I think it would be inhabited.

Capcom queried: 'Don't see anybody waving, is that what you are saying?' Lovell replied: 'I was just kind of curious if I would land on the blue or the brown part.' Borman interrupted: 'You better hope we land on the blue part!' A hint of the isolation which the astronauts were experiencing came when Lovell referred to 'your Earth', then corrected himself: 'It's our Earth as well, of course.'

The next landmark came two hours later when Apollo, having slowed to a relatively snail-like 2,200 mph (3,500 kmh), passed into the lunar sphere of gravitational dominance and began to accelerate once more.

Capcom Jerry Carr passed the news to the crew in nonchalant fashion: 'By the way, welcome to the moon's sphere.' The crew were caught off guard: 'The moon's fair?' Carr repeated: 'The moon's sphere – you're in the influence.' 'That's better than under the influence, hey Jerry?' came the reply. The event was marked in much more serious fashion at Mission Control where a spokesman informed the awed press: 'It's an historic moment. For the first time men are literally out of this world, under the influence of another celestial body.'

Two earlier course corrections had been cancelled since they were unnecessary, but the extra velocity imparted by ejection of waste water necessitated a minor burn 20,000 miles (32,000 km) out from the moon in order to slow the craft by about 1.4 ft per second and lower the pericynthion, or closest approach to the lunar surface, from about 80 miles (130 km) to 70 miles (110 km). This completed, everyone began the checks and preparations for lunar orbit insertion. Unless all systems checked out perfectly, no attempt would be made to enter lunar orbit – the craft would simply be allowed to drift around behind the moon on a direct return to Earth which had been deliberately planned as a safety precaution. Capcom Ken Mattingly informed the crew that all seemed 'Go' so they settled down to a rest period before 'the big day'.

The final approach was marked by Borman's surprising revelation to Mission Control: 'As a matter of interest, we have yet to see the moon.' A startled Capcom asked, 'What else are you seeing?' Anders replied: 'Nothing. It's like being on the inside of a submarine.' With the craft in 'barbecue roll' and orientated away from the moon, the only person to have seen their objective had been Lovell during his navigational stints. The fogging of all but two small windows due to the reaction of window sealant with the space environment also restricted visibility.

By the time the main engine fired, 69 hours into the mission, Apollo 8 was hurtling towards the moon at a speed of around 5,000 mph (8,000 kmh). As the craft disappeared behind the moon and out of radio contact, Jerry Carr called: 'All systems GO. Safe journey, guys.' Anders replied: 'Thanks a lot troops. We'll see you on the other side.' Mission Control settled down to the long nail-biting wait; a tense silence pervaded the building as people stared at the display interspersed by nervous glances across the room at each other. If the engine misfired, everyone knew the two deadly alternatives: the craft could plummet out of control on to the barren wastes on the lunar far side or remain eternally in lunar orbit. The minutes ticked by: 25 minutes and no Apollo – at least the engine must have fired or they would be in touch by now; 34 minutes and Carr began calling, 'Apollo 8, Houston'. Finally, after a seemingly eternal interval, Lovell's voice reverberated around the centre: 'Go ahead, Houston. Apollo 8.' The engine had fired perfectly for four minutes to place them in an elliptical orbit of between 70 and 190 miles (110 to 310 km) above the moon. The three astronauts were closer than anyone had ever been to another world: they were also nearly a quarter of a million miles from home, and had experienced a unique solitude, completely cut off from the billions of inhabitants of planet Earth who at that very moment were preparing for Christmas Day.

During the next four hours and two lunar orbits the crew interspersed their initial reactions to the strange surface visible from the windows with analysis of the burn and orbital trajectory. Lovell described his view: 'The moon is essentially grey, no colour. Looks like plaster of Paris. Sort of greyish sand . . . We're getting quite a bit of contrast as we approach the terminator. The view appears to be good, no reflection of the sun back to our eyes.' Their training immediately paid off as they recognized craters and other features, including potential lunar landing sites, while they swooped low over the rugged grey landscape. TV viewers were given a Christmas treat three hours after Apollo entered lunar orbit – a close-up armchair view of another world. The astronauts took it in turns to describe the scene. Borman began: 'The moon is very bright and not too distinct in this area.' Anders continued: 'The colour of the moon looks like a very whitish grey, like dirty beach sand with lots of footprints in it. Some of these craters look like pickaxes striking concrete creating a lot of fine haze dust.' Lovell concluded: 'As a matter of interest, there's a lot of what appear to be small new craters that have these little white rays radiating from them . . . There is no trouble picking out features that we learned on the map.'

During the third pass behind the moon, the main engine was fired once more to circularize the orbit at about 70 miles. The service module was doing them proud. Each man tried to take turns in snatching a few hours rest while his companions continued the navigational checks and photographic surveys. At one point Frank Borman, a lay preacher in his Episcopal Church in League City, Texas, read out a prayer which he asked Capcom to relay to the congregation:

This is to the people at St Christopher's, actually to people everywhere: Give us, O God, the vision which can see Thy love in the world in spite of human failure. Give us the faith, the trust, the goodness in spite of our ignorance and weakness. Give us the knowledge that we may continue to pray with understanding hearts, and show us what each one of us can do to set forth the coming of the day of universal peace. Amen.

A few hours later, Capcom informed the crew of the reaction to their first lunar broadcast: 'We got the "Interstellar Times" here. Your TV programme was a big success. It was viewed this morning by most of the nations of your neighbouring planet, the Earth. It was carried live all over Europe, including even Moscow and East Berlin. Also in Japan and all of North and Central America, and parts of South America.' More vivid descriptions were radioed back to Earth as the momentous flight continued. Anders described the far side of the moon, never before seen directly by human eyes: 'The back side looks like a sand pile my kids have been playing in for a long time. It's all beat up, no definition. Just a lot of bumps and holes. The area we're over right now gives some hint of possible volcanic action.' The continual activity and lack of sleep began to tell, however. At one stage, Lovell accidentally pushed the wrong button on the computer, partially erasing its memory. Ground control spent many hours checking that he had not wiped out the program which controlled Earth re-entry, and Borman was moved to inform Capcom: 'I'm going to scrub all the other experiments. We are a little bit tired. I want to use that last bit to really make sure we're right for TEI [trans-Earth injection].' Five minutes later he reported: 'Lovell is snoring already.' Capcom replied: 'Yeah, we can hear him down here.'

On the ninth and penultimate lunar orbit, the crew sent their second TV broadcast from the moon. It was late on Christmas Eve in the USA and early on Christmas Day in Europe. Borman introduced the show: 'What we'll do now is follow the trail that we've been following all day and take you on through a lunar sunset. The moon is a different thing to each one of us . . . my own impression is that it's a vast, lonely, forbidding type of existence, a great expanse of nothing that looks rather like clouds and clouds of pumice stone. It certainly would not appear to be a very inviting place to live or work.'

Lovell took over: 'Well, Frank, my thoughts are very similar. The vast loneliness up here is awe-inspiring, and it makes you realize just what you have back there on Earth. The Earth from here is a grand oasis in the big vastness of space.' Anders commented: 'I think the thing that impressed me most is the lunar sunrises and sunsets. These in particular bring out the stark nature of the terrain and the long shadows really bring out the relief.' As they headed towards the dark night hemisphere, Anders introduced a memorable Christmas message for the watching millions: 'For all the people back on Earth, the crew of Apollo 8 has a message that we would like to send to you: "In the beginning, God created the heaven and the Earth . . ." ' and over the gulf of black emptiness came a declaration of faith to be shared by countless millions as each astronaut read from the five verses in the book of Genesis. Borman concluded with, 'And from the crew of Apollo 8 we close with goodnight, good luck, a Merry Christmas, and God bless all of you – all of you on the good Earth.'

The time rapidly approached for the vital main engine burn which would kick Apollo out of lunar orbit on to a return trajectory. Once again the tension mounted; a failure now would almost inevitably consign the crew to a lingering death from suffocation far from home. The craft disappeared behind the moon, leaving ground controllers with nothing to do but cross their fingers and pray. The engine again performed perfectly, firing for three minutes to accelerate the craft sufficiently to break away from lunar gravity. Thirty-seven minutes after loss of signal, Houston reacquired the craft. Lovell broke the good news: 'Please be informed there is a Santa Claus!'

The return trip proved uneventful. Deke Slayton wished the crew a Merry Christmas, adding: 'None of us ever expected a better Christmas than this one.' Borman expressed the astronauts' gratitude: 'Thank everybody on the ground for us. It's pretty clear we wouldn't be anywhere if we didn't have them doing it or helping us out here.' The crew celebrated with a turkey and gravy Christmas dinner eaten with a spoon instead of squeezed from a plastic bag; in Mission Control the fare was cold coffee and sandwiches, though a big Christmas tree appeared to brighten the scene. Two more TV shows were broadcast on the homeward leg as viewers were shown more spectacular shots of planet Earth. Anders summarized their emotions:

I think I must have the feeling that the travellers in the old sailing ships used to have, going on a long voyage away from home. And now that we're headed back, I have that feeling of being proud of the trip, but still happy to be going home . . . We'll see you back on that good Earth very soon.

Two small course corrections put them right on target for re-entry on the morning of 27 December. The return trip took nine hours less than the outward leg due to a fast burn out of lunar orbit. The service module was cast off less than 20 minutes before atmosphere re-entry, exposing the heat shield. The command module was about to undergo the most severe test yet given to any manned spacecraft as it entered the atmosphere at more than 24,600 mph (39,000 kmh). The outer skin glowed orange, a 'real fireball', as the heatshield gradually burned away at temperatures approaching 2,800°C. Radio blackout lasted just over five minutes as the craft was surrounded by ionized particles. The craft gradually decelerated while plummeting Earthwards over the dark Pacific Ocean. An airline pilot reported a long orange-red tail across the night sky; inside the craft, the astronauts were pressed back on their couches by nearly 7G at peak deceleration. The recovery carrier *Yorktown* made radar contact followed by the welcome voice of Jim Lovell: 'We are in real good shape, Houston.' At 24,000 feet (7,500 m) the drogue chutes opened, then the three main chutes lit by a stroboscopic beacon flashing on the capsule. Apollo 8 'hit with a tremendous impact' less than 3 miles from the *Yorktown*. Almost immediately a naval helicopter was over the capsule, dropping flares and waiting for daylight. The command module turned over, leaving the astronauts dangling upside down in their seats for almost five minutes until the flotation bags righted it. One of the crew was seasick as the blistered craft bobbed in 4 ft waves. Nevertheless, the mood was one of elation. One of the helicopter skippers asked if the moon was really made of green cheese. Anders replied: 'No, it's made out of American cheese.'

Dawn broke some 20 minutes after splashdown, enabling frogmen to attach a flotation collar, but another 40 minutes went by before the hatch opened. First into the life raft was Lovell, followed by Anders, and finally the commander. They were lifted one by one into a recovery helicopter using a special 'Billy Pugh net', then taken to the nearby carrier. Borman used the electric razor he had requested, but his companions stepped on to the flight deck proudly sporting their six-day beards. Millions of TV viewers saw the moment of triumph and heard the astronauts praise the carrier's crew and all those who had made the flight possible. A quarter-ton celebration cake had been prepared on the carrier, but the astronauts asked for a breakfast of steak and eggs, 'the same as we had before we left'. Below deck, they were given a preliminary medical and reported 'completely alert and happy: they are not as tired as astronauts have been on previous flights.' President Johnson phoned his congratulations – the last time before his retirement from office. Meanwhile, at Houston there was a scene of unparalleled rejoicing: miniature American flags appeared everywhere, cigars lit up at every console and pandemonium broke loose. It looked as if the USA would meet Kennedy's deadline for a lunar landing after all.

The Apollo 8 mission had, indeed, broken new ground; the bold gamble had paid off. The lunar hardware had been tested in the most exacting manner and not been found wanting. Equally as significant, three men from planet Earth had successfully ventured outside the sphere of influence of their world for the first time, giving millions of people a new perspective on their small 'oasis in the vastness of space'. NASA Acting Administrator Dr Thomas Paine described the flight as 'one of the great pioneering efforts of mankind. It is not the end but the beginning. We are here this morning at the onset of a programme of space flight that will extend through many generations.'

As for the astronauts, Frank Borman spoke to Congress about his hopes for 'an international community of exploration and research' on the moon in the near future, but he did not see himself as part of that community. He was sent on goodwill visits to the Far East and Europe, including a week-long trip to the Soviet Union when he collected medals commemorating dead cosmonauts which would be left on the moon. In 1970 the President sent him on a worldwide tour to seek support for the release of American prisoners of war held by North Vietnam. Later that year he completed an advanced management programme at Harvard Business School and joined Eastern Airlines as a vice president. In 1976 he became chairman of the board, and guided the company through severe financial trouble in the 1980s.

Bill Anders never flew in space again, though he remained involved in space development policies as Executive Secretary of the National

Aeronautics and Space Council. In 1973 he moved to the five-member Atomic Energy Commission, later becoming the first Chairman of the newly established Nuclear Regulatory Commission. There followed a short spell as American ambassador to Norway before he finally left the federal government to join the General Electric Company. Jim Lovell remained on NASA's active list. He was quickly appointed as Apollo 11 back-up commander, then assigned to the ill-fated Apollo 13.

While the Apollo 8 astronauts were enjoying ticker-tape parades before millions of people and embarking on a three-week goodwill tour of Europe to universal acclamation, the crew of Apollo 9 were entering the final stages of a training programme which had begun in December 1966 and been dogged by set-backs and tragedy. The trials of the crew's patience were not over yet: the original launch date passed with the astronauts grounded due to head colds. For an Earth orbit mission like Apollo 9 the change of launch date to 3 March had little significance, but on a lunar mission it would necessitate a complete change of plan, either a month-long delay or a choice of a different landing site. Was it merely coincidence that the crews on the Apollo 7, 8 and 9 missions had contracted minor ailments? The medical staff didn't think so, and cautioned the NASA management over the long, arduous training sessions which they felt made the men run down and more susceptible to infection.

Public and media interest in the forthcoming flight moderated after the spectacular Apollo 8 trip to the moon. Another ten-day Earth orbital mission seemed to lack the danger and excitement of a journey to another world. To those in the know, however, Apollo 9 was a vital step on the road to a lunar landing. Dr Wernher von Braun, head of the Saturn V design team, described the mission as 'at least 25 per cent more difficult' than any previous flight. The main addition to the mission profile was the lunar module, the 'tissue paper spacecraft' as Jim McDivitt called it. This strange-looking craft, christened 'the lunar bug' by the astronauts, had been extensively tested on the ground but had only flown once in space during an unmanned test flight on Apollo 5 during January 1968. If the module failed to pass the exhaustive trials planned in Earth orbit, the lunar landing mission set for 1969 could be jeopardized, perhaps permanently.

The Apollo 9 commander, 39-year-old James McDivitt, and the command module pilot, 36-year-old David Scott, were both Air Force colonels and veterans of the Gemini programme. They were a good, experienced pair, McDivitt the thorough, conscientious pilot who turned down the chance of worldwide acclaim by refusing the Apollo 8 flight and Scott, the guidance and navigation specialist who had participated in the first-ever docking of two spacecraft. Alongside them was a group 3 newcomer, 33-year-old civilian Russell Schweickart. He had served as a pilot in the USAF and Air National Guard in the late 1950s, having graduated with a BSc in aeronautical engineering. Prior to joining NASA he had been involved in research at the Experimental Astronomy Laboratory at the Massachusetts Institute of Technology where he studied upper atmosphere physics and star tracking. He also completed his master's thesis in 1963. To this lively, inquisitive man fell the job of flying a lunar module for the first time. Since two craft would be flying in formation, NASA recognized the need for separate recognition names, so, for the first time since Gemini 3, Gus Grissom's 'Unsinkable Molly Brown', the crew were allowed to name their craft. The management may have had second thoughts when the crew announced their choices: the squat, long-legged LM was dubbed 'Spider', while the CSM became known as 'Gumdrop', a reference to its appearance when first viewed by the astronauts draped in blue tape.

The morning of 3 March 1969 saw everything go like clockwork. Following breakfast the crew suited up and transferred by bus to Pad 39A, entering their cabin over two hours before the scheduled launch. Blast-off occurred exactly on time at 11 am Cape time, the mighty Saturn V lifting the heaviest load to date through the overcast skies and into Earth orbit 11 minutes after launch. Immediately, the crew checked the orbit – an almost perfect 118 mile (190 km) circle – and systems before preparing for the first task of a busy five days ahead of them. The crowded mission profile had been worked out to ensure completion of the major mission objectives even if the flight had to be cut short.

The CM docking probe was extended shortly after reaching orbit, then, 2 hours 45 minutes after launch, Dave Scott fired the explosive bolts which separated the CSM from the third stage of the Saturn V and removed the panels which protected the LM during launch. The craft pulled away to a safe distance, turned and approached nose first, the pilot aiming at a black circle marked on the outer skin of Spider. Despite minor problems with the computer and a thruster, the CM probe was successfully mated to the LM drogue and the two craft were locked together. Once the docking tunnel was pressurized with oxygen, the forward hatch was opened, allowing McDivitt and Schweickart to link the LM to the CM electrical power lines. Scott was now able to extract the LM by backing away, placing the combined craft well clear of the booster stage. This S-IVB was restarted on ground command and fired into an eternal orbit around the sun. NASA officials described the docking manoeuvres as 'smooth and without oscillations'; one more step had been taken towards the moon. The first test firing of the CM main engine proved the stability of the linked craft while lengthening the orbital period. A hectic nine hours in orbit came to an end with the first sleep period: in a change of procedure, the astronauts were all allowed to

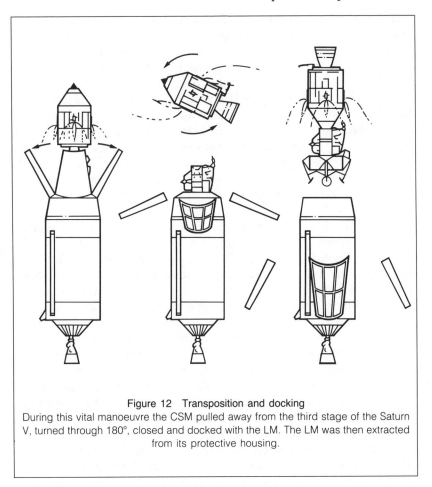

Figure 12   Transposition and docking
During this vital manoeuvre the CSM pulled away from the third stage of the Saturn V, turned through 180°, closed and docked with the LM. The LM was then extracted from its protective housing.

sleep at the same time in order to leave them fresh for the next day.

The second day began with the crew reporting they had slept well, despite a faulty cabin alarm which had caused a few pulses to race. Engineers on the ground had also been busy repairing a tracking computer which registered an inaccurate velocity for Apollo. The main tasks of the day were three more burns of the main engine to test the engine itself, the stability of the two linked craft and the autopilot, as well as changing the orbit and lightening the fuel load. The latter weight reduction was important in making the CSM easier to handle should a rescue mission be required when the Spider was flying independently. All three test burns proved successful, eventually raising the orbital apogee (high point) to 300 miles (500 km). Apart from these successes, there was little for the media to report: the crew were so busy that there were few verbal exchanges and often a complete orbit passed with little contact between Apollo and Houston.

Next day was dedicated to checking out Spider for the first time, but the astronauts' efforts were hampered by space sickness, an ailment familiar to their Soviet colleagues but unknown in American flights until the Apollo series. Sceptics might say that the strange menus did little to help: breakfast that day consisted of fruit cocktail, bacon squares and cocoa. The main victim was Russell Schweickart, a poor traveller at the best of times – he was the only crew member to take a motion sickness pill before launch. The struggle to don his bulky pressure suit in the weightless cabin proved too much for his digestive system. Not surprisingly, perhaps, McDivitt was also feeling rather queasy by this stage.

Eventually, Dave Scott was able to open the hatch to the pressurized tunnel and remove the docking gear. Once cleared, Schweickart was able to squeeze through the 32 inch wide tunnel and into Spider. This action alone could have a peculiar disorientating effect since he was entering a craft in which the 'floor' was facing in the opposite direction to the floor of Gumdrop. Schweickart began to power up the electrical system and check out Spider's environmental control systems, though he still wasn't feeling too good. Like Borman on the previous mission, he took advantage of the private communication link to Houston to inform them of his condition. About an hour after he entered Spider, he was joined by McDivitt, enabling the LM to be sealed off from Gumdrop for the first time as an operationally independent craft. Communications seemed fine, the guidance and navigation system was working properly, so the four landing legs on Spider were deployed, a necessary move even though there would be no landings this time. McDivitt's concern over his partner's health was not eased when Schweickart vomited once more, though he did say that he felt much better after this. The doctors on the ground were puzzled: Schweickart's pulse rate had previously been consistently low, around 70 beats per minute, compared with peak rates of 120 by his two companions. There would have to be serious discussion over the space walk scheduled for the following day.

With the two craft drifting freely, the astronauts in Spider tried brief firings of its thrusters while standing upright at their posts, strapped tightly in position. Happy with their craft, the two men took a break to give a five minute TV broadcast, using a hand-held camera to show viewers the inside of the LM for the first time. Although the pictures clearly showed 'Rusty' Schweickart wearing the portable backpack which lunar explorers would wear and giving a thumbs-up sign, the voice communications almost entirely broke down. The flight director interrupted McDivitt: 'I can see you talking there, Jim. Too bad I can't read your lips.'

A couple of hours later, the astronauts were ready for the major test of

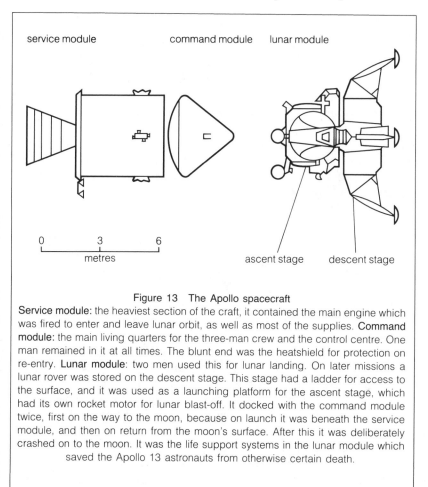

service module   command module   lunar module

0   3   6
metres

ascent stage   descent stage

Figure 13   The Apollo spacecraft

Service module: the heaviest section of the craft, it contained the main engine which was fired to enter and leave lunar orbit, as well as most of the supplies. **Command module:** the main living quarters for the three-man crew and the control centre. One man remained in it at all times. The blunt end was the heatshield for protection on re-entry. **Lunar module:** two men used this for lunar landing. On later missions a lunar rover was stored on the descent stage. This stage had a ladder for access to the surface, and it was used as a launching platform for the ascent stage, which had its own rocket motor for lunar blast-off. It docked with the command module twice, first on the way to the moon, because on launch it was beneath the service module, and then on return from the moon's surface. After this it was deliberately crashed on to the moon. It was the life support systems in the lunar module which saved the Apollo 13 astronauts from otherwise certain death.

the day, a six minute burn of Spider's descent engine with the autopilot controlling it over a wide range of thrust settings, followed by a final manual test. McDivitt was amazed at the accuracy and control which he achieved at this first attempt, just as if he were still operating the simulator. The LM crew proceeded to shut down Spider and return to the waiting Dave Scott. They had spent a mere eight hours inside the LM, but all seemed ready for the most dangerous manoeuvre, the separation of the craft in two days time. Before they could relax, however, another burn of the main engine on Gumdrop had to be completed in order to circularize the orbit. Only then could the astronauts remove their suits and retire for a well-deserved rest. McDivitt warned ground control to be prepared for delays the next day 'because it is really a scramble trying to get suited, and once you get suited you become all tangled up in these hoses'.

The fourth day should have included a two hour space walk during which Schweickart would practise emergency procedures by exiting from Spider's 'front porch' then pulling himself hand over hand along specially provided rails to the open hatch on Gumdrop. Such a procedure would normally only be tried as a last resort if the docking tunnel became blocked and impassable. McDivitt had feared a recurrence of Schweickart's nausea, suggesting instead a simple test of the new life support system with the hatch open but no EVA. A third possibility came to the fore when Schweickart reported feeling much better on awaking: he would be allowed to stand on the porch, taking photographs while trying out the suit and backpack. Covering the pressure suit was a special garment intended to protect him from micro-meteorites and radiation. The backpack, weighing 183 lb on Earth, was designed to provide communications, supply him with oxygen and circulate cooling water through his suit. The only link with Spider was a nylon cord to keep him from drifting off into space.

While McDivitt remained inside, Schweickart floated out through the LM hatch, the door which would one day allow the first men to step out on to the moon. He carefully placed his feet inside the restraints provided on the porch, known to the astronauts as 'the golden slippers', and greeted Dave Scott who had poked his head and shoulders out from Gumdrop's hatch. 'Boy, oh boy, what a view,' exclaimed Schweickart. 'Isn't that spectacular?' agreed Scott. Looking around the black sky, Schweickart spotted the bright star Aldebaran in Taurus: 'You can see that bull's eye. You can really see it at night!' Down below he was able to recognize Baja California. During the 37 minute EVA, Schweickart retrieved some thermal samples from the exterior of Spider, took still and movie pictures of Dave Scott and the two craft, tested the handrails without launching himself away from the porch and generally had a good time. The suit and backpack passed with flying colours, Schweickart's good-humoured banter coming across loud and clear. Back inside Spider with the hatch closed once more, the LM crew gave another ten minute TV show with improved voice quality this time. Then they returned to Gumdrop to prepare for the mission climax next day.

The fifth day of almost ceaseless activity saw the astronauts enter Spider once more; to save time, they were allowed to leave off their helmets and pressure hoses, only donning these when they were ready to pull away from Gumdrop. The docking gear was re-installed in the tunnel, then the hatches were closed. The separation got off to an inauspicious start – when Scott pressed the latch release switch, nothing happened. However, he tried again and breathed a sigh of relief when this time it worked. Scott backed Gumdrop gently away from Spider, then circled around on an inspection tour of the LM before its maiden flight. Satisfied with what he saw, Scott backed away to a safe distance as his colleagues completed their final checks before firing the descent engine for 19 seconds. Spider moved steadily to a distance of 14 miles (22 km) above the passive Gumdrop. Another short burn circularized the orbit, enabling Spider slowly to drop further behind, eventually drifting out of sight more than 100 miles from the quietly watching Dave Scott – the strobe light which was designed to help Scott pick up the other craft had gone out.

The scene was now set for a rehearsal of a lunar take-off: the LM crew had to make up more than 100 miles (160 km) of lost ground, rendezvous and dock with the waiting Gumdrop. The descent stage was successfully jettisoned, pieces of foil insulation flying past the windows in all directions. Now for the crucial firing of the ascent engine, a mere three seconds to create a fast, low orbit which would enable Spider to catch up with Gumdrop. Using the LM's radar and navigation instruments, the crew gradually closed in on Gumdrop while Scott tracked them with his sextant in case an emergency rescue was necessary. Closing to within 100 feet, McDivitt noted that Spider appeared upside down relative to Gumdrop: 'One of us isn't right side up.' Scott commented: 'You're the biggest, friendliest, funniest looking spider I've ever seen.'

The final docking proved the most difficult part of the separation. McDivitt, still strapped in a standing position, had to crane his neck to view the target on the brightly illuminated Gumdrop. The astronauts were relieved when the drogue finally slotted into position, enabling Scott to draw them together in a hard dock. 'OK Houston, we are locked up,' radioed Scott. On hearing the welcome signal tone to indicate docking completion, McDivitt exclaimed: 'Wow, I haven't heard a sound that good for a long time. That wasn't a docking, that was an eye test.' The final manoeuvres had been so difficult that NASA switched the responsibility for them on future flights into the hands of the command module pilot, who enjoyed a much clearer view.

Once the crew had closed down the LM and crawled back into Gumdrop, Spider's ascent engine was fired for the last time by remote control. During a prolonged six minute burst, the friendly Spider was boosted into a high orbit, eventually to burn up in the atmosphere. The thin aluminium skin and non-aerodynamic shape were fine for the vacuum of space or the airless moon, but also ensured that no lunar module would ever fly back to Earth. In five breathless, hectic days, the crew of Apollo 9 had ensured that the LM would be given its certificate of moonworthiness and had achieved every major mission objective. The remaining five days would be much more relaxed as they proved that the command module was equally successful in its role as a home for three men during a 6 million mile voyage around the world.

The second half of the mission was all downhill after the climax of the fifth day, apart from some worried frowns when Gumdrop's main engine failed to fire during a routine burn to lower the orbit. This engine was the only means the crew had of slowing the craft and returning safely to Earth. After a rather tense 90 minutes at Mission Control in Houston, the engineers managed to sort out the problem. Two days later the engine fired perfectly to place Apollo in a suitable orbit from which to initiate re-entry. Houston's only major concern now was a higher fuel consumption than expected, so all unnecessary manoeuvres were banned. This was of little concern to the crew: they were happily carrying out a relaxed programme of landmark tracking, Earth resources photography, navigational alignment checks and general housekeeping. A particularly interesting aspect of their work involved use of four onboard 70 mm cameras which took black-and-white, colour and infrared pictures in an experiment designed to further studies from space of the Earth's land and water resources. They also experimented in using the planet Jupiter as a navigational reference, the first time a planet had been used to guide a spacecraft, and perfected a new way of orientating a spacecraft using the autopilot instead of the inaccurate, time-consuming manual method used previously. At one point they reported sighting the ascent stage of Spider at a distance of more than 750 miles (1,200 km); they had earlier glimpsed a Pegasus research satellite glinting in the sunlight more than 1,100 miles (1,800 km) distant.

The planned 150 orbit mission had to be extended by one when Houston informed the crew on 12 March of bad weather in the Atlantic recovery zone. McDivitt was unconcerned: 'Hey, let's go there.' The site was shifted about 500 miles to the south, a change which meant they would stay in space 97 minutes longer than planned. The main engine fired for the last time as Apollo passed over Hawaii at 11.30 am Cape time on 13 March. Re-entry, as usual, went perfectly so that 30 minutes later the command module dropped into a choppy ocean only 3 miles (5½ km) from the carrier *Guadalcanal*. For once, the craft stayed upright, but nevertheless, the astronauts were not destined to keep their feet dry. During the transfer to a life raft, McDivitt and Scott climbed into a helicopter recovery basket only to be soaked when the chopper's down-draft overturned their raft. Fortunately, they were soon whisked away to a triumphal celebration on the carrier. President Nixon, not entirely truthfully, expressed his admiration in glowing terms: 'The epic flight of Apollo 9 will be recorded in history as ten days that thrilled the world.'

The euphoria which should have filled NASA and the nation after such a successful mission, opening the way to a lunar landing later in the year, was tempered by the attitude of the nation's new president and his administrators. Despite the trumpeting praise of President Nixon, there was an increasing awareness of the great cost of the Apollo programme – each mission cost more than 300 million dollars – so that, even while

Apollo 9 continued on its momentous way, there was talk of Cape Kennedy and its dormitories becoming ghost towns once the moon had been reached. Certainly NASA was under increasing pressure to cut costs at the same time as it was appealing to Congress for funds to initiate new programmes of manned space exploration envisioned for the 1970s and 1980s. Speculation abounded that the final test flight in lunar orbit, Apollo 10, set for May would become the lunar landing attempt. Conservative counsels prevailed, however, as the wiser heads warned of the dangers threatened by the strange mass concentrations (mascons) on the moon which could alter spacecraft orbits, and the problems in lifting an overweight lunar module off the lunar surface. There seemed little point in risking everything simply to save a couple of months and a few hundred million dollars.

The Apollo programme was now in full swing. If Apollo 10 performed to expectations, the lunar landing would follow in July with Apollo 11. The crew for a second landing before the year was out was announced in April. For the astronauts who would carry out this ambitious itinerary there was a continual scramble to be fully prepared, to obtain sufficient practice on the hard-pressed simulators, just as there was pressure on the engineers to check and clear their multi-million dollar charges so that they would not blow up or strand their passengers. For many of the team the Apollo missions marked the end of their association with NASA.

Of the Apollo 9 crew only Dave Scott flew again and had the thrill of actually stepping on to the surface of the moon. Jim McDivitt soon became Manager of the Apollo Spacecraft Programme, retiring from NASA and the Air Force during the run-down of the lunar programme in June 1972. He entered the world of business, eventually becoming a senior vice president of Rockwell International, one of the prime contractors for the Space Shuttle. Russell Schweickart remained with NASA into the late 1970s, serving as back-up commander for the first Skylab mission in 1973, then spending three years at NASA head-quarters in Washington DC. He finally left NASA in 1979 to become Chairman of the California Energy Commission.

The next flight would be the final dress rehearsal for the moon landing. For the first time the lunar module would be test flown close to the lunar surface, the environment for which it was designed, and to that extent it could validly be described as the most dangerous mission yet planned. Apollo Programme Manager George Low summarized the mission objectives: 'The important thing on this flight is to get more LM experience, and also to get that experience at moon distance.' The names given to the craft by the crew once more horrified the NASA Public Affairs Department, though they did have a special significance for the astronauts. The command module was named Charlie Brown while the lunar module was named after his sidekick, the 'World War I flying ace', Snoopy the beagle. In the Manned Space Center at Houston, Snoopy had become the symbol of excellence, with those who did outstanding work being awarded a silver Snoopy pin. For this mission, Snoopy would be required to swap his scarf and goggles for a space helmet. The authorities bore with the astronauts' choice, though determined to ensure that more reverent names would be chosen on the forthcoming historic missions.

The crew who prepared for launch on the morning of 18 May 1969 was one of the most experienced ever launched. The command module pilot was John Young, the cool customer who had already flown twice in space. Mission commander was another rendezvous specialist, Tom Stafford. Since 1966 he had been a lead member of the Apollo mission planning group and a vociferous proponent of the safety first viewpoint concerning the lunar programme. Alongside him in Snoopy would be his partner from Gemini 9, the lunar module pilot Eugene Cernan. He, too, was well prepared since he had served as back-up LM pilot on Apollo 7, even though the lunar module production line had been bedevilled by delays and ensured that his Snoopy was too heavy to risk a lunar landing.

The launch of Apollo 10 was threatened by some technical problems, but these were cured to enable take-off only half a second late – a delay which earned some wisecracks from the assembled reporters – at 12.49 pm. The first stage ignited with orange flame erupting from the base. Within a few minutes it had disappeared through the high cloud which partly obscured the sun. The crew reported that all was well and seemed exhilarated sitting on top of the incredible power of the Saturn V. 'Really a fantastic ride', commented Stafford. 'Just like old times', added Young.

After nearly 12 minutes flight Apollo was successfully placed in a near circular orbit 116 miles (185 km) above the Earth. After the shaking of the launch Young was moved to ask: 'Charlie, are you sure that we didn't lose Snoopy on that staging?' However, apart from a failure in an evaporator, no faults were detected during the intensive checks which they carried out over the next two orbits. Over Australia the third stage engine was restarted for nearly six minutes to boost Apollo on to a moon trajectory. One again, the ride was described as 'rocky' but the crew stuck it out, Stafford calmly reporting, 'We're on our way'. Ground control reassured them: 'It looks good.' In a repeat of the Apollo 9 docking and transposition manoeuvre, John Young swapped couches with Stafford in order to separate from the third stage, pull about 50 feet (15 m) ahead, turn a half-somersault, inch back towards Snoopy still nestled within the stage, dock and finally withdraw the lunar module. The approach to Snoopy and the docking were shown in the first colour TV broadcast from space. Gene Cernan was operating a new 15 lb colour TV camera which gave another dimension to the 19 broadcasts sent back over the next eight days. Two hours later, viewers were treated to a breathtaking geography lesson of the USA as Cernan pointed out the Rockies, Baja California, cloud-covered Alaska and stormy New England. Pointing the camera through a window, he was able to show the remnants of the third stage dropping back on completion of their task.

Further entertainment followed on succeeding days as the crew turned the mission into a public relations officer's dream with their friendly, chatty outlook and varied, sometimes unscheduled transmissions. Between routine housekeeping chores Houston read up their horoscopes, while the crew played back their music selection including, of course, 'Fly Me To The Moon' and 'In My Beautiful Balloon'. In one impressive demonstration, Stafford was seen to move an upside down John Young with little more than a touch of one hand: Young quipped, 'I do everything he tells me'. The high spirits of the crew continued as Charlie Brown carried them smoothly into the lunar sphere of influence, requiring only one of the scheduled course corrections. The only minor irritations came when the thrusters made so much noise that they found difficulty in sleeping, and when the drinking water became overchlorinated causing some to be dumped overboard. Mission control later attributed some of the cheerfulness to the variety of food on board, for, in addition to the usual freeze-dried items, they were provided with individually wrapped bread – a far cry from the controversy over the corned beef sandwich on Gemini 3 – and ham, chicken or tuna salad fillings. Interestingly, there was no problem with space sickness. On this mission, the crew had been asked to spin their drinking water a day before they used it in an effort to remove any traces of hydrogen gas, a possible cause of space sickness according to NASA medical experts.

On the fourth day Apollo 10 swung around the moon for the first time. As on Apollo 8, the crucial lunar orbit insertion burn had to take place behind the moon and out of contact with the Earth. Mission control radioed: 'Everybody here says God speed.' Tension mounted as signal was lost and silence reigned for 35 minutes, but Apollo reappeared right on schedule. Houston breathed a sigh of relief as the Capcom reported: 'AOS [acquisition of signal]. It was a very good burn.' Stafford called in: 'You can tell the world that we have arrived. The guidance was absolutely fantastic.' The main engine had fired for nearly six minutes to slow the craft into an elliptical orbit 69 miles by 196 miles (110 × 315 km) above the moon. Two orbits later, the engine fired again to circularize the orbit at around 69 miles (110 km) altitude.

Almost immediately the crew launched into descriptions of the lunar surface. They differed with the Apollo 8 crew in the precise colour of the moon: rather than 'grey and white like pumice stone' they reported it was 'brownish–grey' while John Young described the far side this way: 'It is shades of black and white and browns in there.' He spotted a couple of features which resembled volcanoes and were 'all white on the outside but definitely black inside' while Stafford said the ridges crossing

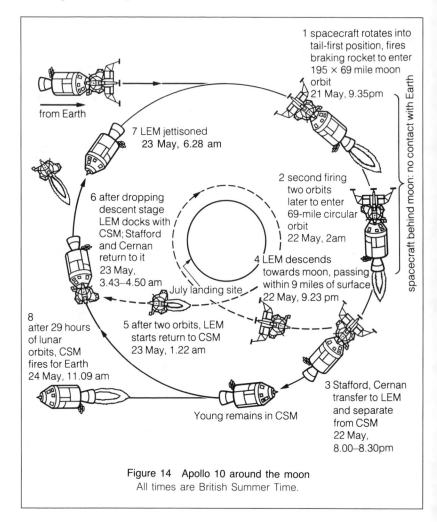

1 spacecraft rotates into tail-first position, fires braking rocket to enter 195 × 69 mile moon orbit
21 May, 9.35pm

spacecraft behind moon: no contact with Earth

from Earth

7 LEM jettisoned
23 May, 6.28 am

2 second firing two orbits later to enter 69-mile circular orbit
22 May, 2am

6 after dropping descent stage LEM docks with CSM; Stafford and Cernan return to it
23 May, 3.43–4.50 am

4 LEM descends towards moon, passing within 9 miles of surface
22 May, 9.23 pm

July landing site

8 after 29 hours of lunar orbits, CSM fires for Earth
24 May, 11.09 am

5 after two orbits, LEM starts return to CSM
23 May, 1.22 am

Young remains in CSM

3 Stafford, Cernan transfer to LEM and separate from CSM
22 May, 8.00–8.30pm

Figure 14   Apollo 10 around the moon
All times are British Summer Time.

the relatively dark, flat mare regions went 'straight down just like the Canyon Diablo in New Mexico'. When the TV camera was unstowed on the third orbit, armchair viewers on Earth were amazed by the first colour scenes transmitted from another world. Mission control described the pictures as 'fantastic – unbelievable' when the sandy brown desert swept across the screen, revealing circular craters of all sizes and black shadows etched into the rugged highlands as Apollo approached sunset. The primary landing site on the Sea of Tranquillity was shown as fairly smooth and free of boulders though pockmarked with a scattering of small, shallow craters.

Thursday, the day of the descent towards the lunar surface, began with a series of unexpected problems. Gene Cernan discovered the first surprise when he made a preliminary check-out of the tunnel and the LM: 'Would you believe we have been living in what you might call snow for three days, and we found out where the rest of it is – it is in our good friend Snoopy.' Flakes of mylar insulation from the docking tunnel had come loose earlier in the flight, necessitating an emergency cleaning session by the astronauts. Stafford crossed into Snoopy to help with the clean-up while Cernan continued his preliminary checks. Ground control suggested the crew turn on their fans and use wet towels to mop up the remaining fragments. Young got some in his mouth and complained: 'It tastes like fibreglass.'

Everyone's fears proved justified when the astronauts tried to close the hatch on Snoopy, prior to separation. This could not be achieved until the oxygen pressure in the tunnel was lowered, but Young reported: 'I am not able to vent the tunnel.' One of the vents had become blocked by insulation particles. The astronauts had to carry out a back-up procedure which bled excess oxygen into Snoopy instead of pumping it back into Charlie Brown. Further cause for concern arose when ground control informed the crew of a list in the alignment of the two craft; Snoopy seemed to have slipped about three degrees out of line, probably during docking. There was a possibility that an attempt to undock might shear the docking latches which held the craft together, thereby making redocking impossible. With time running out before Apollo was due to disappear behind the moon and separate into independent craft, Capcom informed the crew that separation should be possible as long as the list did not exceed six degrees. With that, Apollo was on her own, leaving the controllers to wait anxiously for reacquisition of signal.

Thirty-six minutes later, the Madrid tracking station picked up two craft flying in formation as they rounded the east limb of the moon. As Snoopy stood ready for action with its legs deployed, Young showed TV audiences the weird-looking craft in its home environment. Some 25 minutes later, after 'station-keeping' only 50 feet apart as they circled at 3,000 mph, Young showed Snoopy pulling away on the first leg of the long descent towards the surface. Communications had noticeably become shorter and more businesslike as the critical burn approached. At one point, Cernan was heard to comment: 'There are so many things to do in such a short time.' The astronauts counted down two minutes to go, but Houston corrected them to one minute, with the reassuring words: 'Big Brother is watching you.' Should the descent engine burn for too long, there was a danger of crashing into the moon unless the crew could fire the engine again to regain a safe altitude – everyone was more than aware that such a manoeuvre had never been tried before. As Snoopy shrunk to a tiny dot against the arid background, Cernan told Young: 'Have a good time while we're gone, baby.' Stafford added: 'Adios. We'll see you back in about six hours.' John Young would spend those apparently interminable hours drifting alone around the barren,

inhospitable moon, more isolated than any man since Adam. 'You have
no idea how big the thing gets when there is no more than one man in it,'
he radioed. 'You have no idea how small it looks when you are as far
away as we are,' replied Cernan.

Snoopy's engine did not let them down. A perfect 27 second burn sent
them swooping to within 9 miles of the surface, a frustrating case of so-
near-and-yet-so-far which did not reflect in Cernan's whoop of exhilaration: 'We're down among 'em, Charlie!' When asked for a report
on the burn, he replied, 'Yeah. Soon as I get my breath!' The crew
alternated their quick-fire descriptions of what they saw as the moon
sped beneath them. Cernan spotted the Earth rising above the lunar
horizon: 'Ah, Charlie we just saw Earthrise and its got to be
magnificent.' The Sea of Tranquillity seemed very rocky: 'There's
enough boulders around here to fill up Galveston Bay.' However, the
proposed landing site was 'pretty smooth, like wet clay, like a dry river
bed in New Mexico or Arizona'. Much of the communications from
Snoopy were lost amid a storm of static, though John Young did his best
to keep Houston informed. Another annoying failure occurred when the
automatic cameras jammed, so failing to photograph one of the main
landing sites.

The communication problem was particularly worrying since Houston
lost the stream of coded data which was normally fed straight into its
computers to give an accurate update on the craft's condition. This
omission nearly proved fatal as Snoopy swung down towards the moon
for the second time and the crew prepared to jettison the descent stage
before returning to Charlie Brown. As the descent stage first refused to
budge, then shot off into space, the ascent stage gyrated wildly for
around eight seconds until Stafford took manual control and stabilized
the craft. 'Son-of-a-bitch!' exclaimed Cernan at the sudden, unexpected
movement. 'There's something wrong with the gyro . . . I don't know
what the hell that was, babe . . . I'll tell you there was a moment there,
Tom.' Stafford and everyone else listening in didn't need Cernan to
complete the sentence – it had been a damn close run thing. Controllers
on the ground saw their heartbeats race, but they decided to carry on
regardless: 'Let's worry about it when we make this burn . . . OK
Charlie, I think we got all our marbles . . .' Later analysis showed that a
switch had inadvertently been left in the wrong position due to an
omission on the check-list. As a result, the back-up guidance system had
come into operation, causing Snoopy to swerve wildly as it sought out
the CSM. Once the crew switched to the primary guidance system, no
further problems were encountered.

More than three hours after initiating the descent, Snoopy's ascent
engine was fired to start the LM on the long, difficult journey back to
Charlie Brown, still circling more than 300 miles (500 km) ahead and
above. It took another two and a half hours to bring Snoopy within

40 miles of its target as Stafford and Cernan tracked Charlie and calculated the thruster firings needed. Another hour passed as they edged ever closer, the final approach and docking taking place behind the moon with Snoopy passively waiting for Young to complete the link. Appearing once more around the lunar limb, a relieved Stafford informed Houston: 'Snoopy and Charlie Brown are hugging each other.' In a delighted control centre, a large cartoon appeared showing Snoopy kissing Charlie Brown: the accompanying balloon read, 'Smack. You're right on target, Charlie Brown.' Cernan was equally delighted: 'Man. We is back home. That rendezvous was the best thing we ever had.' They had been inside Snoopy, standing upright, for 12 hours and had flown the LM independently for eight of those. No wonder Cernan commented, 'Man, I'm glad I'm getting out,' as he struggled through the tunnel to greet John Young. Equipment was transferred back inside Charlie Brown, including the faulty camera, then preparations were made to jettison Snoopy on the next pass behind the moon. Soon after separation, Earth stations commanded Snoopy's engine to fire it into an eternal orbit around the sun. Cernan watched it leave with regret: 'I feel sort of bad about that, because he's a pretty nice guy. He treated us pretty good today.'

That was not quite the last they saw of Snoopy. After a long and well-deserved rest, Stafford awoke next day to a bit of a scare when he peered from the window to see the descent stage lingering a little too close for comfort. Mission control spent some time in anxious calculations before reassuring him that Snoopy would gradually drift away long before they were due to fire the engine to return. Most of the day was spent in final descriptions of the lunar surface and in tracking landmarks. Two scheduled TV broadcasts were cancelled to give them time to recover, but, following another rest period, the crew sent back one more spectacular show from the moon. A minor irritation came from some fibre particles still floating around the cabin; they reported them getting under their suits and causing an itch, as well as affecting their noses and throats.

Two orbits after the TV show, Apollo passed behind the moon for the last time; the crew had completed 31 revolutions in a memorable 61½ hour survey of Earth's natural satellite. The faithful main engine fired for 2 minutes 44 seconds, three seconds longer than scheduled, to kick Charlie Brown out of orbit and on to Earth trajectory. Anxious controllers uncrossed their fingers as they picked up a burst of static followed by Stafford reporting: 'Houston, we are returning to Earth. The burn was absolutely beautiful and Geno has the report and we got a fantastic view of the moon now.' Once the trajectory had been checked, Houston informed the crew: 'You're coming right down the fairway.' A delighted flight director informed the press, 'It's downhill from here on in.'

And so it proved. The happy crew sent back more spectacular colour shots of the moon, and played back a tape recording of Dean Martin singing 'Going Back To Houston.' Later broadcasts showed the beautiful blue and white Earth as Charlie Brown sped back home. One surprise for viewers came when the camera turned on the crew: for the first time in space a crew had shaved off their week-long beards. The astronauts had overcome the hazard of weightless whiskers by using tube shaving cream and wiping the razors with wet towels which were then put in waste bags. The 55 hour return journey allowed plenty of time for the crew to relax, though they did carry out more navigational checks and star sightings in addition to the usual housekeeping. On Sunday morning, Tom Stafford's thoughts turned to home; he asked Capcom Joe Engle to apologize to the members of his local church 'for being out of town for church today'. He then requested his pastor to read out a prayer that morning: 'Oh Lord, our Lord, how excellent is thy name in all the Earth, who hast set thy glory above the heavens . . .'

Once again, the flight path was so accurate that only one minor course correction was necessary, a few hours before re-entry. The service module was jettisoned, exposing the heat shield which would protect the crew from the intense frictional heating caused by a record re-entry speed of nearly 25,000 mph (40,000 kmh). It functioned perfectly during the roller-coaster descent as the helpless crew were pressed back into their couches with up to 7G of deceleration. The heavy investment by American TV companies paid off as viewers in the USA saw live colour pictures of a starlike capsule dropping beneath its three giant parachutes as rays from the rising Pacific sun were reflected from its charred exterior. Apollo 10 dropped heavily into a near calm sea, behaving well to the end by remaining in a stable, upright position. The astronauts left Charlie Brown 35 minutes after splashdown, and were then flown the final 4 miles (6 km) to the carrier *Princeton*, changing on the way into light blue overalls. The crew emerged on the flight deck cheerfully waving and stretching their limbs. A red and blue carpet and a giant cake awaited them, as well as the now-traditional embroidered baseball caps.

The reception was predictably rapturous, justifiably so. President Nixon congratulated the men for their 'magnificent achievement', expressing the widely held opinion that the excellent TV pictures had been one of the mission highlights. 'It gave people the feeling that they were right along,' said the President. He invited the astronauts and their wives to dinner at the White House at an early date. The crew were found to be fit and well, apart from some minor skin irritation. Stafford paid tribute to the vast team of ground staff who had made their exploits possible: 'We thank all on Earth who helped our tremendous team effort, and we hope we have added to man's knowledge.' Gene Cernan sounded a patriotic note: 'It's great to come back to the greatest country in the world.' From fellow naval commander John Young, an uncharacteristic humourous quip: 'I'd like to thank a big part of the Navy for coming to pick up a small part of the Navy.'

All three astronauts went on to even more distinguished careers with NASA. Stafford was assigned as head of the astronauts in June of that year, moving to the position of Deputy Director of Flight Crew Operations two years later. In 1975, shortly before leaving NASA, he commanded the first, and so far only joint American-Soviet space flight, the Apollo–Soyuz Test Project. John Young visited the moon again as commander of Apollo 16, this time setting foot on the world he had so frustratingly viewed from his seat 69 miles up on Apollo 10. Eugene Cernan also returned to the moon, commanding Apollo 17 and achieving his ambition of stepping out on to its surface.

In May of 1969 NASA administrator Dr Thomas Paine was telling the Houston press conference:

> This is an historic day . . . While the moon has been the focus of our efforts, the true goal is far more than being first to land men on the moon, as though it were a celestial Mount Everest to be climbed. The real goal is to develop and demonstrate the capability for interplanetary travel. With some awe, we contemplate the fact that men can now walk on extra-terrestrial shores.

With hindsight, we can see that President Nixon's goals were rather different, but in those optimistic summer days the sky was the limit. As Apollo 10 splashed down after 8 days and 3 minutes in space, a large sign appeared in the Houston control centre. It read, '51 days to launch'.

8

One Giant Leap

The First Moon Landings, 1969

Mankind's greatest adventure, the first moon landing mission, began from Pad 39A at Cape Kennedy on the morning of 16 January 1969. Three men from Earth set out to become the first humans ever to set foot on another world, the culmination of a programme set in motion by President Kennedy eight years and 25½ billion dollars earlier. The sun shone brilliantly in a clear blue sky as the 8,000 VIPs and 2,000 journalists from all over the world packed into the viewing area 3½ miles (5½ km) from the launch site. An estimated 1 million people had gathered in the vicinity to glimpse history in the making, with countless millions glued to their TV sets around the globe. The media had been feeding the public a stream of information for weeks so that every intimate detail of the flight programme and the crew was familiar to this immense audience as they followed the countdown with bated breath.

The commander of Apollo 11 was seated on the left couch in the command module Columbia, ready to pull the abort handle if necessary. Although a civilian, 38-year-old Neil Armstrong was described by his colleagues as 'probably the best jet test pilot in the world', a man used to living with danger and cool, calm and collected in a crisis. Since childhood his primary interest had been flying – he got his student pilot's licence on his sixteenth birthday – and he took advantage of a Navy scholarship to further his flying career. In 1951 he was assigned to a jet fighter squadron in Korea where he flew 78 combat missions and won three medals, once escaping from behind enemy lines after being shot down. During the late 1950s Armstrong had the opportunity to fly almost every kind of high-performance airplane and at the same time do research in aerodynamics at Edwards Air Force Base as a civilian working for NASA. He graduated to flying the X 15 rocket plane in 1960, flying in it seven times to a maximum altitude of 39 miles (63 km).

After his selection as an astronaut, he served as back-up command pilot for Gemini 5 before commanding Gemini 8. He had piloted that craft to safety after a jammed thruster had threatened disaster and, as recently as 1968, he had baled out of a lunar module trainer only seconds before it crashed. He was in the classic mould of the American hero – a man of action rather than words.

The man in the centre couch was USAF colonel Edwin 'Buzz' Aldrin, the same age as Armstrong and picked as his companion in the lunar module, Eagle. He, too, had caught the flying bug from his father, but he attended West Point Military Academy before gaining his pilot's wings in 1952. In Korea he flew 66 combat missions, downing two MiG 15 aircraft. His doctoral thesis at Massachusetts Institute of Technology was on 'Guidance for Manned Orbital Rendezvous', an essential aspect of the Apollo 11 mission. Aldrin's proudest moment came when NASA adopted his concept for space rendezvous, and when he was selected as an astronaut in October 1963, one year later than Armstrong, he had the opportunity to try out his ideas. During the flight of Gemini 12 the radar had failed, and Aldrin's own back-up computations had been used to enable him and James Lovell to rendezvous with the Agena target rocket. Buzz – his old schoolboy nickname stuck with him into adulthood – was a perfectionist, deeply involved in everything he did. His wife, Joan, once described his character as 'a curious mixture of magnificent confidence, bordering on conceit, and humility'. As lunar module pilot Aldrin was originally intended to be the first man on the moon, but this plan was later changed. His main role was to check the systems and advise Armstrong, the command pilot of Eagle during its descent.

The third member of the team had the lonely and frustrating task of command module pilot, destined not to see the moon landing or set foot on the moon. Should anything go wrong with Eagle, it would be his craft that would have to attempt a rescue, assuming that the LM was still in lunar orbit. Lieutenant-colonel Michael Collins was a 38-year-old Air Force officer. Born in Rome, he had lived all over America as well as in Puerto Rico before entering West Point. He joined the USAF, eventually becoming a test pilot at Edwards Air Force Base. His application to become an astronaut in 1962 was rejected, but he tried again successfully the following year. A modest, but determined, character, Collins described his main achievement as being 'good at handball'. After Gemini 10 his career was threatened when he was dropped from the Apollo 8 mission when a loose disc in his neck was found to be pressing on his spinal cord. His physical performance and reflexes steadily deteriorated during 1968, so that doctors informed him that his only chance of regaining his health and coveted flight status was to undergo two dangerous operations. Collins took the chance, and his courage paid off. As Deke Slayton, director of flight crew operations, explained, 'He got well so fast he could almost have made Apollo 8. But we didn't know that when he got to surgery, and we couldn't risk it.' His failure to fly on Apollo 8 meant that he, and not James Lovell who replaced him on that mission, joined the Apollo 11 prime crew once his complete recovery was confirmed.

The mission on which these three men were about to embark was the most dangerous ever undertaken in the eight years of manned space-flight. The rocket designer Wernher von Braun saw it this way: 'When we wheel out one of the rockets to the launch pad, I find myself thinking of all those thousands of parts – and all built by the lowest bidder – and I pray that everyone has done his homework.' As the Apollo 1 launch pad fire had demonstrated, his worries were justified; the Saturn V alone consisted of some 12 million working parts. Wherever possible, the principle of redundancy was followed, whereby a back-up system or component was available if anything went wrong with the primary. Nevertheless, this was not always possible and, as the astronauts well knew, a 0.1 per cent failure rate still meant that 12,000 parts would malfunction during the mission. In particular, the LM only had one engine with which the crew could blast-off from the lunar surface, while the CSM had one lone rocket engine with which to make course corrections, enter and leave lunar orbit, and return safely to Earth.

Such were the many possibilities for disaster in the mission that there was a real fear that the responsibility placed on the three men might affect their judgement should a crisis arise. Accordingly, NASA administrator Tom Paine made a point of telling them not to hesitate to abort the mission should any serious trouble arise; their safety was paramount. He even promised to give them an early second chance if the mission had to be aborted. This must have flitted through Armstrong's mind as he lay with his hand poised over the abort handle, but all three seemed relaxed as the countdown continued without a hitch.

A white mist of liquid oxygen shrouded the tail of the towering Saturn V as the astronauts settled into their horizontal couches, so closely packed that their elbows touched. More than 300 feet (100 m) above the ground, they could see nothing of the activity taking place all around them as they spoke to ground control and carried out their routine checks. Outside, TV and film helicopters swarmed around the Cape, revealing miles of messages written in the sand of the nearby Florida beaches, wishing Apollo 11 'Good Luck'. Thousands of people thronged the area, many having camped out all night to ensure a glimpse of the historic launch. Among the VIPs taking their seats in the main viewing area were Hermann Oberth, one of the leading German rocket pioneers, Charles Lindbergh, the first person to fly the Atlantic Ocean solo, and ex-President Johnson, a long-term supporter of the Apollo programme.

At 9.32 am Cape time, an estimated 1 billion people all over the world watched orange flames belch from the base of the Saturn V as it rose, imperceptibly at first, from the billowing clouds of smoke around the launch pad. The exhaust briefly obscured the rocket from view as it struggled to clear the tower. As the Saturn V gradually overcame its battle with gravity, the spectators could see it break through the high layer of thin cloud, followed a short time later by a tiny puff of smoke as the first stage, spent of fuel after only 2½ minutes, cut off at an altitude of 41 miles (66 km) and a speed of 5,400 mph (8,650 kmh). The retrorockets slowed the first stage prior to the separation from the second stage, which now became the workhorse. Soon afterwards, Armstrong confirmed that the escape tower had separated – they were 'looking good'.

Michael Collins later compared the launch with his previous flight aboard the Titan 2:

It was, I thought, quite a rough ride in the first 15 seconds or so . . . I don't mean that the engines were rough, and I don't mean that it was noisy. But it was very busy – that's the best word. It was steering like crazy . . . It was all very jerky and I was glad when they called 'Tower clear', because it was nice to know that there was no structure around when this thing was going through its little hiccups and jerks. But the jerkiness quieted down after about 15 seconds.

The second stage carried the astronauts to an altitude of 110 miles (176 km) and a velocity of more than 14,000 mph (22,400 kmh). This stage was then jettisoned, and the third stage completed the acceleration to Earth orbital speed of 17,400 mph (27,850 kmh) less than 12 minutes after lift-off. Armstrong reported 'shutdown' right on schedule. They were now weightless, orbiting with their feet pointing at the stars.

Following successful navigation and equipment checks, the astronauts were able to relax a little, removing their gloves and helmets. Over western Australia they were given the go-ahead for TLI (trans-lunar injection) which would boost them on their way to the moon. At 2 hours 44 minutes into the mission, on their second orbit, the crew felt a sudden return to normality as the third stage engine fired for 5 minutes 47 seconds, giving them a firm push in their seats. Specially equipped jet transport planes and the tracking ship *Mercury* filled the communications gap in the Pacific Ocean. As Apollo 11 soared out of Earth orbit, even Armstrong sounded a little animated: 'We have no complaints with any of the three stages on that ride . . . It was beautiful.'

The next vital manoeuvres came half an hour later, once the crew had swapped seats; Armstrong was now in the middle, Aldrin on the left and Collins in the command module pilot's couch on the right. First came the gentle separation of the CSM at a snail-like ½ mph (1 kmh) to a distance of about 100 feet (30 m) from the third stage before an equally cautious 180 degree turn, aided this time by the computer. Collins then edged in, nose first, towards the LM, which still nestled inside the adapter section of the third stage. The protective panels were jettisoned, enabling Collins to line up the cross-wires and insert the CSM probe into the drogue of Eagle. The 12 docking latches snapped shut, holding Eagle and Columbia firmly together. The tunnel hatch was removed, the electrical power link to Eagle was connected, and then Collins reversed thrust, gingerly withdrawing the LM from its adapter to a safe distance. It was the first time the two craft had flown together fully fuelled and loaded. Finally, ground control ordered the third stage to vent its remaining fuel, thereby sending it on a separate path around the moon and into solar orbit. Collins was happy with his efforts: 'I thought it went pretty well, Houston, although I expect I used more gas than I've been using in the simulator.'

It was now more than four and a half hours into the mission and, with their main chores completed, the astronauts were relieved to remove their cumbersome pressure suits. They had to help each other wriggle free; Collins later said they were 'like three great white whales inside a small tank'. The discarded suits were carefully folded and stored in bags under the centre couch, leaving the men dressed in white two-piece nylon jump suits. They at last had time to describe their earlier view of the Earth: 'We had the entire northern part of the lighted hemisphere visible, including North America, North Atlantic and Europe and northern Africa. We could see that the weather was good just about everywhere. There was one cyclonic depression in northern Canada . . . Greenland was clear and it appeared to be we were just seeing the ice cap.'

Later, Armstrong's voice came over the radio as the flight controllers changed shifts to wish Dr George Mueller (head of manned spaceflight) a happy birthday.

It was now time to settle down for the long cruise to the moon. A minor course correction was cancelled, but the craft had to be put into a 'barbecue mode' to prevent the sun overheating any part of Apollo. A brief TV broadcast was sent back some ten hours after lift-off, but the crew were more than ready when their first sleep period arrived three and a half hours later; in fact, Collins, the hardest worked of the three, put his head down half an hour early. Each switch was carefully positioned, the windows were covered, and each man found a place to bunk down – Armstrong and Aldrin were in light mesh bags under the couches, while Collins was strapped into the left hand couch with a miniature headset taped to his ear in case ground control called with an important message. The spacecraft was revolving about three times every hour, and the Earth was rapidly shrinking so that, already, it hardly filled one window, even though its pull continually slowed the craft as it sought to escape Earth's gravity.

The next three days saw Apollo slow down to a relatively sedate 2,000 mph (3,200 kmh) before they entered the gravitational realm of the moon. All of the time, the craft was in sunlight, surrounded by a black void studded with unblinking stars which acted as navigational beacons. Much of their work was 'housekeeping' – purging fuel cells, charging batteries, dumping waste water, changing carbon dioxide-removing canisters, preparing food, chlorinating drinking water and numerous other minor tasks. The trajectory was so accurate that only one of the planned four mid-course corrections was needed. The regular TV broadcasts were a great success as they demonstrated the effects of weightlessness and gave guided tours around their home.

During one broadcast Michael Collins decided to display their food cabinet; the food was plentiful and wide-ranging, each astronaut having his own menu for the day, selected before launch, with a 'snack pantry' for bites between meals. Capcom Charlie Duke played the role of introducer: 'Eleven, Houston. We see a box full of goodies there, over.' Collins seized his opportunity:

> We really have them, Charlie. We've got all kinds of good stuff. We've got coffee up here in the upper left and the breakfast items and bacon in little small bites, beverages like fruit drink and over in the centre part we have, oh all kinds of things. Let me pull one out here and see what it is. Would you believe you're looking at chicken stew here? All you have to do is [add] hot water for five or ten minutes. Now we get our hot water out of a little spigot here with the filter on it that filters any gasses that may be in the drinking water out, and we just stick the end of this little tube in the end of the spigot and pull the trigger three times and then mush it up and slice the end off it and there you go. Beautiful chicken stew.

There were more than 70 items on the menu in the form of 'freeze-dried re-hydratable, wet-pack and spoon-bowl foods'. In addition, the snack pantry contained 75 drinks and over 100 food items including strawberry cubes, spaghetti and meat sauce and peanut cubes. There were 72 slices of rye and white bread, with a variety of sandwich spreads. Cleanliness was ensured by means of small, wet-wipe cleansing towels. Certainly Mike Collins appreciated the efforts of the NASA chefs; at one stage, he informed Capcom Bruce McCandless, 'My compliments to the chef. That salmon salad is outstanding.'

The crew also seemed to enjoy their weightless environment. Buzz

Aldrin summed up the general feeling: 'I'll tell you, I've been having a ball floating around inside here, back and forth up to one place and back to another. It's just like being outside [in EVA], except more comfortable.' Later, during a TV show, he expanded on his views:

> There's plenty of room for the three of us and I think we're all willing to find our favourite little corner to sit in. Zero G's very comfortable, but after a while you get to the point where you sort of get tired of rattling around and banging off the ceiling and the floor and the side, so you tend to find a little corner somewhere and put your knees up, or something like that, to wedge yourself in, and that seems more at home.

The only concern arose when the Soviet Union introduced a factor which was out of their control and could not have been foreseen in any flight plan. Three days before Apollo 11 was launched, a Soviet robot craft, Luna 15, was sent towards the moon in an apparent attempt to steal some of the USA's glory by soft-landing on the moon, obtaining soil samples and returning them to Earth without endangering any human lives. As astronaut Frank Borman explained, 'The trouble was, they had our trajectory but we didn't have theirs.' The likelihood of the two craft colliding was minimal, but NASA officials were taking no chances. Borman had recently been on a week-long visit to the Soviet Union, returning with two medals commemorating the dead cosmonauts Yuri Gagarin and Vladimir Komarov which were to be left on the moon by the crew of Eagle, alongside medals for the three astronauts who died in Apollo 1. He was, therefore, asked to contact Mstislav Keldysh, the director of the Soviet Institute of Sciences, in order to obtain a reassurance that there was no danger. The Soviet Academician was able to supply the missing information on the orbit of Luna 15, together with the welcome news, 'In case of further change in the orbit of the probe you will receive additional information. The orbit of probe Luna 15 does not intersect the trajectory of Apollo 11 spacecraft announced by you in flight programme.' Luna 15 remained in lunar orbit until shortly before Eagle blasted off from the moon's surface. Only then were descent manoeuvres attempted, but the craft went out of control and crashed on the moon.

On the third day the crew checked over the LM during a TV broadcast which lasted more than 90 minutes; everything seemed fine. Capcom Charlie Duke told them, 'Eleven, you got a pretty big audience. It's live in the US, it's going live to Japan, Western Europe and much of South America. Everybody reports very good colour and appreciates the great show.' Their spirits were high as they approached the point when the moon would begin to accelerate them again. They played music tapes as they ate, with Michael Collins celebrating 'Because it's a special occasion today, Houston. This is the third anniversary of Gemini 10.'

Their sleep period was delayed for a while as they discussed a strange L-shaped object, apparently a piece of space debris, which appeared outside the window; Collins said he heard it 'thump' against the outside of Columbia. The tension was mounting, and the men slept more fitfully as the moon loomed larger. After seven hours, Aldrin awoke and spoke to Capcom Ron Evans: 'Good morning, are you planning a course correction for us this morning?' Evans had some good news: 'That's negative. Midcourse number 4 is not required. We were going to let you sleep in until about 71 hours, if you'd like to turn over.' Aldrin did not need telling twice: 'OK I'll see you at 71 hours.'

When they did arise from slumber the astronauts gazed upon the shadowy moon – 'an eerie sight' as Armstrong described it. There was a brilliant corona around the moon as it hid all of the sun apart from its gaseous halo. Armstrong now became more eloquent: 'The view of the moon that we've been having recently is really spectacular. It fills about three-quarters of the hatch window, and of course we can see the entire circumference even though part of it is in complete shadow and part of it is in Earthshine. It's a view worth the price of the trip.' Michael Collins commented: 'It's a real change for us. Now we're able to see stars again and recognize constellations for the first time on the trip. The sky's full of stars. It looks like it's the night side on Earth.' Although the cabin was darkened, Collins reported, 'The Earthshine coming through the window is so bright you can read a book by it.' It was also bright enough to reveal the large crater Tycho and other features with a 'very marked three-dimensional effect', according to Armstrong.

However, they could not sit and marvel at the remarkable view if the lunar landing was to go ahead. Their preparations for the lunar orbit insertion burn were lightened by Charlie Duke's news bulletin: 'Even *Pravda* in Russia is headlining the mission and calls Neil 'The Czar of the Ship'. I think maybe they got the wrong mission.' From then on, Armstrong had to endure a fair amount of ribbing from his colleagues over his new title. At 75 hours 41 minutes into the flight, Apollo 11 disappeared behind the moon; there would be no contact with ground control for at least 24 minutes, but, if the burn successfully braked the spacecraft, the interval would be ten minutes longer. Capcom McCandless signed off: 'All your systems are looking good going around the corner and we'll see you on the other side.'

The control room at Houston began to fill with astronauts as the minutes ticked slowly by. Many had arrived to witness Apollo's successful entry into lunar orbit, or the moment when everyone's dreams would become nightmares. Out there in the cold silence of space, some quarter of a million miles from home, Michael Collins was firing the service module engine for a duration of just over six minutes – if anything went wrong, they were on their own.

The time passed the 24 minute mark with no signal. The control room

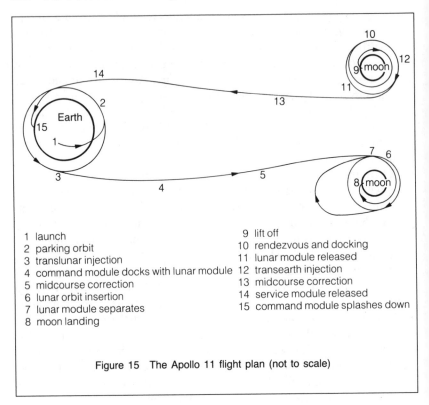

1 launch
2 parking orbit
3 translunar injection
4 command module docks with lunar module
5 midcourse correction
6 lunar orbit insertion
7 lunar module separates
8 moon landing

9 lift off
10 rendezvous and docking
11 lunar module released
12 transearth injection
13 midcourse correction
14 service module released
15 command module splashes down

Figure 15   The Apollo 11 flight plan (not to scale)

was eerily quiet as the clock crept around another ten minutes, then, exactly on schedule, McCandless heard the voice of Collins: 'Read you loud and clear, Houston.' When asked for a report on the burn, Collins replied, 'It was like perfect.' Apollo was now in a highly elliptical orbit which brought them to within 70 miles of the lunar surface (113 km). There was time available once more for descriptions of the moon before the second burn to circularize the orbit. Armstrong commented: 'It's very much like the pictures [from Apollos 8 and 10], but like the difference between watching a real football game and watching it on TV – no substitute for actually being here.'

Two hours after Apollo 11 went behind the moon for the first time, it disappeared once more at the start of the second revolution. There was no reason for concern on this occasion. Further descriptions of the surface, combined with some spectacular TV pictures, followed after they re-appeared. The crew showed a high degree of familiarity with the lunar geography, particularly the landmarks which they would see close-up during the landing attempt. Collins was moved to remark: 'The Sea of Fertility doesn't look very fertile to me.' At 78 hours and 58 minutes since launch, Aldrin said farewell for the third time as they crossed the rugged landscape along the terminator and plunged into darkness.

During this third pass behind the moon, the service module engine was fired for a brief 17 seconds to change the orbit to a near circle: the effect of the lunar gravity would produce a circular orbit without any further use of fuel before the end of the mission. The next task was to check over Eagle; using the LM's power for the first time, Aldrin's voice came across loud and clear. The check-out period lasted for more than three hours, and the crew were ready for their deserved rest period at the end of the fifth pass behind the moon, though they had dropped a little behind schedule as Collins explained: 'We haven't chlorinated the water yet and we haven't changed the lithium hydroxide. We're just still finishing up dinner.'

The rest period was planned for nine hours, but none of them slept very soundly at first. This was hardly surprising with the historic landing attempt due the next day, Sunday 20 July. Collins was the last to succumb, but eventually the flight surgeon was monitoring heart beats down in the 40s. It was quite a struggle for Collins to awake when the alarm call came; in Houston, it was the early hours of Sunday morning. The crew would not have another rest period for more than 22 hours. They ate their last breakfast together for the next two days as the new team of controllers, led by veteran flight director Gene Kranz, took over in Mission Control. Capcom Ron Evans informed them that 'Church services around the world today are mentioning Apollo 11 in their prayers. President Nixon's worship service at the White House is also dedicated to the mission, and our fellow astronaut Frank Borman is still in there pitching and will read the passage from Genesis which was read on Apollo 8 last Christmas.'

The astronauts now had to get down to the main business of preparing for the landing attempt which was little more than seven hours way. The LM crew donned their water-cooled undergarments, while Collins put on his pressure suit. It was just after 8 am Houston time as Apollo 11 went behind the moon to begin its tenth revolution. While they were out of contact, Aldrin moved into Eagle and switched on the descent stage batteries prior to activating all of the systems. He was joined by Armstrong soon after reacquisition of signal, and the airwaves were soon full of technical language between Eagle and Columbia, and the two craft and ground control. Armstrong, then Aldrin, went back into Columbia to don their bulky pressure suits, but the checks continued apace. Before Apollo went 'over the hill' for the eleventh time, the probe and drogue had been refitted in the tunnel and the hatches were about to be closed.

There still remained checks on the LM guidance system, the control thrusters, the descent propulsion system and the rendezvous radar. If all systems were 'go' then ground control would give the crew the green light for undocking. There were no problems at all, enabling the LM crew to deploy Eagle's landing gear. Charlie Duke finally reported:

'Apollo, Houston. We're go for undocking.' Collins replied from Columbia: 'Roger. There will be no television of the undocking . . . I'm busy with other things.' Once again they entered the 45 minute communication blackout as they began the twelfth pass behind the moon, 99½ hours into the mission. The two craft successfully undocked during the blackout, so that, when Duke asked, 'How does it look?' Armstrong replied, 'The Eagle has wings.' The separation of a few yards was increased gradually by an eight second fire of Columbia's thrusters while the latest computer updates were fed to the LM crew by mission control in preparation for the Eagle's descent orbit insertion which would take place on the next pass behind the moon. Collins commented, 'I think you've got a fine-looking flying machine there, Eagle, despite the fact that you're upside down.' Aldrin replied: 'Somebody's upside down.' Collins bade them farewell: 'OK, Eagle . . . You guys take care.' 'See you later,' said Armstrong.

By the time the descent manoeuvre began, the two craft were already 2 miles (3½ km) apart. The LM descent engine was fired for the first time for about 30 seconds, braking Eagle so that it was overtaken by Columbia. Eagle's orbit was changed so it would be carried to within 10 miles (16 km) of the lunar surface; this had already been practised by Apollo 10. However, the final descent would only be given the go-ahead if everything remained perfect. The control room in Houston was packed with NASA administrators, engineers and astronauts, all peering through the glass screen for the latest information on the mission's progress. Every one of them was filled with a tense excitement which became almost unbearable as the clock showed less than half an hour to the beginning of powered descent.

Columbia appeared first beyond the lunar limb. Capcom Charles Duke asked the question on everyone's lips: 'How did it go? Over.' Collins gave the welcome news: 'Listen, babe, everything's going just swimmingly. Beautiful.' Just two minutes behind Columbia came the lunar module. There were now 17 minutes to the powered descent initiation (PDI), but a threat to the landing appeared in the form of communications break-up with Eagle. They improved sufficiently for Kranz to order continuation of the mission, so Duke sent up the message, 'You're go for PDI.' The LM approached its lowest orbital point with the crew, firmly restrained in a standing position, ready for ignition. Communications remained highly variable, but the descent engine was successfully fired at 102 hours 33 minutes into the mission.

For the first part of the descent Eagle flew with its head down, then Armstrong turned the craft right around so that it was flying 'heads-up' to enable the landing radar to lock on and feed data on altitude and velocity to the onboard computer. Aldrin read out the data to Armstrong as the descent continued, with the commander ready to assume control at any time. Mission control could hear the deadpan voice of Aldrin: 'Hang tight; we're go: 2,000 feet.' Ground control was aware that they would probably overshoot the target area, but the computer began flashing danger signals '1201' and '1202' warning of an overload. Aldrin did not speak again until Eagle had descended to 750 ft (200 m). The readings came at regular intervals: '600 feet [altitude], down at 19 [feet per second] . . . 540 feet, down at 15 . . 400 feet, down at 9 . . . 8 forward . . . 330, 3½ down.'

The controllers in Houston then heard the readings tell of a sudden surge forward at up to 80 feet per second, which was definitely not in the flight plan. However, to everyone's relief, the figures settled down to a more expected pattern: 'Down at 2½, 19 forward . . . 3½ down, 220 feet . . . 11 forward, coming down nicely, 200 feet, 4½ down . . . 160, 6½ down . . . 9 forward . . . 100 feet.' By this stage, Eagle was close enough to the surface to stir up dust which tended to obscure the ground, but a red warning light broke in on the watching controllers; only 5 per cent of Eagle's descent fuel remained. Mission rules decreed that Eagle had to abort the landing if it was not down on the surface within the next 90 seconds. It seemed that everyone at Houston was praying, 'Get it down, Neil, get it down.' Failure now would be a heartbreak.

Aldrin's voice came through: 'Light's on.' Then he continued as normal: 'Down 2½ . . . forward, forward. Good. 40 feet, down 2½. Picking up some dust. 30 feet, 2½ down. Faint shadow.' He could see the shadow of one of Eagle's footpads. 'Four forward, drifting to the right a little.' Capcom Charlie Duke announced 30 seconds until abort. It was now or never. 'Forward . . . drifting right . . . contact light. OK, engine stop. ACA out of detent. Modes control both auto, descent engine command override off. Engine arm off. 413 is in,' said a highly satisfied Buzz Aldrin. They had made it with 20 seconds to spare. 'We copy you down, Eagle,' came the voice of Duke. Armstrong uttered the words which resounded around the world: 'Houston, Tranquillity Base here. The Eagle has landed.'

The control centre was a scene of uncontrollable joy. Charlie Duke's voice said it all, as he spoke for all those around him: 'Roger, Tranquillity, we copy you on the ground. You've got a bunch of guys about to turn blue. We're breathing again. Thanks a lot.' Michael Collins, alone in Columbia, had been unable to witness the landing, but he was able to listen in, and his response was equally enthusiastic. 'Fantastic!' was all he could manage to utter.

Gene Kranz immediately ordered a status check to show that all was well with the Eagle before confirming that the crew could prolong their stay. Meanwhile, the long-awaited news was being flashed all over the world, that two men from the planet Earth had at last achieved one of mankind's oldest ambitions and set down on another world.

As the dust blown up during the landing slowly settled and cleared, Armstrong felt he owed mission control an explanation for the unorthodox descent manoeuvres: 'Houston, that may have seemed like a very long final phase. The auto-targeting was taking us right into a football field sized crater, with a large number of big boulders and rocks for about 1 or 2 crater diameters around us, and it required . . . flying manually over the rock field to find a reasonably good area.' Charlie Duke re-assured him: 'It was beautiful from here, Tranquillity,' though many agreed with Lieutenant-General Samuel Phillips, director of Project Apollo, when he stated, 'In my opinion we damn near didn't make it. I could be wrong. Neil might see it differently. But I think we came awfully close to having to abort.' One problem with Armstrong concentrating on spacecraft manoeuvres was that he was unable to identify their precise touchdown point. Ground control sent up information on landmarks to help, but it was some time before they calculated that they had drifted about 1,000 ft (300 m) downrange and 330 ft (100 m) to the left of the target.

Ten minutes after touchdown, Aldrin began to describe the alien environment which surrounded them:

> We'll get to the details of what's around here, but it looks like a collection of just about every variety of shape, angularity, granularity, about every variety of rock you could find. The colours are pretty much depending on how you're looking relative to the zero phase point. There doesn't appear to be too much of a general colour at all; however, it looks as though some of the rocks and boulders, of which there are quite a few in the near area, it looks as though they're going to have some interesting colour to them.

Capcom Duke sounded as if he was still over the moon: 'Roger, Tranquillity. Be advised there's lots of smiling faces in this room and all over the world. Over.' Armstrong replied, 'There are two of them up here.'

Michael Collins was still listening in as he passed overhead in Columbia. 'And don't forget one in the command module . . . And thanks for putting me on relay, Houston, I was missing all the action.' Duke seemed to know how the third astronaut, orbiting far above the site where history was being made, must be feeling: 'Rog. Columbia . . . Say something. They ought to be able to hear you.' Collins was at last able to congratulate his two colleagues: 'Tranquillity Base, it sure sounded great from up here. You guys did a fantastic job.' Armstrong acknowledged the essential, though unspectacular, job that Collins was carrying out: 'Thank you. Just keep that orbiting base ready for us up there now.'

Armstrong now expanded on the limited information so far reported:

You might be interested to know that I don't think we noticed any difficulty at all in adapting to one-sixth G . . . out of the window is a relatively level plain cratered with a fairly large number of craters of the five- to 50-foot variety and some ridges, small, 20 or 30 feet high, I would guess, and literally thousands of little one- and two-foot craters around the area. We see some angular blocks out several hundred feet in front of us that are probably two feet in size and have angular edges. There is a hill in view, just about on the ground track ahead of us, difficult to estimate but might be half a mile or a mile.

Collins broke in, 'Sounds like it looks a lot better now than it did yesterday at that very low sun angle. It looked rough as a cob then.'

The two astronauts were scheduled for a four hour rest period after their first three hours on the lunar surface, but they requested a postponement so that they could prepare for the moon walks; they were too eager and excited to benefit from a rest period at this stage of the proceedings. Houston agreed, and put back the rest to combine with a second rest period scheduled for the pre-launch session. Before eating their first meal on the moon, Aldrin paused with a request: 'I'd like to take this opportunity to ask every person listening in, whoever, and wherever they may be, to pause for a moment and contemplate the events of the past few hours, and to give thanks in his or her own way.' Aldrin bowed his head with a brief silent prayer of gratitude.

Up above, while his two colleagues ate, Collins was still trying to locate Eagle through his sextant with the help of ground control; the module's exact position remained unknown. He reported to Capcom McCandless, 'No joy. I kept my eyes glued to the sextant that time, hoping I'd get a flash of reflected light off the LM, but I wasn't able to see any of him in my scan areas that you suggested.'

It took about two hours for Armstrong and Aldrin to prepare for their moon walks: the electrical check-out took longer than planned and, apart from the other systems checks, a considerable amount of time was taken up with donning the awkward backpacks and depressurizing Eagle's cabin. Each backpack was designed to keep a man alive on the lunar surface for up to four hours; it provided the oxygen to keep the suit inflated and to enable the astronaut to breathe, an oxygen purification and ventilation system, a system to maintain the flow of cool water through the tubes in the undergarment, and a communication system. Although its weight on the moon was only about 20 lb (44 kg) it was necessary for one astronaut to help the other to put it on. Only then could the helmet and gloves be worn, and the cabin be depressurized. Once the men were exposed to the lunar vacuum, the suit was their only protection; if it was ripped open accidentally by catching on the LM or during a tumble, the result would probably be death. Buzz Aldrin once described the possible consequences:

If there were such an accident, I'm not sure that we really know what would happen. We used to think that there would be a foaming up of the blood, and it would boil. We don't think so now; after all, the fluids in the body are not exposed to the outside. All we know is that with a big loss of pressure you would lose consciousness in a short period of time, a matter of seconds.

It was past 9.30 pm Houston time when Aldrin declared, 'The hatch is opening.' Armstrong turned and edged carefully backwards on his hands and knees with Aldrin giving a stream of instructions in order to help him through the small hatch, only 32 in (81 cm) square. At last, Armstrong was through: 'OK Houston, I'm on the porch.' The first-ever excursion on another world was about to begin, about six and a half hours after Eagle landed, and 109 hours 19 minutes into the mission. As Armstrong began his slow descent, he pulled a lanyard which deployed the equipment stowage assembly just to the left of the ladder, including a small TV camera. This heralded another historic 'first' as the fuzzy, ghostlike black-and-white images were picked up by the Australian tracking station and relayed to millions of screens all over the world. As McCandless remarked, 'There's a great deal of contrast in it . . . but we can make out a fair amount of detail.'

Armstrong paused at the bottom rung: 'I'm at the foot of the ladder. The LM footpads are only depressed in the surface about 1 or 2 inches. Although the surface appears to be very, very fine-grained, as you get close to it, it's almost like a powder.' A quarter of a million miles away, an estimated one billion people watched in awe as Armstrong lifted his left leg away from the ladder: 'I'm going to step off the LM now . . .' Then came the immortal words: 'That's one small step for a man, one giant leap for mankind.'

Some time afterwards, back on Earth, Armstrong explained his choice of words:

I had thought about what I was going to say, largely because so many people had asked me to think about it. I thought about that a little bit on the way to the moon, and it wasn't really decided until after we got on to the lunar surface. I guess I hadn't actually decided what I wanted to say until just before we went out.'

Neil Armstrong remained near the foot of the ladder.

The surface is fine and powdery. I can pick it up loosely with my toe. It does adhere in fine layers like powdered charcoal to the sole and sides of my boots. I only go in a small fraction of an inch, maybe an eighth of an inch, but I can see the footprints of my boots and the treads in the fine sandy particles.

He began to survey the immediate landing zone:

There seems to be no difficulty in moving around. As we suspected, it's even perhaps easier than the simulations at one-sixth G that we performed . . . on the ground. It's actually no trouble to walk around. The descent engine did not leave a crater of any size. There's about one foot clearance on the ground. We're essentially on a very level place here. I can see evidence of rays emanating from the descent engine, but very insignificant amount. OK Buzz, we're ready to bring down the camera.

Aldrin watched as Armstrong took the camera. The commander was ready to move further from Eagle: 'OK, it's quite dark here in the shadow and a little hard for me to see if I have good footing. I'll work my way over into the sunlight here without looking directly into the sun.' Aldrin was becoming slightly impatient to join his colleague: 'OK, going to get the contingency sample now, Neil?' This initial sample of surface material was planned in case they had to leave the moon in a hurry. Armstrong was soon busy: 'This is very interesting. It's a very soft surface, but here and there where I plug with the contingency sample collector, I run into a very hard surface, but it appears to be very cohesive material of the same sort. I'll try to get a rock in here. Here's a couple.' Aldrin was eagerly watching: 'That looks beautiful from here, Neil.' Armstrong continued: 'It has a stark beauty all its own. It's like much of the high desert of the United States. It's different, but it's very pretty out here.' The contingency sample was carefully placed in a Teflon bag which Armstrong tucked into a pocket just above his left knee.

'Are you ready for me to come out?' asked Aldrin, perhaps a little frustrated at being a spectator for more than 15 minutes. Armstrong cleared the handrail, then replied: 'All set. OK, you saw what difficulties I was having. I'll try to watch your PLSS [portable life-support system] from underneath here.' He then carefully guided Aldrin through the hatch 'with an inch clearance on top', until he rested on the edge of the porch. Aldrin paused before beginning to descend: 'Now I want to back up and partially close the hatch, making sure not to lock it on my way out.' Armstrong agreed: 'A particularly good thought.' Aldrin continued, 'That's our home for the next couple of hours and we want to take good care of it.'

Armstrong continued to guide his companion as he began to back down the ladder, but Aldrin was enjoying the experience: 'It's a very simple matter to hop down from one step to the next.' When he reached the bottom rung, Armstrong gave a warning: 'That's a good step, about a three-footer.' As he stepped back and turned, Aldrin exclaimed, 'Beautiful view!' Armstrong agreed: 'Isn't that something? Magnificent sight out here.' 'Magnificent desolation,' was Aldrin's version.

The two men were testing various methods of moving around in their bulky pressure suits in conditions where they weighed one-sixth of their Earth weight. Aldrin's continuous commentary showed his exuberance with his new environment:

> The rocks are rather slippery . . . My boot tends to slide over it rather easily . . . About to lose my balance in one direction and recovery is quite natural and very easy . . . And moving arms around, Jack [Schmitt], doesn't lift your feet off the surface . . . Got to be careful that you are leaning in the direction you want to go . . . You have to cross your foot over to stay underneath where your centre of mass is. Say, Neil, didn't I say we might see some purple rocks?

Armstrong asked: 'Find the purple rocks?' Aldrin confirmed: 'Yes, they are small, sparkly.'

The astronauts hopped back to Eagle and unveiled the plaque attached to the descent stage which would remain on the moon. Armstrong spoke:

> For those who haven't read the plaque, we'll read the plaque that's on the front landing gear of this LM. First there's two hemispheres, one showing each of the hemispheres on Earth. Underneath it says, 'Here Men from the planet Earth first set foot upon the Moon, July 1969 AD. We came in peace for all Mankind.' It has the crew members' signatures and the signature of the President of the United States.

In the midst of the ceremonial duties, Collins and Columbia came riding over the horizon: 'How's it going?' McCandless gave him the good news: 'The EVA is progressing beautifully. I believe they are setting up the flag now . . . I guess you're about the only person around that doesn't have TV coverage of the scene.' Collins was philosophical: 'That's all right. I don't mind a bit. How is the quality of the TV?' The astronauts had moved the camera from the LM and set it up on a tripod some 40 ft (15 m) from Eagle so that it would cover the area of maximum activity. McCandless told Collins: 'Oh, it's beautiful, Mike, it really is . . . They've got the flag up, and you can see the Stars and Stripes on the lunar surface.' The flag, 3 ft by 5 ft, was specially designed with a support along its top edge so that it would remain unfurled in the lunar vacuum.

Aldrin and Armstrong then received what NASA described as 'the longest long-distance phone call in man's history' as President Nixon spoke to them from a quarter of a million miles away in Washington.

The two men had been on their EVA for around an hour, and it was now time to concentrate on the more mundane scientific tasks of the mission. The most important scientific part of the landing was the gathering of rock and soil samples. Armstrong used a special aluminium scoop with an extension handle and a pair of aluminium tongs to obtain representative samples of the wide variety of rocks and the dark powdery lunar soil. Core tubes were also hammered into the surface to obtain soil profiles, but Aldrin found problems with this technique: 'I hope you're watching how hard I have to hit this into the ground to the tune of about 5 inches . . . It looks almost wet.' The samples were stored in two aluminium boxes which were vacuum-sealed before they were closed, in order to prevent contamination. The two men collected rocks that weighed 48 lb back on Earth, the most expensive rock specimens ever obtained. Tests in laboratories showed they were formed three billion years ago, before all but the oldest rocks on Earth.

Early in the EVA Aldrin deployed the solar wind experiment, a 4 ft long rectangle of foil-like material which was suspended on an aluminium staff which he forced into the surface. Atomic particles from the sun, unaffected by the moon's negligible magnetic field, were trapped in this device; it was later rolled up and stored in one of the sample boxes.

The two main experiments were set up some distance from Eagle to enable them to survive the blast from the ascent stage when the astronauts' lunar expedition came to an end. The seismometer was intended to detect moonquakes, a valuable method of determining the nature of the moon's interior. The instrument began sending information back to Earth straight away, and soon proved sufficiently sensitive to pick up the vibrations caused by the astronauts' footsteps. It survived for one lunar day and night before the electronics overheated. About 10 ft (3 m) closer to Eagle, Aldrin set up the laser reflector, a small aluminium frame which contained 100 tiny prisms designed to reflect light from a laser beam directed on to it from Earth. The round trip of the light could be timed to an accuracy of one billionth of a second, and, since the velocity of light was accurately known, the distance of the Sea of Tranquillity could be calculated to within 6 inches. At first, no light could be detected when the experiment began, but as the sun moved lower in the lunar sky, the first signals were received at Lick Observatory in California; the experiment worked.

Most of the astronauts' activities were visible on TV screens, and viewers were entertained by shots of the astronauts going through their paces, including a method of locomotion which was termed the 'kangaroo lope'. The men were able to cavort around without excessive heavy breathing, apart from a time when Armstrong loped 200 ft to photograph the smaller of the two craters which obstructed the landing. They were able to manage without the rest periods which had been set aside for them, coping so well with the alien conditions that the duration of the EVA was extended by around half an hour. The only time their heartbeats raced alarmingly was when Armstrong's pulse reached 160 a minute as he was loading the boxes of rock samples into the LM. Besides the black-and-white TV camera, the men obtained highly detailed coverage of their activities and of the lunar surface with still and motion pictures from three additional cameras.

Two hours 25 minutes after Neil Armstrong had begun the EVA, McCandless was beginning to worry about the men outstaying their welcome on the moon, and he broke in several times to hurry them along: 'Neil and Buzz. Let's press on with getting the close-up camera magazine and closing out the sample return container. We're running a little low on time.' A few minutes later, Aldrin had climbed the ladder and was back inside the LM, waiting for his colleague to pass up the sample boxes. Ten minutes later, the samples and equipment were all on board, and two very dirty astronauts were ready to seal themselves in once more: 'OK, the hatch is closed and latched.' The cabin was slowly repressurized and Eagle's systems activated.

The astronauts had been without rest for about 19 hours, and without food for the last seven hours, but they could not relax yet. Once the LM was pressurized, they had to remove their backpacks, attaching their suits to the Eagle's oxygen supply, and collect all dead weight, such as the redundant backpacks, prior to dumping it on the lunar surface. It was two and a half hours later when they depressurized the LM for the second time, opened the forward hatch and deposited their rubbish on the moon. Still there was no rest, as Houston sent up a number of questions relating to what the men had seen and done during their lunar excursion. It was past 3 am in Houston, the TV had been switched off, but the scientists still wanted more from the astronauts.

At 114 hours 50 minutes into the mission, Owen Garriott concluded the question-and-answer session: 'Thank you, and I hope this will be a final good night.' Exhausted and exhilarated as they were, it was hardly surprising that they had difficulty going to sleep, especially with the LM loaded with boxes of samples – it was not the most comfortable or spacious of homes at the best of times. Armstrong rigged up a harness on which to rest his feet and lay across the ascent engine cover; Aldrin had a more comfortable spot on the floor in front of the forward hatch. He later explained some of the problems they had to endure:

> The thing which really kept us awake was the temperature. It was very chilly in there. After about three hours it became unbearable. We had the liquid cooling system in operation in our suits, of course, and we tried to get comfortable by turning the water circulation down to minimum. That didn't help much. We turned the temperature control on our oxygen system up to maximum. That didn't have much effect either. We could have raised the window shades and let the light in to warm us, but that would have destroyed any remaining possibility of sleeping. The light was sometimes annoying because when it struck our helmets from a

side angle it would enter the face plate and make a glare which reflected all over it. Then when we entered a shadow, we would see reflections of our own faces in the front of the helmet, and they obscured anything else that was to be seen.

At one point, Armstrong found light from the brilliantly lit Earth bathing the Eagle's cabin. At least Collins was sleeping soundly high above; the loneliness did not affect him, even though, as he later wrote, 'not since Adam has man known such solitude as Mike Collins when he passes around the back of the moon.' Collins did, however, admit to a feeling of terror should something go wrong with Eagle's lift-off from the moon, with the result that he would be the only crew member to return to Earth.

Armstrong and Aldrin awoke with less than three hours remaining before the crucial burn of the ascent engine. Ground control sent up a stream of instructions and suggestions, many intended to prevent a repeat of the overflow alarms emitted by the computer during the descent. Astronaut Jim Lovell came on the air to voice the feelings of everyone at Houston: 'Eagle and Columbia, this is the back-up crew. Our congratulations on yesterday's performance, and our prayers are with you for the rendezvous.' There was still doubt over the exact position of Tranquillity Base, but the error was insufficient to affect the rendezvous of the two craft.

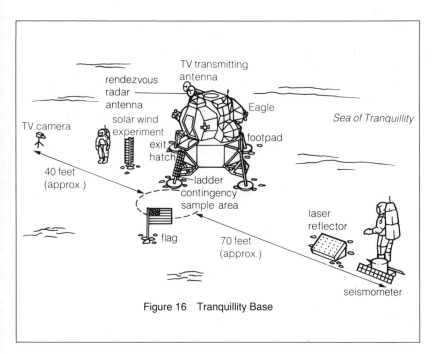

Figure 16   Tranquillity Base

The Eagle had been on the moon for a little over 21½ hours when the countdown reached its nail-biting climax – never before had a manned spacecraft been launched from the surface of another world. No one needed reminding that, if the lift-off failed, the crew would be doomed to a long, lonely wait for death. It was 12.54 pm at Houston as Buzz Aldrin's voice completed the countdown: '9, 8, 7, 6, 5 – abort stage, engine arm ascent, proceed. That was beautiful. 26, 36 feet per second up. Be advised of the pitchover. Very smooth . . . Very quiet ride.' The Eagle had wings once more. The ascent was perfect. Flight operations director Chris Kraft said he felt as if some 500 million people around the world were helping to push Eagle off the moon and back into orbit.

The burn lasted for just over seven minutes and successfully lifted Eagle into an elliptical orbit of 10 × 52 miles (17 × 84 km). The two astronauts had left behind the first human rubbish heap on another world: apart from Eagle's descent stage, there was more than a million dollars' worth of equipment together with a number of more formal mementoes. The laser reflector and seismometer were designed for automatic operation after Eagle departed, but the cameras, backpacks, containers and other equipment had simply been discarded in order to save weight and space. In addition, the crew left an Apollo shoulder patch commemorating the three astronauts – Gus Grissom, Ed White and Roger Chaffee – who died in the Apollo 1 launch pad fire in 1967, and medals honouring the two dead Soviet cosmonauts, Yuri Gagarin and Vladimir Komarov. There was also a small silicon disc on which were etched, in microscopic characters, messages of goodwill from leaders of 73 nations.

The two astronauts stood side by side during the ascent, and remained in this cramped position for the next three and a half hours as they gradually manoeuvred Eagle towards her target and a reunion with Columbia. They were able to see the lunar surface fall away rapidly beneath them, and recognize the topographic features with which they had become familiarized during their training. Aldrin filmed the complete ascent. Shortly after attaining orbit, both craft disappeared behind the moon; it was Columbia's twenty-fifth revolution and more than 125 hours into the mission. As Eagle approached the high point in its orbit, a short burn on Eagle's reaction control thrusters circularized the orbit. Columbia came around the moon three minutes ahead of Eagle, and the astronauts were able to confirm that all was going well – the craft were now only 120 miles apart (190 km) and Eagle was gaining all the time. A series of short burns on the thrusters gradually raised Eagle's orbit in the fashion which had been practised since the Gemini missions had introduced the technique with just this in mind. The craft disappeared from sight behind the moon once more. There was little that Houston could do or say to help; the expertise of the men orbiting the moon would decide the outcome.

The craft reappeared only a few feet apart and aligned ready for the final docking. Collins edged Columbia towards Eagle, then spoke in surprise: 'That was a funny one. You know, I didn't feel it strike [lock together] and then I thought things were pretty steady. I went to retract there, and that's where all hell broke loose. For you guys, did it appear to you to be that you were jerking around quite a bit during the retract cycle?' Armstrong replied that a thruster fired on Eagle for a fraction of a second, long enough to cause a sudden yawing motion and temporary loss of control. Collins came back: 'Yeah, I was sure busy there for a couple of seconds.' At 4.40 pm in Houston, the crew could be heard calling themselves 'Apollo' again, despite the 'horrible squeal', as Collins called it, on the radio.

Collins wasted no time in clearing the tunnel by removing the hatch, the probe and the drogue. However, he had to wait a short time in order for air pressure to build up in the tunnel so that no dust would be sucked into the command module environmental control system. Meanwhile, Armstrong and Aldrin vacuumed all lunar dust from their suits and their sample boxes before they transferred to Columbia. The two men were re-united with Collins as he floated in the tunnel during the pass behind the moon; after shaking hands and greeting each other, it was down to the serious work. Collins later described what happened: 'They passed the rock boxes through to me, and I handled them as if they were absolutely jampacked with rare jewels, which in a sense they were.' Armstrong's comment was: 'It's nice to have a place to sit down.'

When they reappeared, communications had improved and the work was proceeding smoothly as they prepared to jettison the ascent stage of Eagle: 'Houston, this is Columbia, reading you loud and clear. We're all three back inside, the hatch is installed. We're running a pressure leak check. Everything's going well.' The astronauts spent the front pass checking their systems and aligning the two craft ready for the separation. As they neared the blackout time at the beginning of the twenty-seventh lunar revolution, Capcom Ronald Evans commented: 'You're looking great. It's been a mighty fine day.'

The astronauts were so fast and efficient that the next Capcom, Charlie Duke, was able to tell them: 'It looks like you guys are so speedy on us that we're thinking of moving up jettison time.' Everything had gone so smoothly that the explosive charge was fired about 90 minutes ahead of schedule, at 130 hours 15 minutes into the mission. Collins reported the results: 'A fairly loud noise, and it appears to be departing – oh, I would guess several feet per second . . . There she goes. It was a good one.' A short time later, the Columbia's thrusters were fired for seven seconds to boost the craft and increase the gap between it and Eagle. By the scheduled time for the trans-earth injection (TEI) in five hours, Eagle would have drifted 23 miles (37 km) behind Columbia and 1 mile (2 km) above it, thereby minimizing the chances of a collision.

The next revolution was very quiet, apart from a news bulletin giving details of world reactions to their flight, and, perhaps more significant for the three men, the reactions of their wives. Joan Aldrin summed up their feelings: 'After the moon touchdown I wept because I was so happy . . . But the best part of the mission will be the splashdown.' The astronauts enjoyed a meal while contemplating the events of the past 36 hours; yet another notable first had been chalked up by this mission – the *New York Times* had given the landing banner headlines, the largest headlines ever used in the history of the newspaper.

The twenty-ninth revolution was taken up with preparations for the boost out of lunar orbit on to a path that would return them to Earth. Instructions were read and copied down by the astronauts, the computer was updated and the craft was re-aligned yet again to ensure maximum accuracy. However, the critical test would come when the main engine was fired – a failure would ruin all that had been achieved so far. Before passing behind the moon for the thirtieth time, Duke gave the crew the all clear: 'You are go for TEI.' Once again, the vital burn took place during radio blackout, and all the ground controllers could do was cross their fingers and wait. The engine was scheduled to fire for two and a half minutes, accelerating Columbia to an escape velocity of 5,900 mph (9,400 kmh) . . . the ten minute interval from the burn to acquisition of signal seemed an eternity. As they re-appeared, Charlie Duke asked, 'How did it go? Over.' Neil Armstrong confirmed the ground readings: 'Time to open up the LRL (lunar receiving laboratory) doors, Charlie.' Duke happily replied: 'Roger. We got you coming home.'

The return leg would take about 60 hours, but, with the major objectives accomplished, the crew were able to relax for the first time, particularly since the TEI had been so accurate that only one minor course correction would be needed. As they left the barren moon behind, Deke Slayton gave them a call: 'This is the original Capcom. Congratulations on an outstanding job. You guys have really put on a great show up there. I think it's about time you power down and get a little rest . . . I look forward to seeing you when you get back here. Don't fraternize with any of those bugs en route, except for the *Hornet*.' Armstrong was pleased with the good news: 'OK. thank you, boss. We're looking forward to a little rest and a restful trip back, and see you when we get there.' A number of additional questions and instructions followed before the crew were able to settle down for their first real rest period for two days – they slept soundly for more than nine hours, and awakened past noon, Houston time, on Tuesday, 22 July.

The barbecue roll was halted before the main business of the day, the mid-course correction. The reaction control system thrusters on Columbia were fired for about 11 seconds to place the craft on course for a splashdown in the Pacific Ocean two days later. There followed the first of two TV broadcasts during the homeward journey. Armstrong displayed the sample boxes which were the real purpose of the mission: 'They're vacuum-packed containers that were closed in a vacuum on the lunar surface, sealed and then brought inside the LM and then put inside these fibre glass bags.' Aldrin discussed their food and made a ham sandwich which he floated across the cabin to a colleague: 'These bite-size objects were designed to remove the problem of having so many crumbs floating around in the cabin . . . I think we've discovered that we could progress a good bit further than that – to some of the type meals that we have on Earth.' Collins demonstrated the effect of weightlessness on a liquid by filling a teaspoon with water, then turning it upside down with the water still in the spoon. He then showed viewers their water 'tap':

That's not really the way we drink. We really have a water gun which I'll show you . . . This cylindrical thing on the end of it is a filter with several membranes. One allows the water to pass but not any gas; the other allows gas to pass out but not any water . . . And of course all we do to get it started is pull the trigger. It's sort of messy . . . It's the same system that the Spaniards use to drink at a wine stand at the bullfights, only I think [that] would be more fun.

The astronauts played some taped music before retiring for the night, including some topical items such as 'Everyone's Gone to the Moon' and an ancient album of Neil Armstrong's entitled 'Music Out of the Moon'. Eight hours passed before the crew were roused. All systems were operating perfectly, though there was news that two tropical storms were meandering around the Pacific and presenting vague threats to the recovery zone. Much of the time was spent in light-hearted banter with ground control, with references to at least one of their fellow astronauts: 'We've been doing a little flight planning for Apollo 12 up here . . . We're trying to calculate how much spaghetti and meatballs we can get on board for Al Bean.' Garriott replied: 'I'm not sure the spacecraft can take that much extra weight'.

The final TV show was broadcast that evening. Each of the men gave a short speech relating his view on the epic adventure. Armstrong introduced the programme:

A hundred years ago, Jules Verne wrote a book about a voyage to the moon. His spaceship, Columbiad, took off from Florida and landed in the Pacific Ocean after completing a trip to the moon. It seems appropriate to us to share with you some of the reflections of the crew as the modern-day Columbia completes its rendezvous with the planet Earth and the same Pacific Ocean tomorrow. First, Mike Collins.

A moustachioed Collins appeared on the screen:

This trip of ours to the moon may have looked to you simple or easy. I'd like to assure you that that has not been the case . . . We have always had confidence that all this equipment will work, and work properly, and we continue to have confidence that it will do so for the remainder of the flight. All this is possible only through the blood, sweat and tears of a number of people. First, the American workmen who put these pieces of machinery together in the factory. Second, the painstaking work done by the various test teams during the assembly and re-test after assembly. And finally, the people at the Manned Spacecraft Center, both in management, in mission planning, in flight control, and, last but not least, in crew training. This operation is somewhat like the periscope of a submarine. All you see is the three of us, but beneath the surface are thousands and thousands of others, and to all those I would like to say: thank you very much.

Next came Buzz Aldrin:

I'd like to discuss with you a few of the more symbolic aspects of the flight of our mission, Apollo 11 . . . we've come to the conclusion that this has been far more than three men on a voyage to the moon, more still than the efforts of a government and industry team – more even, than the efforts of one nation. We feel that this stands as a symbol of the insatiable curiosity of all mankind to explore the unknown . . . In retrospect, we have all been particularly pleased with the call signs that we laboriously chose for our spacecraft – Columbia and Eagle. We've been particularly pleased with the emblem of our flight, depicting the US eagle, bringing the universal symbol of peace from Earth, from the planet Earth to the moon, that symbol being the olive branch. It was the overall crew choice to deposit a replica of this symbol on the moon. Personally, in reflecting the events of the past several days, a verse from Psalms comes to mind to me: 'When I consider the heavens, the work of Thy fingers, the moon and the stars which Thou hast ordained, what is man that thou art mindful of him?'

The commander completed the crew's observation:

The responsibility for the flight lies first with history and with the giants of science who have preceded this effort. Next, with the American people, who have, through their will, indicated their desire. Next, to four administrations and their Congresses for implementing that will; and then to the agency and industry teams that built our spacecraft – the Saturn, the Columbia, the Eagle and the little EMU, the space suit and backpack that was our small spacecraft out on the lunar surface. We would like to give special thanks to all those Americans who built those spacecraft, who did the construction, design, the tests and put their hearts and all their abilities into those craft. To those people tonight, we give a special thank you, and to all the other people that are listening and watching tonight, God bless you. Good night from Apollo 11.

However, before the crew could retire for the night, news came through of a change in the position of the recovery zone. Charlie Duke informed Collins that the planned splashdown area was being buffeted by heavy seas and thunderstorms, so the target which Collins had to aim at was moved 250 miles (400 km) downrange. He did not sound very pleased, though he did realize the necessity for the change. Duke reassured him: 'Mike, you get your chance at landing tomorrow. No go around.' Collins replied, 'You're going to let me land closer to Hawaii, too, aren't you?' There were 13 hours to splashdown as the crew settled down to sleep in a weightless environment for the last time.

When they awoke soon after 6 am on Thursday morning, Houston time, Columbia was approaching a crescent Earth at around 7,000 mph (11,000 kmh) and accelerating all the time. The astronauts ate their breakfast while listening to the morning news bulletin from Ronald Evans. They were feeling a desire for the comforts of home after eight days in space without a shower or even a proper wash. As Collins admitted later, they were dirty and smelly, and the cabin was little better with its bags of rubbish and urine stashed under the seats in the lower equipment bay. At least they were able to shave with a safety razor, though tissues had to be used to wipe the face.

With less than one hour to re-entry, and travelling at more than 17,000 mph (27,000 kmh), the service module was separated from the command module and moved to a safe distance by firing its thrusters. It would burn up as it entered the atmosphere at more than 27,000 mph (43,000 kmh), leaving only Columbia to return safely out of the giant structure which blasted off eight days previously. Evans informed them: 'You're still looking mighty fine down here. You're cleared for landing.' Aldrin commented: 'Gear is down and locked.' They were still looking fine as Columbia shaved the upper atmosphere and went 'over the hill', heat shield facing forward as their sole protection against the searing temperature of nearly 3,000°C.

As the craft swooped over the dark Pacific, the crew were pressed back in their couches by acceleration forces reaching 6.5G. They watched the tail of flame surround them with its orange–yellow core as friction burned away their heatshield, just as its designers had intended. Visibility was poor on the recovery carrier *Hornet*, with low cloud in the early morning twilight. The men on board saw a brief glow in the sky before it disappeared behind the clouds. When radio contact was re-established, the drogue parachutes, followed by the three giant orange and white main parachutes, dropped the capsule gently into the rolling ocean about 13 miles (21 km) from *Hornet* and 950 miles (1,500 km) southwest of Honolulu.

Columbia landed upside down, but this was soon remedied by inflating the three flotation bags. As *Hornet* and the recovery helicopters raced towards the scene, Collins assured everyone that all was well: 'All crew excellent. Take your time.'

Now came the laborious safety precautions designed to ensure that no organisms, harmful or otherwise, were brought back from the moon and released on an unsuspecting world. Half an hour after splashdown, the hatch was opened so that one of the frogmen could throw in three biological isolation garments. Once these were donned, the astronauts climbed out into a rubber boat where they were sprayed with disinfectant and scrubbed each other down with an iodine solution. They were then whisked away by helicopter to the *Hornet* while their capsule was decontaminated and then sealed. The men looked like aliens from some science fiction movie as they walked down the steps and across the deck, waving to the cheering onlookers. Their first resting place was a specially converted holiday trailer known as the 'mobile quarantine facility'. They would remain inside until they reached the Lunar Receiving Laboratory in Houston. At mission control, the staff cheered and waved small flags as they watched the successful culmination of all their efforts.

The astronauts were relieved to be able to remove the stiflingly hot isolation garments, then luxuriate with their first shower and proper shave for more than a week. After some 40 minutes in the MQF, they were greeted by a proud President Nixon as they peered through a small window in the rear of the mobile unit.

Three days later, the MQF was flown into Ellington Air Force Base where the men were able to see their wives for the first time since their return, though separated by the inevitable window. Within hours, they were installed in the Lunar Receiving Laboratory at Houston where they were to be closeted with Columbia and 20 doctors and officials. They were given every comfort and convenience while undergoing a series of debriefing sessions and medical checks. Neither the men nor any of the creatures exposed to lunar material suffered from contamination, thereby proving the absence of life forms, dangerous or otherwise, on the Sea of Tranquillity. They also proved to be in good health, though Armstrong and Collins had lost a few pounds in weight, and Armstrong had a little fluid in his right ear from the stress of re-entry. As for the precious lunar samples, the scientists expressed their delight with the selection brought back. Dr Robin Brett, a NASA geologist, described the scene on 26 July as the white-clad scientists clustered to peer through a window as a technician opened the first sample box in the high-vacuum chamber. 'When we opened that first box of moon rocks, the hushed, expectant atmosphere in the Lunar Receiving Laboratory was, I imagine, like that in a medieval monastery as the monks awaited the arrival of a fragment of the true cross.' The initial reaction was one of disappointment – one scientist observed, 'What we saw was not much different from a bag of charcoal. The rocks were so covered with dark grey dust that no one could tell a thing about them.' However, as test results came in, it became clear that the rocks were more ancient than all but the oldest rocks on earth. The lunar soil was composed of irregular glassy fragments, indicative of high temperature conditions long ago, a conclusion confirmed by the basaltic nature of the rocks. They seemed broadly similar to terrestrial volcanic rocks, though the elements were present in different proportions.

The men were released from quarantine by Dr Berry on 10 August to their families and the eagerly waiting media. There followed the usual hectic round of press meetings, parades and public engagements. On one day alone, they flew to New York for a three hour ceremony in the City Hall and a reception at the United Nations, experienced a tickertape parade along the route to Kennedy Airport, a flight to Chicago for a civic reception and a speech to 15,000 in Grant Park, a flight to Los Angeles and a state dinner to which the President had invited them on the *Hornet*. The succeeding days saw a triumphal motorcade through Houston, returns to their home towns, a television interview, speeches before a joint session of Congress, and the presentation of a piece of moon rock to the Smithsonian Institution. During September the pace never slackened as they completed a whistle-stop tour through six continents. But the adulation had to end sometime, and none of the three would be content simply to bask in their fame for the rest of their lives.

Michael Collins had already made up his mind never to fly in space again, an intention which he made public during a TV interview on 17 August. He later explained:

The flight itself is wonderful. Other parts are tedious, annoying and frustrating drudgery. During the six months before our flight I had more than 400 hours in simulators . . . I really thought that if I had to spend another two years locked up in those things I'd go bughouse . . . I can only stay 'up' on those things for a certain amount of time. I've been more or less 'up' for five years now, and it's difficult to sustain that pace . . . If I thought I were leaving Deke, or the programme, in the lurch, I wouldn't quit. But Deke has got a whole lot of new guys who haven't flown and who want to fly. Let them have their crack at it. And then there is the matter of having a normal family life with these peculiar working schedules we have. You can do both jobs, space flight and family man, but I don't think you can do a really good job of both at the same time.

In January 1970 Collins resigned from NASA to become Assistant

Secretary of State for Public Affairs. The following year he joined the Smithsonian Institution as Director of the National Air and Space Museum, remaining with the Institution for the next nine years. Today, he maintains his links with aerospace development as president of the Vought Corporation in Virginia.

Buzz Aldrin also felt he had to search for new challenges. He left NASA in June 1971 and became the first astronaut to return to active duty with the USAF with the capacity of commander of Edwards Air Force Base Test Pilots School. The restlessness remained, causing him to retire from the Air Force in March 1972 in order to set up a company of research and engineering consultants. His interest in this sector and in technological innovation remains to this day – he has been an adviser during development of the space Shuttle, and a consultant to companies involved in computers, videos, cable TV, radiology and cockpit instrument displays. He has also been a long-term director of the Mental Health Association; he has himself suffered bouts of depression which have necessitated hospital treatment.

Neil Armstrong, too, was destined never to fly in space again. In October 1971 he left NASA to take up the post of Professor of Engineering at the University of Cincinnati. He held this position until 1980, and later became chairman of Caldwell International in Ohio. For a man whose name is still a household word throughout the world, Armstrong succeeded remarkably well in drifting out of the limelight – until he joined the Challenger inquiry in 1986.

It was he who summed up the feelings of the Apollo 11 crew at the President's dinner, three weeks after their return from their momentous voyage: 'We hope and think . . . that this is the beginning of a new era, the beginning of an era when man understands the universe around him, and the beginning of the era when man understands himself.' With hindsight, this self-enlightenment seems an imaginative dream which is further than ever from realization, yet it cannot be doubted that the 15 year period of space exploration since Apollo 11 has seen an explosion in our knowledge of the universe unprecedented in human history. The words Buzz Aldrin spoke perhaps reflect the true significance of that trail-blazing mission: 'This has been far more than three men on a voyage to the moon . . . This stands as a symbol of the insatiable curiosity of all mankind to explore the unknown.'

After Apollo 11 nine more Apollo missions were on the drawing board, one more in 1969, three each in 1970 and 1971, and two in 1972, but what could they do that was different, would justify the expenditure and ensure continued public and government support for what most people saw as nine repeats of Apollo 11? There were some ambitious plans for a follow-up to the moon programme, named the Apollo Applications

Programme, which envisaged Apollo hardware being used in construction of an Earth-orbiting workshop and scientific laboratory, but President Nixon was showing signs of balking at the cost of such a project. Soon after coming to office the new President had approved further funding for Apollo, enabling the space agency to go ahead with plans to introduce some useful scientific experiments on later missions while extending the length of stay on the lunar surface and providing a means of lunar transport, but the overall NASA budget for 1970 was cut, and a further sign of the times was the cancellation in June 1969 of the Air Force's pet project, the Manned Orbiting Laboratory (MOL), a saving of 1.5 billion dollars on the defence budget over the next four years. (Two months later, NASA accepted 7 of the 15 MOL astronauts who were now redundant after two to four years of training. None of these flew on Apollo missions, although Gordon Fullerton, Henry Hartsfield, Robert Overmyer and Donald Peterson were appointed to Apollo support crews, and all seven eventually achieved their ambition by flying on the Shuttle more than ten long years later.)

Although the financial pressures were building, in one sense at least there was less pressure on NASA's staff; President Kennedy's deadline had been met so there was no longer any need to continue the hectic pace which had preceded Apollo 11. If that mission had failed, Apollo 12 would have had to pursue the elusive goal with a September 1969 launch, but now there was time to assimilate the information gained and to apply the hard-won knowledge in order to turn Apollo 12 into a more meaningful mission. As before, the landing site would be a flat, obstacle-free plains region, a site chosen more for safety reasons than for any special scientific interest. However, there were increased elements of risk: Apollo 12 would be the first lunar mission to leave a free-return trajectory on the final stages of the coast to the moon since this was the only way to reach the Ocean of Storms some 950 miles (1,500 km) west of Tranquillity Base. In addition, the crew would remain on the moon for more than 30 hours, during which they would carry out deployment of scientific instruments and complete two EVAs which would both exceed in duration the Apollo 11 EVA and take the astronauts well away from the safety of their spacecraft. A special point of interest would be an excursion to visit the unmanned Surveyor 3 craft which had soft-landed on the moon two and a half years earlier and now sat decaying in its final resting place.

The crew chosen to further man's knowledge of Earth's only natural satellite comprised three naval commanders with varying degrees of spaceflight experience. Mission commander was the ebullient Charles 'Pete' Conrad, a veteran of two Gemini flights including a successful rendezvous and docking with a target Agena. Alongside him once more as command module pilot was Richard Gordon, a colleague from Conrad's test pilot days at Patuxent River and his partner on Gemini 11.

This time, however, Gordon would have to forgo EVA, waiting patiently during a lonely sojourn in the command module. The lunar module pilot who would accompany Conrad to the surface was a rookie, 37-year-old Alan Bean. He, too, had attended the Navy Test Pilot School after graduating in aeronautical engineering from the University of Texas and spending four years with a jet attack squadron in Florida. Bean joined the astronaut corps in 1963, the same year as Gordon, but was unfortunate enough to miss out on Gemini. His partner on the Gemini 10 back-up crew, Clifton Williams, had been killed in a plane crash two years previously while training for Apollo with the prospect of flying as lunar module pilot on Apollo 12. Now, as Bean stood in that position instead, he did not forget his dead friend – there were four stars on the Apollo 12 badge, one for each of the astronauts and an extra one to commemorate C. C. Williams. Bean and Conrad were to combine to become one of the liveliest, most proficient teams in the history of space exploration. The only competition between them seemed to be over who had the least hair and who could give the broader smile; but when it came to eating there was only one winner – Al Bean was the champion spaghetti eater on the entire astronaut staff if his colleagues were to be believed.

The astronauts needed a sense of humour during the launch of Apollo 12 at 11.22 am Cape time on 14 November 1969. The previous day had seen a mixture of sunshine and thunderstorms, and as the astronauts waited in their cabin for the final countdown, threatening weather began to close in again. They knew that a delay longer than three hours would mean a two-day postponement while flight plans were revised for a different landing site. They also knew that the viewing area was packed with VIPs, including President Nixon, and that a failure to launch on time would be a blow to American prestige. As the final seconds ticked away, black rain clouds enveloped the Cape, virtually obscuring the launch pad from the spectators 3 miles (5 km) distant. Pete Conrad, settled back in his couch for the blast-off, saw rain water seeping behind the protective shroud and running down the command module windows and prepared for a rough ride. Launch Operations Manager Paul Donnelly conferred with the new Director of Launch Operations, Walter Kapryan, and with the rule book; the verdict was unanimous – 'go' for launch. It was a decision that nearly killed three men in what would have been the most embarrassing fireball in the history of manned spaceflight.

The Saturn V ignited on schedule, lifting Apollo 12 into the dark blanket which hovered only 800 feet (240 m) above the ground. The President and his entourage, sheltering under umbrellas or simply getting soaked to the skin, had a brief 20 second glimpse of the world's most powerful rocket as nature reminded the leaders of the human race they still had a long way to go before they could control the elements.

Shortly after disappearing from sight, the crew saw a bright flash outside the craft, then the caution and warning panel lit up like a Christmas tree. Conrad reported there were so many lights on that the crew could not read them all. Sixteen seconds after the first flash, all hell broke loose as the rocket approached maximum acceleration: 'OK, we just lost the platform, gang. I don't know what happened here. We had everything in the world drop out.' For a while, there was the distinct possibility of the first abort during a launch as a massive power surge tripped the circuit-breakers, shutting down the main electrical system for a few seconds. Using their back-up batteries, the crew set about restoring order in their cabin as they continued to head out over the cold Atlantic. Back at the Cape, worried controllers scanned their instruments for any danger signs, but the rocket seemed to be functioning normally. Second stage separation occurred on schedule, much to Conrad's relief, as he reported: 'We are weeding out our problems here. I don't know what happened – I'm not sure we didn't get hit by lightning.' Soon after, he was able to give a reassuring progress report: 'We have reset all the fuel cells, we have all the buses back online, and we'll just square up the platform when we get into orbit.' Fortunately, although the guidance platform on board the command module was no longer properly aligned, the system which controlled the rocket was functioning normally; a failure in this would have sent the Saturn careering out of control like a bucking bronco. Gordon expressed the crew's relief and satisfaction with the booster as they neared orbit: 'Got a little vibration of some kind, but she's chugging along here minding her own business.'

With the emergency apparently over, the irrepressible Conrad radioed to Houston: 'That's one of the better sims [simulations], believe me!' Capcom replied: 'We had a couple of cardiac arrests down here too, Pete.' 'We didn't even have time for that up here,' retorted Conrad. He added drily, 'I think we need to do a little more all-weather testing.' Apollo 12 reached Earth parking orbit, 118 miles (190 km) above the Earth, allowing the crew an opportunity to check all systems and follow the complicated procedure necessary to realign their guidance platform. Apart from a minor error in the guidance computer on the third stage, all seemed in remarkably good order, enabling Apollo to blast out of orbit during the second revolution. Vision through the windows was partly obscured by ice formed during the launch, but TV viewers were still able to see colour pictures of the successful docking manoeuvre as Conrad linked the command module, Yankee Clipper, with the lunar module, Intrepid. Soon after, the third stage was jettisoned. Conrad watched it disappear: 'It reminds me of some guy standing back there with a water hose just spraying it in any old direction.'

Houston conferred with the crew about the launch and its conse-quences. Mission control radioed: 'Your theory that it was probably lightning that did it – that looks about the best idea.' Conrad replied: 'I guess the other thing that we were thinking about, maybe not lightning so much, just unstable air. We were a pretty big piece of electricity builder going through there, so we might just have discharged ourselves.' Capcom said they were also thinking along those lines. Since their trajectory was so accurate, a planned course correction was cancelled, leading Houston to suggest a preliminary check-out of Intrepid instead. Gordon was happy to convey the crew's agreement: 'OK, it sounds good. We really don't have any place to go tonight so we don't mind working late.' Conrad and Bean spent an hour inside the LM powering up its systems and checking that no damage had occurred during the bumpy launch. On returning to the CSM, Houston noticed power consumption was one amp higher than expected, so the two astronauts were obliged to open up Intrepid once more. It turned out that a small interior lamp was not switched off; compared to the malfunctions earlier in the day, this latest breakdown was almost laughable. After more than 22 hours without sleep, it was an exhausted crew that turned in that night. Not surprisingly, they were allowed a ten hour rest period followed by three days of light duties.

The cruise to the moon gave plenty of opportunity for the cheerful, chatty crew to give TV viewers the almost routine guided tours of the craft and views of Earth. They were given a 'first', however, when the crew broadcast the scenes from inside Yankee Clipper as the computer-controlled engine burn slowed the craft to send it into a non-return trajectory. The slight vibrations could be seen on the screen as Conrad commented: 'That was nice to get an extra touch of G.' The commander made it clear that the crew had cast the anxious moments of the launch into the backs of their minds: 'We've been sitting here chuckling about it.' The overall impression was of a happy ship under the jovial, laid-back leadership of Pete Conrad: 'We are trying all these things we didn't have in Gemini, like toothpaste and shaving – we are really having a ball up here.' Capcom replied: 'Roger. All dressed up and no place to go.' Back came the reassuring comment: 'Oh, we're going someplace. We can see it getting bigger and bigger all the time.' The only event out of the ordinary came when the astronauts reported being followed by an unknown tumbling object, but Houston decided it was probably only the Saturn's third stage. 'OK, we'll assume it's friendly anyway.' Capcom confirmed: 'Roger, if it makes any noises it's probably just wind in the rigging.'

The third day saw a TV transmission from inside Intrepid as Conrad and Bean checked out the frail lunar module and showed viewers the interior fittings. Pictures were so clear that dust particles floating in the oxygen atmosphere were clearly visible. The two men were back with Gordon in less than an hour. With the next course correction cancelled, all was ready for the lunar approach.

The astronauts' fourth day began with a bosun's pipe playing

'Sweepers Man Your Brooms'. Navy commander Gordon responded: 'All persons accounted for, sir.' Once again, men from planet Earth swept into the moon's shadow and gave what might be a last farewell to their relatives, friends and workmates listening helplessly some 230,000 miles distant. Once again came the eerie silence as ground controllers and spectators alike glanced at their watches, counting the seconds to reacquisition of signal. And once again, the main engine of the CSM fired perfectly for nearly six minutes behind the moon to bring them back safely around the moon's east limb. 'Hello, Houston. Yankee Clipper with Intrepid in tow has arrived on time . . . I guess like everybody else that just arrived, we are all three of us plastered to the windows . . . but for the Navy troops it doesn't look like a very good place to pull liberty, though.'

Shortly before the crew fired the main engine again to put Yankee Clipper into a roughly circular orbit, they sent back a colour broadcast of the surface accompanied by an almost schoolboy-like exuberance: 'Boy, it's beautiful! Look at that crater! Wowee!' Al Bean's commentary was rather more staid and down-to-earth:

The black is about as black as you've ever seen in your life. It just doesn't have any hues or anything to it. It's just solid, straight, dull black. The moon is just sort of very light concrete colour. In fact, if I wanted to look at something I thought was the same colour, I'd go and look at my driveway.

They were soon clearing the tunnel and giving Intrepid a final once-over. Everything was ready for the next day's descent to the Ocean of Storms. Houston was concerned at the threat of high radiation levels due to increased solar activity, but regular monitoring inside and outside the spacecraft suggested that danger level would not be reached. The only obstacle to a sound night's sleep was a biomedical sensor attached to Conrad's chest which he said had caused 'a bunch of blisters'. The doctors agreed to let him move it to a less sensitive area.

Shortly after 7 pm Cape time on 18 November, Conrad and Bean wriggled through the narrow tunnel into Intrepid. Once the LM was powered up and sealed, Gordon reinstalled the docking mechanism in the tunnel. With its legs deployed and turned at right-angles to the orbital path, Intrepid was ready to leave her mother ship. Gordon acted as TV cameraman, pointing the camera through a window, as Intrepid finally separated on release of the docking latches at 11.16 pm. 'You're on your way, Intrepid,' radioed Gordon. 'I'll be watching you.' Behind the moon, Intrepid's descent engine fired for 28 seconds to reduce speed and initiate the drop towards the surface. 'We had a great burn,' reported Conrad as Intrepid reappeared around the limb on her long, gentle glide down across the front side.

Ground tracking stations picked up Intrepid's beacon signal, allowing very accurate measurements and course corrections to be fed into the computer as it guided Intrepid over the hazardous terrain. Conrad followed their progress intently: 'I sure hope you got us lined up right, Houston, 'cause there's sure a big mountain in front of us right now.' He need not have worried. As the LM swung into an upright position ready for the final stage, the astronauts could see the lunar landscape below, almost identical to the scenery they had studied so frequently in the simulators. 'I think I see my crater,' shouted Conrad. 'There it is! Oh my gosh! Son of a gun! Right down the middle of the road.' Switching to manual control, Conrad excitedly listened to Bean as he read off the altitude and rate of descent. Finally, there it was in the middle of a dust cloud on the Ocean of Storms, a pinpoint landing right beside the target, Surveyor crater: '18 feet coming down at 2. He's got it made. Come on in there . . . contact light.' 'A typical Navy landing,' came the congratulatory message from Capcom. Mankind's second landing on the moon was clocked at 1.54 am Cape time on 19 November.

Conrad and Bean moved rapidly through the initial check-out of Intrepid. All was well, so the astronauts peered through the windows to describe their strange environment. Conrad told Capcom: 'Man, oh man, Houston. I'll tell you I think we're in a place a lot dustier than Neil's.' He described the landing zone as 'sort of like an undulating plain.' He went on, 'I'm sure that some of these rocks have different colours and different textures, but from here in the spacecraft . . . they all appear to be of the same material and they all appear to be pure white.' Al Bean summed up their feelings: 'I can't wait to get outside.' For once, however, they dropped behind schedule in the struggle to don their large, ungainly backpacks. It was not until 6.39 am that Conrad opened the hatch and moved on to the porch. Pulling on the lanyard, he successfully deployed the Equipment Stowage Assembly, a special shelf built into the descent stage where the experiments and colour camera were stored. Pictures received on Earth showed a ghostly white figure slowly backing down the ladder to the bottom rung, 3 ft above the surface. To Pete Conrad, at 5 ft 6½ in one of the smallest astronauts, that seemed a mighty long way down to Intrepid's footpad:

> Whoopee! Man, that may have been a small one for Neil, but that's a long one for me. I'm going to step off the pad . . . Oh, is that soft. Hey, that's neat. I don't sink in too far . . . Boy, that sun's bright. That's just like somebody shining a spotlight on your hands. I can walk pretty well, Al, but I've got to take it easy and watch what I'm doing. Boy, you'll never believe it. Guess what I see sitting on the side of the crater. The old Surveyor. Does that look neat! It can't be any further than 600 ft (200 m) from here. How about that?

And so it continued. The happy astronaut bounded around as if he was on his favourite holiday beach, humming to himself and letting out the occasional war whoop as he collected the contingency sample of rocks and 'soil'. These samples were passed up to Bean in case a quick getaway was called for, then the second astronaut crawled backwards through the hatch and down the ladder. Conrad called him over to observe the nearby Surveyor, but Bean found walking harder than he expected: 'Boy, you sure lean forward.' Conrad replied lightheartedly, 'I don't think you're gonna steam around here quite as fast as you thought you were.' Bean began to look closely at the barren surface: 'Hey, you can see some little shiny glass, yeah, glass, in these rocks . . . you can also see some pure glass if you look around.'

Jolted by the realization of how quickly time was passing, Conrad reminded his partner of the work ahead: 'Hustle, boy, hustle. We've got a lot of work to do.' They set up the umbrella-shaped antenna which would relay colour TV pictures back to earth, but then came the lone failure of the session; Bean carried the camera from Intrepid to set it up on a tripod, but when he pointed it towards the LM, the screens

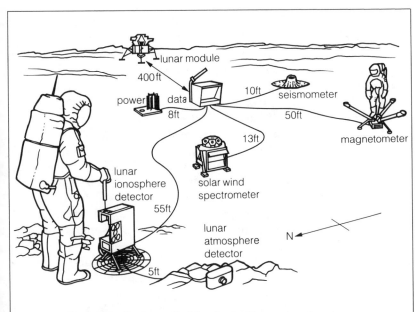

Figure 17   Deployment of the Apollo 12 experiment package

Science began to play a more important role in Apollo 12. The major innovation was a small nuclear-powered generator, linked to the various experiments. The magnetometer was to determine whether there was a lunar magnetic field, while the seismometer measured tiny moonquakes to give information on the lunar interior. Other instruments measured whether the moon had any atmosphere or ionosphere, and detected the high energy particles from the sun known as solar wind.

appeared blank. He struggled to correct it for a few minutes, even resorting to banging, shaking and turning it upside down, but to no avail. Engineers told him he had probably inadvertently turned it directly towards the brilliant sun and burned it out. The American flag was set up with little difficulty, but deployment of the experiment package proved more frustrating. The site they chose lay about 400 ft (130 m) northwest of Intrepid on the far side of a small crater, sufficiently distant to afford protection from Intrepid's blast during lift-off. The dust proved a continual nuisance, coating everything: 'There's no way to handle all this equipment with all the dust on it. Every time you move something, the dust flies . . . goes way up in the air and comes in and lands on you.' The Lunar Atmosphere Detector toppled over on its side several times; the aluminium skirt on the seismometer persisted in curling up at the edges until Houston suggested weighing it down with some dirt; and the fuel rod for the nuclear generator stuck in its graphite cask, forcing Conrad to spend ten minutes gently tapping it before it could be extracted and inserted into the power unit.

The experiments deployed at last, the astronauts headed in a northwesterly direction to the lip of a large crater named Shelf. They described the panorama: 'We're looking down at this big crater and it looks rather old and it has bedrock at the bottom . . . there are some big boulders resting inside the rim.' On the return leg, they followed a route slightly to the east which brought them across two unusual hillocks. Bean told Houston: 'I don't know what they are. They're just sort of mounds.' Warning the geologists not to take him too literally, he went on: 'It looks like a small volcano only it's just about 4 feet high and about 5 feet across at the top. It slopes down to a base with a diameter of 15 to 20 feet.' The geology field trip continued as they filled their collection bag with more rock specimens. Most of these were picked up using special tongs, but one block was too large for this, so one man pushed it over to the other who then lifted and stowed it. Among the specimens was a piece of pure glass as well as some small black 'beads'.

On arrival near Intrepid, Bean sank the core tube some 32 in (1 m) into the surface, much deeper than had been possible at Tranquillity Base. Advised that he had plenty of oxygen left, Conrad and Bean took their time stowing the samples before returning to the LM. They also attempted to remove the clinging dust which gave their white clothes a grubby, soiled appearance: 'Man, are we filthy. We need . . . a whisk broom.' They agreed to try to brush each other down, though without much success. Conrad closed the hatch 3 hours 56 minutes after he had first ventured forth. Already, the crew had extended by half the time spent by men exploring the moon, and had collected some 50 lb of rocks and soil. Mission control was particularly pleased with the fitness of the men – only towards the end of the EVA had their heart rates begun to rise, peaking at around 150. Conrad's only complaint was that he had wet feet, caused by condensed water from his cooling system collecting in his boots.

The mission profile allowed nine hours sleep after the men had eaten, then recharged their backpacks with fresh oxygen, cooling water and batteries. In an advance on Apollo 11, hammocks had been provided, but the crew were too excited to sleep on, waking after only five hours and immediately pressing Houston for an early start to the second excursion. This time they would travel further afield, visiting sites of potentially high scientific interest, including the old Surveyor. Soon after 11 pm Cape time, Conrad stepped on to the moon for the second time, followed shortly afterwards by Bean. Both set out to retrace their steps, equipped with collection bags, 70 mm cameras strapped to their chests and small maps clipped to their wrists, and a tool kit. Conrad loped past the seismometer, commenting that he felt like a giraffe running in slow motion. Mission control told him: 'Pete, we're watching you down here on the seismic data. Looks as though you're really thundering right by it.' While he checked out the experiments, Bean found a 'dandy' extra grapefruit-size rock' near Head crater. Rejoined by the commander, they tried activating the seismometer by throwing a rock down the crater side. Glass beads and fragments lay around them; one rock appeared 'shining very, very bright and clean like a ginger ale bottle'.

Moving around the western rim of Head crater, Conrad kicked the dark surface and uncovered a lighter, cement-coloured soil. This discovery prompted them to dig a trench for samples. Heading south, they skirted Bench crater, where they described what seemed like bedrock on its floor with a small, apparently once molten, central peak. Yet another very different crater, named Sharp, was visited next as they turned west. Only 40 ft (10 m) wide, Sharp had a white rim of soft material raised about 2 ft above the general level, while surrounding it was a ray pattern. Bean trenched the area for samples and a core tube sample was taken. Doing a U-turn, the men began the second half of their trek, following instructions from Houston. They were beginning to find their arms were tiring from carrying the tools and rocks, despite the aid of one-sixth gravity. Bean, parched from his exertions, told Houston: 'I'll tell you one thing, I'd go for a good drink of iced water now.' More core and rock samples were taken near tiny Halo crater. To ease the task of bending over to pick up rocks in their rigid pressurized suits, Bean tried holding on to Conrad's backpack to steady him – unknown to Houston Conrad had already fallen over several times, fortunately without damaging his suit.

The astronauts now entered Surveyor crater, cutting across the gently sloping wall with little difficulty. The robot craft showed few signs of wear from its two and a half year sojourn, though its surface now had a brownish tint. Imprints in the surface showed that it had bounced after touchdown. The mirror had warped slightly, but Bean was able to wipe

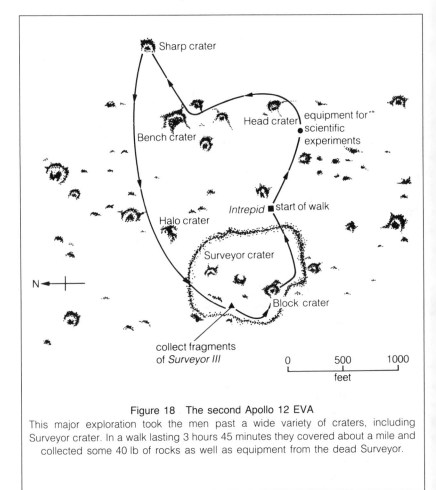

Figure 18   The second Apollo 12 EVA
This major exploration took the men past a wide variety of craters, including Surveyor crater. In a walk lasting 3 hours 45 minutes they covered about a mile and collected some 40 lb of rocks as well as equipment from the dead Surveyor.

off some of the dust coating. Using a cutting tool, they dismantled two pieces of tubing, though the piece originally intended as a target proved too thick. The men also cut off a length of cable, the craft's trenching scoop and its camera. With Intrepid gazing down on them from the dark horizon, the two enthusiastic explorers made one more visit, photographing and sampling Block crater, a small impact hollow within Surveyor crater. Around 2 am on 20 November, after some three hours' exploration, they arrived back at Intrepid loaded with about 80 lb of rocks and equipment. Conrad was sorry it was all over: 'It's really a shame, Houston. We could work out here for eight or nine hours.'

One final photograph of the area swept clean by the engine exhaust, then on with the loading of the treasures. 'Hi-ho, hi-ho,' sang Conrad as the precious cargo was lifted on board. It was nearly 3 am by the time the hatch was closed. Left behind on the surface was some 15 million dollars worth of redundant equipment; it was only later that the astronauts realized they had left a fully exposed colour film outside as well. The men were tired and dirty, but elated with the success of their moon walks: 'Man, is it filthy in here. We must have 20 lb of dirt and all kinds of junk. Al and I look like a couple of bituminous miners. But we're happy.' Conrad at last admitted that he had fallen over during the EVA: 'It's no big deal.' He added that a vigorous one-handed push was sufficient to regain an upright position.

After a fairly relaxed six hour session of eating, cleaning out the LM, checking systems and chatting to ground control about the EVA, Intrepid was ready to return to mother. At 9.26 am, Conrad pushed the 'engine start' button to blast off the Ocean of Storms after 31½ hours on the surface. 'Lift off and away we go!' yelled Conrad. 'Harbour master has cleared you into the main channel,' came the calm voice of Capcom. The initial orbit proved too high, so the crew had to fire the thrusters to drop down a little, but from then on it was strictly routine. Behind the moon, the lower, faster Intrepid closed to within 125 miles (200 km) of Yankee Clipper. The mother ship was in their sights once more. During the next revolution the gap closed steadily until mission control advised Gordon: 'Stand by to receive the skipper's gig.' 'Aye-aye, sir,' responded Gordon. He was pleased and relieved to be receiving company again after 19 solo orbits. Turning the camera out of the Clipper's window, he showed TV viewers the amazing sight of Intrepid growing from a tiny speck against the cratered background. With the separation only half a mile, Gordon called up Conrad: 'Hey, Pete, how can you look so good when you look so ugly?' 'I don't know,' replied Conrad cheerily, 'you look pretty good yourself.' The transmission continued right up to the final docking when the roof of Intrepid filled the screens. 'Attaboy, we're in,' came Conrad's voice as the docking latches closed. 'That was cool, wasn't even a ripple.'

It took two hours to unload Intrepid of her booty, but Gordon refused to allow the filthy pressure suits inside his clean command module, so the two lunar explorers had to strip off and float back in a state of undress. Intrepid was then cast off and her thrusters fired to send her crashing into the moon. Scientists received quite a surprise when the Apollo 12 seismometer registered shock waves for 55 minutes, far longer than any vibration recorded on Earth. 'It was as though one had struck a bell in a church belfry a single blow and its reverberation had continued for 55 minutes,' said one expert. Clearly, the internal structure of the moon was very different from that of its nearest neighbour.

A day of lunar observations remained before the crew could head for home, continuing the programme of surface photography carried out by Gordon during his long stay in orbit. Then, at the end of the forty-fifth lunar revolution after nearly 90 hours circling the moon, the trans-Earth injection burn was successfully completed in that strange, silent period when communication was impossible. Swinging back into view, mission control received the welcome confirmation: 'Houston. Apollo 12's moving home.' TV pictures showed the rapidly receding moon with the apparently rough terrain thrown into sharp relief by the black, elongated shadows. At last the period of excitement and constant activity was over; mission control allowed the weary adventurers a welcome 12 hours sleep period. By the time they awoke, Yankee Clipper was already under the Earth's gravitational influence as they accelerated homewards. Further TV sessions, one featuring a question-and-answer sequence with the press, and long sleep periods marked the uneventful return. The only minor problems came with an excess of water from the fuel cells and their old enemy, the lunar dust. Conrad complained: 'The spacecraft is so loaded with dust that we've got to clean the screens in here every 2 or 3 hours. Our suits look like we've been wallowing around in graphite.' The doctors prescribed decongestant tablets to ease their nasal irritation. The commander's message to the captain of the recovery carrier *Hornet* was more typical of his normal exuberance: 'Tell the skipper of that ship to have it right on target because Dick's gonna be driving this thing right down the middle of it.'

The final approach was marked by 'a fantastic sight' as the Earth moved across the face of the sun. Al Bean described the unique spectacle:

> This has got to be the most spectacular sight of the whole flight. The sun is eclipsed now and what it has done is to illuminate the entire atmosphere all around the Earth. All you can see is a sort of purple-blue with some shade of violet. It's a heck of a time to be without a 70 mm colour film, I'll tell you. I'll try to get it on my 16 mm . . . We can see clouds sort of on the dark part of the earth . . . The clouds appear sort of pinkish-grey and they're sort of scattered all the way around the Earth.

Two minor course corrections put Yankee Clipper right on target for her Pacific splashdown. Three and a half hours after observing the eclipse, the service module was jettisoned, exposing the Clipper's heat shield. Streaking into the atmosphere at just the right angle, the CM skipped to aid braking. Only 11 minutes after beginning re-entry, the craft was spotted swinging in the breeze beneath its three giant parachutes. In rough seas and a 15 knot wind, the craft hit the water hard and turned turtle: it was probably just as well all three men were Navy personnel. The flotation bag soon righted the craft as helicopters dropped frogmen and hovered expectantly overhead. Yankee Clipper had fulfilled her commander's boast; she had splashed down only 3 miles (6.5 km) from *Hornet*. The full biological isolation garment was considered unnecessary this time, so the crew were able to clamber into the life raft wearing their flight overalls though their faces were still obscured by the germ filtering masks. Once on the *Hornet*, they were immediately whisked to the mobile quarantine facility where they remained until they transferred to the Lunar Receiving Laboratory at Houston.

The Apollo 12 crew spent 16 days in quarantine before being released to their families and world acclaim. Conrad remained with NASA and the Navy until December 1973, having flown in space for a fourth time as commander of the Skylab 2 mission earlier that year. Alan Bean also participated in Skylab as commander of the third mission in 1973, setting a world endurance record of 59 days in space, then acted as back-up commander for the Apollo–Soyuz collaboration in July 1975.

Richard Gordon continued to be involved in the Apollo programme after his initial success, though he never set foot on the moon and he never flew in space again. After serving as Apollo 15 back-up commander, he was assigned to head the Advanced Programmes section of the Astronaut Office where he worked on the design and testing of the Shuttle. Unable to settle to this new role, he resigned from NASA and the Navy in 1972 to take charge of the New Orleans Saints Professional Football Club as executive vice president. In 1977 his knowledge as a chemistry graduate was utilized when he went to work for a Texas company involved in the discovery and exploitation of oil and gas resources. In 1978 he moved into the engineering field, becoming president of Astro Systems and Engineering in Los Angeles.

The amazingly successful mission of Apollo 12 following so soon after Apollo 11 seemed to confirm public and government opinion that flights to the moon and back were now more or less routine. The loss of colour pictures from the lunar surface was considered a disaster by the NASA public relations department, who were anxious to maintain public interest and support, and thereby keep the pressure on the President and Congress to continue funding. From a scientific viewpoint, the 'ringing of the bell' during the crash of Intrepid on to the surface had been an unexpected bonus and a sure incentive to send more and better equipment to as many varied sites as possible. The pinpoint landing next to Surveyor demonstrated perfectly the ability to send missions to more rugged, inhospitable regions. The question was, would the American administration support the continued exploration of the moon while Congressmen cried out for an end to such 'wasteful' expenditure and for increased spending on health and welfare programmes? There was a sign of the times when a militant atheist took out a suit to prevent astronauts sending back prayers on future space missions. What a difference a few months made!

9

'We've Had a Problem'

Apollo 13, 1970

The successes of the Apollo programme had led to widespread indifference among the public and the government as preparations went ahead for Apollo 13 at the end of 1969 and the beginning of 1970. Neither the President nor Congress were prepared to accept the necessity of funding eight repeats of the Apollo 11 mission at a cost of around 400 million dollars per mission. The fact that the later Apollo missions were scheduled to carry a greater variety of scientific experiments and explore a wider area made little impression on a government seeking budget cuts or on a public becoming blasé about lunar voyages. NASA inevitably came off worse in the annual budget negotiations, so that one, and then another two Apollo landings had to be scrubbed from the list.

The atmosphere among space administration staff seemed to rub off on the forthcoming Apollo 13 – even the number 13 led some to gloom and despondency. The preparations were hardly incident-free: liquid oxygen drained from the rocket during a test, formed an oxygen-rich atmosphere and caused sparks from car ignition systems to set the vehicles alight, while a helium tank on the lunar module repeatedly reached a higher pressure than expected. More significantly, one of the two tanks which supplied oxygen for the CSM's fuel cells and cabin atmosphere refused to drain properly, despite numerous attempts. The only way to empty the tank was to turn on its internal heaters and carry out numerous pressure-cycle operations. Under pressure to meet a launch date of 11 April, the engineers decided – fatefully – to retain the tank since the drainage problem did not seem to affect its performance.

Only five days before the launch date, the 'jinx' spread to the crew. In August 1969, the prime crew of James Lovell, Fred Haise and Thomas Mattingly was announced with a back-up crew of John Young, John Swigert and Charles Duke. Already fate had intervened since Lovell, a

Gemini veteran, had been on the Apollo 8 back-up crew with Armstrong and Aldrin when Michael Collins was grounded for medical reasons. As a result, Lovell flew on Apollo 8 and missed a place in history when his companions on the back-up became the prime crew for the first lunar landing. Now it was the turn of Thomas Mattingly when he was removed from the Apollo 13 crew on 6 April because Dr Berry found that he had no immunity to German measles and had been in contact with Charles Duke just before the latter had come down with the disease. There was no possibility of a straight switch between the prime and back-up crews, so Jack Swigert was drafted in as a last minute replacement as command module pilot. This was hardly an ideal situation for Swigert or the two men who would be his intimate companions in a small, overcrowded spacecraft for a week, with no experience of working together in critical situations. In order to see how the men got on together in stressful conditions, the three were put through a rigorous series of simulations in the short time remaining. Lovell and Haise were happy with their new partner, causing Swigert to make hasty arrangements for being away from home for the next nine days.

Captain James Lovell, 42 years of age and an astronaut for nearly eight years, was the veteran commander of Apollo 13. He had already been in space three times and, having viewed the moon from close quarters on Apollo 8, was unlikely to be overawed by setting foot on another world. His lunar module pilot was Fred Haise, a country boy from Biloxi, Mississippi, well known to Lovell since he had been the LM back-up on Apollos 8 and 11. As a pilot he was hard to beat, with experience flying jet aircraft for the US Marine Corps, US Navy, Oklahoma Air National Guard and USAF before he became a civilian research pilot with NASA. He became an astronaut at the same time as the new command module pilot, Jack Swigert, in April 1966. Now 38 years old, Swigert was the only member of his astronaut group to have seen service in Korea, but he too was a civilian when he joined NASA, having previously worked as a test pilot with North American and Pratt and Whitney. So far he had only worked on the support crews for Apollos 7 and 11, and his first space flight must have seemed several years away when he was appointed Apollo 13 CM back-up. Then came the German measles episode and Swigert's opportunity to jump the queue of astronauts.

Apollo 13 set off from Pad 39A at 2.13 pm on 11 April 1970 without the public interest and acclaim accorded to her illustrious predecessors; such was the price of success. The message from ground control was, 'Good Luck. Head for the hills', a reference to their lunar destination of the Fra Mauro uplands. The Saturn V rocket turned temperamental, however, when one of the second stage engines cut off more than two minutes early. 'What's the story?' asked Lovell, to which ground control could only answer that they did not know the reason for the cut-off but the mission was still 'Go'. The four remaining engines were fired for 34 seconds longer than planned to help compensate for the shortfall in thrust, then the third stage engine fired for an extra 9 seconds so that Apollo 13 entered Earth orbit as if nothing untoward had happened, though 44 seconds behind schedule. Despite the extended burn, there was still enough fuel for the trans-lunar injection by the main engine. All three astronauts had remained calm during the crisis: Lovell's pulse reached 116, while his companions' registered a maximum of 102. When passing over the Cape, 98 minutes after launch, the three men completed the first TV broadcast of the mission, then prepared for the next crucial stage.

Apollo 13 now performed perfectly, as the craft was boosted out of Earth orbit, the CSM, Odyssey, separated from the third stage and the lunar module, named Aquarius, was successfully docked and extracted from the housing on the third stage by Swigert as Haise pointed the camera through the window in a live broadcast of the event. Approaching six hours into the mission, the third stage was set on a collision course with the moon; three days later it impacted on the Ocean of Storms, causing an intense seismic signal which was picked up by the nearby seismometer left by Apollo 12's crew.

The ship drifted along on her 'free-return' course until the evening of the second day when a short engine burn sent her towards lunar orbit. Once again, this critical manoeuvre was watched by TV viewers, a sign of the confidence that pervaded everyone concerned with the Apollo programme. The lighthearted atmosphere was exemplified by Jack Swigert's revelation to Capcom Vance Brand that he had found a sign under one of the computers saying, 'My name is Hal'. 'I wonder how that got there,' he said. 'Just remember you have to be good to Hal,' replied the Capcom.

The flight controllers, organized on four shifts, were relaxed and confident, as the flight entered its third day. The craft was performing well, rolling in barbecue mode at three revolutions an hour and so accurately on course that the final mid-course correction was cancelled. The flight director, Milton Windler, told the press, 'We've had no hardware problems at all', a comment he would soon regret. The astronauts continued their programme of routine checks, TV broadcasts, meals and sleep, relaxing to music provided by on-board tapes. The only crisis came when Swigert remembered that he had left in such a hurry that he had forgotten to file his income tax return before 15 April, the final date. 'That isn't too funny,' he moaned. 'Things started happening real fast down there and I do need an extension.' Capcom Joe Kerwin reassured him: 'We'll see if we can get the agent out there in the Pacific when you come back.'

There were other minor problems, however, which later assumed a much greater significance than seemed proper at the time. Swigert had been having trouble reading the gauge for one of the oxygen tanks: the reading had risen off the scale, causing ground control to make frequent requests for a 'cryogenic stir' to stir up the oxygen. There were also worries over helium pressure in the lunar module descent stage, again continuing the hitches first experienced before launch. Lovell also reported low pressure in one of the hydrogen tanks. 'It might be interesting that just after we went to sleep last night we had a master alarm and it really scared us. And we were all over the cockpit like a wet noodle.'

At 55 hours into the flight, the crew of Apollo 13 were making another of their scheduled TV broadcasts as they opened up Aquarius for the first time. Lovell opened the broadcast in the manner of an experienced TV compere: 'What we plan to do for you today is start out in the spaceship Odyssey and take you on through from Odyssey in through the tunnel into Aquarius.' In fact, few networks were taking the programme live due to a drop in the ratings, so it was a small audience who watched as Lovell panned the camera to show the grey interior of the command module. The guide for the tour of the lunar module was Fred Haise, since he had few tasks to tackle until Aquarius was powered up when they entered lunar orbit. Cameraman Lovell followed the white, ghostly figure of Haise through the tunnel into the LM. There he demonstrated some of the equipment to be used on the moon, including a drinking bag which he and Lovell would have inside their helmets. Lovell commented, 'So if you hear any funny noises on television during our moon walk, it is probably just the drink bag.'

Lovell returned to Odyssey in order to focus on the third crew member, seated on the middle couch and confronted by control panels on three sides. He was too busy to do more than smile at the camera initially, but a few minutes later he took over the camera so that Lovell appeared on the screen for the first time. The commander demonstrated their selection of music, including, of course, 'The Age of Aquarius'. Finally, at 55 hours 47 minutes, following a suggestion from the ground, Lovell brought the broadcast to a close. 'This is the crew of Apollo 13 wishing everyone there a nice evening, and we're just about ready to close out our inspection of Aquarius and get back for a pleasant evening in Odyssey. Goodnight.' Ten minutes after they went off the air the astronauts' hopes for a pleasant evening were suddenly shattered. Lovell later described the situation:

Fred was still in the lunar module, Jack was back in the command module in the left-hand seat, and I was halfway in between in the lower equipment bay wrestling with TV wires and a camera – watching Fred come on down – when all three of us heard a rather large bang – just one bang. Now before that . . . Fred had actuated a valve which normally gives us that same sound. Since he didn't tell us about it, we all rather jumped up and were sort of worried about it; but it was his joke and we all thought it was a lot of fun at the time. So when this bang came, we really didn't get concerned right away . . . but then I looked up at Fred . . . and Fred had that expression like it wasn't his fault. We suddenly realized that something else had occurred . . . but exactly what we didn't know.

It was 10.11 pm Cape time, and one hour earlier at Mission Control in Houston, Texas. The crisis had begun five minutes earlier, unknown to either the crew or Mission Control. An amber warning light had flashed on, indicating low pressure once more in a hydrogen tank in the service module. As before, a message was transmitted to Jack Swigert: '13, we've got one more item for you when you get a chance. We'd like you to stir up your cryo tanks.' The CM pilot duly obliged, throwing the four switches that would set the fans in motion in the hydrogen and oxygen tanks. In the service module, attached to the rear of the command module, the wires in oxygen tank 2 were almost bare of insulation – later investigations suggested that this was the result of accidental overheating during the prolonged emptying of the oxygen tank more than two weeks earlier. Sixteen seconds after the operation started, an arc of electricity jumped across two wires, causing a fire in the oxygen tank and a rapid increase in oxygen pressure. This went unnoticed for some time since the hydrogen pressure warning light overrode the warning light system for oxygen pressure. Meanwhile, leaking oxygen allowed the fire to spread through the whole interior of Bay 4 in the service module. The oxygen, together with gases given off by burning insulation, finally blew out the weakest part of the bay – the panel on the craft's outer hull – causing the whole spacecraft to vibrate violently with a loud bang which was audible to the three astronauts. About two seconds later Swigert noted that a master alarm was sounding in his earphones, and that an amber warning light signalled a power drop in main bus B. Swigert slammed the hatch shut behind Haise as he emerged from the tunnel, then returned to his seat to inform Mission Control: 'OK, Houston, we've had a problem.'

Jack Lousma was unprepared for the sudden announcement. 'Say again please.' Swigert repeated: 'Houston, we've had a problem. We've had a main bus B undervolt.' Haise added, 'And we had a pretty large bang associated with the caution and warning there.' The urgency in the men's voices made the flight controllers sit up to attention. No one knew exactly what had happened: the telemetry from the craft could not directly relay the fact that an oxygen tank had blown up, it could only send back a confusing pattern of pressure and temperature readings. The reason for the loss of power was not immediately obvious either; in fact, the explosion had closed oxygen supply line valves to fuel cells 1 and 3 which had then ceased to operate. Without all three fuel cells functioning normally, the moon landing was automatically cancelled. The seriousness of the astronauts' position gradually dawned as Lovell reported that the oxygen pressure gauge for tank 2 was now reading zero and the pressure in tank 1 was noticeably dropping. Not only was the oxygen vital for the fuel cells to make electricity, but it was the source of nearly all the water and the cabin oxygen consumed by the men. It began to look increasingly likely that the crew of 'unlucky' 13 would die from suffocation in a dark, cold spacecraft drifting out of control in the vast emptiness of space.

Lovell described his feelings at the time:

> When you first hear this explosion or bang . . . you don't know what it is . . . then I looked out the window and saw this venting . . . my concern was increasing all the time. It went from 'I wonder what this is going to do to the landing' to 'I wonder if we can get back home again' . . . and when I looked up and saw both oxygen pressures . . . one actually at zero and the other one going down . . . it dawned on me that we were in serious trouble.

Lovell floated out of his seat to peer through the window. He saw a white, wispy cloud surrounding the service module. He calmly reported to a disturbed Jack Lousma: 'It looks to me that we are venting something. We are venting something out into space.' This sighting explained why the spacecraft kept veering off course, even allowing for the loss of half the attitude thrusters due to the failure of main bus B. The thrust from the leaking gas was counteracting the manual manoeuvring being practised by the commander. It was also threatening to push the craft into an attitude that would cause the guidance system to lock, thereby leaving Apollo 13 without its most important navigational aid.

Chief flight director Gene Kranz set inquiries in hand immediately. 'OK, let's everybody think of the kind of things we'd be venting. GNC [guidance and navigation control officer], you got anything that looks that normal in your system?' The reply was negative. 'OK, now let's everybody keep cool. We got LM still attached, let's make sure we don't blow the whole mission', was Kranz's message to his team. It soon became abundantly clear that the presence of the LM was the one saving grace in a potentially fatal crisis, for it, alone, was unaffected by whatever was disabling the CSM.

The situation on board Odyssey continued to deteriorate. Communications with mission control were poor, and sometimes ceased altogether as the craft continued to wobble. Two fuel cells were dead – an unprecedented state of affairs – and one oxygen tank was registering zero. The CSM's electricity would last only as long as oxygen continued to reach it from the other tank, but the power output from the surviving fuel cell was steadily dropping, as was the oxygen in the second tank.

Kranz ordered the astronauts to begin powering down the CM in order to reduce the strain on the main bus A, but he was still hoping somehow to salvage the moon landing. Fred Haise and his companions were more realistic, as Haise later explained:

The ground may not have believed what it was seeing, but we did. It's like blowing a fuse in a house – the loss is a lot more real if you're in it. Things turn off. We believed that the oxygen situation was disastrous, because we could see it venting. The ground may have been hoping there was an instrumentation problem, but on our gauges we could see that the pressure was gone in one tank and going down in the other, and it doesn't take you long to figure out what happened.

The reason for the reluctance to believe the evidence was later given by a NASA engineer: 'Nobody thought the spacecraft would lose two fuel cells and two oxygen tanks. It couldn't happen.' Jack Swigert supported this view: 'If somebody had thrown that at us in the simulator, we'd have said, "Come on, you're not being realistic." '

The astronauts had been boosting the power output from main bus A by connecting it to the re-entry battery, a back-up supply designed to last for up to ten hours. However, as the power drain continued, they were told to disconnect it; the battery would be their sole source of electricity during re-entry, and without it they were as good as dead. A further precaution followed when the oxygen surge tank in the CM was isolated, so ensuring an oxygen supply during the period of re-entry. Ground control then closed the valves which linked fuel cell 3 to the oxygen tanks in the vain hope that it was the fuel cell, not the tanks, which was leaking. Once closed, these valves could not be opened again, so the moon landing was finally cancelled. The closure made no difference to the fall in oxygen pressure, and it was, at last, clear to everyone that the LM would have to act as a liferaft over the next four days.

A number of problems immediately surfaced as this plan was closely studied. Although some tests of the LM had been carried out in Earth orbit by the crew of Apollo 9, there had never been any practice or simulation of a mission with the whole crew relying solely on the LM. Nevertheless, contingency plans for such an emergency had been laid long ago by NASA engineers. More significant, perhaps, was the fact that the LM was designed to keep two men alive for up to 50 hours, whereas it was now required to provide life support for three men over a period of 84 hours. With the CSM main engine not to be relied upon, all course corrections would have to be made by the LM descent engine, including the vital burn that would return Apollo 13 to an Earth rendezvous trajectory.

The crew began charging their re-entry battery to ensure that it would be in prime condition when it was needed. Jack Lousma contacted Fred

Haise 1 hour 45 minutes after the crisis began: 'We have a procedure for getting power from the LM we'd like you to copy down.' Haise replied, 'OK, Jack. About how long is it?' The Capcom reassured him: 'It's not a very long procedure, Fred. We figure we've got about 15 minutes worth of power left in the command module. So we want you to start getting over in the LM and getting some power on that.'

Meanwhile, mission control contacted the manufacturers of the major systems which comprised Apollo 13, so that their top specialists were immediately available for consultation and advice. A spacecraft analysis team was set up to keep in constant touch with engineers at Grumman Aerospace in Bethpage, New York, the prime contractors for the LM, and at North American Rockwell in Downey, California, who were the CSM prime contractors. These and their subcontractors, notably those responsible for the LM's environmental control sub-system – Hamilton Standard Division of the United Aircraft Corporation at Windsor Locks, Connecticut – were tied into a trans-continental link-up of simulators, computers and experts assembled with the sole objective of returning the crew of Odyssey safely to Earth.

The first priority was to align the guidance system on Aquarius before the corresponding system on Odyssey was switched off, a power-consuming but essential task. Lovell explained this later: 'Without knowing exactly which attitude the spacecraft is in, there's no way to tell how to burn or how to use the engines of the spacecraft to get the proper trajectory to come home.' Within an hour of Haise entering Aquarius, the CM was shut down completely. Now that the crew were totally reliant on the LM for survival, the question arose concerning the quickest and safest way to get the men back. The idea of simply reversing course was soon squashed; the LM's descent engine did not have the power to carry out such a manoeuvre, and a failure would cause the craft to crash into the moon. Yet, assuming the burn to place Apollo in a free-return trajectory once more was successful, the time that would elapse before re-entry would be an unacceptable 100 hours. Ways would have to be found to stretch the consumables on board the LM while reducing the duration of the flight. Fortunately, it soon became clear that there was enough oxygen in the LM to last the required four days, though electricity and water were very much on the borderline.

The flight controllers eventually agreed to two burns, the first to place Apollo on course for Earth, and the second, after the craft passed behind the moon, to boost Apollo on its way. At 61½ hours into the mission, and nearly six hours after the bang, the LM descent engine was fired manually by Lovell for 30 seconds. The first step on the long road home had been successfully taken and everyone felt a little better. Lovell commented that his main concern was at least to get into Earth's atmosphere since it would be better to burn up like a meteor than not to come back at all.

Although there would be little sleep for the flight controllers that night as they debated, argued and tried to reach some conclusions concerning the remainder of the flight programme, the astronauts were given a chance to rest, and, if possible, relax. There were no seats in the LM, with standing room only for two men, so while two crew members remained in Aquarius, the third moved into Odyssey and tried to catch some sleep. The rota began at around 4 am when Lovell sent Haise to his 'bed' on the couch in the dark, silent command module. Fatigue and stress were affecting all three by now, so that Lovell found it very difficult to control the craft's attitude by manual control of the LM's thrusters. His task was made even more difficult since the LM was positioned at one end of the combined craft and its thrusters were not designed to control a spacecraft with a mass several times that of Aquarius; furthermore, the dashboard display for the guidance platform had been switched off in order to conserve power. It was hardly surprising that he sometimes forgot which direction the craft was drifting, or that he once turned the craft completely around.

One of the major worries was the water supply; the LM supplies were very limited. Without consulting Mission Control, Lovell instructed Swigert to transfer some of the drinking water from the command module to Aquarius. This, too, proved something of a problem, since the hose fittings in the two modules did not match, so Swigert had to resort to using plastic drink bags, with the inevitable spillages. Meanwhile Lovell finally placed Apollo in the correct attitude – sideways on to the flight path – with the help of astronaut Charles Duke, the back-up LM pilot, who was working in a lunar module simulator at the Houston control centre. However, Aquarius was not designed to roll the craft automatically in order to prevent overheating, so Lovell was told he would be reminded by the Capcom to rotate Apollo through 90 degrees once every hour.

When the morning came, the leading NASA executives involved in the manned spaceflight programme assembled at the rear of the control room to debate the type of burn Apollo 13 would make at around 8.30 pm Houston time that evening, two hours after it had passed behind the moon. The first proposal was to jettison the SM in order to reduce the craft's mass, then fire the LM's engine at full throttle so that the return trip would only take a day and a half. This idea was rejected on the grounds that it left no margin for error. A second proposal involved a fast burn which would give them a Pacific splashdown in the vicinity of the recovery ships, but it, too, was rejected on the grounds that the SM would have to be jettisoned, thereby exposing the heat shield on the command module to the bitter cold of space. The proposal eventually adopted envisaged a slower burn, which would take 24 hours longer, with the advantages that the SM would remain attached and a Pacific splashdown would occur.

After six restless hours trying to sleep, Haise was aroused by Lovell. All three had a meal, then Lovell and Swigert retired for some well-deserved rest to the command module bedroom. They, too, had difficulty sleeping; shafts of sunlight entered the windows as the craft rolled on its axis. Lovell suggested pulling down the shades, but this simply cut out the sun's heat, and with no electrical power available, the temperature dropped markedly. Lovell eventually gave up trying to sleep and returned to Aquarius where he discussed ways of checking the alignment of the guidance platform – normal methods of star sightings were impossible due to the cloud of bright particles which kept the craft company, so ground control substituted a sun check. Despite Lovell's grumbles, he and Haise managed to hold the craft steady long enough in order to align the platform.

The crew were so busy they had little time to notice how rapidly they were approaching the moon. Apollo entered the moon's shadow 76½ hours into the mission. Lovell had already swivelled the craft into position for the subsequent burn, so they were able to relax a little as they passed out of contact with mission control and swept to within 160 miles (254 km) of the lunar surface. Eight minutes after losing radio contact, Apollo's crew watched the sun rise above the lunar horizon, enabling Haise and Swigert to grab their cameras for souvenirs of the world on which they were destined never to set foot. After 25 minutes of radio silence, Apollo 13 re-appeared around the limb of the moon – the astronauts were on their way home at last.

The tension mounted during the next two hours as everyone prepared for the crucial boost which would cut the return time by nearly ten hours. Capcom Vance Brand informed the crew that the third stage had impacted on the moon, to which Haise wryly replied that he was glad to hear that something had worked right on the flight. Swigert, now redundant like his command module, watched over their shoulders as Haise and Lovell prepared for the burn. At 79 hours 27 minutes into the mission, Lovell manually fired the LM descent engine for a total of 4 minutes 23 seconds, gradually increasing the thrust in three carefully planned and executed stages. Mission control then gave instructions to the crew to attempt the 'barbecue' roll again, a task which proved just as difficult as previously. Meanwhile, elation turned to dismay and disbelief on the ground as tracking stations showed that Apollo 13 was moving further and further off course – something seemed to be venting. The astronauts were not immediately informed; they had been psychologic-ally boosted by the successful burn, but they were tired and increasingly irritable. One of them complained to Brand as he read up yet more suggestions, 'Hey, we've gone a hell of a long time without any sleep.'

At about 10.30 pm Houston time on 14 April Brand at last told them they could begin to power themselves down, but they were still awake an hour later. Swigert, in particular, was worried – although he had little to do at present, he would be responsible for the re-entry. He anxiously inquired of Brand, 'You think main bus B is good, don't you?' The Capcom re-assured him: 'That's affirm. We think it is, but we want to check it out anyway. We think you guys are in great shape all the way around. Why don't you quit worrying and go to sleep?' A wistful voice came back over the loudspeaker, 'Well, I think we just might do that. Or part of us will.'

Haise took over as his two companions tried to get some rest. There was further alarm soon after when he reported some debris floating past the window: 'I just heard a little thump down in the descent stage and saw a few snowflakes coming up from that way.' Hurried checks by ground control could find no new fault, much to everyone's relief. Later, as the carbon dioxide level rose higher than on any previous flight, Haise switched to a secondary lithium hydroxide canister as a short-term replacement before the next primary canister was brought into use.

Lovell was unable to sleep, so after about three hours in the command module he left Swigert and returned to relieve Haise. Capcom Jack Lousma was surprised: 'Gee whiz. You got up kind of early didn't you?' Lovell explained: 'It's cold back there in the command module.' Lousma replied that they were going to leave him for a longer rest 'because we figure you're pretty worn out', but he accepted Lovell's restlessness and briefed him on the carbon dioxide situation. He also suggested that the astronauts might wear their spacesuits to combat the falling temperature – now below 10°C – but they declined, feeling that they would be too cumbersome. There were blankets in their emergency landing pack, but these were left buried under a mound of other stores and equipment in the command module.

The morning of Wednesday, 15 April saw the astronauts piecing together an adapter which would enable the lithium hydroxide canisters in the command module to remove carbon dioxide from the LM's atmosphere. Following instructions sent up by mission control, a 'mailbox' was constructed to cleanse the air by sucking it through a hose taped to a canister. Although not the standard of equipment usually installed on spacecraft, the important point was that it worked; the carbon dioxide level was never a worry again.

Lovell tried to rest once more, but only slept for 'four or five hours'. The afternoon was spent preparing for yet another crucial engine burn to correct Apollo's drift off course; if it failed, the craft would skip off the Earth's upper atmosphere and be lost for ever. This manoeuvre had to be made before the build-up of pressure in the helium tank on the LM reached the point when it automatically vented into space. Everyone hoped that this new venting would not push Apollo off course again.

At 10.31 pm Houston time, the LM descent engine was fired yet again for a 15 second burst intended to slow the craft by just 7 feet per second so that it would hit the narrow re-entry corridor through the atmosphere.

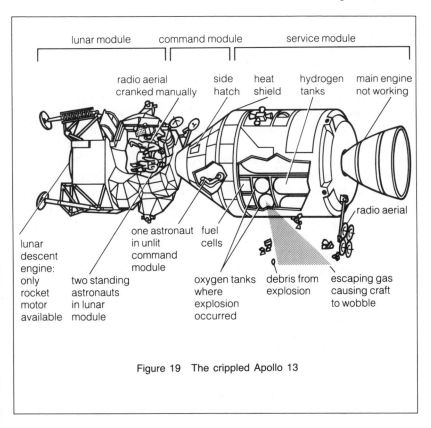

Figure 19   The crippled Apollo 13

Lovell recounted his feelings at a post-flight press conference:

When the ground read out the procedure to us, I just couldn't believe it. I thought I'd never have to use something as way out as this. And here I was on Apollo 13, using this very same procedure [which I had helped work out on Apollo 8]. Because it was a manual burn, we had a three-man operation. Jack would take care of the time. He'd tell us when to light off the engine, and when to stop it. Fred handled the roll manoeuvre and pushed the buttons to start and stop the engine.

Lovell himself had to keep the telescope cross hair just touching the 'horns' of the crescent Earth by manually firing the attitude thrusters. Successfully completed, the LM was powered down again so that it used only 10–12 amps of current. Lovell stayed on watch while the other two rested.

The helium tank blew off its surplus pressure at about 2 am Houston time on Thursday 16 April. It was supposed to be non-propulsive, but Lovell and Capcom Jack Lousma agreed that they would hate to see a propulsive release. Lovell commented: 'I noticed a lot of sparklies going out.' He then had to struggle to set up the thermal roll once more; he was so bleary-eyed that at one stage there was confusion over whether the Earth or the moon was passing across the window. When Lousma asked about his shipmates, Lovell described their strange sleeping arrangements: 'It's sort of humourous. Fred's sleeping place now is in the tunnel, upside down, with his head resting on the ascent engine cover. Jack is on the floor of the lunar module with a restraint harness wrapped around his arm to keep him down there.' This was hardly surprising, since the temperature in the CM was now down to about 3°C. The walls and windows of Apollo were dripping with condensation, Haise was coming down with a kidney infection and Swigert, without the lunar overshoes available to the others, still had wet feet after two days. None of them had more than 12 hours sleep in the 3½ days between the explosion and splashdown.

The doctors were worried about the effects of fatigue on their efficiency. As Apollo neared Earth, there were fewer problems about electricity, though the water supply was a worry – the rationing forced by the freezing up of the command module water tank meant that the crew were drinking only 6 oz (170 g) a day, less than one-fifth the normal minimum requirement. This, too, was a cause of growing concern, for dehydration can cause those affected to make uncharacteristic errors.

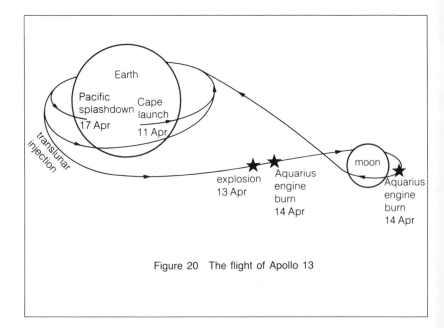

Figure 20   The flight of Apollo 13

The last 24 hours were the longest that any of the flight controllers or the crew had ever experienced. Lovell's thoughts, like those of everyone else, were on the re-entry. He asked Joe Kerwin if the re-entry checklists were ready to be sent up; the reply was that they still required a final check in the simulator. While waiting for this to be completed, the astronauts transferred any essential film and equipment from the LM to the CM, together with extra ballast to take the place of the 100 lbs of moon rocks which they would normally have been carrying. Meanwhile, Haise moved out as much as possible of the 'garbage' which had been stored in Odyssey. When he had finished he told Capcom Vance Brand, 'Boy, you wouldn't believe this LM right now! There's nothing but bags from floor to ceiling.' Brand tried to sound lighthearted: 'It's kind of a cold winter day up there isn't it? Is it snowing in the command module yet?' Haise replied, 'No – no, not quite. The windows are in pretty bad shape . . . every window in the command module is covered with water droplets. It's going to take a lot of scrubbing to get that cleared off.'

At about 6.30 pm Houston time, Vance Brand informed Lovell that the re-entry checklist was ready at last, starting with Swigert's command module, their only means of returning safely to Earth. 'He'll need a lot of paper,' warned Brand. There followed a series of delays over a lack of copies for the flight engineers to follow, causing Lovell to cut in sharply: 'We just can't wait around here to read the procedures all the time up to the burn! We've got to get them up here, look at them, and then we've got to sleep!' Eventually, astronaut Thomas Mattingly, fresh from the simulator, read up the checklist, a task that took nearly two hours.

By that time the first battery on the CM had been fully charged by the LM, and the second battery was recharged during the next two and a half hours. The descent stage of Aquarius ran out of water soon after, but the crew were able to switch over to the ascent stage supply immediately. Ground control confirmed that they wanted to make a minor course adjustment about five hours before re-entry, just to make sure that Apollo hit the entry corridor – there was a fear that separation from the LM might push Odyssey off course.

The early morning hours of Friday 17 April showed more signs of stress and fatigue among the shipwrecked crew. Jack Lousma was reading up a number of checklist alterations to Haise and Swigert when Lovell interrupted: 'OK, Jack, this is Jim. I just want to make sure that any . . . of the changes to the checklist that come up, you make sure that they're absolutely essential.' Deke Slayton, the Director of Flight Crew Operations, decided it was time to intervene: 'I know that none of you are sleeping worth a damn, because it's so cold, and you might want to dig out the medical kit there and pull out a couple of dexedrines apiece.' Lovell replied, 'Fred brought that up. We might consider it.' None of them were very happy about the idea of using such a powerful stimulant unless absolutely essential. Their use was put off until a couple of hours before splashdown, but the men were so exhausted by then that they had little effect.

Lousma radioed up to the shivering astronauts, 'Wish we could figure out a way to get a hot cup of coffee up to you. It would taste pretty good now wouldn't it?' Lovell sounded miserable: 'Yes, it sure would. You don't realize how cold this thing becomes in a passive thermal control mode that's slowing down . . . the sun is simply turning on the engine of the service module. It's not getting down to the spacecraft at all.' Lousma could only try to give moral support: 'Hang in there. It won't be long.'

Soon after, however, he was able to give them some welcome news. 'OK, skipper, we figured out a way for you to keep warm. We decided to start powering you up now.' With nine and a half hours still to go before splashdown, Lovell wanted reassurance: 'Sounds good. And you're sure we have plenty of electrical power to do this?' Lousma replied in the affirmative. The temperature in Aquarius crept up, eventually enabling the crew to switch on the window heaters; by 5 am Houston time, they were basking in a cabin temperature of 16°C. Even Odyssey was warming up, as Swigert commented: 'It's almost comfortable.' Much to his relief, main bus B, the first to fail after the explosion, was found to work perfectly. He began to defrost the CM's thrusters.

Five hours before splashdown the LM thrusters were fired for 23 seconds in order to increase Apollo's speed by 3 ft (1 m) a second and ensure an accurate splashdown near the recovery ships. Half an hour later, the damaged service module was jettisoned by firing the explosive charges which broke the links and giving a gentle push with Aquarius' thrusters. The astronauts had to carry this out without help from the ground, but as Swigert later related, 'The procedure went well – we used a push–pull method . . . Jim and Fred were in the LM and using the translation controller to give us some velocity . . . when Jim yelled "Fire!", I jettisoned the service module and it went off in the midst of a lot of debris, which is usual.'

Swigert quickly grabbed his camera to snap the source of all their problems before it drifted out of range. Lovell and his companions saw for the first time the damage which had disabled their craft and cancelled the lunar landing: 'There's one whole side of that spacecraft missing! . . . Right by the high-gain antenna, the whole panel is blown out, almost from the base to the engine.' They could see debris hanging out from the jagged gash in the module, though they were unable to see exactly what had been damaged. Their description caused immediate concern at Mission Control because the ruptured oxygen tank had been near the heat shield of the CM and there were fears it could be damaged.

Swigert and Haise were too busy powering up the command module to worry about such a possibility. Having had to rely on his companions during most of the flight, Swigert was happy now to be working in Odyssey once more. The most difficult task proved to be aligning the CM's guidance platform; the cloud of 'little white fluffy objects' surrounding the craft obliterated the stars when Swigert sought them in his sextant. Eventually, he succeeded by lining up the stars Altair and Vega on the shaded side of the craft with the light off. The alignment transferred from the LM's guidance platform turned out to be exactly correct.

Lovell now manoeuvred Apollo to the attitude required for the safe separation from Aquarius. Switching to automatic control, he informed Capcom that he was going to bail out of the LM. Before closing the hatch, he looked back at their liferaft; it looked like 'a packed garbage can'. Safe inside the CM, and using its batteries and oxygen, the astronauts prepared to jettison Aquarius. The explosive charges were fired 90 minutes before re-entry to send the LM on its final journey. Mission Control radioed, 'Farewell, Aquarius, and we thank you.' Lovell added, 'She was a good ship.' Haise kept taking photographs until Aquarius was a mere speck of light.

Ground control seemed happy with the trajectory and the craft's status as the moment of truth approached. Swigert acted as the crew's spokesman: 'I know all of us here want to thank all you guys down there for the very fine job you did.' Capcom Joseph Kerwin gave them a smooth talk-down, prompting Swigert to compliment him, 'You have a good bedside manner, Joe,' to which Kerwin replied, 'That's the nicest thing anybody has ever said.' Nine minutes later, Odyssey skipped into the upper atmosphere, subjecting the crew to a welcome of more than 5G. All around the world people watched their TV screens or listened to the radio; many, including some of the ground controllers, offered up a prayer for the astronauts' safety.

The heat shield withstood the 4,000°C inferno as it glowed white hot. The men, who had left their spacesuits behind on Aquarius, were not let down by their craft when they were most dependent on it. Meanwhile, the flight controllers were becoming concerned; three minutes passed, then another 30 seconds and still no word. Kerwin gave them a call, but there was no reply. Everyone began to despair. Then came the sound of Jack Swigert's voice: 'OK, Joe.' In full view of the cameras on board the carrier *Iwo Jima*, first the drogues and then the three main parachutes opened. It was one of the most accurate splashdowns of all time, and the fastest recovery. The clapping and cheering in Mission Control were echoed all over the world. NASA administrator Thomas Paine voiced everyone's feelings: 'There has never been a happier moment in the United States space programme. Although the Apollo 13 mission must be recorded as a failure, there has never been a more prideful moment.'

President Nixon declared the following Sunday a national day of prayer and thanksgiving. Church bells rang out across the country as the President proclaimed: 'Never have so few owed so much to so many.' The three men who were the centre of all this attention were tired and unshaven, but relieved and delighted to be home. Medical checks revealed them to be in worse condition than any previous astronauts, hardly surprising in view of their ordeal. The President flew to Houston to present the ground crew with America's highest civilian award, the Medal of Freedom, then repeated the ceremony in Hawaii with the three astronauts.

More significant, perhaps, was the post mortem, which began on 21 April. It took over two months to reach a conclusion concerning the cause of the aborted mission. The Review Board recommended modifications to the command and service module systems which would eliminate potential dangers in spacecraft which used pure oxygen. All Teflon insulation in the oxygen tanks was replaced by stainless steel, while a fourth oxygen tank, located away from the others and easily isolated, was added to the service module. The inquiry and the modifications meant that the Apollo 14 mission was put back from October until February 1971, and rearranged so that it would land in the Fra Mauro uplands which had been the objective of Apollo 13.

James Lovell left NASA and the US Navy in March 1973 to join Bay-Houston Towing Company, later becoming an executive in a telecommunications company. He made four space flights, holding the individual endurance record of 715 hours 5 minutes until the advent of Skylab. Fred Haise and Jack Swigert never flew again in space either. Haise stayed with NASA until 1979, operating as back-up commander for Apollo 16 and later as technical assistant to the manager of the Space Shuttle project. He is now an executive with Grumman Aerospace. Swigert worked on the Science and Astronautics Committee of the House of Representatives before he resigned from NASA in July 1978. On 2 November 1982, he was elected a Congressman for Colorado, but he died of cancer on 27 December 1982 – the first American astronaut to die of natural causes.

10

The Mountains of the Moon
The Lunar Exploration Missions, 1971–2

The near disaster to Apollo 13 forced NASA to reassess the safety of its spacecraft and to redesign the command and service modules, resulting in an improved craft after a 5-month delay. The long interval between Apollo missions allowed the Soviet Union to regain some prestige while, at the same time, stirring up the old controversy over the role of man in space. In September 1970 the Soviet robot craft Luna 16 succeeded in obtaining the first sample of lunar material not handled by astronauts; and a further blow to NASA followed in November when Luna 17 carried the first mobile lunar robot to a soft landing in a shallow crater on the Sea of Rains. The Lunokhod ('moon walker') trundled down a ramp and began to wander over the lunar surface using cameras for eyes. The ground controllers successfully navigated the craft around all obstacles and up slopes of more than 20 degrees during the 11 month life span of this remarkable vehicle. Before it ground to a halt in the freezing temperatures of its eleventh lunar night, Lunokhod had roamed over 6½ miles (10½ km) of lunar terrain, transmitted thousands of photographs, carried out various soil analyses and detected solar flares which Soviet scientists said would have caused a radiation hazard for any astronauts.

NASA tried to defend its policy, though not entirely convincingly, arguing that astronauts were more flexible and adaptable to unexpected situations than robot machines. Astronaut Pete Conrad was quoted as saying that if his Apollo 12 mission had been unmanned, a 25 million dollar lunar experiment package would have been a complete loss. The American space administration was also at pains to point out that its astronauts would be more mobile in future, starting with a small handcart on Apollo 14 and extending to a lunar 'buggy' on subsequent missions.

There was controversy over the crew selected for the third lunar landing, for it was the least experienced of any Apollo crew. The man appointed to command the mission was the oldest man to go into space up to that time, 47-year-old Navy captain Alan Shepard. Since he made history by becoming America's first man to go into space almost ten years earlier, Shepard had been grounded with an inner ear disorder, unable even to fly an aircraft unaccompanied. Despite being given an important desk job as Chief of the Astronaut Office, Shepard was determined to fly a spacecraft once more, possibly because he felt he had never received the recognition he deserved after risking his life on the first Mercury mission. Undoubtedly, his position in the NASA hierarchy gave him access to the people who decided future policy, and his selection as Apollo 14 commander followed hard on the heels of his successful ear surgery and restoration to flight status in May 1969. Nevertheless, Shepard was not the kind of man to sit back and relax, and during the months of training he worked hard to reach a peak of mental and physical fitness. He was not particularly amused when the white room technicians gave him a walking stick before boarding the command module, named Kitty Hawk.

Accompanying Shepard were two men who had never flown in space before, 40-year-old Navy commander Edgar Mitchell and 37-year-old Air Force major Stuart Roosa. Both men had been selected by NASA in April 1966, and had worked together as support crew for Apollo 9. Mitchell, the lunar module pilot, had a distinguished academic record, with degrees in industrial management and aeronautical engineering, and a doctorate in aeronautics and astronautics from Massachusetts Institute of Technology. After assignments to jet squadrons in Okinawa and on board aircraft carriers, he became a research project pilot, passing out top of his class at the Aerospace Research Pilot School. Before joining NASA, he acted as manager of the Navy field office for the proposed Manned Orbiting Laboratory which was planned to follow Apollo. He was the back-up LM pilot on Apollo 10 in 1969. Roosa had been a test pilot at Edwards Air Force Base prior to astronaut selection. He had graduated in astronautical engineering while flying jet fighters, then moved up to being a maintenance test pilot before attaining every pilot's dream of serving at Edwards. There was no question concerning the experience and dedication of the Apollo 14 crew but the fact remained that the three men had just 15 minutes of space time between them.

The launch of Apollo 14 seemed destined to be a repeat of Apollo 12's. The weather at the Cape on 31 January 1971 was squally with thunderstorms lingering in the vicinity. Thousands of VIPs and spectators, the largest number for any flight in the series apart from Apollo 11, had to endure a 40 minute hold in the countdown as the flight director applied the new launch regulations introduced after Apollo 12 – there would be no lightning strikes this time. The crew cheerfully endured the first major delay in an Apollo launch, and were eventually given the go-ahead after a weather aircraft reported a clearing in the weather. The final eight minutes of the countdown proceeded on schedule, enabling Apollo 14 to launch into an overcast sky at 4.03 pm Cape time with lightning flashes illuminating the horizon for American Vice President Spiro Agnew and his guests. They were treated to little more than half a minute of the Saturn's fiery trail before it disappeared into the dark grey blanket overhead, but, to everyone's relief, the rocket performed perfectly in placing the craft in Earth orbit. Shepard's biomedical sensors failed to work, but doctors reported peak pulse rates of 132 for Roosa and less than 90 for Mitchell.

In order to minimize the changes necessitated by the launch delay, Houston decided to bring everything forward by 40 minutes. Apollo 14 was blasted out of Earth orbit at the time planned on the original schedule, but it was not long before the entire lunar landing was in jeopardy yet again. The transposition and docking manoeuvre had never caused any major problems, so everyone was amazed when Roosa guided the probe on Kitty Hawk into the conical drogue on top of Antares, the lunar module, but the latches refused to catch. The TV pictures sent back by the crew showed everyone in Mission Control what was happening as Roosa backed off and tried unsuccessfully a second time. Three more attempts failed as Roosa tried to drive home the probe at faster approach speeds. Things were getting serious; without the docking there could be no lunar landing. Gene Cernan, the back-up commander, discussed the problem with Roosa and Shepard and agreed they should try for a hard dock which would activate the 12 main latches rather than the three small capture latches on the probe. At the sixth attempt, a relieved crew reported they had docked at last. The anxiety at Houston remained, however. What if the mechanism was faulty and prevented Antares from docking with Kitty Hawk on its return from the lunar surface? The mission director, Chester Lee, ordered a detailed TV examination of the assembly. A close examination by the astronauts revealed deep scratches inside the cone, indicating that the spring catches had not retracted, but no malfunction in the equipment. This led to the suggestion that ice might have formed around the catches during the wet launch. Mission control decided to continue as if nothing had happened. The moon landing was back on. The unexpected extra work had taken its toll on the men, however, leading them to ask for an early rest period.

The cruise to the moon over the next three days was marked by long periods of relative silence, leading to the men being dubbed 'the silent crew'. One observer commented: 'I don't see how it could be much quieter whether they are asleep or awake.' A ten second engine burn ensured that Apollo 14 would speed up sufficiently to make up the 40 minutes lost by the launch delay, with a second mid-course correction the next day putting the craft on the right trajectory. Gerry Griffin, one of the flight directors, assured pressmen: 'As far as the bird is concerned, it's going to get to the moon at the right time.' One unusual series of observations which occupied the crew concerned bright flashes observed by the Apollo 13 astronauts during the dark sleep periods. Scientists believed they were caused by high energy cosmic rays impacting on the eye, though they were generally believed to be harmless. Mitchell reported: 'You first see them as being very bright, followed by a more sustained type of light like a halo.'

Although the docking difficulty seemed to have been resolved, further minor problems dogged the mission. A rest period was rudely interrupted by Capcom Gordon Fullerton when controllers reported a high oxygen flow rate, the signature of a cabin leak. Ed Mitchell found the solution by tightening a valve on the waste disposal system which had apparently been overworked! Another worry, and a potential threat to the moon landing, was related to a slight voltage drop in one of Antares' batteries. Mitchell was again deputed to check it out, and his test showed no apparent malfunction. The Fra Mauro uplands still beckoned.

Swinging behind the moon for the first time, the main engine fired for just over six minutes to place the craft in lunar orbit. On the second revolution, the engine fired once more to place Apollo into a highly elliptical orbit instead of the usual circular orbit. The crew experienced the strange sensation of rising to an altitude of 68 miles (109 km) then dropping to within 11 miles (17 km) on each orbit, the lowest a complete Apollo craft ever flew. The crew, staring out at the bleak wilderness, became more communicative. Shepard commented: 'Well, this is really a wild place up here. It has all the greys, browns, whites, dark craters that everybody's talked about before. It's really quite a sight . . . Everything is really clear up here. Really fantastic.' Mitchell added: 'I think the best description that comes to mind, we mentioned this when we first looked at this thing, is that it looks like a plaster mould that somebody has dusted with greys and browns but it looks like it's been moulded out of plaster of Paris.' As for the vital descent burn, performed by Kitty Hawk and designed to conserve fuel on the lunar module, Stuart Roosa delightedly commented: 'Well, that turned out to be a piece of cake.' Mitchell found difficulty adjusting to the unaccustomed low altitude: 'As a matter of fact we're below some of the peaks on the horizon, but that's only an illusion . . . it certainly is an unusual sensation flying this low.'

The separation of Antares was successfully completed on the twelfth revolution, drawing the comment from Roosa, 'Boy, you sure look mighty pretty out there'. On passing inspection, Antares was left to move ahead as Roosa fired Kitty Hawk into a circular 68 mile orbit behind the moon. The lone flight of Antares was soon interrupted by an unsettling situation for Shepard and Mitchell, however, as a computer fault flashed on to the display keyboard. The abort system had somehow been activated, probably through some minute specks of dirt in the abort switch; had this occurred once powered descent had been initiated, the entire landing would have been aborted. It was some time before mission control came up with a way of bypassing the computer blockage to prevent such a disaster.

All was well as Antares came around the front side on its fourteenth revolution, the fragile craft was slowed to drop the short distance to the surface. Things became a little hairy during the 11½ minute descent when the landing radar cut out, leaving the men 'blind' as they flew over the threateningly rugged terrain. Mitchell urgently flicked the radar on and off: 'C'mon radar! Get the lock on . . . Phew, that was close.' Once again, a landing abort had been averted by the skin of their teeth. Shepard finally settled in a depression just to the east of the target zone with the LM tilting at an eight degree angle. A relieved Mitchell called in: 'We're on the surface.' Shepard described the scene: 'A stark place . . . pockmarked by craters . . . the sky is completely black.' Surface colour varied between 'mouse brown and mouse grey'. The brilliant morning sun shone low above the horizon on to the weird craft covered in gold foil. It was 4.17 am Cape time on 5 February 1971.

Shepard's first sortie on to the surface was delayed for about 45 minutes while a poor communications link on his backpack was dealt with. Eventually, with everything in working order, the commander struggled through the narrow hatch and backed down the ladder, pausing only to deploy the equipment in the storage bay. Watching the colour pictures in Mission Control, Capcom Bruce McCandless told him: 'Not bad for an old man.' A restrained, but nevertheless jubilant, Shepard replied: 'OK, you're right. Al is on the surface, and it's been a long way, but we're here.' Despite all the trials and disappointments of the past eight years, Shepard had reached his goal, becoming the fifth man to step out on to the moon and by far the oldest to achieve that distinction. Only six minutes later, he was joined by Ed Mitchell and the two set to work collecting rock and soil samples, and deploying the large number of scientific experiments. The Apollo Experiment Package included six research instruments powered by a small nuclear power pack, similar to those carried by Apollo 12. Other equipment was set up to reflect laser beams directed at it from Earth, to collect solar wind particles, and to measure the lunar magnetic field. An innovation on Apollo 14 was the use of a rickshaw-like handcart which was stored on the descent stage. No longer would the astronauts have to struggle to carry tools and bags full of samples.

For nearly five hours the men could be seen scampering around the landing site, leaving a trail of deep footprints in the grey dust as they wandered up to 1,000 feet (300 m) from Antares. A checklist on Shepard's wrist detailed around 200 tasks which they were required to complete, making them the busiest moon walkers yet. Of course, there were the usual frustrations when equipment fell over or failed to work to full capacity. In particular, there was some disappointment when a seismic experiment did not go according to plan. Three detectors called geophones were laid out in a line on the surface to pick up vibrations from four small mortar bombs to be fired on command from Earth after the departure of the astronauts. However, Mitchell also carried a 'thumper' which contained 21 explosive charges which he was to set off in an attempt to activate the geophones. Over a 30 minute period he tried to complete the task, but at least five of the charges failed to detonate as he struggled to pull the trigger device on the thumper, and the first four explosions were not received because one geophone was resting on its side.

The busy work schedule and the clinging dust eventually took their toll as the men became tired and their heart rates rose – Shepard's was reported to have reached 120. A planned walk to the Doublet craters was cancelled because of this and the fact that they had run out of time. It was revealed by mission control that Mitchell's suit was leaking oxygen at a higher than normal rate, though it was not considered dangerous. Preparing to enter Antares, Shepard tried to brush down his colleague: 'You're a mess.' Mitchell replied: 'It looks like you've been wallowing in mud.' Closing the hatch behind them, the men were happy to accept their ten hour rest period. Meanwhile, at home on Earth, Shepard's wife sidestepped the media fascination over her husband's age: 'They can't call him old man Moses any more. He has reached the Promised Land.'

The second moon walk was to take the astronauts nearly a mile from the lunar module on the first attempt at hill climbing on another world. The target was Cone crater, a large hollow excavated by an impact during the moon's early history. Despite their efforts of the previous day, the two astronauts awoke early to prepare for their expedition. 'OK, we're up and running,' came Shepard's cheery voice. He was true to his word, as the men exited from Antares about two and a half hours earlier than originally planned.

Trundling their cart along beside them, TV viewers saw them head off towards the black horizon leaving the first wheel marks on the virgin lunar soil. The early stages were easy-going as they crossed a shallow depression and stopped to take the first soil and rock samples, and the first readings with their portable magnetometer. By the second stop, the trek was beginning to get more strenuous, causing Mitchell to turn up the ventilation on his suit and mission control to advise a short rest. They were standing in 'a large boulder field with more numerous boulders than we've seen in the past'. The men were beginning to sample the material blasted out of Cone crater during its birth billions of years ago.

The absence of landmarks began to cause disorientation. 'Get the map and see if we can find out exactly where we are,' said Mitchell. At least they could still see Antares sitting alone on the cratered plain behind them: 'The old LM looks like it's got a flat [puncture] over there.' The slope became steeper as they set off once more, and the soft dust slowed their progress until Mitchell reported: 'The soil here is getting firmer.' The handcart was proving something of a liability; when tools began to bounce off, one man had to follow behind to retrieve them, then the wheels tended to stick in the fine dust, forcing the men to carry the cart. Fortunately, even loaded down with tools, rock samples and camera, it still only weighed about 25 lb on the moon.

The terrain became ever rockier as the men progressed in short bursts followed by enforced breathers. Heart rates began to race with the target not yet in sight – Shepard's pulse was reading 150, Mitchell's was 128. Their hopes rose each time they crested a rise, only to be dashed again as another obstacle confronted them. 'You know the rim, we haven't found that yet,' lamented Shepard. Houston tried to help them sort out their exact position, but there was very little to go on – there was no point in. trying to take compass readings in the weak, localized magnetic field of the moon. Shepard summarized their situation for Capcom Fred Haise: 'Our positions are all in doubt now, Fredo. We've got a ways to go yet. Perhaps you can say what it is you want. I'd say that the rim is at least 30 minutes away. We're approaching the edge of the boulder field here from the south Flank.' After a brief consultation, Haise told them to continue for a little longer. Mitchell, already complaining because of a

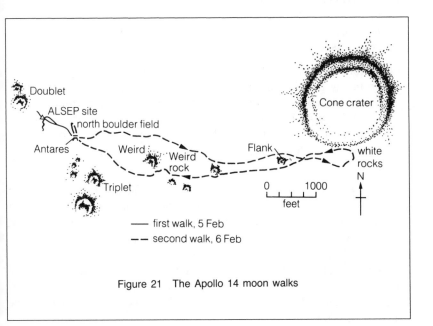

Figure 21   The Apollo 14 moon walks

broken cable and some loss of mobility in the right hand glove of his spacesuit, was moved to comment: 'It's further than it looks.' An equally frustrated Shepard replied, 'That's the order of the day.'

As the walk entered its third hour, mission control was closely monitoring their oxygen and water supplies. The men headed for some 'almost white rocks' then paused for another consultation. They were standing 'right in the midst of a whole pile of very large boulders . . . a pretty darn rugged boulder field area.' Unknown to them, they had wandered around the rim of Cone crater without realizing; their target was less than 200 feet away when Shepard decided it was too risky to continue. Mitchell tried to argue for a continuation: 'Aw, gee whiz, let's give it a whirl.' But the more conservative counsel of the commander prevailed. Mitchell could not hide his disappointment: 'I think you're finks.' So the men had to be satisfied with photographs and samples of the large white rocks which littered the surface.

The return journey was largely downhill and so considerably faster than the outward trek, speeded still further by the cancellation of many of the allotted tasks. This time there was no navigation difficulty; the friendly sight of Antares beckoned them onwards. Even so, they described the landscape as 'country so rolling and undulating, with rises and dips everywhere, that you can be going by a fairly good size crater and not even recognize it'. Shepard was later to liken the experience to trying to find your way around the Sahara Desert. Back at Antares, the crew realigned the radio antenna and TV camera, collected the solar wind experiment which would return with them to Earth and prepared to load the fruits of their labours.

The normally staid Shepard then surprised everyone by producing his party piece as the climax to the show.

> Houston, while you're looking that up you might recognize what I have in my hand as the handle for the contingency sample return and just so happens I have a genuine six iron on the bottom of it. In my left hand I have a little white pellet that's familiar to millions of Americans. I drop it down. Unfortunately, the suit is so stiff I can't do this with two hands, but I'm going to try a little sand trap shot here.

Using a special golf club head provided by Jack Harden, a former professional at Houston, Shepard became the first man to swing at a golf ball on the moon! 'Hey, you got more dirt than ball that time,' chided Mitchell. As a second ball lifted into the black sky, Capcom joined the criticism: 'That looked like a slice to me, Al.' Shepard tried once more, sending the ball 'straight as a die . . . miles and miles and miles'. The second walk lasted 4 hours 35 minutes, bringing the total lunar EVA time to 9 hours 25 minutes, considerably longer than either of the previous missions.

Preparations for take-off went smoothly, so that Antares was able to lift her crew towards the mother ship orbiting overhead after a record 33½ hours on the lunar surface. Using a new series of rendezvous manoeuvres, Antares arrived in lunar orbit much closer to Kitty Hawk than had been the case on earlier missions. The result was a fast rendezvous completed within one revolution. Roosa televised the final docking as the two craft closed. Anxiety over the docking mechanism proved unnecessary as the Kitty Hawk captured Antares at the first attempt. Capcom radioed congratulations: 'Beautiful. There's a big sigh of relief around here.' Curiosity remained, however, about the docking mechanism, so the crew were told to retrieve it and stow it on board Kitty Hawk for later examination.

Antares was soon jettisoned and fired by remote control to her final crash-landing on the moon, creating shock waves which were picked up by the seismometers left by Apollos 12 and 14. The crew were allowed a lie-in after successfully firing the main engine behind the moon for the return to Earth. It was 11 hours before Mitchell broke the silence: 'We've had no medication. We are all in excellent shape, so just tell the surgeon to sit back in his chair and have a cup of coffee.' The return home included the usual TV broadcasts, but with a difference. Alan Shepard, who had been a military man for 26 years, spoke about the Vietnam War and ended with a plea for peace: 'It is our wish tonight that we can in some way contribute through our efforts in the space programme to promote a better understanding of peace throughout the world.' Alas, there were few people who heard his plea, for, such was the decline in public interest that few American TV networks showed the broadcast.

A new departure for this mission was the in-flight experiments carried out on Kitty Hawk. Stuart Roosa had performed some advanced photographic and radar assignments on his lonely vigil while his companions were on the moon, although one special topographic camera broke down. Now it was time for all three men to try out four zero-gravity experiments in front of the TV cameras to study the effects of weightlessness on prospective future materials and manufacturing processes in space. NASA was looking ahead, beyond Apollo, to the days of the Earth-orbiting space laboratory. Of particular interest were the experiments dealing with the manufacture of vaccines and metals in space. Another experiment had a more immediate significance; the CSM was tested to see whether it could supply oxygen at a sufficiently high flow rate for in-flight EVA activity on future Apollo missions. Unexpected drops in oxygen and water tank pressures after about an hour brought the experiment to an end, though it was later deduced that the abrupt fall had simply been caused by a urine dump during the test period.

The crew conducted a press conference with newsmen from around the world as they neared home once more. The final question concerned Shepard's golf prowess. He said the longest shot went about 400 yards, '. . . not bad for a six-iron'. Meanwhile, Mission Control warned the crew to expect a chilly homecoming – temperatures in Houston were below freezing. Mitchell commented: 'Have you moved Houston to the North Pole already?' He added that it was 'a very comfortable 71 degrees' in his cabin. Only one minor course correction was required on the way back as Kitty Hawk headed for a perfect splashdown in the south Pacific, about 900 miles south of Samoa. The only worry involved the faulty docking mechanism; the crew had to make sure it was carefully stowed and could not break free during descent. They also had to be prepared to take over the controls in case the extra weight caused a steeper angle of re-entry than was safe. Neither fear materialized as Kitty Hawk floated down within sight of the recovery ship, *New Orleans*. The craft remained upright after hitting the water, though one parachute refused to separate. Shepard reported: 'We're in very good shape here.' Then the crew transferred to a liferaft wearing clean coveralls and face filter masks before being whisked to the recovery ship by helicopter. They had the unusual experience of splashing down on Wednesday then crossing the International Date Line to land on the *New Orleans* on Tuesday. The helicopter was wheeled to a lift and lowered to the hangar deck below where the crew were greeted by a Navy band playing 'The Stars and Stripes'. Shepard led his companions down some steps and into the mobile quarantine facility, the cheers and applause of the crew ringing in their ears. The Apollo 14 crew had the distinction of being the last crew to endure the three week quarantine period; the only living organisms brought back from the moon had been discovered inside the camera retrieved from Surveyor 3 by the Apollo 12 crew, survivors from the construction of the robot craft years earlier.

The Apollo 14 flight ended amidst a wave of apathy and virtual indifference. Even the President remained aloof, merely contacting Deke Slayton and asking him to pass on good wishes and congratulations. The landing vied for attention in the world's media with more down-to-earth news stories, and the record booty of 96 lb of moon rocks was insufficient to stir the public imagination. Alan Shepard summarized the situation better than anyone: 'Apollo 14 has shown we have reached maturity in the manned space programme.' Three months later, his own maturity was emphasized when he attended an unveiling ceremony marking the tenth anniversary of his first flight into space. It was too much to expect that he would ever experience space a third time at his age. Promoted to rear-admiral, Shepard resumed his former duties as Chief of the Astronaut Office until he retired in 1974, entering the world of business. He is now president of the Windward Company in Deer Park, Texas.

Neither of his Apollo 14 colleagues flew again in space. Mitchell and Roosa worked together again on the Apollo 16 back-up crew, with Roosa also serving on the Apollo 17 back-up crew. Mitchell left NASA before the Apollo programme concluded at the end of 1972. Only later was it revealed that he had carried out an unofficial extra-sensory perception experiment during the Apollo 14 mission with four friends on Earth. Analysis of the results suggested that ESP was a possibility even at the tremendous distances involved in space travel. Mitchell is now Chairman of the Board with Forecast Systems of Utah and Florida. Roosa left NASA in 1976 after working on the Space Shuttle programme. He went into market research related to the Middle East, then into real estate investment, before becoming president and owner of Gulf Coast Coors beer distribution company in 1981.

Information transmitted by the Apollo seismometers suggested a world with a deep, rigid crust, but the questions remained – was the interior of the moon now cold and inactive, or was there a hot, massive core similar to that found within the Earth? NASA had three missions remaining in which to answer the questions and tried very hard to sell these missions to a public enthralled by space adventure spiced with the hint of danger but less than enthusiastic about obscure scientific theories. The hard sell succeeded in attracting nearly a million people to southern Florida on the morning of 26 July 1971 for the launch of what President Nixon described as 'the most ambitious exploration yet undertaken in space.'

The crew selected to double as scientist–geologists had spent more than a year preparing for the longest and most dangerous landing to date, taking part in numerous field trips to mountain and desert in order to become familiar with the bewildering array of rocks and minerals that would confront them, and to learn the scientific procedures of observation, description and sampling. It was a far cry from the air force and test pilot backgrounds of all three crewmen. Mission commander was David Scott, now 39 and taking part in his third space mission, having previously gained rendezvous and docking experience on Gemini 8 and Apollo 9. Alongside him were two newcomers to spaceflight, both selected in April 1966. The senior of the two was lieutenant colonel James Irwin, 41, the man who would accompany Scott to the mountains of the moon. He had graduated in naval science from the US Naval Academy in 1951, later gaining a master's degree in aeronautical engineering from the University of Michigan. After ten years in the USAF, Irwin moved to Edwards Air Force Base in California to become a qualified test pilot. Lucky to survive a serious jet crash in 1961, he went on to reach the top of the tree by graduating from the Aerospace Pilot School in 1963 and being posted to Headquarters Air Defence Command. His early work for NASA after astronaut selection involved thermal vacuum testing of the lunar module, moving on to serve on the Apollo 10 support crew and the Apollo 12 back-up crew.

The man who would carry out the most intensive scientific programme from lunar orbit while awaiting the return of his colleagues was 39-year-old major Alfred Worden. He had graduated from the US Military Academy in 1955, moving around various Texan air bases before serving with a jet fighter squadron at Andrews Air Force Base in Maryland. In 1963, Worden gained his master's degree in the same disciplines and at the same university as Irwin. His career prospects were further enhanced by attendance at Randolph Air Force Base Instrument Pilots Instructor School, the Empire Test Pilots School at Farnborough in England, and finally the Aerospace Research Pilots School at Edwards, where he became an instructor. Within a year, he had become an astronaut, later serving on the Apollo 9 support crew, and the Apollo 12 back-up crew with Irwin and Scott.

All three men showed their awareness of the new role to be played by Apollo 15 by choosing the name Endeavour for their lunar module in honour of the ship which carried Captain Cook on his first voyage of exploration 200 years earlier. As Scott said, 'Cook made the first purely scientific expedition in history and ours is the first extensive scientific expedition to go to the moon.' Equally fittingly, the crew named their command module Falcon after their US Air Force mascot. Among the articles carried to the moon by the crew were a wooden fragment of the sternpost of Cook's ship, courtesy of the Marine Museum in Rhode Island, and a real falcon's feather.

Apollo 15's take-off weight was well above that of its predecessors. Contributing to this was a major new piece of hardware – the battery powered lunar rover. This vehicle folded for storage in the side of the LM, but could be deployed to its full 10 ft length by the astronauts pulling a few lanyards. The four-wheel car was designed to be functional rather than sporty, with wire mesh tyres aimed at maximum grip on the dusty surface, large wheelguards to disperse the dust, two folding seats and seat belts to hold down the spacemen. Each wheel had a separate electric motor and mechanical braking system, there were independent steering systems linked to a handle rather than a steering wheel which enabled it to turn full circle within its own length, and top speed was around 9 mph. Most important from the astronauts' point of view was the fact that it took much of the hard slog out of moon exploration, enabling them to travel much further from their craft than earlier crews and retrieve rock samples for stowage in the 'boot', where the tools were carried. TV pictures could be broadcast to Earth by an umbrella-like directional antenna whenever they stopped, and voice communications were relayed to Earth by a smaller antenna. The astronauts' work was also facilitated by an improved, more flexible pressure suit which enabled them to crouch down or even touch their toes, and a backpack which

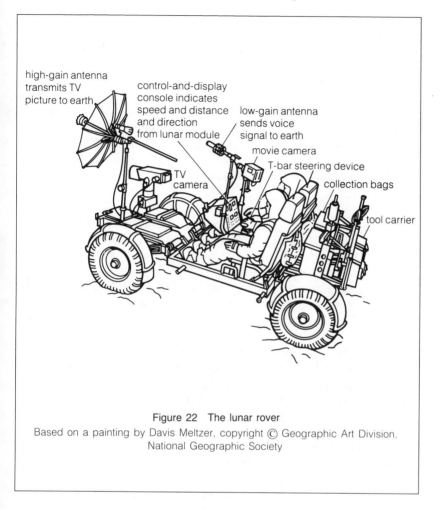

high-gain antenna
transmits TV
picture to earth

control-and-display
console indicates
speed and distance
and direction
from lunar module

low-gain antenna
sends voice
signal to earth

movie camera

T-bar steering device

collection bags

tool carrier

TV
camera

Figure 22   The lunar rover
Based on a painting by Davis Meltzer, copyright © Geographic Art Division,
National Geographic Society

supplied sufficient oxygen and water for up to seven hours EVA for any excursion. However, it was the time factor which prevented the men from wandering all over the lunar surface; if their backpacks failed, the emergency supplies carried on their backs only lasted for 75 minutes – if they weren't inside Falcon within that time limit, they would be dead.

The Apollo programme had attracted its fair share of wrath from the gods, and Apollo 15 was no exception with the launch pad being struck 11 times by lightning during June, but their anger seemed to have abated as Monday 26 July dawned hot, humid but sunny. The astronauts were briefly glimpsed by spectators when they boarded the transfer bus to Pad 39A. At the other end, they headed for the towering white giant which had been uprated to carry the extra weight, Worden pausing to shake hands with his proud father. At 9.34 am Cape time, Apollo 15 blasted off amid awed silence in what was described as 'the most spectacular and trouble-free launching ever seen at Cape Kennedy'. Unlike the cloud-shrouded launches of Apollos 12 and 14, the watching multitude were treated to a wonderfully clear view as they followed the white trail into the blue beyond.

Apollo entered Earth orbit almost as planned, and the by now familiar routine of trans-lunar injection, transposition and docking was happily trouble-free. The only unusual departure was the immediate move to an orbit of no-return in order to attain the correct lunar orbit inclination and to save fuel on the CSM; Apollo 15's landing site was the furthest from the lunar equator of all the Apollo sites.

The cruise to the moon was largely uneventful apart from the usual minor malfunctions. Most worrying was a warning light which suggested that two valves on the main engine were 'in an unusual position'. There was a possibility that the engine could be activated early by a short circuit, a dangerous possibility which threw doubt over the lunar landing. Tests by the astronauts proved inconclusive, but the short was eventually located after Scott used the back-up electrical system to make the necessary mid-course correction while the primary system was proved to be reliable for manual firings. A master alarm sounding in the cabin sent everyone's pulses racing on the second day as a complete electrical failure seemed likely, but it proved to be a false alarm when all readings reverted to normal. The only other reason for concern was a broken glass cover on a range-altimeter in Falcon, but mission control assured the crew it still functioned normally. The rest of the four-day cruise to the moon was taken up with the usual household chores and TV broadcasts. Only when they were nearing their objective did Scott report a water leak from a chlorine port, but this was soon dealt with by using the plumber's standby, a spanner from the tool kit. Scott told Capcom: 'Our trusty LMP [lunar module pilot] came up with an interesting analogy relative to our last event. He wondered if the original Endeavour had ever sprung a leak like that.' Capcom replied: 'OK, that's a good question. We'll put our historians out to check that one. Hey, what did you do with all that extra water? Stick it overboard or drink it, or what?' Scott came back: 'Oh no. We've got a bunch of towels hanging up in the tunnel right now. It looks like somebody's laundry.'

Prior to their disappearance behind the moon, the bay door was fired free from the scientific instrument module (SIM) in the service module, paving the way for the first intensive scientific survey of the moon by an orbiting Apollo. The instruments which would be turned towards the lunar surface included a laser altimeter which would operate in conjunction with a mapping camera to record the surface topography, a panoramic camera to take high resolution stereo pictures of the moon, and two spectrometers to record the chemical composition of the tenuous lunar atmosphere and the rocky surface. These experiments would ensure that Worden never became bored on his lone voyage around that desolate world. Meanwhile, the first priority was to enter lunar orbit. Capcom Joe Allen wished them bon voyage: 'Gentlemen, everything looks perfect down here and all we can say is "Have a good burn".' For the sixth time, everyone on Earth had to endure the suspense of the sustained silence as an Apollo craft fired its main engine to decelerate for lunar capture. For the sixth time the burn was successful as Scott broke radio silence: 'Hello, Houston. The Endeavour's on station with cargo, and what a fantastic sight.' For the first time the crew became excited as they attempted to describe the sights: 'I'll tell you, this is absolutely mind-boggling up here . . . hard for the mental computer to sort out.' Passing over features never observed by previous crews, the geological training already began to pay off. Worden commented: 'After the King's training, it's almost like I'd been here before.' (The 'King' was the crew's nickname for one of their teachers, Egyptian-born Farouk El-Baz.)

Less than one hour after lunar orbit insertion, the third stage of their Saturn booster crashed into the Sea of Serenity, activating the seis-mometers left by earlier Apollo missions. Little over three hours later, the craft was lowered to an elliptical orbit which brought them within 8 miles (13 km) of the lunar surface. Capcom asked the crew: '15, does it look like you are going to clear the mountain range ahead?' The reply was equally lighthearted: 'We've all got our eyes closed; we're pulling our feet up.' The Apennine mountains where they were scheduled to land the next day were smooth and rounded, unlike any major mountain ranges on Earth. 'We are over the mountains and they are just incredible. The mountains jump out of the ocean in great relief.'

As the men slept in preparation for the busy day ahead, the irregular gravitational pull of the 'mascons' beneath the lunar seas slowly altered the orbit. On awaking, the crew were informed of the minor re-adjustment that would be required. The short thruster firing did the trick, but the separation of Falcon and Endeavour behind the moon proved a headache: 'OK, Houston, we didn't get a sep, and Al's been checking the umbilicals now on the probe.' Worden found a loose power connection in the tunnel, enabling Falcon and Endeavour to move apart 25 minutes later than planned. One hour later, Worden bade farewell to his colleagues as he fired the CSM into a higher, circular orbit. Scott and Irwin sped twice more around the moon before braking over the Apennines; it would be the steepest descent in Apollo history. Sixty feet (20 m) above the surface, the blast from the engine began to churn up the lunar dust, engulfing the craft and blinding the pilot. Gingerly feeling his way down 'on the gauges', Scott brought Falcon down with an 'abrupt jar' only a few hundred feet from the target with more than 100 seconds of hover-time remaining. It was 6.16 pm Cape time on 30 July. Scott's jubilant voice came over the speakers: 'OK, Houston. The Falcon's on the plain at Hadley.'

The men threw the switches to turn their tiny craft into their temporary home and prepared for their three-day stay. A couple of hours passed before Scott poked his head out of Falcon's top hatch to survey the scene. The ancient name given to the site was 'Marsh of Decay', but this was far from the reality which Scott portrayed as he spent half an hour describing the view. Falcon's perch was tilted by about 11 degrees because a footpad had come to rest in a small crater, but Scott had a fine view:

> All of the features around here are smooth. The tops of the mountains are rounded off. There are no sharp, jagged peaks or no large boulders apparent anywhere . . . It's a gently rolling terrain completely around 360 degrees, hummocky much like you saw on 14 . . . Mount Hadley itself is in shadow.

Turning clockwise, using a map and sun compass to take bearings and taking a complete panorama with his camera, he continued:

> To the east of Hadley Delta, again I can see smooth surface . . . but there appear to be lineaments running, dipping, through to the north east . . . there are definite linear features . . . The craters on the side of Hadley Delta are rather few.

Already there was plenty for the listening geologists to theorize and argue about. Capcom transmitted everyone's delight: 'Superb description, Dave . . . You've answered everyone beautifully. Outstanding.'

The astronauts' first reconnaissance over, they settled down to an evening meal of cold tomato soup and cold hamburger followed by about five hours sleep. The hammocks proved quite comfortable, but with their minds whirring and imaginations running riot, not to mention the mechanical noises which pervaded the spacecraft, sleep was not easy to achieve. The men had foreseen this latter problem, and had practised sleeping in an altitude chamber while a tape recorder played the sounds produced by a working spacecraft. They soon dropped to sleep in their 'portable Leaning Tower of Pisa'. Their slumber was interrupted an hour early, however, with Mission Control reporting oxygen leaking from Falcon. The trouble was soon dealt with by tightening a urine dump valve on a tube which carried urine from their spacesuits to the outside tanks.

The hatch was opened at 9.13 am Cape time on Saturday 31 July. Dave Scott deployed the colour TV camera as he carefully backed down the ladder, enabling millions of viewers to watch him become the seventh human to step on to the surface of the moon. Scott gave a short speech: 'As I stand out here in the wonders of the unknown at Hadley, I sort of realize there's a fundamental truth to our nature. Man must explore. And this is exploration at its greatest.' He was joined within a few minutes by Jim Irwin, and once the camera was installed on a tripod they proceeded to unwrap their lunar buggy. This took a lot more time and effort than had been envisaged, then Scott disappointedly told Capcom Joe Allen: 'I don't have any front steering, Joe.' He would have to drive using the rear steering only. Once the TV and radio communications were set up and the tools loaded on the rear, they set off on the first cross-country drive over the hazard-strewn dusty wilderness. Scott told Houston: 'I'll have to keep my eye on the road. This is really a rock'n'roll ride.' Irwin confirmed: '[It's a] combination of bucking bronco and a rowboat in a rough sea.'

The first few minutes were taken up by continuous comments about

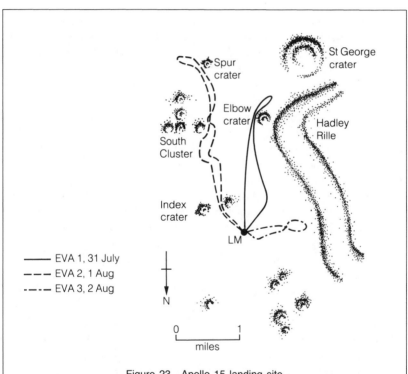

Figure 23 Apollo 15 landing site
Apollo 15 landed in the most rugged terrain yet explored by astronauts. The landing site was a basin bounded on three sides by mountains rising up to 15,000 ft above the heavily cratered plain. To the west was enigmatic Hadley Rille, a strange dry valley averaging 1,000 ft deep and nearly one mile wide. Utilizing the lunar rover to full advantage, the astronauts made three excursions to investigate the rille and the foothills of the nearby Apennine Mountains.

their new automobile: 'OK, Joe, the rover handles quite well. We're moving at, I guess, an average of about 8 kmh (5 mph) . . . It feels like we need the seat belts, doesn't it Jim? . . . The steering is quite responsive, even with only rear steering . . . There's no accumulation of dirt in the wire wheels.' Scott agreed with Allen's description: 'It sounds like steering a boat, with the rear steering and the rolling motion.'

Bouncing along happily, they headed south past small craters with strange names like Rhysling and Pooh before reaching their first geological stop on the east side of Elbow Crater. In a mere 25 minutes, they had covered more ground than the walking explorers on Apollo 14 had covered in their entire day. As they collected samples, took pictures and described geological specimens, Houston commanded the TV camera to scan the area. Slowly panning across the unique scene, the camera showed the two men near the rover, then moved on to a large field of boulders with two smooth mountains in the background, finally showing the edge of Hadley Rille sharply etched in bright sunlight and black shadow. Irwin described this mysterious gash in the rolling plain: 'I can see several large blocks that have rolled downslope . . . I can see the bottom of the rille. It's very smooth. I see two very large boulders that are right on the surface there . . . I'd say [the bottom] is flat . . . I'd estimate maybe 200 m (660 ft) wide on the flat area of the bottom.'

Houston urged them on as time was passing. When asked about the performance of the buggy, the men were very happy, though they admitted it was 'sporty driving' and told of the rear end breaking away in a 180 degree skid when making a downhill turn too fast. Another difficulty not normally encountered on Earth was the seatbelt operation: 'In one-sixth G we don't compress the suits enough to be able to squish down and get the seatbelts locked without a certain amount of effort.'

Eleven minutes later they reached stop 2 on the south side of Elbow Crater. For the next hour they fulfilled a geologist's dream as they scampered around searching for unusual specimens. Irwin was ecstatic about the scenery: 'This is unreal. The most beautiful thing I've seen.' TV viewers were able to appreciate his point as the camera pointed into the nearby rille. Scott showed them a partly buried rock which had 'glass on one side of it with lots of bubbles'. From nearly a quarter of a million miles away, geologists in Houston were able to advise him that it was probably 'fresh' – no older than three and half billion years.

Mission control informed the men that stop 3 would have to be omitted due to the earlier delays; the deployment of the experimental package, scheduled for the final hours of the EVA, took precedence over the geology field trip. Using their gyroscopic navigation system, they turned around on a northward track and soon came across their wheel tracks. 'Somebody else has been here,' quipped Irwin. After logging more than 6 miles on their drive, they returned to the vicinity of Falcon loaded with six bags of rocks, four more of soil and two double core tube samples. Already four hours into the EVA, another two and a half hours of hard work lay ahead as they opened the doors in the LM to deploy the Apollo Lunar Surface Experiment Package (ALSEP). Irwin observed: 'Boy, it's going to be hard to keep the dust off the ALSEP.' He carried the two pallets, weighing nearly 300 lb at one G, to a safe site about 350 feet from Falcon while Scott drove the rover. Irwin's job was to deploy the £10 million atomic-powered package, which included the usual seismometer, solar wind experiment, magnetometer and ion detector. Scott couldn't resist joining in, however. 'I'll give you a demonstration here, Joe. Got the TV on this pallet?' Winding himself up like an aborigine throwing a boomerang, Scott hurled the small platform into the black sky, then, unable to stop his own momentum, he whirled around in a flurry of dust, desperately trying to remain upright. Capcom Joe Allen, on witnessing this drunken dance, was moved to ask: 'What was that a demonstration of, by the way? It started out to be of gravity and wound up being of centrifugal force, I think!'

Scott's main task, using a special jackhammer drill to penetrate the ground surface to a depth of 10 ft for insertion of two heat-flow probes, also ran into difficulties. Below the soft surface layer he ran into soil denser than anything he had encountered on Earth. 'I tell you one thing,' he panted, 'the base at Hadley is firm . . . Boy, that's really tough rock.' Leaning his full weight on the drill to prevent it from turning him around, he eventually gave up after penetrating only 5½ ft. The second hole proved just as difficult, forcing him to inject the probes as far as he could and hope the result would be worthwhile. The men's exertions had led to faster than normal oxygen consumption so Joe Allen passed up instructions to terminate EVA half an hour early. A super-sensitive laser reflector was set up, the samples were loaded and the tired astronauts climbed thankfully back into Falcon after more than six hours of almost constant activity. On removing their helmets, they commented on a strange odour, like gunpowder, which they attributed to lunar dust in the air. Scott's fingers were sore because his glove tips were too tight; this would result in blackened fingernails by the time he returned to Earth. Another minor irritation was the need to mop up water from a leaking water spigot, but the universal opinion was expressed by Capcom Allen: 'Real fine day's work up there, guys. Why don't you take the rest of the day off?' The men were more than happy to oblige.

Up above, the lone voyager, Al Worden, had also been busy 'trying to be three places at once in here'. He could not communicate directly with Falcon, though he could see it clearly as he passed overhead. Houston encouraged him to maintain his vigil: 'In general all orbital experiments are working very well and we have some very happy principal investigators . . . they say "keep up the good work".'

The second EVA was delayed for more than an hour as the astronauts searched for more spilled water and carried out the mopping up chores.

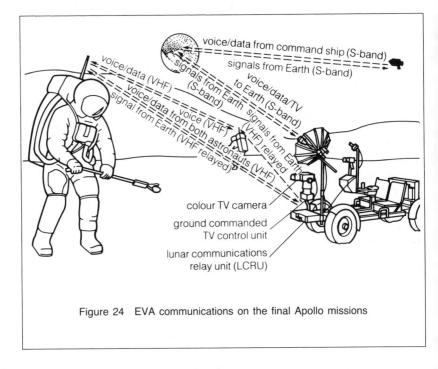

voice/data from command ship (S-band)

signals from Earth (S-band)

voice/data from command ship (S-band)

voice/data (VHF)

signals from Earth (S-band)

to Earth (S-band)

voice/data/TV

voice/data/TV to Earth (S-band)

voice/data from both astronauts (VHF relayed)

signals from Earth (VHF) relayed

voice (VHF)

signals from Earth (VHF) relayed

signal from Earth (VHF relayed)

colour TV camera

ground commanded
TV control unit

lunar communications
relay unit (LCRU)

Figure 24   EVA communications on the final Apollo missions

Emergency repairs to a broken antenna on Irwin's backpack were also necessary before the men could exit once again, along with the replacement of water in his backpack to remove air bubbles. Scott, however, was clearly looking forward to the trip to the foot of the towering Apennines as he cheerfully commented: 'Down the ladder to the plains of Hadley.' To their delight, the front steering on the rover also worked when they tried it: 'You know what I bet you did last night, Joe. You let some of those Marshall guys come up here and fix it didn't you?' Ironically, once they got under way Scott decided he preferred driving on only one system.

Pushing south past a variety of small craters for a distance of some 3 miles (5 km), the men stopped at the base of Mount Hadley Delta. Scott was moved to exclaim, 'My, oh my! This is as big a mountain as I ever looked up.' Once he had a chance to look around, he told Capcom: 'Joe, this rover is remarkable. I'm telling you we have climbed a steep hill and we didn't even really realize it.'

Resting on a 15 degree slope with a magnificent view all around, they set to work on trying to unravel the history of these grey, lineated mountains so different from anything found on Earth. Two rock specimens particularly excited their interest. Irwin discovered the first: 'Come and look at this . . . This is the first green rock I've seen.' A little later, a second green rock was found, causing doubts to set in – was it simply an optical illusion? It was not until scientists examined it in the lunar receiving laboratory that the pale green glass formed by sudden melting and cooling was confirmed. Scott's turn soon came: 'Guess what we just found! I think we found what we came for!' He held up a piece of milky-white crystalline rock about the size of a fist and weighing half a pound. Scott excitedly speculated that he might be holding an ancient fragment from the lunar interior, a theory confirmed by scientists on Earth when they dated it as at least four billion years old. 'This really is a gold mine,' enthused Scott as he and his partner worked like beavers. At one point, they were so excited that they dropped a sample bag, but this was quickly noticed in Houston so that they were able to retrieve it.

The EVA was into its fifth hour by the time the astronauts completed their work in the vicinity of Spur Crater and began their return journey. As on the previous day, Scott's oxygen consumption was higher than Irwin's, but Mission Control was so pleased with their progress that the 30 minutes it had deleted from the activity after the first day's excess oxygen usage were restored. After what they modestly called their 'nice trip', Scott tried once more to drill a hole the full 10 ft beneath the surface. Again he soon ran into difficulties and was advised to give up. A final sampling session and the erection of the American flag preceded the climb back into Falcon. The EVA had lasted for more than seven hours and had carried the men over nearly 8 miles of the Apennine foothills. One jubilant NASA official described it as 'the greatest day in the history of scientific exploration'.

Six hours sleep allowed the men to recuperate ready for the final excursion. The extended traverse of the second EVA had put activities behind schedule, however, so the third excursion was limited to between four and five hours – they did not want to miss the flight home! Once more, they headed for the edge of enigmatic Hadley Rille, though taking the shortest route by driving due west in the 'trusty old rover'. The first task, however, was to extract the core drilled the previous day. Scott began to fume as he struggled with a vice designed to pull out the tubes. First, he vented his spleen on Mission Control: 'Are you guys that interested in this thing, Houston?' Then he turned on Irwin: 'You can be doing something useful instead of just standing.' At last the reluctant core was withdrawn, and Scott was mollified by Allen telling him how important the core was: 'Boy, that sounds good. I don't feel like I wasted so much time.'

The trip to the rille gave the astronauts a respite from their physical workouts and once more delighted the geologists. Scott described the far wall of the gorge in intimate detail as the TV camera relayed the scene, clearly showing the various layers of supposed lava flows topped by occasional slumps of loose debris. Lumps of bedrock were hammered free to supplement the verbal and visual evidence before they left the rille for the last time. The sunlit mountains to right and left made a lasting impression on the two explorers. Irwin commented: 'Dave, I'm reminded about my favourite Biblical passage from Psalms: "I'll look unto the hills from whence cometh my help." But of course, we get quite a bit from Houston too.'

Back at Falcon, the men performed the closing ceremonies of their epic mission. Scott used a small franking machine to cancel the first and second stamps of a new US Postal Service issue commemorating American achievements in space. He then proceeded to conjure up the falcon's feather in his left hand while holding his geological hammer in his right hand: 'I guess one of the reasons we got here today was because of the gentleman named Galileo, a long time ago, who made a rather significant discovery about falling objects in gravity fields and we thought, where would be a better place to confirm his findings than on the moon?' Releasing both objects simultaneously, viewers could see the feather slowly accelerate and touch the surface at the same time as the hammer, confirming the theory put forward by the great Italian scientist some 350 years earlier. Scott drove the buggy to its last parking site about 300 ft (100 m) east of Falcon and turned the camera back towards the LM. It would televise for the first time the spectacular blast off from the Marsh of Decay. Beside the rover he placed a small replica of a 'fallen astronaut' with a plaque which carried the names, now totalling 14, of Soviet and American astronauts who had died serving their countries. The EVA closed soon after, nearly five hours after it began. It was time to clear out the LM and prepare for take-off. Joe Allen told them: 'OK, troops, explorer hats off and put on the pilot hat.' Scott replied: 'We sure are ready to do some flying.' After 67 hours on the lunar surface, the Falcon ascent stage took wing admist a brief flurry of debris, ending the longest stay on the moon up to that time. The occasion was marked by a short tape playback from the LM of 'Off We Go into the Wild Blue Yonder'.

Overhead, Al Worden in Endeavour had modified his orbit to prepare for boarders. The Falcon reached lunar orbit right on the nose, enabling a relatively straightforward catching up procedure during the next lunar revolution. Two hours after take-off, the craft were reunited. 'Hard dock. We got you,' came Scott's voice. 'Welcome home,' was Worden's simple reply. It took some three hours to transfer the large number of samples and close up both craft ready for jettisoning Falcon. This manoeuvre had to be postponed, however, when a routine check on air pressure in the tunnel suggested one of the hatches was not sealed properly. The astronauts had to re-open the tunnel, check the seals then close the hatches again. This was successfully achieved, though it took up an entire revolution. At last Falcon was 'go' for separation. The craft was deliberately crashed into the moon, an impact which registered on all three active seismometers located by Apollo missions.

Although the bulk of the work was now over, Endeavour remained in lunar orbit for another two days continuing the programme of orbital experiments. The óne break from routine was the launch of a mini-satellite from the SIM bay shortly before Endeavour blasted out of orbit. Only 31 in (76 cm) long, this hexagonal satellite, powered by solar energy, was designed to operate for up to a year, measuring nuclear particles and magnetic fields and enabling scientists to gain more information on the gravity anomalies on the moon which caused spacecraft to drift away from their planned orbits.

The homeward journey for Apollo 15 began right on schedule at 5.23 pm Cape time on 4 August. The crew were glued to the windows as they saw the world with which they had become so familiar shrinking in size as it once more resolved into a disc. The main event of the return leg was Worden's spacewalk the next day to retrieve two cassettes of film from the SIM bay. Watched by Jim Irwin from the open hatch, Worden moved easily hand-over-hand three times to the rear of the service module, using the hand rails provided. The entire EVA lasted only 38 minutes, but it was a major highlight of the mission for the CM pilot who had been closeted in the cramped cabin for 10 days. He was particularly taken up with the sight of Irwin framed against a backcloth of the nearly full moon. 'Jim, you look absolutely fantastic!' he exclaimed.

The last two days were taken up with observations of cosmic ray flashes, star navigation sightings, a press conference and a small course correction to ensure Endeavour hit the atmosphere at just the right angle. Fifteen minutes before re-entry, the service module was separated and the blunt heat shield turned to absorb the blistering heat caused by atmospheric drag. Everything functioned normally until the final stages, then, just as the three giant main parachutes appeared to be opening properly, one of them collapsed. The watching recovery ship warned the crew to expect 'a hard landing', but the craft remained upright after hitting the rolling Pacific some 335 miles north of Honolulu. Scott quickly reported that all three men were fine, and within six minutes of splashdown they were surrounded by frogmen. A triumphal greeting awaited them on the recovery carrier *Okinawa*.

Apollo 15 was an undoubted success: the crew left behind a powerful array of scientific instruments which would send back invaluable information for years to come; they brought back more than 10,000 photographs of which more than 1,100 were taken on the moon; also returned were around 100 documented soil and rock samples weighing 173 lb and including an 8 ft long core which showed in intimate detail the last billion years' events on that almost unchanging world. David Scott summarized the crew's attitude at their first public press conference after landing:

We went to the moon as trained observers in order to gather data, not only with our instruments on board, but also with our minds. And I'd like to quote a favourite statement, which I think expresses our feelings since we've come back: 'The mind is not a vessel to be filled, but a fire to be lighted.'

The spectacular pictures of the strange lunar terrain certainly fired people's imaginations all over the world. Two years later, Scott would write: 'Frequently I reflect upon those three most memorable days of my life. Although I can reconstruct them virtually moment by moment, sometimes I can scarcely believe that I have actually walked on the moon.'

Only five days after splashdown, the crew of Apollo 15 were announced as the back-up for the final Apollo mission scheduled for the end of 1972. However, within a year the sweet taste of success had turned sour. NASA discovered that all three men had been involved in an undercover operation to gain a massive profit from the sale of 100 unauthorized envelopes which they had franked on the moon. To handle the profits, said to exceed $150,000, the astronauts had broken NASA's conduct code by setting up a trust fund for their children. Further controversy arose over the appearance of duplicate statuettes like the 'fallen astronaut' left on the moon. Officials had no choice but to reprimand the heroes of a year before and take away their flight status. It was a sad smear on the integrity of the astronaut corps.

Irwin left NASA almost immediately, resigning from the space agency and the Air Force on 31 July 1972. Since then, he has been the chairman of the board and president of High Flight Foundation, a non-profit organization which he founded 'to share his faith in God and serve humanity, through speaking engagements, publications, retreats and training activities'. He has devoted recent years to a search for Noah's Ark on Mt Ararat. Worden moved to Ames Research Center in California where he could use his knowledge of aeronautical engineering and instrumentation. After a year, he was promoted to chief of the systems studies division. He resigned from NASA and the Air Force in September 1975, becoming vice president of Irwin's High Flight Foundation and president of the Americans For The National Dividend organization. He now has his own energy management services company in Florida and is a director of a Michigan energy management organization as well as maintaining his religious work. David Scott remained with NASA until 1977. He was first moved to Johnson Space Center as an Apollo programme technical assistant and was soon involved in mission operations for the joint Apollo–Soyuz flight planned for 1975. He was promoted to Dryden Flight Research Center at Edwards Air Force Base, first as deputy director, then as director, a post for which he was eminently well qualified and which he held for two years. He formed his own advanced technology company in 1977, specializing in research and development, and project management.

The flight of Apollo 16, the penultimate lunar mission, would visit another highland area, the Descartes plateau, and in most respects would be a copy of its predecessor. The more leisurely pace of the Apollo programme as it entered its final year allowed minor modifications to both the spacecraft and the pressure suits during the nine-month interval. It was just as well, for several accidents during assembly would have caused problems on a tighter schedule; at one point the entire rocket had to be wheeled back from the launch pad to the assembly building when a command module fuel tank ruptured. Another unplanned factor entered the scene when one of the crew, Charles Duke, went into hospital with pneumonia. Three days later, the launch date was put back from 17 March to Sunday 16 April 1972.

The weekend and the fine weather ensured a large turnout for the launch at 12.54 pm – there was still no spectacle on Earth like the fiery ear-shattering blast-off by a Saturn V, and people flocked to the Florida coast for the awe-inspiring event which would soon be confined in the pages of the history books. Among the VIPs present were Vice President Spiro Agnew and King Hussein of Jordan, proving that Apollo still had its value as a symbol of American leadership in the fields of science and technology, even though the administration had decided to terminate its funding.

Commanding the mission was Navy captain John Young, now 41 and on his fourth space mission, only the second man to reach this landmark. Alongside the phlegmatic Young – his usual exclamation of excitement was a laconic 'That was something' – was a man of an entirely different mould. The lunar module pilot, 36-year-old Air Force lieutenant-colonel Charles Duke, was an excitable extrovert who would ensure that there were few dull moments during the forthcoming exploration. He had already led a distinguished career before astronaut selection in 1966, having gained a bachelor's degree in naval sciences before being commissioned in the Air Force in 1957, passing his flight training with distinction, then spending three years as a fighter interceptor pilot in West Germany. He gained a master's degree in aeronautics in 1964, graduated from the Aerospace Research Pilot School the next year and stayed on there as an instructor until he became a group 5 astronaut. Although Apollo 16 was his first space flight, he had gained considerable experience on the Apollo 10 support crew and the Apollo 13 back-up crew. Perhaps the most meaningful recommendation, however, came from Neil Armstrong, who specifically requested that Duke held the Capcom hotseat during the Apollo 11 landing phase.

The command module pilot, 36-year-old Navy lieutenant-commander

Thomas Mattingly, was also on his first mission, having joined the astronaut corps at the same time as Duke. Their fates seemed intertwined, for Mattingly had been due to fly on the ill-fated Apollo 13 but had been grounded when doctors found he had no immunity to German measles after Duke, with whom he had worked closely, had become infected during a visit to a friend's home. The flight assignments were rescheduled to allow him to fly as soon as possible, and so he and Duke lined up alongside each other on Apollo 16. Mattingly had graduated in aeronautical engineering in 1958, joined the Navy and become a qualified pilot in 1960, serving in carrier-borne attack squadrons for five years before moving to the Air Force Aerospace Research Pilot School where Charlie Duke was already serving. At NASA, he served on the Apollo 8, 11 and 12 support crews, also working on development and testing of the Apollo spacesuit and backpack.

The outbound leg proved the maxim that nothing related to spaceflight could be taken for granted; once again, minor irritations kept both ground controllers and astronauts on their toes. The third stage began to leak helium gas used for attitude control while Apollo was still on its first Earth orbit, causing Houston to warn the crew to be prepared for their craft to assume attitude control. Fortunately, the problem was not sufficiently serious to affect the trans-lunar injection burn over the Pacific two and a half hours into the mission. Transposition and docking followed in the usual way, except that John Young reported a cloud of bright particles blowing past the window. The apparent venting continued after extraction of the lunar module, named Orion. Capcom Gordon Fullerton assured the crew that it was probably flakes of thermal paint which had been added to give increased protection against overheating, but as a precaution Young and Duke were told to open up Orion for a quick examination. The TV pictures showed nothing amiss, so the LM was closed once more until the next scheduled check-out the following day. Charlie Duke was not especially pleased with the situation, however, for it was his job to attempt navigational star sightings through the cloud of shiny white particles.

Navigation became even more difficult on the second day out from the Cape when a master alarm sounded twice, sending shivers up the crew's spines. Each time, the navigation platform locked for around 18 minutes as the result of an intermittent short circuit, effectively preventing any spacecraft manoeuvres until it was overcome. The potential threat to the craft should the fault recur during critical phases of the flight was very serious, causing engineers at Massachusetts Institute of Technology to work on a computer program which would prevent a mission abort. Then communications problems arose when flight controllers reported difficulty in switching antennas by ground command as the spacecraft rotated three times an hour in its 'barbecue roll'. Even Charlie Duke got in on the act when he complained that his pressure suit was too tight and asked permission to slacken the laces; Houston felt the situation would right itself in the lunar vacuum and refused.

The crew did have time to conduct some experiments en route, including a repeat of a fluid separation test carried out on Apollo 14 in order to evaluate methods of producing vaccines in space. The detection of cosmic rays was extended beyond mere visual recognition by an emulsion detector held in front of the face, keeping a permanent record of cosmic ray tracks. Only one mid-course correction was necessary. The experiments on board the CSM, Casper, were readied as the craft accelerated towards the moon, and shortly after 3 pm Cape time on Wednesday 19 April Apollo 16 disappeared behind the moon's west limb for the first time.

John Young's voice greeted Houston after the usual agonizing wait: 'Hello, Houston. Sweet 16 has arrived.' The elliptical orbit swept the crew to within 67 miles (107 km) of the lunar surface, an experience which enthralled the veteran Young, the first man to orbit the moon twice, as much as his companions. After only two revolutions, the main engine fired once more to lower the craft into the orbit from which descent would begin, a mere 12 miles (20 km) above the objective. The initial enthusiastic survey over, it was time for a meal and a rest before the excitement of the next day.

The morning saw Young and Duke enter Orion for the final check-out, full of high hopes for what lay ahead. They began 40 minutes ahead of schedule, but the time advantage was gradually eroded as problems arose. First, the steerable S-band antenna refused to swivel properly, causing considerable extra work for Duke as he manually updated the onboard computer. Then the pressure of propellant in the thrusters began to rise out of control, forcing the crew to transfer liquid from the propellant tanks to make room for the leaking helium gas. At last, undocking could go ahead and as they came from behind the moon on revolution 12, Orion informed Houston, 'We're sailing free'. The two craft kept station a few yards apart while Young and Duke prepared for descent, and Mattingly checked out Casper prior to raising and circularizing the orbit. Then came drama on the thirteenth pass behind the moon, out of contact with ground control. Only minutes before Orion's descent was due to begin, Mattingly noticed oscillations in the motor which moved the main engine nozzle to ensure that the thrust was directed at the correct angle. The primary system seemed to be normal, so he assumed the trouble lay in the back-up system. Nevertheless, the problem was so serious that the moon landing was automatically postponed; if the primary system now failed, there would be nothing on which to fall back, and no way of getting home. As soon as Casper appeared in view again, Mattingly passed on the bad news and the engineers feverishly began to analyse the problem. Time was not on their side; if the descent was not begun within ten hours or five revolutions, Orion's orbit would have altered too much and its supplies would have depleted sufficiently to abort the landing. Meanwhile, Mattingly was told to close up once more on Orion and await developments. On board Orion, the crew was only able to wait helplessly and speculate on the future. Duke told Houston: 'John and I have been talking about if we get to land this thing, we'd like to probably think about going to sleep first and then we'd get in a full EVA tomorrow.' Houston agreed that the suggestion made sense. But would there be a landing?

For nearly three hours the astronauts circled the moon in dread of the order to abort and return home; there was even some talk of an Apollo 13 type of crisis. However, tests at Houston and at the main contractors' plants all over the country confirmed that no structural damage would occur during a normal firing. Chris Kraft, Director of the Manned Spacecraft Center, announced the decision to continue the lunar landing as the two craft came around the moon for the fifteenth time.

The necessary flight plan alterations were passed up to the crew, and it was clear that the delay had not made the landing any easier: Orion's thrusters had been fired during the long wait for news, using vital propellant which might be needed during the descent from an altitude of 12 miles (20 km), the highest yet. To make matters worse, the main steerable antenna was still not working in the yaw direction, thereby limiting communications with ground stations. Nevertheless, the descent and landing went unbelievably smoothly with only minor course corrections being punched into the computer by the crew. Charlie Duke could hardly contain his excitement:

> Coming in like gangbusters . . . I can see the landing site . . . We're right in, John . . . Man, there it is – Gator, Lone Star . . . Perfect place over there, John . . . A couple of big boulders . . . Contact . . . Stop. Well, old Orion is finally here, Houston. Fantastic! All we have to do is jump out of the hatch and we've got plenty of rocks . . . Man, it really looks nice out there . . . I'm like a little kid on Christmas Eve.

It was 9.23 pm Cape time on Thursday 20 April. Orion had arrived on the Descartes plateau nearly six hours late and come to rest close to a sizeable crater. In the distance were the rolling slopes of the lunar highlands.

Following a brief description of the terrain visible through the windows, the crew powered down the LM to minimum level, prepared a meal, then began an early rest period prior to the first EVA which had been postponed to the next day. Up above, Tom Mattingly was safely installed in a circular orbit and beginning his schedule of orbital surveys

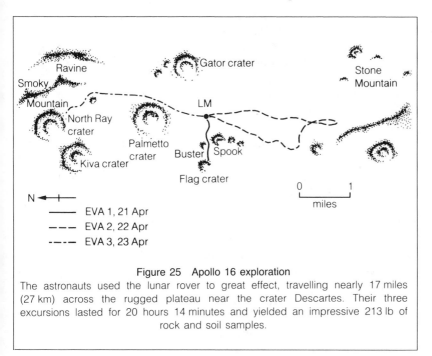

Ravine

Smoky Mountain

North Ray crater

Kiva crater

Palmetto crater

Buster

Spook

Flag crater

Gator crater

LM

Stone Mountain

N ◄──┼──

0 ────── 1
miles

─────── EVA 1, 21 Apr

─ ─ ─ EVA 2, 22 Apr

─ ·─ ·─ EVA 3, 23 Apr

Figure 25   Apollo 16 exploration
The astronauts used the lunar rover to great effect, travelling nearly 17 miles (27 km) across the rugged plateau near the crater Descartes. Their three excursions lasted for 20 hours 14 minutes and yielded an impressive 213 lb of rock and soil samples.

after successfully firing the main engine. The optimism of the ground crew seemed to be vindicated. Their worries had not completely evaporated, however, causing the introduction of further flight plan alterations. To reduce the consumption of supplies and cut to a minimum the reliance on main engine burns, the post-landing orbital duration would be reduced by one day, despite the loss of information from the orbital experiments.

The next morning, Orion's cabin was depressurized and the first EVA began. John Young backed down the ladder to become the ninth man on the moon: 'Here you are, mysterious and unknown Descartes highland plains. Apollo 16 is going to change your image.' The historic moment was not recorded for posterity, however; the antenna problem prevented reception from the TV camera. Young was followed only five minutes later by the eager Duke, though his enthusiasm was dampened somewhat by a leak from a faulty valve on his orange juice bag: 'I've already had an orange shampoo with the helmet on. I wouldn't give you 2 cents for orange juice as a hair tonic.'

They deployed the lunar rover, put up the flag, installed a TV camera on a tripod and set up an ultraviolet camera. John Young was amazed at the 30 ft deep crater only 10 ft from Orion: 'Any place else around here and we'd have landed on a great big slope.' While he took a series of exposures with the TV camera, Duke was left to carry the ALSEP experiments to a safe distance from the lunar module. The bulky packages were harder to carry than he thought, and one of them slipped from his grasp: 'It hit the dirt like a bomb, but the experiments appear to be intact.'

Duke began drilling the two holes for the heat sensors, and, to his delight, the first nine foot hole was completed with little effort. Then, as he started on the second, John Young negated all of his partner's efforts by tripping over the line running from the heat flow sensors to the nuclear power source. With loose wires lying all over the dusty surface, it was surprising such an accident had not happened before, but Young was still very upset: 'God almighty! It broke right at the connector. I didn't even know it. Oh, rats. I'm sorry, Charlie.' Despite efforts to revive the experiment, it eventually had to be abandoned. The other experiments were activated without any further mishaps, Duke succeeding in extracting an 8 ft (2.6 m) core sample quite easily. Young set up three geophones, then fired a total of 19 charges from his portable 'thumper' in a successful re-run of the Apollo 14 experiment. A mortar box containing four more charges was also set up ready for remote control detonation after the landing was completed, though one of its four legs would not move into position.

Nearly four hours had elapsed before the men climbed aboard the rover for their first long distance excursion. They travelled about a mile due west to a small crater named Flag, then returned via Spook crater, each time collecting soil and rock samples, and taking magnetometer readings. The trip ended with a spectacular test drive by Young as he put the rover through its paces before a suitably impressed Duke: 'He's got about two wheels on the ground. It's a big rooster tail out of all four wheels, and as he turns he skids. The back end breaks loose just like on snow . . . Man, I'll tell you, Indy's never seen a driver like this!' Young later admitted: 'The thing that really saved us was that no one was coming in the opposite direction!' The mission commander had really enjoyed the experience of one-sixth gravity, occasionally displaying considerable exuberance and humour; at one point, he commented on Duke's considerable height advantage: 'One small step for Charlie is one giant leap for me!'

The second EVA the next day was also scheduled for seven hours, but this time most of the effort would be aimed at geological exploration and sampling. The target was Stone Mountain, a rounded hill to the south of the landing site. It was well named, for the astronauts found the journey rough going as they traversed boulder fields and cratered, undulating terrain. Young was moved to comment, 'It's the roughest ride I've ever been on.' He was in the position of the reluctant passenger suffering helplessly from the 'reckless' driving of Duke. Several times he complained to the over-exuberant driver: 'Charlie, quit pushing this thing around,' and later, 'Charlie, whatever you do, don't hit that brake.' The rover took quite a battering, resulting in the breakdown of the instrument which told the crew the vehicle's degree of roll.

The traverse extended some 750 feet up the slopes of Stone Mountain and 2½ miles (4 km) from Orion. Geologists on Earth gloried in the beautiful pictures returned by the onboard remote controlled camera while the crew raked the soil, overturned boulders, probed the surface, took magnetic readings and collected 82 lb of samples. They were generally disappointed with the results, however; many experts had predicted evidence of lava flows which did not materialize. There was slight consolation when Young picked up a rock and commented: 'This is the first one I've seen I really believe is crystalline.' After 7 hours 23 minutes they were back inside Orion for a well-earned rest. The realities of astronaut life were relayed worldwide in an unintentionally hilarious broadcast when the astronauts inadvertently left their microphones switched on and treated listeners to a catalogue of their digestive problems resulting from an experimental orange juice laced with potassium to help stabilize their heart function.

The third and final EVA was shortened to about five hours in order to give the men time to prepare for lift-off. With the sun now higher in the black lunar sky, Young warned his companion: 'Charlie, it's gonna be hot out there today,' as he climbed awkwardly down the ladder. Once again, they were so eager to start work that they left the LM ahead of schedule, despite Duke having to resort to sleeping pills for the third night in succession. This time, the main targets lay to the north, in particular the spectacular North Ray Crater and the lower slopes of Smoky Mountain. North Ray was so named because the surroundings were covered by a lighter coloured deposit of material ejected by the impact which created the crater. The depression itself was found to be about three-quarters of a mile across and some 600 ft deep. Scattered all around were the largest boulders yet seen on any Apollo mission; one of them was called the 'house rock' because of its giant size. They explored the lower terraced slopes of Smoky Mountain before turning back, but again there was an embarrassing lack of volcanic evidence. So involved were the crew in their work that they pleaded with Capcom Tony England for a brief extension: 'How about an extension, you guys? We're feeling good,' came Duke's voice. England gave the official line: 'We understand. We'd like you to get back in the lunar lander on time.' Young joined in: 'All we were going to do tonight was sit around and talk.' England tried to stall: 'We like to hear you talk.' Duke turned to pleading: 'Ten minutes and we can get all this done. How about ten minutes, Tony? Come on, Tony, pretty please?' The controllers gave way under the pressure. 'OK, we'll give you ten minutes. How's that? Just because we love you,' came the reply from Capcom.

The EVA eventually terminated after 5 hours 40 minutes and 7 miles

(11.4 km) of meandering across the rugged lunar surface. The samples from the third excursion weighed about 90 lb. These and the solar wind experiment were loaded on board Orion before a final TV spectacular, described by Young as an abbreviated 'lunar Olympics'. He explained: 'We're gonna show what a guy could do, like jump flat-footed straight in the air three or four feet.' Sure enough, viewers saw the commander take off in a slow motion vertical leap. Unfortunately, in his enthusiasm to excel, Duke overdid the push and lost his balance, falling heavily on his backpack. Luckily, there was no damage to Duke or the suit, but Young reprimanded him: 'Charlie, that ain't very smart.' An abashed Duke was forced to agree: 'Sorry about that.' Further events followed as they discarded equipment by flinging it long distances like a strange lunar discus competition. Young parked the lunar rover several hundred feet from Orion, then expressed his appreciation: 'Boy, that's a good machine.' With the camera set up to record take-off, they entered Orion to prepare for the launch. Everything went to plan as Mattingly fired his main engine successfully for the second time since separating from Orion and moved into the correct rendezvous orbit. The lunar module blasted off in a cloud of debris at 8.26 pm Cape time, watched by millions of viewers and carrying a record load of samples. Apollo 16 had also broken existing records for total time on the surface, and for total EVA duration. 'What a ride, what a ride!' came the excited cry from the ascent stage as it disappeared into the black beyond.

A little more than two hours later, Orion had caught up with the mother ship and docking had been successfully completed. The lunar module was looking distinctly the worse for wear now; Mattingly described great strips of thermal blanket hanging loosely from the craft after being ripped free during blast-off. Attempts to clean up the dust which filled Orion had to be abandoned when the vacuum cleaner failed, so some dust inevitably floated through to Casper. Mission Control annoyed the crew by altering the schedule so that Orion was jettisoned without the crew having time for a rest after their hectic day. Not surprisingly, a switch was left in the wrong position, resulting in Orion tumbling in orbit and causing Mattingly to make a hurried evasive manoeuvre. Plans to crash the LM on to the lunar surface had to be abandoned. Orion remained out of control in lunar orbit for nearly a year before finally impacting on the moon. Within an hour of separation, the small sub-satellite was released from the SIM bay. The continued unwillingness to use the main engine more than necessary meant that Casper's orbital plane change was omitted, causing the satellite to enter a short-lifetime orbit rather than the year-long orbit originally planned.

Trans-Earth injection took place less than five hours after the satellite ejection; the period of orbital observations had been curtailed by that same fear of an engine malfunction, so Young and Duke had less than a day in lunar orbit before Apollo 16 hurried them off towards home. The engine which had caused so many hearts to flutter and so many flight alterations performed perfectly behind the moon. Mattingly put into words the relief everyone was experiencing: 'Morale around here just went up a couple of hundred per cent.' The crew were at last able to retire for a much-needed extended rest period. Next day, 25 April, saw the highlight of the return trip, a deep space EVA by Mattingly to retrieve magazines of exposed film from the panoramic and mapping cameras which had scanned the moon during his lengthy solo flight. Tethered by a 25 ft (8 m) line which gave him his oxygen and communications, Mattingly inched his way along the side of the craft using the hand and footholds to the scientific instrument bay. Meanwhile, Duke poked his head out of the open hatch to expose spores, watercress seeds, bean embryos and shrimp eggs in an experiment designed to show the effects of cosmic rays on biological specimens. The entire EVA lasted for 1 hour 24 minutes, the longest of the Apollo deep space excursions.

The remainder of the homeward leg was relatively straightforward with the usual cosmic ray eye tests, the onboard press conference and some minor equipment hiccups. The only unusual activity came three hours before landing when the thrusters were briefly fired in a small course correction designed to ensure that debris from the re-entering service module would not head towards the tiny inhabited island of Penrhyn. Splashdown occurred at 2.45 pm Cape time on Thursday 27 April, about 175 miles (280 km) south east of Christmas Island. The spacecraft hit the gentle swell less than a mile from the recovery carrier *Ticonderoga*, the most accurate splashdown yet. Despite the craft turning turtle on touching the water, it was quickly righted, and the value of the bullseye landing was demonstrated when the crew appeared on the carrier's deck only 37 minutes after splashdown, another Apollo record. Residual propellants were retained in the command module for the first time after fears that their ejection may have damaged Apollo 15's parachute lines. It was a costly decision: the craft later exploded at North American's plant in Downie, California, injuring several technicians. Indeed, the final verdict on the mission was summarized neatly by one newspaper: 'Brilliantly successful scientifically, but disappointing from an engineering and operational point of view.'

The crew spent two days on board *Ticonderoga* before flying back to Houston and a hero's welcome. All three remained with NASA beyond the Apollo era, though Young and Duke also continued to work together as a team for the next eight months on the Apollo 17 back-up crew. Unlike many of their fellow astronauts, the Apollo 16 men decided to help the Space Agency over its long period of transition which followed the euphoria of Apollo.

Duke was the first to leave, in December 1975, on completion of the Apollo–Soyuz test mission. Soon after, he resigned from the USAF, though he continued to serve with the Air Force Reserve with the rank of brigadier general. He set up his own investment firm, became president of Southwest Wilderness Art, Inc. and devoted much of his time to lay preaching. Both Young and Mattingly were assigned to the Space Shuttle programme in January 1973. Young, now one of the senior astronauts, was appointed Acting Chief of the Astronaut Office in January 1974, a post which was made permanent the following year. Meanwhile, Mattingly worked on various aspects of the Shuttle flight test programme, culminating in service as back-up commander for the second and third Shuttle missions and as commander for the fourth flight of Shuttle Columbia.

On 23 May, ground controllers fired three of the four mortar charges left at Descartes – signals from the mortar indicated that it had fallen over after the third firing. Seismic waves picked up by the geophones gave valuable evidence to geologists; even more significant were the tremors picked up by Apollo seismometers on 17 July, when a meteorite weighing about a ton crashed onto the far side of the moon. It began to look as though the moon is not entirely dead after all; evidence was mounting for a molten core, possibly with a high iron content. Curiously, the magnetic readings obtained by the crew of Apollo 16 were 10 times higher than any previously recorded. All the evidence from the Apollo lunar experiments suggested that the highlands seemed to hold the vital clues to the early history of the moon and Earth, so it was to the mountains of the Taurus–Littrow region that the final Apollo was directed, and it was no coincidence that among the crew was the only scientist to set foot on the moon, a geologist named Harrison Schmitt.

In 1969, when the back-up crew for Apollo 14 was announced, group 5 astronaut Joe Engle, named back-up LM pilot, had been favourite for a trip to the moon on board Apollo 17 – the well-established practice was for a back-up crew to bypass the next two missions then become the prime crew for the third. Engle seemed to have all the necessary qualifications: he had flown jet fighters, graduated from the USAF Experimental Test Pilot School and the Aerospace Research Pilot School at Edwards Air Force Base, and flown the X 15 rocket plane on 16 test flights, exceeding an altitude of 50 miles (80 km) on three occasions. As a result of these three daring flights, he was already entitled to the rating of astronaut before he joined NASA in 1966. The disappointment must have been considerable when he learned that the LM pilot's seat on the last Apollo would be occupied by Dr Schmitt, though, to his credit, no visible resentment or envy surfaced. Engle patiently waited his turn, conducting essential Shuttle approach and landing test flights in 1977 before finally experiencing the challenge of spaceflight as commander of the second Shuttle mission in November 1981.

Harrison Schmitt had the good fortune to be in the right place at the right time. The 37-year-old bachelor had graduated in science from Caltec in 1957, studied at the University of Oslo in Norway for a year, then undertaken work for the Norwegian and US Geological Surveys before taking up a teaching post at Harvard in 1961. Gaining a doctorate in geology from Harvard in 1964, he moved to the US Geological Survey's Astrogeology Center in Flagstaff, Arizona where he became the project leader for lunar field geological techniques and became an authority on photo and telescopic mapping of the moon. Already, he was helping to instruct astronauts on geological field studies in preparation for the moon visits. In 1965 he was chosen as one of a select group of only six scientist–astronauts, and was able to continue astronaut instruction in lunar navigation, geology and feature recognition while himself undergoing flight training. Once samples began to be returned from the lunar surface, Schmitt participated in analysis of the rare treasures, but his own chance to collect specimens first hand seemed to have gone when the Apollo missions were curtailed, despite assignment to the Apollo 15 back-up crew. However, for once, scientific pressure carried political weight, so Schmitt was moved up to the LM pilot's seat on Apollo 17, possibly the last opportunity for a trained scientist to visit the moon this century. Now he had the chance to rebut the criticism of his selection.

Schmitt's two companions were the usual mixture of veteran and rookie. The commander was Gene Cernan, still only 38 but taking part in his third mission and his second to the vicinity of the moon. Cernan was determined that Apollo 17 would go down in history not just because it was the last but because it was the most successful mission; in true patriotic style, his command module was named America while the LM became Challenger. The man whose job it was to keep America in healthy condition while conducting orbital experiments and continuing a lone vigil was 39-year-old Ronald Evans, another Navy man. Evans had graduated in electrical engineering from the University of Kansas in 1956, then gained a master's degree in aeronautical engineering in 1964 from the US Navy Postgraduate School. He completed flight training in 1957, becoming a carrier fighter pilot with the Pacific Fleet, and eventually a combat flight instructor. Prior to NASA selection, he was flying F 8 Crusaders on combat duty in Vietnam from the carrier *Ticonderoga*, one of only a few astronauts to serve in that bloody conflict. The bald-headed Evans was just as aware of the significance of Apollo 17 as his commander and the entire Manned Spaceflight Center: 'We may be the last crew going up there, but we're going to be the first team, that's for sure.' This sentiment rubbed off on the ground staff; one group of Grumman engineers hung a sign on the service structure at Pad 39A which read, 'This may be our last but it will be our best'.

For about two and a half hours during the darkness of 6–7 December 1972 it seemed as if all of their efforts had been in vain: Apollo 17 refused to budge, threatening a costly and humiliating postponement. More than half a million spectators had flocked to the Cape for the one and only opportunity to observe a night launch by a Saturn V booster. Apart from an hour's delay which had been made up, the countdown had proceeded normally. Expectation rose as the automatic sequence began, then, with only 30 seconds to go, deflation – the countdown stopped. The automatic sequencer had failed to give the command to pressurize the third stage oxygen tank, and, although this had been manually corrected from the firing room, the computer refused to accept the correction, terminating the countdown. It was the first such halt in an Apollo countdown, and resulted in a frantic race against time to get the giant rocket off the ground before the launch window closed.

Ground controllers and engineers at Huntsville, Alabama tried out ways to bypass the automatic command system while the countdown was recycled to T minus 22 minutes. Although there was no apparent danger to the astronauts, the service tower was swung back into position in case an evacuation was necessary. The crew had to endure a seemingly interminable wait strapped in their couches as the clock slipped past midnight. Eventually, the final stage of the countdown resumed, this time with no hold-ups. At 12.33 am Cape time, the spectators stared in wonder at the brilliant burst of flame which lit up the sand bars and lagoons surrounding the launch pad. Spectators had to shield their eyes from the blinding ball of orange light which erupted from the five first stage engines amidst a billowing cloud of white smoke. A trail of fire and vapour appeared across the black sky as the last Saturn V to see active service accelerated in a great arc out across the Atlantic Ocean on its way to the parking orbit 105 miles (169 km) above the Earth. The launch was visible not only to Florida residents – reports of sightings came in from as far away as Cuba, the Bahamas, North Carolina and Alabama. The mighty Saturn V, Wernher von Braun's brainchild and still the mightiest rocket ever constructed, had given countless thousands of spectators the thrill of a lifetime. It was also an unforgettable experience for the crew, as they witnessed the flickering glow at launch then later as the second and third stages were jettisoned. 'Let me tell you, this night launch is something to behold,' commented Cernan.

The main priority for the journey to the moon was to make up for the 2 hour 40 minute delay on the ground. Accordingly, the trans-lunar injection burn was completed 15 minutes earlier than usual, boosting Apollo into a fast non-return trajectory: if the main engine on America failed, there would be no escape this time. The crew were in good spirits, despite a series of master alarms which kept them on their toes but proved to be spurious. Transposition and docking went according to plan, enabling the third stage to be cast off on a collision course with the moon. A minor course alteration on day three put America on line for lunar orbit while further increasing the acceleration. Checks on Challenger showed all was well – after a hesitant start, Apollo 17 was performing beautifully. The crew resorted to sleeping pills, however, a practice which caused ground control some problems next morning. Brass band music, klaxons and alert signals were broadcast to the slumbering astronauts before they were finally aroused after an hour. Capcom asked: 'How do you read us this morning?' to which came the reply, 'We're asleep.' 'That's the understatement of the year!' commented Houston. 'Never let Evans be on watch . . . Our biggest problem this morning is keeping Ron from going back to sleep,' said Schmitt. Houston responded by jokingly threatening to dock a day's leave for their late start, although in fact, there was no real need for urgency.

The crew were presented with a magnificent view of the sunlit moon as they closed in, a view denied to their predecessors because they had to approach in shadow. The SIM bay door was jettisoned prior to final approach, then, at 2.36 pm Cape time, Apollo 17 disappeared behind the moon's west limb. Eleven minutes later, a six and a half minute burn by the main engine slowed the craft into lunar orbit. After a minor communication problem, Cernan's voice came through loud and clear on rounding the moon: 'Houston, this is America. You can breathe easier. America has arrived on station for the challenge ahead.' Understandably perhaps, 'Jack' Schmitt was particularly enthusiastic about the view from his window. During the first orbit he reported a strange sighting on the dark lunar surface: 'It was a bright little flash near that little crater on the edge of Grimaldi. It was a hard little pin-prick of light . . . It means something is still happening on the surface of the moon.' There was little time for such discussions, however, for the second engine firing was due at the end of the second revolution. This burn put the craft into an elliptical orbit whose low point (pericynthion) lay 10 degrees to the west of the landing point. Apollo was now cruising within 17 miles (27 km) of the lunar surface.

Following a lengthy rest period, the crew awoke without too much trouble and prepared for the final descent. Cernan and Schmitt moved into Challenger and powered up the LM. Everything checked out perfectly during the eleventh pass across the front side, enabling Challenger to separate from America as they sailed out of sight once more. Re-appearing for the twelfth time, Evans told Houston: 'We're floating free.' Slowly the gap between the two opened up as Evans fired his thrusters. Schmitt enthused over the view of the landing site: 'I can even see right there where we're going to park this baby down. The way this ridge is coming up, you can almost stick your hand out to grab a

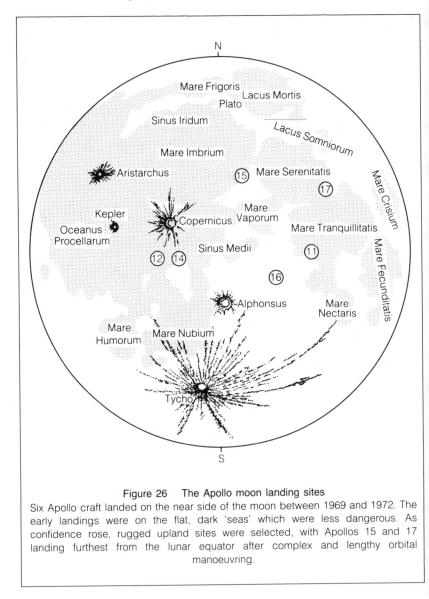

Figure 26   The Apollo moon landing sites

Six Apollo craft landed on the near side of the moon between 1969 and 1972. The early landings were on the flat, dark 'seas' which were less dangerous. As confidence rose, rugged upland sites were selected, with Apollos 15 and 17 landing furthest from the lunar equator after complex and lengthy orbital manoeuvring.

rock.' The pitted Taurus–Littrow valley passed some 12 miles (22 km) below, a dark grey enclave in the rugged lunar mountains.

Back on the far side, Evans fired the main engine to circularize America's orbit. Five minutes later, Challenger was slowed to drop even closer to its target. Within the hour, the main engine on the LM fired yet again to initiate the descent. Swooping ever lower, Cernan was heard to cry out: 'The day of reckoning comes in four minutes, Jack!' The view from the window appeared just as it had so many times in the simulator. Challenger was now below the mountain crests with Cernan in manual control. Dust billowed from the plain as he inched downwards, then shut down the engine to land with a jolt. Challenger had come to rest with one foot inside a small crater, causing the craft to tilt gently backwards. 'The Challenger has landed,' came Cernan's delighted report. 'We is here. Man, we is here!' It was 2.54 pm Cape time on Monday 11 December. The breathtaking view of the rolling valley with the rounded peaks reaching 6,000 feet into the black heavens deeply impressed the two explorers. 'This is the most majestic moment of my life,' said an emotional Schmitt. 'This is something everyone gotta do once in his life.' The surroundings became a topic of excited chatter. 'You can see the boulder tracks. There are boulders all over that massif. We should've hovered around a bit.' Schmitt asked Cernan: 'Who told you this is a flat landing site?' to which the commander replied: 'What do you want, an absolute guarantee? I like it right where we are.'

There was little time for sightseeing, however, since they had to prepare for their first EVA. Just over 4 hours after the landing, Cernan opened the hatch and backed on to the porch. The long black shadow of the LM stretched over the pristine surface – it was not long into the lunar day, a continuous period of sunlight which lasts for 13 Earth days. 'Jack, I wouldn't lower your gold visor until you get on the porch because it's plenty dark out here,' commented Cernan. Minutes later, he became the eleventh man to set foot on the Earth's only natural satellite: 'I'm on the footpad, and, Houston, as I step off at the surface of Taurus–Littrow, I'd like to dedicate the first step of Apollo 17 to all those who made it possible.' Schmitt repeated the commander's progress shortly after, though Cernan felt he had to remind him as he closed the hatch: 'You lose the key and we're in trouble!'

The first task was to unfold and test the lunar rover. 'Challenger's baby is on the road,' called Cernan as the wheels bit into the dust and rubble. With their transport in working order, the astronauts set up the sixth and last American flag on the moon: 'Houston, I don't know how many of you are aware of this, but this flag has flown in the MOCR [Mission Operations Control Room] since Apollo 11. And we very proudly deploy it on the moon to stay for as long as it can in honour of those people who have worked so hard to put us here and to put every other crew here.' The hard work began with the deployment of the nuclear powered ALSEP, a very different package from those previously carried. There were experiments to determine the composition of the tenuous lunar atmosphere, to measure the speed, mass and direction of travel of dust particles derived from impacts, and to try to detect the gravity waves predicted by Einstein's general theory of relativity. In addition, there was a more advanced seismic experiment with four geophones and eight varied charges, and the heat flow experiment which had so unfortunately been destroyed on Apollo 16. A further series of experiments to help determine the nature of the lunar surface and subsoil were also deployed.

It proved to be hard going, particularly the drilling of the two eight foot holes for the heat flow sensors and a third hole for the deep core sample. Cernan's heart rate peaked at around 150 on several occasions, necessitating a higher than expected oxygen consumption. A series of minor delays put the men still further behind schedule so Houston told them to curtail the planned EVA; instead of driving to Emory, they were to stop short at Steno crater, less than a mile to the south. It took a mere eight minutes to cover the ground, and once there they were able to set up one of the explosive charges, rake the soil for small fragments and take two core samples. Unfortunately, Cernan accidentally damaged part of a rear fender on the rover, and on the return leg it completely broke away, causing the two men to be sprayed by a cloud of grey, clinging dust. Not that they were clean up to that time; Schmitt had fallen over

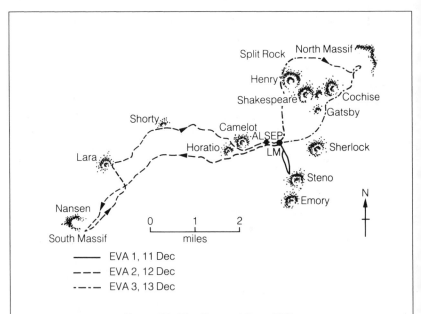

Figure 27   The Taurus–Littrow EVAs

All three EVAs lasted for more than seven hours, a total of 22 hours 6 minutes. During these excursions the astronauts drove a distance of 21 miles (35 km), including 12 miles (19 km) on the second day. The weight of samples collected was also a record, 243 lb, including the celebrated orange soil from Shorty crater. The first EVA was intended to examine Emory crater but had to be curtailed.

several times, and their normal activities had covered their white suits with a film of grime. It was a dirty and weary crew who clambered aboard Challenger after an EVA lasting 7 hours 12 minutes. The rest period was very welcome; it had been nearly 24 hours since their last sleep.

During the night, John Young worked on a way of repairing the broken fender. His solution was not very elegant, but it turned out to be quite functional – four lunar maps were taped together, then fastened to the remaining fender by emergency lighting clips. Once again, the crew were late beginning their excursion, but Cernan set off in high spirits: 'OK, Houston. On this fine Tuesday evening, as I step out on the plains of Taurus–Littrow, Apollo 17 is ready to go to work.' Schmitt added: 'Oh, what a nice day . . . there's not a cloud in the sky – except on the Earth!' The target this time was the slopes of South Massif, the site of an ancient landslide. Schmitt was disappointed by cancellation of some of his planned activities and asked Houston what he was supposed to do. 'Help Gene, I guess,' came the abrupt reply.

Once the journey began, he soon cheered up again, however, as a glittering array of rocks and craters appeared en route:

Occasional craters show lighter coloured ejectas all the way down to half a metre in size. Other craters that are just as blocky have no brightness associated with them. Most of the brightest craters have a little central pit in the bottom which is glass lined.

The number of blocks decreased as they passed Horatio and Camelot, apart from some perched near the crater rims. They collected samples from the light coloured material which had slid down long ago from South Massif, then darker samples from the regular surface. Passing by a steep scarp to their left, the men pulled up at a prominent crater named Nansen for their most extended spell of exploration by foot. Once again, Schmitt was in his element: 'When you look down into the bottom of Nansen, it looks like some of the debris there has rolled off the South Massif and covered up the original material.' They scrambled around taking measurements and collecting samples for over an hour; indeed, so engrossed were the two men that Houston told them they could have a ten minute extension if they cut out a later stop. The LM was some 5 miles distant, easily within reach by rover, but more than two and a half hours away by foot should their transport break down – there were no emergency roadside services close at hand!

After tearing Schmitt away from Nansen, there was little time to spare. A half-hour stop at Lara crater left them only a few minutes at the next crater, named Shorty. This football field size crater had a dark rim with many blocks lying on its inner wall and on the central mound; some geologists, including Schmitt, thought it might be an ancient volcanic vent, but there seemed little evidence to prove this theory. Then, as

Cernan shuffled around the rim, his feet scuffed the surface, revealing a sight he could hardly believe: 'Oh, hey, there is orange soil!' Schmitt cried out in delight, 'Well don't move until I see it.' Cernan explained his discovery: 'I stirred it up with my feet. It's orange. Wait a minute, let me put my visor up. It's still orange.' Schmitt agreed: 'Sure it is. Crazy! I gotta dig a trench, Houston . . .'

It wasn't only Schmitt who went into a state of euphoria; back in Houston, the discovery was greeted with astonishment, then unbelievable excitement by the assembled scientists. The general consensus of opinion was summarized by Dr Robin Brett, chief of geochemistry at Houston: 'One of the most important finds reported from the moon since Apollo 11.' What was not generally agreed was the true significance of the find – was it evidence of water or had it been caused by something else, a fumarole or some other volcanic activity for example? With the arguments raging on Earth, it was time for the astronauts to headback to Challenger, stopping only for a brief look at Camelot. Schmitt was dropped off to adjust one of the ALSEP experiments, then the LM was loaded with the rich haul of samples collected during the past 7 hours 37 minutes, the longest lunar EVA ever.

The final trip was to the slopes of the towering North Massif. It proved to be a real endurance test as they drove across the rugged, pitted surface towards the giant boulders which had been dislodged and snaked their way down the mountain slopes thousands, perhaps millions, of years ago. Driving north for more than an hour, they finally came upon their goal, a giant split rock, now broken into five massive slabs; the astronauts were dwarfed by their discovery. Clambering around the slopes of North Massif proved to be quite an exhausting experience, a task not made any easier by the higher temperatures as the sun climbed higher in the black sky and the dark shadows receded. Even the rover was beginning to suffer; Cernan reported 'a couple of dented tyres' where the wire mesh had given way under the stresses of the rough ride. A number of football sized rocks were collected, then they were on their way once more. Van Serg crater turned out to be something of a disappointment, for no signs of volcanic activity were uncovered, and so the last lunar EVA drew to a close.

Back in the neighbourhood of Challenger, the crew performed their cermonial duties. Speaking to a group of students from 75 countries assembled at Mission Control, they selected and dedicated a rock 'of all sizes and shapes, and even colours, that have grown together to become a cohesive rock' to the young people and their countries. They then swung the TV camera round to show a plaque attached to one of Challenger's legs. Beneath two Earth hemispheres and a central moon map showing the Apollo landing sites were the words: 'Here man completed his first explorations of the moon December 1972, AD. May the spirit of peace in which we came be reflected in the lives of all mankind.' It was signed by the three Apollo 17 astronauts and President Richard Nixon. Cernan added: 'This is our commemoration that will be here until someone like us, until some of you who are out there, who are the promise of the future come back to read it again, and to further the exploration and the meaning of Apollo.' Cernan's final emotional speech before ascending the ladder for the last time concluded with a prayer of thanksgiving: 'We leave as we came, and, God willing, as we shall return, with peace and hope for all mankind. God speed the crew of Apollo 17.'

The final EVA had clocked up another 7 hours 15 minutes, bringing the total surface exploration period to more than 22 hours. On board Challenger was a record haul of 249 lb of rocks and soil, a load which technically put the LM overweight for take-off, but no one was going to ask the men to throw away their hard-won treasures. Redundant equipment was discarded on the surface, then the astronauts settled down for their final sleep period. Up above, Evans fired America's thrusters to move into the rendezvous orbit. His long solo flight was nearing its end, though he had been accompanied by five Pocket mice as part of an experiment named Biocore. These tiny desert mice carried cosmic ray particle detectors implanted beneath their scalps to determine whether such rays can injure cells in the eye or brain.

Lift off came at 5.55 pm Cape time on 14 December, ending a record 75 hours on the moon. TV viewers were treated to spectacular colour pictures as Challenger's ascent stage blasted free amidst a cloud of debris. Shortly afterwards, Houston informed them: 'Your trajectory is right on the money.' An enhanced burn boosted the overweight LM into a perfect orbit, despite a communications problem which meant messages had to be passed through America. Rendezvous was completed on time, though docking needed two attempts after a too-gentle approach by Evans on the first run-in. Once sample bags and equipment had been transferred, it was time to bid farewell to Challenger: 'It seems like an unfitting finish to a super bird, but it's got one more job to do.' The LM was jettisoned prior to its deliberate crash-landing into the South Massif near the recent landing site, the impact being recorded by the orbiting crew as a bright flash.

Two more days were spent in lunar orbit. The astronauts were kept busy monitoring the orbital experiments in the SIM bay and photographing the surface; carried in the service module for the first time were a sounder which bounced electromagnetic impulses off the lunar surface, an infrared radiometer to map temperature on the surface and a far-ultraviolet spectrometer to analyse the tenuous atmosphere, in addition to the panoramic and mapping cameras. On America's seventy-fifth lunar orbit, on the far side pass, the craft was fired out of orbit towards home. The main engine burned perfectly as Cernan acknowledged 24 minutes later: 'Houston, America has found some fair winds and some following seas, and we're on our way home.' TV viewers were treated to their last close-up of the arid moon, including a rare sighting of the moon's far side and its most outstanding feature, the crater and mountain peak of Tsiolkovsky. The crew continued to discuss the significance of Apollo in its wider context: 'We're looking back at some place, I think, we will use as a stepping stone to go beyond some day . . . It has been a beginning. I don't think there ever will be an end, not as long as man is alive and willing.' On the moon, the batteries on the lunar rover succumbed after letting Houston watch only two of the explosions in the seismic experiment, but the ALSEP continued to send back a stream of information.

The next day, 17 December, saw Ron Evans demonstrate his prowess at EVA in a 1 hour 6 minute walk to the SIM bay in order to retrieve three camera cassettes and inspect the equipment bay area. He took the time to pause and take in the spectacular view: 'I can see the moon right behind me . . . down here to the right. Full moon. And off to the left, just outside the hatch down here is the crescent Earth. A crescent Earth is not like a crescent moon. It has horns that go almost all the way around, about two-thirds of a circle.' The blue and white Earth was still 180,000 miles (290,000 km) distant, but growing in apparent size all the time. Monday was filled with routine housekeeping and the usual press conference, though there was some concern over a missing pair of scissors – controllers were unhappy about the possibility of these potentially lethal weapons floating out from behind some crevice during the vulnerable minutes of re-entry. The scissors were never found, and it was eventually concluded that they must have drifted into space during Evans' space walk.

The service module was jettisoned shortly before 2 pm Cape time on Tuesday 19 December. Re-entry began soon afterwards with America accelerating to a velocity of 27,000 mph as it swept across northern Siberia and the north Pacific towards a dawn splashdown. Evans was wearing a special conditioning garment designed to apply pressure to his legs in an experiment to determine the value of such aids in reducing heart stress on return to normal gravity. The parachutes opened on schedule, allowing America to drift into a calm sea 400 miles (640 km) south east of Samoa. The craft hit the water about 4 miles (6 km) from the carrier *Ticonderoga*, behaving beautifully to the end by resting in an upright position. Almost immediately, the helicopters appeared over-head, dropping the frogmen alongside the scorched hull of the ungainly craft. Happy astronaut voices filled the airways as the crew celebrated the successful completion of their historic mission. Schmitt was the first to exit and enter the liferaft and then the helicopter, with Cernan following tradition by being the last to leave his ship.

Less than an hour after splashdown, the crew set foot on the carrier to receive the traditional red carpet treatment. Gene Cernan summarized a mission which was generally agreed to be the most perfect flight ever flown: 'This has been an extremely rewarding 13 days for us – 13 days I hope people throughout the world will be able to share.' The scientific rewards had been tremendous, though there was one disappointment: when analysis of the orange soil was eventually completed, it turned out to be tiny orange glass beads which had been formed by a mighty impact nearly four billion years ago – the orange tint was due to a high concentration of titanium.

The Apollo era was over. In just over four years, the spacecraft had been tried and tested, flown to the moon on nine occasions and actually set down on the surface six times to enable 12 men to walk, and later drive, across its barren wastes. Was it worth the cost in time and money, and above all, lives? For one group of men the answer to that question was never in doubt – to the astronauts, such a challenge was the reason for their existence. But now there were few new challenges on the NASA horizon.

The three Apollo 17 crewmen stayed with the space agency for a few more years, to see out the era of expendable rockets for manned spaceflight, then left the future of space exploration to the younger generation of spacemen and women. Gene Cernan was assigned to help with the planning and development of the Apollo–Soyuz joint mission, acting as senior American negotiator with the Soviet Union. On 1 July 1976 he retired from NASA and the Navy to join Coral Petroleum of Houston as executive vice president. Five years later, he started his own company as a consultant to the energy and aerospace industries. His attachment to NASA has not totally disappeared; he acts as a commentator for ABC–TV during Shuttle missions. Ron Evans was also closely involved with Apollo–Soyuz as back-up command module pilot. He retired from the Navy in April 1976, but remained active on the astronaut list to help with Shuttle development until March 1977. He now lives in Arizona, and is director of space systems marketing for Sperry Flight Systems. Harrison Schmitt was appointed as one of the first Sherman Fairchild Distinguished Scholars at Caltec 1973–5. At the same time, he took on extra duties as chief of scientist–astronauts and NASA assistant administrator for energy programmes. In 1975, he resigned from NASA to run for the Senate in New Mexico. He was elected in 1976, but defeated six years later during an attempt at re-election. He still maintains his links with the American space programme, and has remained a vigorous proponent of an expansionary space policy – in an address to the Planetary Society in Washington in 1984, he warned of the Soviet development of giant booster rockets that would rival the Saturn V, and predicted that the Soviet Union might well send a manned orbital mission to Mars in commemoration of the seventy-fifth anniversary of the Bolshevik Revolution during October 1992.

11

Plunge into Despair

The Salyut Debacle 1968–73

The death spiral of Vladimir Komarov in April 1967 virtually ended Soviet hopes of beating the Americans to the moon. Apollo 7 had proved the spaceworthiness of the main hardware in October 1968, leaving the way open for a circumlunar debut at Christmas. Clearly, there was no way in which the Soviet Union could land a man on the moon before its rival unless Apollo 8 turned into an unmitigated disaster. There was one other possibility, however, which could save some face for the leader of the Communist world – a circumlunar flight before the end of 1968. Should this not be achieved, an alternative propaganda front was prepared in the form of an orbiting space station, named Salyut, something which NASA was not contemplating until the early 1970s. Again, there was a minor snag, however; no Soviet spacecraft had yet carried out orbital rendezvous and docking, manoeuvres which had become commonplace during the American Gemini programme. An enforced lull in manned missions followed upon Komarov's untimely demise, so progress had to be made by automatic craft while the existing Soyuz craft was redesigned and tested.

The first step came six months after Komarov's Soyuz 1 plunged to Earth. Two stripped down Soyuz craft were launched within three days of each other into almost identical orbits. Operating completely under automatic control from Earth, the first craft, designated Cosmos 186, closed on its passive partner, known as Cosmos 188. Within a matter of hours they had successfully docked in the world's first automatic link-up of two spacecraft. After flying together for three and a half hours, they separated over the Soviet Union, a feat recorded by TV cameras and later broadcast to the Soviet people. Cosmos 186 soft-landed on 31 October 1967 followed by its sister craft two days later. Western media speculation that it was the forerunner of a space platform from which a manned craft would be launched towards the moon proved unfounded. The analysts were unaware of the turmoil and disarray in the Soviet camp. Six more months passed before Cosmos 212 and 213 repeated the feat, and once again, it fuelled speculation that a manned mission would follow shortly, leading to 'a Soviet attempt at a moon landing next year'.

Sure enough, Soviet activity did increase, but not quite the way predicted. Late August 1968 saw Cosmos 238, another unmanned Soyuz, successfully complete a four-day running-in mission; the ship seemed ready for another human occupant at last. Less than three weeks later, a Soviet Soyuz-type craft known as Zond 5 was launched towards the moon with life on board: not the long-expected man, but a collection of tortoises, fruit flies and mealworms. Nevertheless, a tape-recorded voice was heard sending back instrument readings, and a splashdown near a recovery fleet in the Indian Ocean, another Soviet first, ensured survival of the inhabitants. Zond 5 had simply looped around the moon and returned to Earth on a ballistic trajectory over the South Pole which had generated more than 10G, much greater than would be acceptable on a manned mission. Soviet scientists explained that they wanted to measure the effects of radiation on living organisms, but most Western observers believed it was the precursor of a manned flight around the moon.

Their beliefs seemed to be justified by the events of October and November. On 26 October 1968, after an 18-month interval, a cosmonaut was launched inside the redesigned Soyuz 3 on a four-day check-out of the new craft's systems. The cosmonaut was Colonel Georgi Beregovoi, already decorated as Hero of the Soviet Union for his exploits in the Second World War, and now, at 47, at that time the oldest man to go into orbit. The son of a railway porter, Beregovoi had begun working at a steel plant on leaving school but took an interest in aircraft modelling and then enrolled at a flying club. Having found his vocation, he moved on to Lugansk Air Force Pilots' School. The war made great demands on his nerve and stamina, but he survived 185 combat missions to become a highly-rated test pilot when peace returned. Beregovoi was already well into his forties when selected as a cosmonaut in 1964, but his ice-cool temperament and experience in a crisis made him ideally suited for a dangerous shake-down mission. But was that all his flight plan required of him?

An unmanned target vehicle, Soyuz 2, had been launched into an almost identical orbit 24 hours earlier. The only logical explanation was that a rendezvous and docking would subsequently take place in order to supplement the experience gained by previous automatic dockings and pave the way for crew transfers on later missions. Sure enough, on his first orbit Beregovoi's craft closed to within 600 ft (180 m) on automatic control before he took over manually; but the docking never came. The manoeuvre was repeated the next day, but there seemed to be no attempt to dock or even match the relative velocities of the two craft. Soyuz 2 made a successful soft landing in the Soviet Union on 28 October, leaving Beregovoi to continue his tests and observations, including terrain and weather photography. In addition, he made regular TV reports which were recorded by tracking stations as he passed over the Soviet Union for subsequent broadcasting. In a new venture for Soviet flights, he was seen to give a limited tour of his craft, including the rest quarters in his cabin. On the fourth day, Beregovoi manually orientated his craft, then initiated the automatic retro-fire system. The parachutes opened perfectly, and there was a suggestion that an improved drag-lift capability in the new design led to the accuracy of the landing on the plains only a few hundred miles from the Baikonur cosmodrome where his flight began.

At a press conference the well-built veteran revealed that his height – 5 ft 11 in – had been more of a problem than his age in making the flight. However, neither he nor the authorities would admit whether there had been an original intention to dock with Soyuz 2. Officially, it was simply a mission to improve techniques in 'space navigation'. Beregovoi retired from active duty soon after his flight, and was appointed to the influential post of head of the Yuri Gagarin Training Centre for cosmonauts at Star City.

Events now seemed to be leading towards a climax. Only eleven days after the safe return by Soyuz 3, another unmanned Zond was launched towards the moon. As Zond 6 sped towards its target, NASA Acting Administrator Tom Paine announced to the world his agency's intention to send Apollo 8 into lunar orbit during late December. If this could be achieved, it would be a major blow to Soviet hopes of getting there first. There was no doubting the purpose of Zond 6: as Tass news agency explained, it was 'intended to perfect the automatic functioning of a manned spaceship that will be sent to the moon'. No attempt was made to enter lunar orbit, but there was one vital difference from the previous Zond flight plan. The craft returned to Soviet soil by utilizing the braking effect of 'two immersions in the atmosphere'. This double skip, rather like the path followed by a flat stone skimming across the surface of a pond, enabled a more gradual deceleration than the more direct ballistic re-entry. As a result, maximum G force was reduced to around 7G, a much more suitable level for a returning cosmonaut. Once again, the Soviet Union had successfully tried out a technique long proposed by the Americans but not yet put into practice. Yet, even while the Soviet press praised the efforts of its space experts, and a manned flight to the moon seemed very much on the cards, one of those experts, Professor Keldysh, was quoted as saying that there was no race to the moon, and there never had been one. It seemed that the Soviet authorities were preparing to salvage what they could from a defeat in the moon race by asserting that their efforts had always been in a different direction. The success of Apollo 8 showed how astute this change of public attitude was.

On 14 January 1969 Soyuz 4 was launched from Baikonur with one man on board, 41-year-old Lieutenant Colonel Vladimir Shatalov. The cosmodrome was blanketed by thick snow with sub-zero temperatures, hardly ideal conditions for a launch, but the mission's importance clearly justified the risk. Shatalov, an aviation enthusiast since his schooldays, and a highly experienced air force pilot, had become a cosmonaut in 1963 and served as Beregovoi's back-up.

He safely reached Earth orbit after only a few minutes, then circularized the orbit during his fourth revolution. He was not alone for long: a companion craft, Soyuz 5, was launched into a similar orbit the next morning as Shatalov cruised overhead. On board were three more cosmonauts. Commander of the new craft was Lieutenant Colonel Boris Volynov, a 34-year-old former fighter pilot who had been recruited to the first group of cosmonauts in 1960 but was now making his first spaceflight. His previous experience was as back-up to Valeri Bykovsky on Vostok 5. Alongside him was another member of the first cosmonaut group, 35-year-old Lieutenant Colonel Yevgeni Khrunov. Born into a peasant family, one of eight children, he studied at an agricultural technical training school before taking up flying. Noted for his skill as a pilot and ability with machines, he became an aircraft and spacecraft design engineer, and served as back-up to Alexei Leonov for the first space walk on Voskhod 2. It seemed highly likely that he would soon be able to take advantage of this training on his current mission. The third crew member was a 34-year-old civilian engineer who had graduated in chemistry before taking an interest in space engineering and joining Korolev's design bureau in 1963. Alexei Yeliseyev and Khrunov had been photographed alongside Komarov before the fatal Soyuz 1 crash, reinforcing suspicions that both men had been assigned to an aborted Soyuz 2 which was scheduled to link with Komarov's craft. Yeliseyev had been assigned to the cosmonaut group in 1966 as part of the new policy of giving such men first-hand experience of their handiwork.

The expected rendezvous and docking materialized on 16 January as Soyuz 4 approached under automatic control to within 300 ft (100 m) of its quarry. Shatalov then assumed manual control to inch his craft towards the passive Soyuz 5. As the probe nestled inside the drogue assembly, Shatalov jokingly shouted, 'I've been hunting all over for you.' One of the receiving crew, probably Khrunov, replied, 'Yes, and now you're raping us!' The comment passed the censors on the early TV replay, but was later cut out. Nevertheless, there was considerable excitement in the Soviet camp now that the first ever docking between two manned craft had been successfully completed – the docking between the Apollo 9 command and lunar modules was still six weeks away. In that irritating manner of theirs, the Soviets had upstaged yet again a feat which had been planned well in advance by their rivals. More was to follow.

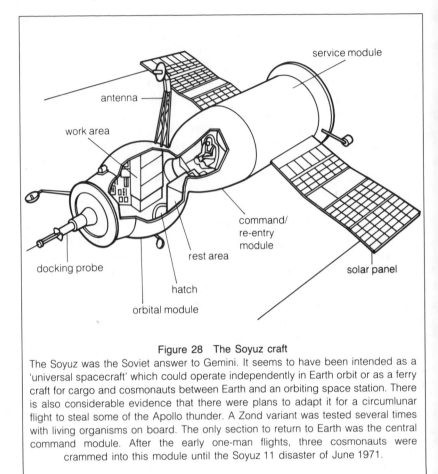

Figure 28   The Soyuz craft
The Soyuz was the Soviet answer to Gemini. It seems to have been intended as a 'universal spacecraft' which could operate independently in Earth orbit or as a ferry craft for cargo and cosmonauts between Earth and an orbiting space station. There is also considerable evidence that there were plans to adapt it for a circumlunar flight to steal some of the Apollo thunder. A Zond variant was tested several times with living organisms on board. The only section to return to Earth was the central command module. After the early one-man flights, three cosmonauts were crammed into this module until the Soyuz 11 disaster of June 1971.

Official statements described the linked craft as: 'The world's first experimental assembled cosmic station with four compartments for the crew, making it possible to perform a large variety of experiments and investigations, and also providing comfortable conditions for work and recreation.' This was undoubtedly a considerable exaggeration, if only because there was no means of internal access between the craft, though there was a telephone link. The answer was to 'walk' through space from one to the other. So, after one orbit in unison, Yeliseyev and Khrunov moved into the orbital module to don their pressure suits. This achieved, they closed the air-lock and depressurized the module before exiting through a side hatch. Utilizing a new backpack for breathing and cooling, together with a series of tethers and special grips on the outside of the Soyuz, the two men slowly swung towards the open hatch on the companion ship. Thirty-seven minutes later, they were greeted by a hastily scribbled 'Welcome' sign as they floated through the hatch into Soyuz 4. Once the oxygen–nitrogen atmosphere was replaced and their suits were removed, the two men occupied the empty seats next to Shatalov, and the delighted crew toasted their feat in blackcurrant juice. It was the first time men had transferred from one spacecraft to another.

Tass trumpeted the news as a triumphant innovation which would lead to eventual replacement of crews on long-term orbital stations or to crew rescue in emergencies. However, the reality was much less glamorous. In a technological sense it was a dead end, for the future lay in the Apollo-type crew transfer which used a docking tunnel for internal access. Neither was the docking a major breakthrough, except in the sense that the Soviets were years behind the Americans in this field and that they had finally succeeded in doing something hardly more advanced than what the Gemini crews had been doing years earlier. The new Soviet emphasis on orbital space stations was significant, however, reaffirming their newly acquired disinterest in sending men to the moon.

The two Soyuz undocked after four hours, prior to a return to Earth. Soyuz 4 brought Shatalov and the two spacewalkers back to a safe soft-landing on the freezing, snow-covered steppes in Kazakhstan. The orange parachute was spotted by hovering helicopters, enabling warm clothes to be rushed to the crew as they emerged. Volynov continued alone in Soyuz 5 until 18 January, broadcasting regular TV trans-missions and taking Earth photographs. A request to extend his mission was turned down on the grounds that the mission objectives had been accomplished, so Volynov tried out the manual steering and orientation controls, turned the craft into its re-entry attitude, fired the retro-rockets as he passed over the Gulf of Guinea on his fiftieth orbit, and sat back while the automatic controls guided the Soyuz safely back to Earth. Once again, the touchdown was so accurate that helicopters were quickly on the scene. The four cosmonauts were feted as the Soviet media went overboard with their praise, but the propaganda value of the missions far outweighed their scientific value.

Further nails were hammered into the Soviet moon race coffin by the quick-fire successes of the Apollo programme in 1969, culminating in the celebrated moon landing by Apollo 11 on 20 July, and the failures of the new Soviet giant booster rumoured to be in its final stages of development. This monolith never materialized: vague reports filtered through to the West of a launch pad explosion in June 1969, followed by two flight test failures in 1971 and 1972 before the project was scrapped. Why the Soviets needed a competitor for Saturn V is still not clear. Design studies for a space station began in January 1969, so it would have been a useful launch vehicle, though rather on the large side, for such a station. However, when the booster was initiated in the early sixties, its main purpose must have been as a potential moon launch vehicle comparable to America's Saturn V. In the event, it proved a miserable failure. Only now has the project been revived as a means of meeting Soviet plans to visit the moon and Mars though, at the time of writing, it has yet to fly.

During the Apollo summer spectacular, the Soviet authorities could only sit idly by as impotent spectators. Official policy now insisted that there had never been any intention to send cosmonauts to the moon, and, though grudging praise was bestowed on the American heroes, it was tempered by continual reminders of the advantages of robot craft over the risky, expensive manned programme. The one attempt to grab some headlines from the historic landing by Apollo 11, a soft-landing by the automatic Luna 15 followed by the return of a soil sample to Earth, failed when the craft crash-landed on the lunar surface. The battered, humiliated Soviet space programme desperately needed another of its morale-boosting blockbusters. Hence the triple Soyuz mission in October which saw three craft and seven men in orbit simultaneously for the first time.

The first craft into orbit, Soyuz 6, was seen in the West as the advance guard for a manned space laboratory, even though a Soviet scientist involved in the mission revealed that the craft had no docking equipment. It was widely expected that two more Soyuz would join Soyuz 6 within the next couple of days for a major space spectacular. A hint of what would follow appeared in an official magazine, *The Week*, which showed drawings of a space station composed of two Soyuz at either end of a tunnel. Speculation was further fuelled by the news that one of the main tasks for the Soyuz 6 crew would be to experiment with welding equipment in conditions of deep vacuum and weightlessness. Pictures released a couple of hours after the launch showed the two cosmonauts entering the ship wearing only thick woollen clothes.

The commander was Lieutenant Colonel Georgi Shonin, a 34-year-old pilot who had served with the Baltic and Northern fleets before cosmonaut selection in 1960. Shonin had recently graduated from the Zhukovsky Air Force Engineering Academy, then served as Volynov's back-up on Soyuz 5. Flight engineer was a 34-year-old civilian, Valeri Kubasov, the man responsible for the Vulkan welding equipment. The son of a shipping-line mechanic, Kubasov did so well at school that he was able to enter the Moscow Aviation Institute without taking the entry examinations. He graduated in 1958 with the diploma of engineer-mechanic, then joined Korolev's space design bureau. During this time, he wrote several books and scientific papers which so impressed Korolev that he used his influence to have Kubasov added to the cosmonaut list. He was accepted in 1966, and received an MSc degree in engineering in 1968. He acted as Yeliseyev's back-up on Soyuz 5.

Right on schedule, Soyuz 7 followed into a similar orbit on

12 October. On board were three more cosmonauts with no spaceflight experience between them. Commander was 41-year-old Lieutenant Colonel Anatoli Filipchenko, a 1963 cosmonaut recruit who continued to fly supersonic aircraft as a test pilot after his selection. Filipchenko, like many cosmonauts, had come from what we would describe as a working class background: he helped his mother at a motor repair plant in the war, then started work as a lathe operator at the age of 15 before enrolling at Cheguyevo Air Force School in 1947. Research engineer on the flight was Lieutenant Colonel Viktor Gorbatko, a cosmonaut since 1960 and a back-up with Khrunov to Alexei Leonov on Voskhod 2 before backing up Khrunov himself on Soyuz 5. Gorbatko was born on a state farm and joined a military flying school at an early age. The odd one out of the crew was civilian Vladislav Volkov, another of Korolev's design bureau engineers who were recruited as cosmonauts in 1966. Volkov at 33 years of age had gained quite a reputation as an all-round sportsman earlier in life, but any ambitions in this field were subordinated to his career in spacecraft design. His role on Soyuz 7, like most of the mission, remained a mystery: at one point in a live TV broadcast, he commented that he was responsible for communications and other tasks, 'but I'm not going to tell you about the other tasks'.

Everything seemed to be going according to plan when the third craft, Soyuz 8, duly joined its two companions on 13 October. No comment was made about the absence of docking equipment on either Soyuz 7 or 8, so experts assumed both of them were equipped for docking. On board Soyuz 8 were two men who had partnered each other nine months earlier during the joint Soyuz 4/5 mission. Overall commander of the triple flight was Vladimir Shatalov, while his flight engineer was Alexei Yeliseyev. The successful and accurate launch of three manned craft within three days spoke volumes for the efficiency of the cosmodrome ground staff. It remained to be seen how well they could cope with the extra communications load, and how well the mission fulfilled its secret objectives. The usual huge gaps in the ground communication network were filled by eight tracking ships scattered around the oceans with links to communications satellites. It was clearly a major effort from which great dividends would be expected. Yet, for reasons unknown to this day, it turned into a damp squib.

During 14 October the predicted link-up did not take place, with Soyuz 7 and 8 closing to an unimpressive 2 miles (3 km). The official news agency, Tass, took delight in reporting, 'the ships, according to aviation tradition, waved their solar panels at each other'. Other reports spoke of Soyuz 6 and 8 experiments: the crews 'observed and photo-graphed cloud formations, the moon and stars at the horizon, and assessed the luminosity of the Earth' while Soyuz 7 practised manual orbital manoeuvres. Could this really be the sum total of the space spectacular on which so much time and effort had been expended?

On 15 October something at last seemed to be happening. Soyuz 7 and 8 made a series of close approaches, observed by the crew of Soyuz 6 from a safe distance. At one time, they approached to within 1500 ft of each other, but the long-awaited docking never came. After station-keeping for about 24 hours, the two craft pulled apart once more – the docking was not to be. Meanwhile, the crew of Soyuz 6 sealed themselves in their re-entry module, depressurized the workshop module and operated their remote control welding apparatus. Three different methods were tried in an attempt to find the best process for future space construction. A few hours later, Soyuz 6 soft-landed in the normal Karaganda recovery area, where the samples were whisked away for analysis. It was heralded in the Soviet press as a major breakthrough, even before the results were known. In effect, it was the only real success of the entire mission.

Soyuz 7 and 8 both returned after five days in space with the Western analysts asking what went wrong. There seemed no logical reason why Soyuz 6 could not have flown independently in order to carry out the welding experiments, unless it was considered a useful propaganda countermeasure to offset the Apollo successes by launching three craft simultaneously. As for Soyuz 7 and 8, why did they not dock? At a post-flight press conference, Shatalov explained that they could have done so, but it was not part of their flight programme. He did, however, add that there had been 'difficulties, as in every spaceflight'. And why the pictures of a dual Soyuz space station released only the day before Soyuz 6 was launched? It seemed an unfortunate piece of timing, most unlike the Soviets, if it had no bearing on the forthcoming mission.

Most of the cosmonauts involved in the disappointing flight remained on the active list, though Georgi Shonin disappeared from the programme. Valeri Kubasov was able to utilize his experience in space welding during the Soyuz 19 link-up with Apollo in 1975 when he used the multipurpose furnace with his American counterpart. Filipchenko made one more Soyuz flight, while Gorbatko was to command two more Soyuz.

Another 8 months passed before the Soviet Union launched the next manned mission. This time, the scientists seemed to have won the argument over policy, for it turned out to be a straightforward long-duration mission with just two cosmonauts and a single spacecraft. The intention soon became clear: the space endurance record at that time was only 14 days, held by the American Gemini 7, while the Soviet record of less than five days had stood since 1963. There was no point in pressing ahead with an orbital space station if people were found to be unable to survive and maintain their health during long periods of weightlessness. The two medical guinea-pigs on board Soyuz 9 were space veteran

Andrian Nikolayev, now 40 years old, and newcomer Vitali Sevestyanov, a civilian engineer who had only been a cosmonaut since 1967. Since his marriage to the world's first spacewoman, and the birth of their daughter, Nikolayev had drifted out of the limelight. It is now known that he was closely involved in the early Soyuz flights, and served as the back-up commander for the triple Soyuz mission in 1969. Sevestyanov had been his partner on the back-up crew, having graduated to the cosmonaut corps as one of the last recruits from Korolev's design bureau. Before joining the cosmonauts, he had been one of their lecturers on rocketry; now he was about to put his theories to a practical test.

Soyuz 9 was placed into a stable orbit suitable for long-duration work 134 miles (220 km) above the Earth following the first ever night launch by a manned craft on 1 June 1970. The orbit was raised by manual thruster firing to an eventual altitude of 155 × 166 miles (248 × 267 km) by the second day. Tass put 'medico-biological research' at the top of the list of mission tasks: it was vital for doctors to learn more about the effects of prolonged weightlessness.

In an attempt to combat the debilitating effects on bones and muscles, the cosmonauts were given a programme of exercises to carry out, but this was not always viable – in one exchange reported by *Izvestiya*, ground control had to chide the men gently for not carrying out their exercises, to which the crew replied that they had no free time. An extra aid to be worn during exercises in order to normalize blood circulation and strengthen muscles was a 'load suit'. The work load provided for the men ensured they were kept fully occupied. Apart from medical tests and biological experiments, they were expected to carry out navigation by stellar alignment, engineering tests on the spacecraft, and Earth observations. Photography of the Earth's land and oceans was intended to pioneer research into forestry, farming, fishing, geology and weather systems. TV relays enabled ground controllers to watch the men at work as well as public relations broadcasts to the people in their homes. Conditions on board were kept as comfortable as possible, with a roughly normal cabin atmosphere and temperature around 22°C; they were also given an electric oven for heating their food and an electric razor for shaving.

The Gemini 7 record was smashed as Nikolayev and Sevestyanov stayed circling the Earth for nearly 18 days. But the cramped conditions and lack of exercise took their toll. It was a weak, exhausted crew which soft-landed on 19 June. Both men had to be carried from their re-entry capsule. During the next four to five days they complained of difficulty in walking and sleeping, and it was more than a week before they cast off the sensation of living in an environment where everything seemed twice as difficult as normal, as though they were in a continual 2G centrifuge. Nikolayev never flew in space again; he moved into the cosmonaut training office and was promoted to major-general. Sevestyanov made one more flight, as flight engineer on board Soyuz 18, a 63-day space marathon spent mostly on the Salyut 4 space station. He was already working as a journalist and science reporter on TV before the 1975 endurance test, a career he has since continued.

A long period of evaluation followed as doctors analysed the results of the Soyuz 9 mission and while work continued on the Salyut space station. The interlude was enlivened by final proof that the Soviet Union had turned away from sending men to the moon and was now following its widely broadcast policy of substituting robots. In September 1970 Luna 16 succeeded where its predecessor had so lamentably failed, soft-landing on the surface, scooping up a soil sample and returning it safely to the eagerly awaiting scientists. The prize was only a few ounces in weight, but it was trumpeted as a major triumph by the Soviet media. Further 'proof' of the superiority of robots came in November when Luna 17 carried a mobile wheeled craft called Lunokhod to the Sea of Rains. For the next 11 months the sturdy little 'moon walker' trundled across the lunar plains, controlled by a team of drivers on Earth. It was an impressive feat, and one that inevitably led to clashes of opinion in the USA where so much emphasis was placed on human explorers. Astronauts found themselves obliged to defend their roles because of this wire-wheeled technological wonder.

The long-awaited Salyut 1 was finally launched on 19 April 1971 into a low Earth orbit of altitude 125 × 139 miles (200 × 222 km).* Weighing about 20 tons on Earth, the world's first operational space station measured 65½ feet long and 13 feet wide. This was sufficiently small to be carried on top of the medium-sized Proton launcher – there was no need for any superbooster. Salyut was divided into several separate modules, three of which were pressurized: the access or transit module allowed entry from a docked Soyuz ferry craft; the principal module contained the living and working quarters; and the equipment module contained the vital heart of the station – its power supply, control panels, communications and life support. Finally, at the rear of the station was the unpressurized service module with the main engines and propellant storage. Electrical power was provided by two pairs of giant solar panels, one fore and one aft.

Four days after the Salyut launch the expected Soyuz 10 lift-off took place, carrying the first occupation crew for the space station. Official Soviet sources were typically cagey about the purpose of the launch, simply stating that the two craft would be carrying out 'joint

---

* The name Salyut (Salute) was apparently chosen as a tribute to Yuri Gagarin, now dead, who had first orbited the Earth almost exactly ten years previously.

experiments', but the third crewman, 38-year-old civilian Nikolai Rukavishnikov, was described as 'test engineer of the orbital station'. Rukavishnikov had spent his early years travelling from camp to camp across the Soviet Union with his stepfather, a railroad surveyor, and his mother. On leaving school he had enrolled in the Moscow Institute of Physics and Engineering, then been accepted in 1957 as one of Korolev's designers. A desire to try out the automatic control systems he designed led to an application to join the cosmonaut corps. He was selected in January 1967, and became a recognized expert on Earth science and solar physics, a very useful man to crew the first space laboratory. He was fortunate in his companions: Shatalov and Yeliseyev were now a veteran team with two flights' experience behind them, including the only manned docking yet achieved by cosmonauts.

All seemed to be going according to plan as Soyuz slowly closed on its objective, though the rendezvous took almost a day to complete. While Soyuz manoeuvred automatically, ground control tried to help by making four minor changes in the Salyut orbit. The space station was first spotted by the crew at a range of about 9 miles (15 km), but automatic procedures continued until the craft were only about 600 feet (180 m) apart. Shatalov assumed manual control for the final run in during Soyuz's twelfth orbit, eventually driving home the probe on the nose of the ferry craft. It had been a worrying and exhausting day; not surprisingly, mission control allowed the crew to begin a rest period once preliminary check-outs proved favourable.

To everyone's amazement, the craft remained docked for only five and a half hours. The crew initiated re-entry procedures after manoeuvring around the space station and checking it for damage; after 30 Earth orbits, Soyuz 10 soft-landed in darkness near Karaganda on the morning of 25 April. The mission had lasted a fraction over two days. Soviet sources claimed that the mission had been fully successful in carrying out the testing of new docking systems and approach techniques. They maintained that no entry to the Salyut had been planned, and certainly the space station's orbit was very low for a prolonged orbital mission. Shatalov explained that, although the flight duration was short, 'it was tense and very great in its tasks', particularly since the final docking manoeuvres had been completed out of range of ground stations. Western experts remained sceptical – what was the point in going to all the trouble of linking with a new space station if you weren't going inside it? One favoured explanation was Rukavishnikov's ill health. The rookie cosmonaut had complained during the flight of 'the unusual and rather unpleasant sensations arising as a result of the increased flow of blood to the head', but he denied that this had any significance at a post-flight press conference, insisting that 'work in a weightless state was a joy'. So if that wasn't the cause, what was? The only alternative seemed to be a technical malfunction. Favourite candidate was the new hatch and docking tunnel system being used for the first time on a Soviet flight. No other explanation has been forthcoming from Soviet officials.

Shortly afterwards Salyut was boosted into a higher orbit to await new visitors. Clearly, it would not be long before another Soyuz crew attempted to enter. Although the American Skylab would not fly for another two years, the Soviets were anxious to grab some of the limelight which had been almost exclusively reserved for the Apollo landings over the past two years. Soyuz 11 was launched at 7.25 am Moscow time on 6 June 1971 with a crew of three and the usual vague statements concerning the flight objective. The only experienced crewman was Vladislav Volkov, flight engineer for the second time on a Soyuz mission. Commander was 43-year-old Lieutenant Colonel Georgi Dobrovolsky, a hard-bitten fighter pilot who had been a cosmonaut since 1963. Test engineer on Soyuz 11 was the second civilian on the crew, 37-year-old Viktor Patsayev, a graduate of Penze Industrial Institute. The balding Patsayev had been a radio researcher and design engineer at the Central Aerological Observatory until his selection as a cosmonaut in 1967.

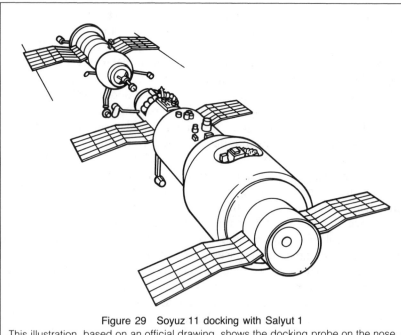

Figure 29   Soyuz 11 docking with Salyut 1

This illustration, based on an official drawing, shows the docking probe on the nose of the Soyuz about to enter the drogue assembly in the transfer section of the space station as they pass over the north Pacific. The date was June 7 1971. The two craft when docked weighed about 25 tons and measured about 90 ft (28 m) long. Note the solar panels; these were replaced by batteries on Soyuz 12 and all subsequent ferry craft.

Rendezvous and docking remained a slow, painstaking operation. The trio manoeuvred their craft to within 4 miles (6 km) of Salyut, allowed their automatic systems to close the gap, then closed the last 300 ft under manual control once more. It was an exhausting period which lasted more than 24 hours, but the two craft were linked at last, air seals were checked and cabin pressures equalized. The historic moment came as Patsayev opened the docking hatch and floated through the tunnel into Salyut, followed soon afterwards by Volkov, the first men ever to occupy an orbiting space station. The Soviet media were understandably jubilant. Tass declared: 'The use of the new cosmic complex, orbital stations with ferry ships to service them, opens wide perspectives for further research and mastery over space in the interests of science and economy'.

The first few days of occupation were taken up with checking the Salyut systems and fully activating the equipment after its long spell in mothballs. A major priority was to raise the decaying orbit once more by firing the Salyut engine, thereby increasing the space station's lifespan. The crew became TV and film stars almost immediately as they somersaulted, joked and showed off their new home to the Soviet public. Behind the lighthearted banter, however, was a serious scientific programme tied to a rigid daily schedule of exercise and work. These three men had the unique task of pioneering a whole new lifestyle, preparing the way for those who would come after and spend many months rather than weeks in such an orbiting laboratory. Apart from a series of simple exercise devices, the crew were provided with 'Penguin' elastic restraint suits designed to prevent weakening of muscles due to lack of use in the weightless environment.

In addition to monitoring the performance of the onboard systems, the crew had to carry out an intensive programme of scientific experiments. These included observations of geology, snow and ice cover, atmospheric processes and weather systems, together with Earth resources surveys of land and sea; astronomical observations using an 'Orion' optical and spectrographic telescope which could operate automatically once it was directed at a particular star; and biological experiments involving both plants and animals. Of particular interest was their 'Oasis' greenhouse which was Patsayev's responsibility and included a variety of crops including flax and marrow-stem kale. Tass related how the first 'sprout' appeared two days after the greenhouse was activated, evidence that it could be possible for future generations of space travellers to cultivate their own food crops on long duration flights. The experiments had another significance for the homesick cosmonauts: 'Our pets give us great pleasure to watch'. Among their 'pets' were a jar full of tadpoles which hatched out from frogspawn and the customary fruit flies.

As the crew headed for a new space endurance record, their spirits remained high, although doctors admitted some signs of fatigue were beginning to show. As Salyut made its one-thousandth orbit on the afternoon of 19 June, the crew relaxed by celebrating Patsayev's thirty-eighth birthday. Soviet viewers were able to watch his colleagues toast him in fruit juice drunk from tubes; they also gave him an onion and a lemon with his normal diet, both smuggled on board without the knowledge of their medical supervisors. By now, two of them had sprouted beards. The endurance record of Soyuz 9 was passed a few days later, this time with the crew in relatively good health. Then, as 30 June dawned, it was all over, after nearly 24 days in orbit. The men transferred their flight log and experimental results back to the Soyuz, ready for the triumphant welcome being prepared in the Soviet Union. The mission had genuinely broken new ground, unlike the staged propaganda spectacles of earlier Soyuz extravaganzas, and would be a well-merited boost for a space programme that had suffered some major setbacks and humiliations in recent years.

At 1.35 am Moscow time, the crew, wearing the usual woollen flight suits and leather helmets, settled back in their couches and initiated retro-fire, having finally separated from Salyut four hours earlier. Dobrovolsky reported to mission control: 'This is Yantar 1 [his call-sign]. Everything is satisfactory on board. Our condition is excellent. We are ready to land.'

As their craft slowed to enter the upper atmosphere, the final separation of the command module from the service module to the rear and the orbital module in front was completed by firing explosive bolts. There followed an eerie silence which remained unbroken throughout the remainder of the re-entry. Everything seemed to function normally as the craft plunged through the atmosphere, the drogue parachute opened, followed by the main chute, and the retro-rockets fired perfectly a few feet above the ground to enable the Soyuz to make what seemed a model touchdown in the Kazakhstan target zone. Helicopters were hovering overhead within seconds. The absence of communication with the crew during the re-entry was a cause for concern, but no one was prepared for the sight which met the doctors as they opened the hatch and peered in.

Inside the cramped capsule the three men hung motionless from the straps which held them on their couches. There was no response to the words of the recovery crew, no motion to clamber out through the hatch for a cheering, backslapping welcome. The awful truth began to dawn, soon to be verified when the doctor found no sign of a heartbeat. All three men had died in some unknown way during the 20 minute re-entry. The official announcement simply stated: '. . . a helicopter-borne recovery group, upon opening the hatch, found the Soyuz 11 crew in their seats without any signs of life. The causes of the crew's death are being investigated.' A special commission under Mstislav Keldysh, President of the Soviet Union's Academy of Sciences, was set up to inquire into the cause of the disaster. While solemn music played on Moscow radio, messages of condolence from all over the world flooded into the capital, including those of many world leaders. NASA's deputy administrator, George Low, realized the significance of this tragedy more than most, having lived through the aftermath of the Apollo 1 fire four years earlier. 'Our hearts go out to their families and to their colleagues,' he said.

American astronauts were also part of the brotherhood; they knew it could easily be their turn tomorrow to suffer a similar fate. Previous requests to attend cosmonaut funerals after the deaths of Komarov and Gagarin had been turned down – the nation's grief was seen as an internal affair by the authorities. This time, with detente in the air and negotiation well under way for a joint manned mission in 1975, permission was given for Tom Stafford to act as one of the pallbearers. Crowds flocked to leave flowers at the dead cosmonauts' feet as they lay in state in Moscow. In a televised state funeral their bodies were cremated and the ashes interred in a niche in the Kremlin Wall alongside Gagarin and Komarov. All three men left widows and children: Dobrovolsky had two daughters, aged 12 and 14; Volkov had a 13-year-old son; and Patsayev left a son of 14 and a daughter of 9. Posthumous awards of Hero of the Soviet Union were made to the three victims.

Speculation as to the cause of death was inevitably rife after the news was released. Some Western experts suspected the heat shield – overheating might have affected the environmental controls, leading to suffocation. Others feared that the weightless barrier had been reached, arguing that overexposure to the debilitating effects of zero gravity had so weakened the men that they were unable to cope with the stresses of re-entry. Several days after the accident, unofficial Moscow sources leaked a story of a hatch that was not properly sealed, allowing the cabin air to escape gradually without the men realizing until it was too late. On 11 July, the commission of inquiry admitted that the cause had been a 'rapid pressure drop occurring inside the descent vehicle' due to 'a loss of the ship's sealing', but the details were not released until 1973, when at the insistence of American scientists who were working on the joint Apollo–Soyuz Test Project, the Soviets revealed the detailed results of their inquiry.

It appeared that the vibrations caused by the explosive bolts which separated the command module from the remainder of the Soyuz had caused a pressure equalization valve to open prematurely. Normally, this valve would only operate near touchdown, but in the vacuum of outer space, it simply resulted in the cabin air escaping into space. If the crew had been wearing pressure suits, this would not have been fatal, but the cabin was very cramped for a crew of three dressed in such suits, and Soviet designers were confident that such protection was unnecessary. This misplaced confidence cost the men their lives. The lack of communication between the men and mission control suggests that decompression was quite rapid, probably a matter of seconds. Even if they had realized what was happening, there was no possibility of doing anything about it, for the valve was so designed that it took nearly two minutes to close manually. A combination of lack of oxygen and pressure drop caused the men to black out in the early stages of re-entry. By the time the atmosphere leaked back into the cabin, it was far too late.

The tragedy had proved once again that nothing could be taken for granted when it came to sending humans into space. Short cuts could have potentially deadly consequences. The first obvious precaution to take with future flights was to install pressure suits for the crew. Such an apparently simple modification meant a complete rethink of the Soyuz concept, however, for there was insufficient space in the craft to accommodate the extra life support equipment as well as the bulky suits. In addition, the entire method of command module separation and equalization of cabin air pressure had to be re-evaluated. The necessary modifications took about a year to complete; an unmanned test took place in mid-1972 with a craft designated as Cosmos 496. When Soyuz 12 finally flew in September 1973, it was with a crew of two due to the reduced cabin space available. Not until the launch of a modified version known as Soyuz T-3 in November 1980 would the Soviets again send up a three-man crew.

The Soviet space programme continued to run into problems. Following the Soyuz 11 tragedy, the Salyut 1 space station was raised into a long-life orbit. There were reports at the time suggesting that a replacement crew was ready to go, but the three deaths altered that. With no reliable ferry craft to carry men up to Salyut, and a shortage of propellant looming, the Soviets decided to cut short the station's life. On 11 October 1971, the world's first operational space laboratory was deliberately slowed and sent to a fiery end over the Pacific Ocean. It had orbited the Earth for 175 days.

Plans to launch a replacement ran into difficulties as the American Skylab launch approached and the Soviets tried desperately to regain prestige by beating their capitalist opponents to the punch as they had done so frequently in the past. No official announcement was made, but intelligence reports confirmed that a Salyut station had been destroyed during a Proton booster failure in July 1972. Worse was to come. Salyut 2 was successfully placed in orbit on 3 April 1973, just five weeks before Skylab was scheduled for launch, but no Soyuz craft went to join it. Instead, after three weeks of vague statements from Moscow, the mission was said to have been completed. Western tracking stations picked up more than 20 pieces of space debris, evidence that the station had begun tumbling out of control and breaking up. Another launch on

11 May, only three days before Skylab, also had all the hallmarks of a Salyut despite the official designation of Cosmos 557. This giant craft also remained unoccupied, eventually crashing back to earth after 11 days – Western experts speculated that the manoeuvring controls or main engine had malfunctioned. The likelihood of two such stations being built at the same time seems small unless one imagines a proposed dual mission. Certainly, two Soviet space stations in orbit at the same time would have drawn much of the sting from the launch of the 100 ton Skylab. Fortunately, no more cosmic deaths resulted from what can only be described as a ruthless attempt at political opportunism.

These failures left the Soyuz ferry with no station to service, so the Soyuz 12 and 13 flights had to be solo fill-ins. Soyuz 12, crewed by an air force colonel, Vasili Lazarev, a qualified doctor, and civilian flight engineer, Oleg Makarov, lasted only two days during a brief test of the new modifications. Three months later, a solar-winged version which would be suitable for a long solo mission such as the forthcoming Apollo–Soyuz Test Project, was sent up for a week-long scientific expedition. Commanding was air force major Pyotr Klimuk, who had been a mere 23 years old when he joined the cosmonaut corps in 1965, and alongside him was civilian flight engineer Valentin Lebedev, who had only been in training for a year. After a successful week of experiments, these two representatives of the new generation of cosmonauts landed on Boxing Day in a snowstorm. Another six months passed without a flight until Salyut 3 was at long last placed in orbit as the second Soviet operational space station.

One further change took place after the Soyuz 11 débâcle. General Nikolai Kamanin, head of the cosmonaut team, was replaced by veteran cosmonaut Vladimir Shatalov. The exact circumstances of Kamanin's forced retirement are unknown, but Shatalov retained this influential position into the 1980s and has become an effective spokesman for the cosmonauts and for the Soviet space programme as a whole. Under his leadership, many retiring cosmonauts have been encouraged to take up positions of responsibility at Star City or Mission Control after completing their stint of up to three missions, lending a further blend of practical experience and wisdom to the theoretical deliberations of the scientific community.

12

Skylab Sets the Record
The Skylab Missions of 1973

The age of innocence was over: manned spaceflight was about to enter a new age of maturity. Perspectives differed geographically, but the effect was the same. In the USA the pioneering flights of the brave guinea-pigs who risked their lives to win the space race were over – the race was considered won. The days of almost unlimited spending initiated by President Kennedy gave way to a more austere, thrifty regime in which every penny of funding had to be justified. NASA now had to produce a programme of research which would be both cost-effective and beneficial to the American economy. The ultimate goal was now a re-usable spacecraft, a winged combination of rocket and glider which would replace the forced wastage of conventional boosters and become the space transportation system of the late twentieth century. In the Soviet Union, too, there was a new realism. The moon race had been lost, and with it the technological race. It was time to push propaganda to the background and forge ahead with a realistic programme of research based on Earth orbital space stations and spacecraft already developed. Progress from now on would follow a logical, pre-determined course towards the economic utilization of near space.

In the USA, the transformation was traumatic. All over the country, engineers, designers and technicians who had enabled the nation to gain its pre-eminent position in space technology were faced with redundancy. The decline had been as sudden as the boom during the hectic expansion of the early sixties. From less than 50,000 workers employed in space developments in 1960, the total number of people in the space programme had risen to more than 400,000 by 1965. Since then, there had been a steep, but consistent, decline of around 50,000 workers a year – even the year leading up to the triumph of Apollo 11 had seen an average of 1,000 workers laid off each week. There was hardly a major aerospace company that was not affected: McDonnell Douglas had been responsible for Mercury and Gemini, and had built stages for the Saturn V and IB; Grumman built the lunar module; Rockwell International constructed the Apollo command and service modules, and the Saturn V second stage; Boeing were responsible for the Saturn V main stage. The west coast factories were particularly hard hit – Boeing had to lay off 64,000 workers in one week in order to remain solvent – but Kennedy Space Center was equally transformed. The old launch pads from which historic missions had begun were left to decay into ruins. House prices collapsed as the whole commercial fabric of the Cape complex and the surrounding area was adversely affected; staff levels at the Cape had been halved since the mid-sixties to match the decline in the number of launches. Only Skylab and Apollo–Soyuz remained in the manned spaceflight pipeline between the last Apollo in 1972 and the first Shuttle flight in 1981 – five major launches in nine years, and a six-year interval without an astronaut entering space. It was certainly a far cry from the heady days of Mercury, Gemini and Apollo.

The astronauts were inevitably affected by these changes. As Deke Slayton was at pains to explain, there was little chance of any mission for most of them except in the very long term; he advised them to look elsewhere for employment, and many of them took his advice. Of the 73 astronauts selected by NASA since the first group attracted banner headlines in 1959, eight had died by the beginning of 1973. Only 35 astronauts remained on the active list, almost all of the remainder having retired to civilian life. Apart from Slayton, only Alan Shepard was left from the original Select Seven. There were four survivors from group 2, but Jim Lovell and Pete Conrad resigned that year, leaving just Tom Stafford and John Young on the active list. It was a similar story with the Apollo heroes of group 3: of the original 14 selected, only Alan Bean, Gene Cernan, Rusty Schweickart and David Scott remained with NASA after the moon landings. The future would be entrusted to the young, untried who had been selected after 1966, and to the new breed of military pilots and civilian mission specialists recruited from 1979 onwards, and including the first women astronauts.

The astronaut resignations were not the only changes to occur in the USA after the Apollo honeymoon had ended. There had been several changes at the top of the NASA hierarchy after James Webb resigned in October 1968; Dr Thomas Paine had been put in the difficult position of Acting Administrator before his position was confirmed in March 1969. A year and a half later, Paine left to work for General Electric, having lost the battle to develop new, pioneering programmes which would lead NASA towards the twenty-first century. During his short reign, the man who had inspired the manned space programme and designed the mighty Saturn V, Wernher von Braun, was moved from his beloved Marshall Space Flight Center to a planning job at headquarters in Washington,

DC. He stayed for nearly two and a half years, arguing hard for expansion of the manned programme, but his pleas fell on deaf ears and it was a tired and disillusioned man who retired from NASA on 1 July 1972. His schemes for nuclear-powered rockets and manned landing on Mars still remain dreams. Meanwhile, another Acting Administrator, George Low, tried to buoy up the floundering ship until replaced by James Fletcher in April 1971. The new man was left with the unenviable task of presiding over the prolonged period of transition, which saw the successful completion of Skylab and the joint Apollo–Soyuz mission, a rare venture into friendship and cooperation with the Soviet Union under the fashionable new policy of detente.

The Skylab programme grew out of a desire to build on the expertise and hardware which would be developed during the moon landings. From the mid-1960s, NASA scientists and engineers began planning some kind of orbital space station and laboratory which could be derived at a relatively low cost from the Saturn rocket and the Apollo command and service modules. The initial Apollo Applications Programme envisaged a very ambitious launch schedule of orbital workshops and manned ferry craft coincidental with the moon landings, but as the national budget was increasingly constrained by the demands of the Vietnam War and an ever more cost-conscious Congress, the plans were cut back to a more realistic level. By 1970, the project had been renamed Skylab and reduced to a single workshop carved from the third stage of a Saturn V rocket. At this stage, NASA had been placed under such financial constraints that planners were obliged to cancel a moon landing in order to find a booster capable of launching the laboratory. Three visiting crews were planned once Skylab was placed in orbit.

The unmanned workshop was launched on 14 May 1973 on top of a two-stage Saturn V, to be followed the next day by a three-man crew riding a Saturn IB. Thousands of spectators felt the earth tremors and heard the tremendous roar that could only be generated by this monumental Saturn V as it slowly lifted off in what seemed a perfect launch. Hardly a minute passed before triumph turned to despair. Under building atmospheric drag, the meteoroid shield was torn away from the lab's exterior, causing a loose aluminium strap to anchor around an unopened solar wing. Minutes later, the blast from a staging rocket tore loose the shield and the solar panel to which it was attached. Telemetry indicated a perfect orbit 271 miles (442 km) high, but, although the X-shaped panels on the solar telescope mount deployed properly, one main panel was lost and the second failed to deploy fully. Skylab's power supply had been cut by half in one stroke. Another threat to the entire programme soon became apparent as the missing shield allowed the intense solar radiation to raise cabin temperature to around 40°C. It was

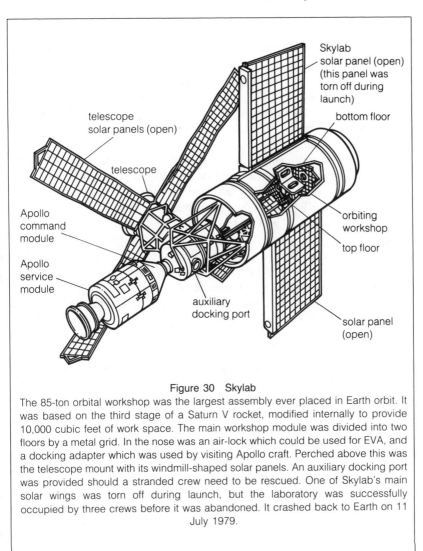

Figure 30   Skylab
The 85-ton orbital workshop was the largest assembly ever placed in Earth orbit. It was based on the third stage of a Saturn V rocket, modified internally to provide 10,000 cubic feet of work space. The main workshop module was divided into two floors by a metal grid. In the nose was an air-lock which could be used for EVA, and a docking adapter which was used by visiting Apollo craft. Perched above this was the telescope mount with its windmill-shaped solar panels. An auxiliary docking port was provided should a stranded crew need to be rescued. One of Skylab's main solar wings was torn off during launch, but the laboratory was successfully occupied by three crews before it was abandoned. It crashed back to Earth on 11 July 1979.

only a matter of time before the station's supplies deteriorated and the cabin became too hot for occupation. The first manned visit was postponed until someone could come up with a way out of the jam.

As engineers grappled with the problem of rigging a sunshade for the overheated Skylab, ground controllers tried a wide assortment of orbital alignments in order to reduce the surface area exposed to the sun while maintaining a reasonable power level from the forward telescope array.

They eventually compromised by placing Skylab at a 50 degree angle to the sun, thereby stabilizing the temperature at about 40°C and reducing the risk of poisonous gases being produced by decaying plastics inside the lab. While this was going on, the engineers hit upon the idea of a parasol made of nylon, mylar and aluminium foil which could be furled to fit inside the confined space of an Apollo command module but opened out to act as an external sunshade for Skylab. Another piece of equipment hurriedly put together was special cutters for the attempt to free the jammed solar panel.

With only hours to spare, the Apollo rescue craft was loaded with the tools and equipment for the most dangerous and ambitious repair yet attempted in space. The crew assigned to carry out the repairs to the 250 million dollar laboratory had been able to fit in a minimal amount of training for their task, and had to be prepared to deal with unexpected snags through their own initiative. In command of this most demanding of missions was Gemini and Apollo veteran, Pete Conrad. The wisecracking Conrad would need all his experience and sense of humour if he was successfully to reactivate the station and guide his two rookie companions through the forthcoming trials and tribulations. Fellow Navy pilot Paul Weitz, selected by NASA in 1966, and medical doctor Joseph Kerwin, an astronaut since 1965, occupied the other seats in the command module as the Saturn IB blasted off from the Cape on the morning of 25 May. One of Kerwin's main tasks was to monitor the adaptation of the crew to weightlessness over a scheduled 28 days, but it was obvious to all concerned that their return would take place much sooner unless they could achieve the necessary repairs. Conrad, as ever, remained optimistic. 'We can fix anything!' he declared as they roared off in pursuit of the ailing Skylab.

Even he must have had his doubts, however, as the disabled station drifted into view some seven hours after launch. 'As you suspected, solar wing 2 is gone completely off the bird. Solar wing 1 is, in fact, partially deployed . . . the gold foil has turned considerably black in the sun.' It soon became clear that a metal strap had snagged the remaining panel, so preventing its opening fully. The crew were optimistic that they could dispose of this debris. They pulled in for a soft dock on to the adapter module while they ate a quick meal and discussed the forthcoming stand-up EVA. The plan was for Weitz to stand in the open hatch and either cut through the metal strip that had twisted around the wing or pull it free. For the next hour, Weitz pulled and prodded with his 'pruning hook' while Kerwin held grimly on to his legs and Conrad manoeuvred the Apollo, but to no avail. The panel stubbornly refused to deploy. Returning to the docking adapter, more frustration followed as the capture latches failed to catch hold of the drogue. Repeated attempts at conventional docking led to further failures and a torrent of colourful language from Pete Conrad. As a last resort before backing off for the

(Previous page) **Return from the lunar surface**
The Apollo 11 lunar module ascent stage rises from the Sea of Tranquillity to rendezvous with Michael Collins in the command module Columbia. The lunar module Eagle carries Neil Armstrong and Buzz Aldrin, the first men to set foot on another world. In the background is the half-illuminated disc of the Earth.

(Right) **Apollo splashdown** The Apollo 15 command module Endeavour touches down in the Pacific Ocean on 7 August 1971. On board are astronauts David Scott, Alfred Worden and James Irwin. Impact was a little harder than usual since one of the three parachutes failed to function properly. The men were picked up by helicopter and flown to the prime recovery ship, USS Okinawa, 6½ miles away

(Below) **The mountains of the moon** Apollo 15 astronaut James Irwin salutes the Stars and Stripes beside the lunar module Falcon and the Lunar Roving Vehicle. The flag was deployed from a horizontal bar in the absence of a lunar wind. The lunar module descent stage, the lunar rover and the flag remained behind on the moon. In the background is Mount Hadley Delta.

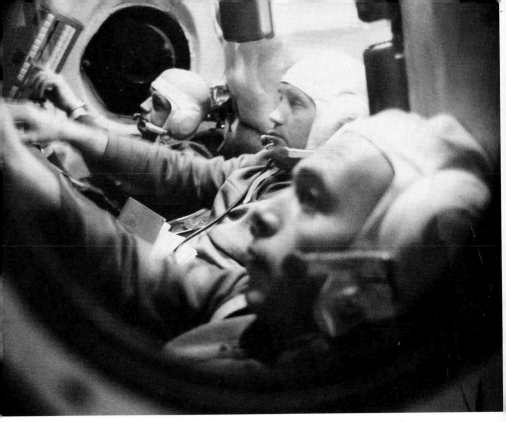

The crew of Soyuz 11 The three cosmonauts of Soyuz 11 made history when they became the first crew to transfer to an orbiting space station (Salyut 1) on 7 June 1971. After nearly 24 days in space, the crew returned to Earth but the lack of pressure suits to save space in the cramped cabin proved fatal: they were killed by cabin depressurization. Left to right: Flight Engineer Vladislav Volkov, Commander Georgi Dobrovolsky, and Test Engineer Viktor Patsayev.

Skylab The largest space station so far placed in orbit, and the only US station to date, was the 100 ton Skylab. This picture, taken by the Skylab 3 crew prior to docking, shows the main docking port with the X-shaped telescope solar array above. The station was nearly disabled when the right hand main solar panel was ripped off during launch. Also visible through the support struts of the telescope mount is the parasol deployed by the Skylab 2 crew. Far below is the Amazon River valley of Brazil.

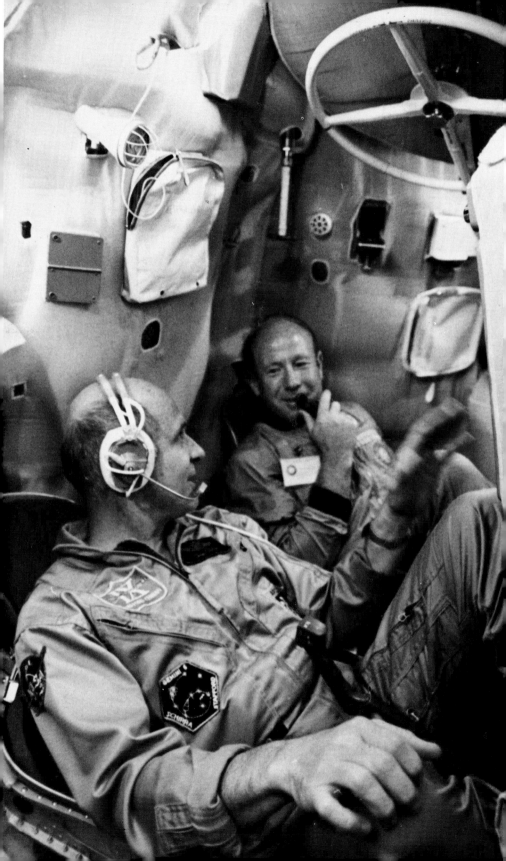

(Left) **Apollo–Soyuz Test Project** The only joint American-Soviet space mission to date took place in July 1975. Using a specially designed docking adapter, a Soviet Soyuz craft was able to dock with the Apollo 18 command module and crew transfers followed. Here the Apollo commander, 45-year-old Thomas Stafford, chats with his Soviet counterpart, 41-year-old Alexei Leonov, in a mock-up Soyuz craft at the Yuri Gagarin Training Centre near Moscow on 9 July 1974.

(Right) **The first Shuttle launch** On 12 April 1981 the world's first re-usable spacecraft was launched from Cape Canaveral in Florida. On board the Space Shuttle Columbia were astronauts John Young and Bob Crippen. The Shuttle's engines are supplied with liquid oxygen and liquid hydrogen from the giant external tank, painted white on this first mission. Also visible is one of the two solid rocket boosters which give the extra thrust necessary to lift the 100-ton Shuttle.

(Below) **Shuttle flight deck** Fish-eye view of Columbia's flight deck with Commander John Young on the left and Pilot Bob Crippen on the right. The Shuttle is fitted out with duplicate controls, though most manoeuvres are controlled by the five onboard computers. The crew can call up flight status information on four TV displays.

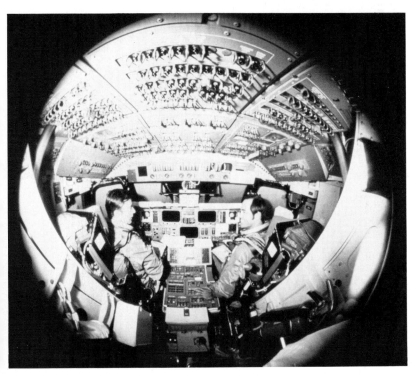

The Shuttle in orbit The Shuttle Challenger photographed with the cargo bay doors open during the seventh Shuttle mission in June 1983. This first picture of the Shuttle in orbit was taken from a camera on board the Shuttle Pallet Satellite (SPAS). This small satellite was retrieved by the Shuttle remote arm after about 9½ hours of free flight.

Portrait of the ill-fated STS 51-L crew Crew members were: front row, left to right, pilot Michael J. Smith, commander Francis R. (Dick) Scobee, mission specialist Ronald E. McNair; back row, left to right, mission specialist Ellison S. Onizuka, Teacher In Space participant Sharon Christa McAuliffe, payload specialist Gregory Jarvis, mission specialist Judith A. Resnik.

Shuttle satellite launch The Tracking and Data Relay Satellite (TDRS-A) is launched from Challenger's cargo bay during mission STS–6 on April 5 1983. The satellite was to be boosted to a geostationary orbit 22,300 miles above the equator by its Inertial Upper Stage. Data from the Shuttle and various unmanned satellites were to be relayed to Earth by three of these satellites located at 120 degree intervals around the earth, but TDRS–A remains the only one in orbit to date.

Training in zero gravity Simulating working conditions in zero gravity back on Earth is a major problem for flight planners. Spacesuited astronauts and cosmonauts use large water tanks for long duration training, but true weightlessness can only be simulated by aircraft flying parabolic curves. Here Svetlana Savitskaya (centre) practises with Alexander Serebrov (left) and Leonid Popov (right).

(Right) **Working in space** In July 1984 cosmonaut Svetlana Savitskaya made history by being the first woman to work outside a spacecraft during her second trip into space. During a 3½ hour spacewalk, the 35-year-old flight engineer used a general purpose hand tool for cutting, welding, soldering and spraying operations. Note the main solar panel and its two auxiliaries attached to the Salyut 7 space station.

(Below) **The Mir space station** The Soviet Union's third generation space station is seen during assembly. Launched on 20 February 1986, the 20-ton Mir has been described as a core module to which other specialist scientific workshops can be docked. Mir has six docking ports, five of them in a spherical transfer module, here seen just to the right of centre.

night, the crew tried a back-up procedure which entailed depressurizing the command module, removing the drogue from the docking tunnel, and attaching a by-pass lead to the switches of the docking mechanism so that the initial soft dock was no longer needed. It was nearly 15 hours since launch when Conrad's joyful voice came across the radio: 'We got a hard dock out of it!' With Apollo firmly attached to Skylab's nose, the crew were at last able to settle down for a sleep in the cramped cabin.

Next day the men tested the air in Skylab for toxic fumes, found a negative response, and decided to enter the station. Weitz took pole position, finding a cool 10°C in the adapter a welcome surprise. A thorough check-out found the relatively spacious adapter 'clean as a whistle', so the way was clear to float through the air-lock into the main workshop. Wearing a gas mask in case the oxygen–nitrogen atmosphere had been contaminated, Weitz conducted a quick survey of the giant workshop. He reported that the temperature was 'a little bit warm . . . like in the desert' but that everything seemed in good shape. Obliged to retreat temporarily from the workshop while further tests were conducted, the crew spent the time preparing a meal and admiring the spectacular view. At last they were given the go-ahead for permanent occupation of the laboratory. The men wasted little time in screwing together the rods for the handle of the parasol, then pushed the assembly through a small air-lock and opened the parasol. Although it was a little crumpled, the area shaded was sufficient to cause a temperature reduction almost immediately. By 4 June the temperature was down to a comfortable 24°C, but the power problem worsened as some of the battery packs refused to take more than a 50 per cent charge from the solar array. Something had to be done to deploy the jammed main panel.

Conrad passed on to ground control the crew's solution to the problem: a space walk to free the panel. Without such a venture, the station would never be able to achieve its full working potential, and mission controllers admitted that the next two missions would 'probably not be possible'. On the other hand, there was the grave threat of sharp metal edges which might cause a fatal accident if they punctured a pressure suit, plus the fact that there were no hand rails or footholds which an astronaut could use as anchors near the jammed panel. Ways of carrying out such a delicate operation were practised in a water tank at the Marshall Space Flight Center in Huntsville, Alabama, with the best solution passed up to the crew by teleprinter. On 7 June, almost halfway through the scheduled four-week mission, the astronauts were ready to make the attempt.

As soon as the air-lock was depressurized, the hatch was opened to allow Conrad and Kerwin to assemble the 25 foot pole with its bolt cutter attachment. By this time, Skylab had passed into the Australian night, so the crew sat back and waited. Breathing through 60 ft umbilicals and continually checked by Weitz inside the station, the men began the next stage of operations once daylight returned. Kerwin managed to clamber on to the struts of the solar telescope, then, extending the pole towards the aluminium strip tangled around the panel, he tried to clamp the bolt cutters on to it. In the absence of a decent foothold, Kerwin found himself floundering and expending a lot of energy without achieving any success. As his heart rate soared to 150 Conrad advised him: 'Cool it for a minute, Joe.' The breather had the desired effect, for a combined effort finally succeeded in positioning the cutters.

Now came the most dangerous part of the entire operation. Conrad moved hand over hand along the pole to the solar wing, stringing a tether behind him. Once the tether was attached to the panel, they had to pause once more for the night pass over the Indian Ocean. With daylight returned, Kerwin pulled with all his might on the lanyard which would close the jaws to cut the strap, but it remained stubbornly in place. 'Man, I'm really pulling,' gasped Kerwin. Suddenly the strap parted, causing both men to burst into relieved laughter. But the boom was still not fully deployed so Conrad moved closer, bracing himself on Skylab's outer skin with the rope dangling over his shoulder. Like some prize tug-of-war, they both heaved with all their might, until suddenly the panel broke free and locked in position. The repairmen were yanked off their perches by the unexpected jerk, and Conrad later confessed that he never did see the panel deploy as he was cartwheeling at the end of his tether. Mission accomplished after more than four hours of EVA, the delighted space walkers headed back inside. A loud 'Whoopee!' from Conrad was heard across the radio link. Full deployment of the vital panel was achieved by ground controllers pitching Skylab towards the sun to defrost the hydraulic fluid. In one magnificent stroke, the station's power supply had been increased by 3 kilowatts and the operational life of the laboratory had been assured.

With Skylab back in full operating condition, the crew were able to settle into their experimental routine and enjoy the home comforts of the giant workshop, though minor repairs were needed to various pieces of equipment. The space station was a self-contained home with adequate provisions for three long-term crews including air, food, water and clothing. Almost a ton of food was stored in containers and freezers, much of it in cans with pull-off tops. The men could cook and heat their food as they sat at the wardroom table by using the three small overhead ovens and the hot plates on the table, and each man had his own water dispenser to reduce the risk of infection. Three sleeping bags hung on the wall, enabling the men to zip themselves in and sleep like bats with their weightless arms tucked out of harm's way. Washing was mostly by means of dampened cloths, though each astronaut was allowed use of the shower once a week. Showers were restricted due to the limited water supply, though only three quarts were needed on each occasion. To prevent water spreading throughout the station, the men were obliged to seal themselves inside a cylindrical shower curtain which was extended upwards by the occupant to fit a ring on the 'ceiling'. The main problem was drying everything afterwards without allowing water droplets to escape, an awkward, time-consuming business which put some Skylab astronauts off the whole idea. Other useful hygiene facilities included a zero-gravity wash basin and a toilet. To wash his hands an astronaut had to insert them through rubber cuffs into a small wash basin sealed off from the cabin. The toilet was suction-operated, propelling all body wastes into disinfected bags which were then wrapped, carefully weighed and labelled for later analysis on Earth. Dirty laundry was simply disposed of through an air-lock into a large tank.

The men found adaptation to their new environment a pleasurable experience. Conrad informed anxious doctors on the ground: 'Mobility around here is super. Nobody has any motion sickness.' To prove the point, they displayed their gymnastic expertise on a live TV broadcast with a series of effortless somersaults, cartwheels and backflips. Another form of recreation tried out by the crew was space darts, but their performance was disappointing. Kerwin commented: 'The darts didn't work worth a darn.' He added: 'The paper airplanes are rather promising though.' Flight surgeon Kerwin was very pleased with the physical response of the crew to the stresses of the record-breaking 28-day space voyage, noting only a predictable lowering of blood pressure and accelerated heart rates. By the end of the mission, he gave his medical scorecard: 'Right now the score is man 3, space 0.' The worse problem he had to deal with was sore throats – sound carried poorly in the low pressure oxygen–nitrogen atmosphere, so they had to shout to be heard. Even whistling proved something of an effort.

During the four-week mission a number of records were broken as Conrad became the overall space-time leader, then the American endurance record was passed and finally the all-time record of the ill-fated Soyuz 11 crew. More significant, however, were the scientific results the mission yielded. Despite the early setbacks, ground controllers reckoned that the crew obtained about 80 per cent of the planned solar observations and completed 11 out of 14 planned Earth resources surveys. In addition, all 16 medical experiments were conducted, and they even found time to operate five experiments contributed by students. Particularly exciting for physicists were the results from the eight solar telescopes clustered on Skylab's nose. Five of these operated at X-ray and ultraviolet wavelengths beyond the range of Earth-bound observatories, a sixth concentrated on the glowing corona which formed a halo around the sun, while the other two observed changes on the solar disc. On 15 June these instruments swung into action for the first detailed study of a solar flare, an unpredictable violent outburst which had puzzled scientists for years. They took so many pictures with the solar array that Conrad and Weitz had to leave the air-lock to retrieve the exposed film on 19 June. During the 90 minute EVA, the 'fix it' crew succeeded once again, using what Capcom called 'the old Army technique' – a stuck battery relay on the solar array was freed with a hefty blow from Conrad's wooden hammer. Not all activities proved quite so straightforward. At one stage tempers snapped as a result of the old tendency of ground controllers to pile on the chores. 'About two or three times now you got us doing things where we got 89 pieces of gear out, and you got us running all over the spacecraft,' scolded Conrad. It did not take long for Mission Control to get the message.

At the end of the four-week stint the crew packed their blistered Apollo craft with film cassettes, blood and urine samples, then turned off the air conditioners and lights to save power. As a precaution against blackouts during re-entry through blood pooling in their legs, the crew wore inflatable pressure suits which they could activate if necessary. Their remarkably successful flight ended on 22 June with splashdown near the carrier *Ticonderoga* some 830 miles (1,330 km) south west of San Diego. The crew remained in their spacecraft to reduce the risks of transferring to liferafts and helicopters in their weakened state. Once the capsule was winched aboard, doctors were prepared to lift them out, but the men insisted on walking, albeit a little unsteadily, on to the deck. The only person who suffered at all during re-entry and recovery was Kerwin, who felt obliged to inflate his pressure suit after feeling a little faint, then forgot his naval background by becoming seasick in the rolling command module. All three men had lost weight – between 4 and 8 lb – and exhibited significant losses in heart volume, red blood cells and bone calcium which caused some doubts among doctors about the upper limit of human space endurance. The returning crew were more positive, simply suggesting more rigorous exercise by their successors.

As for the triumphant Pete Conrad, his fourth mission was his last. He resigned from NASA and the Navy at the end of the year to take up a career in civilian aviation. He is currently a senior vice president in the Douglas Aircraft Company. Both his companions remained with NASA: Weitz stayed on the active list, eventually going on to command the first flight of the shuttle Challenger, while Kerwin became a senior NASA official and currently holds the post of space and life sciences director at Houston.

The second visiting crew for Skylab was launched on 28 July 1973, along with a menagerie which included two minnows, 50 frogs' eggs, six mice, 720 fruit fly pupae and two spiders nicknamed Arabella and Anita. Commander of this unusual crew was Apollo 12 moon walker Alan Bean, making his second flight since joining NASA in 1963. His science pilot. Owen Garriott, was a cheerful, moustachioed electrical engineer who had gained a doctorate from Stanford University before astronaut selection in 1965. He was an expert in ionospheric physics and electromagnetic theory. Third crew member was stocky Marine fighter pilot Jack Lousma, an astronaut since 1966. The crew did not contain a qualified physician, but all three had trained as paramedics.

Even before the eight hour chase of Skylab was completed, the alarm bells were ringing on the mission. During CSM manoeuvres Bean reported spotting 'some kind of sparklers'. At the same time, ground control noticed a pressure drop in one of the four thruster assemblies, so Bean had to complete docking with this assembly shut down. Six days later, a second assembly sprang a leak, instantly causing concern that the craft was fatally flawed. Houston immediately ordered a red alert, hurriedly initiating preparations to launch an Apollo rescue craft modified to take five men and piloted by Vance Brand and Don Lind. In the event, the leaks proved to be unrelated, and engineers succeeded in finding a way to steer the original craft with only two operational sets of thrusters, so the rescue mission was thankfully mothballed.

The faulty thrusters were not the only source of misery for the incoming crew. By the end of the first day in space all three were feeling the effects of spacesickness: Lousma had swallowed a sickness tablet even before the rendezvous was completed and was now vomiting at irregular intervals, unable to eat. His companions were faring little better, with bouts of dizziness and nausea which restricted their capacity to work. Bean lamented to Houston: 'We're just not as spry up here right now as we'd like to be. None of us has been able to eat all our breakfast. There's a desire to just lie still for a while.' Not until the fifth day did they feel back to normal. 'Food tastes a lot better today,' commented a more cheerful Lousma. Less fortunate, however, were some of their passengers: the Pocket mice and fly pupae colonies both died after the controls on their environment capsules short-circuited. The minnows also seemed unhappy with their new environment, swimming around in small loops, but those hatched from eggs on board the station swam normally straight away. This observation led Mission Control to ask Garriott if this might suggest a way of selecting future astronauts, but the science officer declined to comment. The most successful creature to adapt was Arabella: after only a couple of days she was spinning webs comparable to those made on Earth. 'It seems she learned very rapidly in zero G without the benefit of any previous experience,' commented Garriott. Sister Anita repeated the delicate construction work when given her chance later.

With everything shipshape at last, the crew settled down to an intensive Earth resources survey, starting with a 6,000 mile sweep from the Pacific, across the USA to the Gulf of Mexico and the Amazon Basin. Tying in with a small armada of fishing boats deployed in the Gulf, Skylab's instruments were tested for their ability to predict fish abundance through sea temperature or colour. The crew were so dedicated to this aspect of their work that by the end of the mission they had completed 39 Earth resources passes instead of the planned 26. Observations included the life cycle of severe storms, soil moisture and potential subterranean water sources in Africa, the eruption of Mount Etna and assessments of soils, crops and mineral deposits. On each occasion the crew had to tilt Skylab away from the sun so that its instruments pointed directly at the target areas, but the data were particularly valuable since Skylab's orbit, inclined at 50 degrees to the equator, enabled the crew to observe sections of 33 countries apart from the USA.

One of their first tasks on joining up with the station should have been the deployment of a second sunshade over the deteriorating parasol erected by their predecessors, but their sickness forced a postponement. The space walk eventually took place on 6 August, a record-breaking six and a half hour excursion by Garriott and Lousma. In a procedure similar to raising a sail on a ship, the men first assembled two 55 foot (17 m) rods and positioned them horizontally over the workshop so that they could be attached to the forward telescope mount at one end and a baseplate at the other end. They then pulled two lanyards to extend the folded shade along the rods so that it covered the now redundant parasol. Although it resembled a gold-coloured concertina more than a sunshade, the sail had the desired effect on cabin temperatures. This achieved, the efficient space walkers moved on to load film canisters in the telescope cameras, install panels to assess micro-meteorite damage, and inspect the thrusters of the suspect Apollo ferry. As a reward for their efforts, the crew were allowed to sleep in the next morning before beginning their intensive solar observations.

Once again the crew showed great dedication, putting in between 12 and 14 hours a day in search of the sun's secrets. Altogether they spent more than 300 hours on solar astronomy compared with the planned 200 hours, snapping an astonishing total of 75,000 pictures of the sun and its elusive features, such as coronal holes, bright spots, flares, coronal bubbles, sunspots and giant tongues of hot gas, 'prominences'. They were rewarded in the first week of September by a sudden burst of violent activity which produced 31 flares in a single day. One tremendous flare sent out such a stream of energetic particles that it disrupted radio communications on Earth and produced a superb display of aurorae in the polar skies. Fortunately, the astronauts were low enough to be shielded by the Earth's magnetic field.

As the mission approached its climax, doctors were surprised and delighted to observe that the crew's physical condition had stabilized. Much of the credit for this was given to the one-hour work-out that the men completed each day, with the small exercise bicycle taking most of the strain. So well adapted and efficient did they become that they asked for extra assignments and found no difficulty coping with only six or seven hours sleep each day. So happy were the NASA physicians that there was no hesitation in giving the go-ahead for two more space walks towards the end of the mission. On 24 August Garriott and Lousma stepped out of the air-lock to plug in a new set of gyroscopes they had brought with them from Earth – the original system had been giving a lot of trouble – and replace film cassettes. The third EVA occurred on 22 September, enabling Bean to have a turn. He and Garriott spent more than two and a half hours outside the station unloading film cassettes, retrieving strips of material and cleaning an occultation disc fitted to the coronagraph. As on previous occasions, the men were so inspired by the view that they were reluctant to come back inside.

With their eight-week stint of 59 days, 852 orbits and 21½ million miles (35 million km) coming to an end, Bean optimistically inquired of mission control whether they could have an extension, but they were turned down. They had already broken just about every record in the book. Splashdown took place in the Pacific 230 miles (370 km) south west of San Diego on 25 September. During re-entry the men wore the pressurized trousers to prevent pooling of blood in their legs. The partially disabled command module functioned perfectly following special re-entry instructions from the ground, but turned turtle after hitting the choppy ocean. Despite their rough ride and 40 minute wait before they could be winched aboard the carrier *New Orleans*, the men smiled and waved to waiting cameramen. Rather unsteady on their feet, they were whisked below decks to undergo immediate medical checks: for the first time there were no welcoming speeches to be endured. The results showed that 'in some tests the men look better than the first Skylab crew'. The way was open for an even greater challenge to the hostile environment occupied by the battered Skylab.

Alan Bean continued with NASA until he retired in 1981 to take up full-time painting of space landscapes. Garriott applied his practical experience in a number of senior administrative posts before returning to the active list and taking part in the first Spacelab mission of the Shuttle in 1983. Jack Lousma teamed up again with Bean on the back-up crew for the Apollo–Soyuz Test Project, and eventually returned to space as the commander of the third Shuttle flight.

An inventory of supplies remaining on board Skylab suggested to programme planners that the third crew would be able to occupy the station for at least as long as Bean's highly motivated trio. Skylab seemed in reasonable shape, and the medical barriers had been pushed back, so the new crew were given a target of 84 days in orbit. Just to make sure their food would last out, the NASA nutritionists provided them with special high-energy food bars to supplement the existing stocks. Also crammed inside their command module was extra film and camera equipment with which to survey a newcomer to the Solar System, Comet Kohoutek, potentially the most brilliant comet of the century. But while the crew were raring to go, their ageing hardware was beginning to show its fragility. Hairline cracks in the Saturn's tail fins were discovered during routine checks. With the distinct possibility that these could cause the rocket to break up in flight, all eight fins had to be replaced. Further cracks in the staging between the first and second stages had to be patched up, and buckled fuel tanks carefully refuelled. This, added to the six-day delay in launch, was not confidence-inspiring for crew, who were all on their first space trip.

This unusually inexperienced crew was headed by a 41-year-old Marine lieutenant colonel, Gerald Carr, an experienced fighter pilot with a master's degree in aeronautical engineering from Princeton University. He was working on tactical data systems when his assignment to NASA came in 1966. During the following years he served on two Apollo support crews and helped test and develop the lunar rover. His science officer was 37-year-old civilian engineer Edward Gibson, a research physicist who worked for the Philco Corporation after obtaining his doctorate in 1964. Third crewman was Air Force lieutenant colonel William Pogue. He was an astronaut from the traditional mould, having begun his flying career with combat duty in Korea, joined the Thunderbirds in the mid-fifties, attended the Empire Test Pilots' School in England and been assigned as an instructor at Edwards Air Force Base prior to joining NASA in 1966.

The rookie crew were happy with their morning launch on 16 November 1973, the thirtieth such occasion at the Cape, exhibiting only minor rises in heart beat. Docking proved something of a problem, eventually succeeding at the third attempt, but the crew were looking forward to entering their new home. 'She looks pretty as a picture,' said Carr warmly. To avoid the illness suffered by their predecessors, all three took medication during the approach, but the unfortunate Pogue was still struck down with nausea. His affliction was not relayed to ground control, and was only picked up on a transcript of their onboard conversation, an omission which led to a mild reprimand from Alan Shepard and an apology from the mission commander. The crew restrained their enthusiasm and decided to spend the first night inside the Apollo.

After the workaholic second crew ground controllers had become used to an efficient crew operating like some well-oiled machine, and failed to make allowance for the acclimatization required by the newcomers. Finding their way around the station took some time, especially since equipment was not always located in the expected places, and the task of unloading the command module's thousands of items, then stowing them safely away, proved a trying experience. Already the crew were complaining they had too much to do and too many new tasks, especially medical experiments, which had been inadequately rehearsed after the decision was made to extend the duration of the mission. Gibson later commented: 'The first seven or eight days were not something I would want to go through again.'

On 22 November Gibson and Pogue earned praise from Mission Control when they repaired a jammed antenna on the radiometer-altimeter during a record-breaking 6 hour 34 minute space walk. It was a tricky job, partly because the antenna was in an awkward location, and partly because it entailed removing six screws before installing a new electrical switching box. While Pogue completed the repair, Gibson had the unenviable task of hanging on to his feet to hold him steady. Unfortunately, while they were conducting their other duties outside the station, they forgot to put filters on six of the cameras, an error which lost some valuable Earth resources data.

The next day gave no respite for the beleagured crew. One of the three gyroscopes which adjusted Skylab's attitude failed and had to be shut down. Although the remaining two could still operate the station, it meant much tougher flight planning, and more use of valuable thruster propellant in manoeuvring the craft. The remainder of the mission was spent with the constant threat of a second gyroscope failure, like a guillotine threatening to drop at any moment. Earth surveys suffered badly while ground controllers struggled to find a way of using less fuel, resulting in the crew being told to concentrate on solar astronomy and medical experiments.

The crew continued to show an alarming lethargy as Christmas approached, achieving less concrete results in the time allotted than their predecessors, and continually looking forward to their rest periods. Even the sun remained in quiescent mood, while Comet Kohoutek refused to put on the expected fireworks display. Two welcome diversions came on 18 December when the sun unexpectedly erupted with the largest prominence seen in 20 years, and Skylab was joined in orbit by the Soviet Soyuz 13, the first time both superpowers had representatives in orbit at the same time. Although the crews were unable to communicate directly with each other, and were never within sighting range, the Americans showed considerable interest in the welfare of their rivals. 'Tell them to drop by for coffee some time,' radioed Carr. Cheered by carols broadcast from Earth, the crew gave a live broadcast on Christmas Eve to TV viewers across the nation. They were in a meditative mood prior to their next space walk, speaking in favour of peace on Earth and the brotherhood of man.

Christmas Day saw Carr and Pogue step outside for a close-up view of the comet, though they were unable to see it themselves because it was so close to the sun. With Pogue 'holding me like a sausage under his arm', Carr also used his screwdriver to pry open a stuck filter wheel on an

X-ray telescope, replaced film cassettes and retrieved particle collection samples. Following the seven hour EVA the men were given a taste of home. Against the background of a Christmas tree fashioned from food containers, the crew opened the presents from their families which had been secretly stowed away before the launch.

As the comet reappeared from behind the sun, Carr and Gibson ventured out once more on 29 December to focus in on the elusive object. For astronomers and public alike, the comet had proved a damp squib and a great disappointment, but the astronauts were filled with wonder at the sight that greeted them. 'Hey, I see the comet!' shouted Gibson. 'There's the tail. Holy cow!' As they snapped away for nearly three and a half hours, they described in detail the comet's spike and tail. 'The spike is very evident. It is not 180 degrees out from the tail, but more like 160 degrees. It is yellow and orange . . . just like a flame . . . and there is a diffuse amount of material which goes out and joins up with the tail.' Gibson went on to call it 'one of the most beautiful sights in creation I've ever seen'.

The problem of crew morale had still not been solved, however, despite attempts to 'back off' and give the crew more time to relax and use their own initiative. So far this crew had been earning the reputation of the most ill-tempered ever to orbit the Earth. They found everything wrong with the station and its equipment – the toilet was too complex to operate, the towels were too rough, the clothes were boring and colourless, the food was bland and tasteless, and so on. Exercise on the bicycle and treadmill was a drudge which they tried to avoid, and even shaving become too onerous, so the men simply grew beards. On one Sunday, usually regarded as a rest day, the crew downed tools and refused to follow instructions from the ground – an incident which Carr later termed 'the first space mutiny'. In a final session intended to clear the air once and for all, frank communication took place between the crew and Houston on 30 December. Carr commented that 'the problem we had at the beginning was we started too high'. He went on: 'A guy needs some quiet time to just unwind if we're going to keep him healthy and alert up here. There are two tonics for our morale – having time to look out the window, and the attitude you guys take and your cheery words and your occasional bits of music.' Pogue summarized his viewpoint: 'I'm a fallible human being. I cannot operate at 100 per cent efficiency. I am going to make mistakes.' Such confessions were rare among the cool, hard-nosed flier fraternity, but the message got home to Capcom Richard Truly and his companions. From now on the work schedules would be adjusted and the crew would have more time to relax: it would be a working partnership instead of a management–worker situation. From this date, the crew's performance showed a marked upturn so that by the end of their mission their results exceeded the original targets.

A few days later the crew told listeners on Earth of their changed attitudes towards themselves and life in general, an 'almost spiritual' transformation. Pogue summed it up when he explained: 'My attitude towards life is going to change, towards my family it's going to change. When I see people, I try to see them as operating human beings and try to fit myself into a human situation instead of trying to operate like a machine.' They threw themselves with a will into their experiments and the exercise regime as if taking second wind, and although still error-prone and bugged by niggling equipment problems, they set out to prove they were as competent as any previous crew. Despite the continual threat posed by the erratic gyroscopes, flight directors and doctors agreed to extend the mission bit by bit, enabling the astronauts to pass the Skylab 3 record of 59 days on 12 January, and the accumulated flight time record for space voyagers on 25 January. Previous all-time record-holder Alan Bean congratulated the crew: 'You did it the hard way – all in one shot.'

Gibson's long vigil with the solar observatory at last paid off on 21 January as he focused in on a suspiciously active region on the solar disc. Flares were notoriously unpredictable on the ever-changing surface of Earth's star, but the astronaut took a gamble by initiating a rapid-fire photographic sequence, despite being low on film. To his delight, a small bright spot developed before his eyes into a full-grown flare, the first observation of a solar flare's life cycle and the answer to a solar physicist's prayer. The vital pictures and information were retrieved on 3 February during the fourth space walk of the mission. Carr and Gibson spent more than five hours outside the station, also recovering materials and collection devices used to discover the effects of the hostile space environment. In an optimistic gesture they even left behind a fresh micro-meteorite experiment which could be picked up by some future space walkers.

As the men began closing down the station, Skylab was hanging on by the skin of its teeth: the second gyroscope was on its last legs, the shower no longer worked, the cooling unit on Gibson's EVA suit was leaking, and even one of the sunshades had split. On 8 February they put together a 'time capsule', a bag containing samples of food and drink, unused film, camera filters, clothing and electronic equipment, which might someday be collected by a visiting astronaut. To lengthen Skylab's life in orbit, the command module thrusters were fired in order to raise the altitude of the station. Controllers in Houston expressed optimism that this would be sufficient to keep Skylab aloft for perhaps nine years, by which time the Shuttle would be operational. In the absence of funds to develop a new space station, Skylab would be the only vehicle of its type available to NASA in the foreseeable future.

Stowed aboard the command module were the results of nearly three months' work in zero gravity. These included some 20,000 pictures of

Earth, 19 miles (30 km) of magnetic tape, and 75,000 telescopic images of the sun, as well as some of the most detailed pictures of a comet ever obtained. Despite the problems during the first half of the flight, the crew had still conducted four space walks, including a record-breaking seven hour jaunt, had completed 39 Earth survey passes rather than the 30 originally planned, and had used twice as much film on solar observations as expected. In addition, they had produced improved crystals and new metal alloys using the small electric furnace, the forerunner of future space factories, they had completed the test runs on a manned manoeuvring backpack which they flew around the workshop, and had obtained more medical information than either of the previous crews.

With Skylab's main systems shut down, the crew squeezed into the command module for return to Earth. Ground control radioed to Apollo: 'Say goodbye for us. She's been a good bird.' Gibson agreed with Capcom Bob Crippen: 'It's been a good home. You can tell Al Bean and the guys that they did a great job putting that sail up. Hate to think we're the last guys to use it.' However, even re-entry did not go smoothly on this trouble-torn mission. A helium leak in one of the thruster loops meant that the crew had to use the other thruster system. They were also warned to put on their oxygen masks immediately they smelled anything unusual. In their haste to complete re-entry procedures, the crew mistakenly shut off the automatic control circuit used during the return to Earth. As Carr later confessed, 'Our hearts fell and our eyeballs popped' when they discovered the error. Forced to resort to the manual back-up system, Carr was able to guide Apollo back through the atmosphere to a Pacific splashdown. The command module condemned the frustrated astronauts to an uncomfortable wait as they dangled upside down in their couches, but after nearly an hour, they were winched aboard the recovery carrier.

Surrounded by helping hands, the three heroes moved awkwardly to the waiting seats. After nearly three months without a shave, Pogue and Carr were almost unrecognizable to their wives, though Gibson remained his normal clean-shaven self. Carr described the main lesson of their 34 million mile epic: 'That man can live a normal existence in space and that he can accomplish a great many things that can't be done on the ground.' For veteran space physician Charles Berry, the repercussions were startling: 'From what we know today, there is no medical reason to bar a two-year mission to Mars'. Only too aware that this was a dream with little chance of becoming reality for decades to come, NASA administrator James Fletcher came to a more prosaic conclusion: 'In a very real sense, Skylab can be considered a turning point – for while it was still basically an experimental space station, it nevertheless possessed many qualities and ingredients that will characterize operational missions of the future.'

As for the futures of the new endurance record holders, none of them ever experienced the trials and joys of zero gravity again. Carr worked on a NASA design support team until he retired in 1977 to become a senior consultant to a California research company. Gibson resigned from NASA in 1974 to do research on the Skylab solar physics data he had so painstakingly gathered. He spent a year in West Germany helping in the design of the European Spacelab, then returned to NASA from 1977 to 1980 as Chief of Scientist–Astronauts. He is currently a senior manager for TRW Inc., working on efficient ways of extracting energy from natural resources. Pogue left NASA to head his own consultancy to aerospace and energy firms. In 1985 he wrote a book about his Skylab experiences entitled *How Do You Go to the Bathroom in Space?* By that date the aged space station was long gone, obliterated as it crashed to Earth in July 1979. Scientists had not allowed for increasing friction between the station and the upper atmosphere as the sun reached the peak of its 11 year cycle of activity, and with the Shuttle still unavailable for a salvage attempt, the battered giant was doomed to a short, though spectacular, death ride. Its replacement would have to wait until well into the 1990s, leaving the field of orbital laboratories open for the Soviets to develop unopposed.

13

The $500,000,000 Handshake

The Apollo–Soyuz Friendship Mission, 1975

On Thursday 17 July 1975 the Iron Curtain was breached as millions of people worldwide saw the green-suited, bulky frame of Soviet cosmonaut Alexei Leonov float through the tunnel from his craft and cross into an American spacecraft. The ensuing warm handshake and cordial greetings were the end result of a space pact signed three years earlier by President Nixon and Soviet Prime Minister Kosygin at the height of detente. Tax payers in both countries had provided perhaps half a billion dollars to witness the first, and so far only, space rendezvous between the two superpowers. As viewers on both sides of the world could clearly see, there was no doubting the genuine warmth of the greeting, but already the Nixon–Kosygin initiatives were beginning to be regarded as outdated and misguided. The warmth generated by the Apollo–Soyuz joint mission would prove inadequate to prevent revival of the Cold War in the years to come.

The story began in 1970 with an informal meeting in New York between former NASA administrator Dr Thomas Paine and Soviet space scientist Anatoly Blagonravov. Both sides were well aware of the dangers of manned spaceflight, having suffered fatalities during the space race of the 1960s. Such problems were also highlighted in the contemporary imagination by a Hollywood movie entitled *Marooned*. American scientists amazed their Soviet counterparts by relating the plot, in which an astronaut stranded in space was finally rescued by a cosmonaut in another craft. Joint working groups had already hammered out the general details by the time the agreement on the peaceful exploration of space was signed by Nixon and Kosgyin in May 1972. The space rendezvous mission was named the Apollo–Soyuz Test Project (ASTP).

Nixon's foreign policy achievements reverberated around the world in the early 1970s: the SALT 1 missile treaty with the Soviet Union, the

Vietnam peace treaty, open dialogue with Communist China and the search for peaceful cooperation with the Soviet Union, which included ASTP. However, behind the scenes all was not well, and the bubble burst when Nixon was forced to resign in disgrace in August 1974 as a result of the Watergate scandal. It was left to his vice president, now President Gerald Ford, to watch over the final stages of his former chief's work.

The years of preparation did not prove easy. Although NASA was well used to revealing all as part of the freedom of information clause in its charter, the Soviets found it difficult to break away from their routine secrecy and suspicion. At times, the Americans found the only way to make progress was to dig in their heels and stand firm. On one occasion the man who was to command Apollo 18, Tom Stafford, threatened to withdraw from the mission unless facilities were granted for a thorough inspection of launch facilities and spacecraft at the Baikonur cosmodrome. As usual, the Soviet team consulted behind the scenes with higher authority, and next day expressed surprise that such an inspection had ever been in doubt! The visit went ahead, though the Americans were not allowed to carry cameras and were carefully driven away from any sites of strategic significance, including the mission control centre.

Nevertheless, regular exchange visits every few months by crews, engineers, technicians and administrators began to bear fruit. One basic difficulty to be resolved at an early stage involved the different atmospheres used in the two craft. Normal Soyuz cabin atmosphere consisted of the familiar oxygen–nitrogen mixture at a sea level pressure of 14.7 psi. However, in Apollo the atmosphere was pure oxygen at a pressure only one-third that in Soyuz. Any transfer of men from the Soviet craft to Apollo would result in a painful, and possibly fatal, attack of the 'bends' caused by nitrogen bubbles in their body tissues. Agreement was finally reached on the need for an air-lock in which the atmosphere could be adjusted for either crew during transfer from one craft to the other. Equipment was also built into Soyuz so that air pressure could be reduced to 10 psi during the period of joint flight.

Before crew transfer could take place, the two craft would have to rendezvous and dock successfully. Despite the considerable experience both sides had built up in this department, there was the major disadvantage that the two systems and methods were incompatible. Soyuz usually orbited at around 120 miles altitude, much lower than Apollo. The Soyuz could dock using both manual and automatic procedures, but was less manoeuvrable, while Apollo had much more advanced radar and navigation systems, an onboard computer and a much greater fuel capacity but could only dock using manual guidance. In the end, the fuel supply took precedence, so Apollo was designated the chase craft and Soyuz the passive craft. To make the dark Soyuz easier to see without altering its thermal characteristics too drastically, the craft was fitted with a transponder for Apollo to home in on during rendezvous and signal lights, as well as a new coat of paint – green and white. The only major new piece of equipment, the docking module, was developed and constructed by NASA. Although attached to the forward end of Apollo, the module had to be designed so that either craft could approach and actively dock with the other. This had to be possible while at the same time leaving a clear transfer tunnel through the module for the crews. Design of the docking mechanism had to start virtually from scratch because the existing probes, drogues and latches on the two craft were of different dimensions.

As American experts became familiar with the rival technology, they suddenly realized how successfully they and their government had been duped during the sixties. Far from being a potential design break-through, the Soyuz was found to be a relatively primitive craft, comparable in many ways to the Mercury craft built in the USA nearly 15 years earlier. Instead of a sophisticated computer, the cosmonauts were restricted to use of various rolls of punched tape with fixed instructions on the necessary manoeuvres; engine burns had to be timed by stopwatch and re-entry was controlled only by spinning the craft during descent. The many debacles with Soyuz flights had done nothing to ease fears that astronauts' lives would be in danger should the Soyuz malfunction again. Some Americans were openly scathing about the entire mission. 'Plain, goddamned brute force engineering' was how one critic summed up Soviet space technology, while another quipped, 'Our arms around their shoulders and their hands in our pockets'.

Against this background, the Soviets hastened to reassure their new partners that the rocket which aborted during the launch of Soyuz 18 in April 1975 was from an older batch than the booster being prepared for ASTP. Less easy to dismiss was the Soyuz 11 tragedy which killed three cosmonauts. Not unreasonably, NASA officials insisted on a full account of the reasons for the fatalities. For once, the Soviets were completely frank and open, revealing both the causes and the subsequent modifications which had been flight tested by the unmanned Cosmos 496 and 573 as well as Soyuz 12. In order further to reassure both themselves and the sceptical Americans, a Soyuz craft adapted to ASTP specifica-tions was sent into orbit on 2 December 1974. On board was the second ASTP prime crew, Colonel Anatoli Filipchenko commanding and Nikolai Rukavishnikov sitting alongside as flight engineer. For both men, now well into their forties, it was their second space mission. Filipchenko had previously commanded Soyuz 7 in the triple mission more than five years earlier, while Rukavishnikov had participated in the Soyuz 10 flight to the first Salyut space station. The six day Soyuz 16 mission proved much more successful, although Soviet fears and American doubts were not immediately dispelled. The first information concerning the launch to reach NASA came one hour after the event, and resulted in ASTP flight director Glynn Lunney and his associates being aroused from bed to attend mission control in Houston. NASA tracking stations noted that the initial Soyuz orbit was too low but were reasssured when the craft succeeded in raising its altitude to the planned ASTP value.

Soyuz 16 carried the same equipment as would the ASTP craft, Soyuz 19, the following summer. Its life support system was modified to supply four men instead of the normal two, and the ability to lower and raise cabin pressure was also demonstrated. On its nose was 'an imitation docking ring' which simulated the Apollo docking mechanism. This was eventually jettisoned after numerous tests. Modifications to the Soyuz control system also passed out with flying colours. The one regret of the Americans was that no direct voice communication between the crew and Houston had been possible.

The crew landed safely on the frozen steppes of Kazakhstan to complete an almost perfect rehearsal. Confidence soared, only to be dented once more when the Soyuz 18 rocket failed. Such embarrassments were not easily explained away, although the Soviets pointed out that a second rocket and crew would be ready and waiting on the day of the ASTP Soyuz launch.

One more concession from the Soviets was necessitated by ASTP. For the first time, they were obliged to reveal the identities of the crews prior to launch. In May 1973, the first prime crew for Soyuz was announced: the Soviet Union was clearly selecting the most prestigious crew possible. In command would be Alexei Leonov, the first person to walk in space in 1965. The 41-year-old air force colonel had continued in training during the ten-year interval, graduating from the Zhukovsky Air Force Academy in 1968 then preparing for a mission to Salyut which never materialized. He had built up a reputation as a space artist with his portrayals of the space walk, with his drawings, used on Soviet stamps, and with an exhibition of his work opening in Moscow during May 1975. His astronaut companions described him as a 'very funny guy' as well as a very competent and hard-working pilot. The Soyuz flight engineer was another Soviet pioneer, 40-year-old civilian Valeri Kubasov, the first man to carry out experimental welding in space. This experience would be invaluable in performing the metal and materials processing experiments scheduled for the present mission. In contrast to the ebullient Leonov, his companion was a shy, quiet man equipped with an 'encyclopaedic' knowledge of spacecraft designs and celestial mechanics. Leonov described him as an 'absolutely first-class engineer'.

Greeting the two cosmonauts in space would be one of the most unusual crews ever launched by NASA. The Apollo commander was 44-year-old US Air Force brigadier Thomas Stafford, one of the most experienced astronauts. Stafford had worked his way up the ladder to fly on two Gemini missions and command Apollo 10 to within a few miles of the lunar surface. After that success, he had become head of the astronaut group and therefore responsible for the selection of Apollo and Skylab crews. Further responsibilities followed in June 1971 when he became Deputy Director of Flight Crew Operations under Deke Slayton. The Soviets seem to have considered him a man of 'high spirits', though there was no doubting his calibre in an emergency. Curiously, Stafford's Houston boss, Donald 'Deke' Slayton was also assigned to ASTP. Selected as a Mercury astronaut back in 1959, Slayton had suffered the frustration of watching his comrades blaze their trails across the heavens and the headlines while he was grounded for a minor heart irregularity. Assigned a desk job with overall responsibility for crew selection and training, Slayton gained a reputation as a tough but fair task master and became an indispensable part of the NASA decision-making team. Yet he never abandoned the dream of flying in space one day. In March 1972, following a 'comprehensive review of his medical status' by NASA's Director of Life Sciences and the Federal Aviation Agency, he regained full flight status. This was too late for Apollo or Skylab, and the Space Shuttle was still at least seven years away, by which time he would be too old. ASTP was the one and only opportunity remaining for the 51-year-old former test pilot, and he jumped at the chance. There was apparently little resentment at his jumping the queue for the prized position of docking module pilot. The third seat would be occupied by 44-year-old Vance Brand, an astronaut since 1966 but participating in his first flight. Although he had worked as a flight test engineer and test pilot for Lockheed Aircraft Corporation prior to selection by NASA, Brand had left the Marine Corps in 1957 and so was the third civilian out of the five ASTP prime crew members. After serving on the Apollo 15, Skylab 3 and 4 back-up teams, Brand was given the consolation of ASTP. He was luckier than many of his group 5 astronaut contemporaries, who had to wait patiently for the Space Shuttle to give them their first opportunity in spaceflight.

Training for the joint mission was unlike that for any other flight before or since. Every few months the astronauts and cosmonauts would visit each other's country for training and familiarization with the spacecraft and new equipment. Housing for the Americans was specially built at Star City north east of Moscow: the Americans assumed it was bugged and acted accordingly. They soon found that if they needed anything, the most efficient way of passing on the request was within listening distance of the unseen microphones! One of the more serious aspects of the mission was the language difficulty. A single failure to communicate clearly during any part of the mission could jeopardize the entire project, so from early 1974 the crews underwent intensive teaching in the relevant language. Their teachers accompanied them in the gymnasium and at the briefing sessions as well as in the classroom, until both teams were fluent in the vocabulary needed for everything from eating and drinking to the most complicated spacecraft manoeuvre. As a precaution, however, all flight documents and primary ground and space controls were labelled in both Russian and English. After all, there was always the chance that the Soviets might misunderstand Stafford's version of Russian, christened 'Oklahomski' by the cosmonauts! Between working sessions, the crews had some time for relaxation. In Star City, the evenings almost invariably turned into vodka drinking parties in which the astronauts inevitably came off second best. During their visits to Texas, the cosmonauts attended numerous cocktail parties and ceremonies, displaying a way with words which the Americans could only stand back and admire. More than one Texan was forced to revise his opinion of Soviet Communists after watching Leonov, sporting a stetson hat, sing a number with a Country and Western band and gyrate with a belly dancer at the County Fair. Their favourite recreation in America, however, was to spend as much time as possible hunting, waterskiing and racing fast cars.

The final countdown to the mission was marked by a surprise climbdown by the Soviets. Having formerly insisted that no Americans could attend the Soyuz launch at Baikonur, the authorities relented and invited the American Ambassador, Walter Stoessel, and his wife; William Shapley, a NASA administrator; and American scientist Egon Koebner. Unlike the visiting astronauts, they were flown to the cosmodrome in daylight, were able to speak with the cosmonauts and see the launch pad for themselves. The Americans were surprised by the lack of restriction: such an interview and visit to an active launch pad would not have been possible in the USA, where there were strict precautions against infection or accident.

Final good wishes from the leaders of both countries and from Tom Stafford had been passed to the cosmonauts before they ascended to the tower to clamber into their craft on 15 July 1975. By the time the squat white arrow ignited amid a thunderous roar, the American crew were sleeping in their quarters, so the perfect launch was recorded for them to watch later. 'Let's go!' came the cry from Leonov: 530 seconds later, flight director and former astronaut Alexei Yeliseyev announced, 'The craft is in orbit.' As the solar panels extended like wings on either side of Soyuz, control passed to the Moscow centre. Already, another milestone in spaceflight had been achieved: for the first time a Soviet launch had been broadcast live to about 100 million Soviet viewers along with countless millions worldwide.

The tension now switched to Cape Canaveral. The Soviets had come through in style, and the pressure was on the Americans to do likewise. As Soyuz entered its second orbit, the three astronauts were roused. After the traditional breakfast of steak and eggs, it was time for the final medical and suiting operations. By now, the Soyuz crew had equalized pressure in both modules of their craft, removed their pressure suits and dried them prior to storage. An orbital correction raised their altitude to 140 miles (232 km) as the astronauts waited patiently in their craft for the final countdown. 'Looks like it's a good day to fly,' came Stafford's voice over the intercom. There had been thunderstorms all week, but today the sky was a brilliant blue. At 3.50 pm Cape time, seven and a half hours after the Soyuz launch, the biggest crowd of spectators and pressmen to gather at the Cape for many years cheered as the Saturn IB lifted slowly and majestically from Pad 39B. For many space buffs, it was a moment of nostalgia: no more would the Florida landscape reverberate to the roar of von Braun's babies. The Saturn era was over.

The final flight of Apollo began perfectly. As staging was completed without a hiccup, even the normally taciturn Slayton was moved to exclaim: 'Man, I tell you, this is worth waiting 16 years for.' The astronauts were safely in orbit, only 100 miles (160 km) above the Atlantic, below and behind their Soviet counterparts. It was the lowest ever orbit by a manned American craft, giving the crew an unusually rapid bird's eye view of the surface. 'The thunderheads seemed to reach a quarter way up to us,' recalled Deke Slayton later.

By now, partial depressurization of the Soyuz had been completed, in readiness for the three guests, enabling the cosmonauts to begin their onboard biological experiments with micro-organisms, fish eggs and fungi. One problem had arisen with a TV antenna intended to broadcast scenes from the interior of Soyuz. Troubleshooters on the ground worked out a way of repairing the connection using sticking plaster from the first aid box, but it proved only partially successful and resulted in the crew starting their sleep period 90 minutes late.

Meanwhile the first major task of the Apollo crew, the transposition, docking and withdrawal of the docking module from its housing in the Saturn's second stage, had been successfully accomplished. Apollo was now free to commence the Soyuz chase through a series of short main engine burns. The routine was broken, however, by a series of minor mishaps. When Stafford reported he had discovered a 'super Florida mosquito' flying around the cabin, it was the cue for much hilarity and wise-cracking. Slayton delivered a 'scientific report' on the giant stowaway, suggesting they should bring it home and award it a pair of astronaut wings. More serious problems included a helium bubble which had to be purged from the fuel lines, a slight overheating of the command module, and a partial blockage in the urine waste duct. The difficulty which prevented the Americans from beginning their sleep period on schedule was, however, an old friend from the lunar missions. Vance Brand entered the tunnel to stow the docking probe and drogue, but found he could not remove the assembly with the correct tool because incorrectly installed electrical wiring was in the way. There was

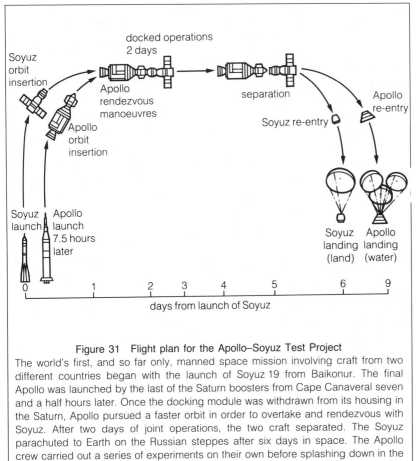

Figure 31   Flight plan for the Apollo–Soyuz Test Project

The world's first, and so far only, manned space mission involving craft from two different countries began with the launch of Soyuz 19 from Baikonur. The final Apollo was launched by the last of the Saturn boosters from Cape Canaveral seven and a half hours later. Once the docking module was withdrawn from its housing in the Saturn, Apollo pursued a faster orbit in order to overtake and rendezvous with Soyuz. After two days of joint operations, the two craft separated. The Soyuz parachuted to Earth on the Russian steppes after six days in space. The Apollo crew carried out a series of experiments on their own before splashing down in the Pacific three days later on 24 July 1975.

no panic in mission control: it turned out that engineers were very familiar with this particular assembly since it had caused a lot of trouble during the Apollo 14 mission and had been returned to Earth for a thorough inspection. Flight director Neil Hutchinson told the crew to get some sleep while his team analysed the problem on the ground.

The overnight investigation proved fruitful the next morning. Brand followed his instructions on how to clear the wire and remove the probe, completing the job within half an hour. Breakfast had not been quite as straightforward: a plastic bag of strawberries burst, and Stafford spilt some orange juice, causing him to quip, 'We'll have a psychedelic spacecraft when we get back.' As Apollo gradually moved towards her quarry, the two crews continued routine housework and experiments. Slayton checked over the docking module, encountering some trouble in extracting a frozen sample for a medical vaccine (electrophoresis) experiment and a stuck door on the small electric furnace. Both craft were able to send back TV pictures, while Leonov and Kubasov were delighted to exchange greetings with their Soviet comrades in the Salyut 4 space station.

With their days still not synchronized, the Soyuz crew went to sleep some five hours earlier than the Americans. Unfortunately for the Apollo crew, fate seemed to have decreed that they would not enjoy their full sleep period. Stafford had found a niche in the command module, Brand in the tunnel and Slayton in the docking module. After about 5 hours, the master alarm sounded. As so often in the past, it turned out to be a false alarm, but they decided not to try sleeping again. As Brand commented, in a classic understatement, 'That sort of alarm sure wakes up a crew . . . and makes them very alert.'

The historic link-up was scheduled for the mission's third day. Failure now would negate all that had gone before. An altitude correction put Apollo on to a rapidly overtaking path, like a runner on the inside track. Soon afterwards, Vance Brand told Capcom Richard Truly, 'OK Dick. We've got Soyuz in the sextant.' Still more than 200 miles (320 km) apart, the Soviet craft appeared as 'just a speck, hard to distinguish from the stars'. Like a rider jockeying for position, Stafford gradually guided Apollo up and ahead of the target over the next few orbits. Voice communication was established, then radar contact. Both craft were now within range of the ATS-6 communications satellite, orbiting 22,300 miles (33,680 km) above the Earth. From this vantage point, the satellite could broadcast messages to nearly half the globe, enabling communications links for about 50 minutes in each 90 minute orbit. It was yet another historic first in the story of manned spaceflight.

With the cosmonauts, now suited in case of emergency, watching and waiting, Apollo braked and closed in. Dramatic TV pictures showed the ant-like Soyuz drifting just a few hundred yards away. Then came the news the five spacemen were waiting for: 'Moscow is Go for docking. Houston is Go for docking. It's up to you guys. Have fun.' The mission control centres in both countries were packed with excited onlookers, including the ambassadors as political representatives. Using manual control, Stafford edged towards the open petals on the nose of Soyuz. '5 metres, 3 metres, 1 metre,' came the count, then a jubilant cry in Stafford's best Oklahoma–Russian accent, 'Contact!' Leonov replied, 'Capture' in well-practised English. Stafford concurred: 'We also have capture. We have succeeded. Everything is excellent.' The Soviet commander was clearly delighted: 'Soyuz and Apollo are shaking hands now . . . We're looking forward now to shaking hands with you on board Soyuz.'

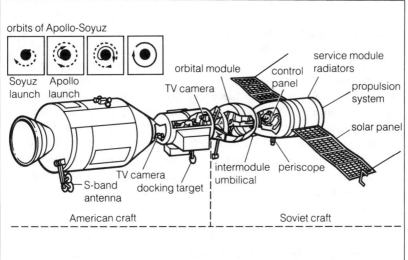

orbits of Apollo-Soyuz

Soyuz launch Apollo launch

orbital module

TV camera

control panel

service module radiators

propulsion system

solar panel

TV camera

S-band antenna

docking target

intermodule umbilical

periscope

American craft

Soviet craft

Figure 32 Apollo–Soyuz link-up
Soyuz was designed as a ferry craft between the ground and orbiting space stations. Introduced in 1967, it consisted of three modules: the orbital module was used for both work and rest; the descent module contained the main controls and crew couches and was the only section to return to Earth; the service module contained the power, communications and propulsion systems. Apollo was designed as a lunar mission craft. Also introduced in 1967, it consisted of two modules: the command module was the crew area, while the large service module at the rear contained the main spacecraft sub-systems. The docking module was designed by NASA to be compatible with both craft. Soyuz was about 23 ft (7 m) long and weighed about 6 tons. Apollo was 34 ft (10 m) long and weighed about 13 tons. The docking module was 10 ft (3 m) long, 5 ft (1½ m) wide and weighed 2 tons.

The final insertion of Apollo's probe into the Soyuz docking ring had taken place above the Atlantic on the Soviets' thirty-sixth and the Americans' twenty-ninth orbits. They would remain in each other's embrace for the next two days. However, when the hatches were opened to the docking module 20 minutes after docking Stafford reported a smell 'like burnt glue, something like acetate' and a burning sensation in the eyes, so the crew prepared to close the hatch and don their oxygen masks. Hasty check-outs in Houston found nothing out of order, leading to the conclusion that some Velcro had been scorched earlier in the day by stored samples from the furnace.

The unpleasant smell slowly cleared, enabling Stafford and Slayton to shut themselves in the docking module and alter the atmosphere to match that inside Soyuz. The Soviets had removed their pressure suits, but the battery of instructions from the ground and the rather chaotic condition of their cramped cabin led to increased tension in relations between the crew and the ground. After all the protocol that had been drummed into them, the five men seemed uncertain how to proceed once Stafford cracked open the hatch and peered into the Soyuz cabin. 'Looks like they got a few snakes in there too,' joked Stafford as he saw the TV cables and power lines floating in the background. 'Come in here,' he called to the unusually reticent Leonov. Viewers across the world watched as the commander's rotund face, partly obscured by his helmet, appeared at the tunnel entrance. Hands from east and west finally grasped in the time-honoured gesture of friendship. 'Glad to see you,' came the Russian–English greeting. It was 3.19 pm Cape time, 10.19 pm Moscow time: 140 miles below them was the beautiful Dutch city of Amsterdam.

According to the schedule, the Americans were to accept Soviet hospitality on this first meeting, so four men squeezed themselves around a small green table in the Soyuz orbital module to hear congratulatory messages from Chairman Brezhnev and President Ford; Vance Brand remained behind in Apollo as duty supervisor. Slayton loaned his headset to Leonov so that the Soviet commander could speak with the President and give his formal reply. When asked if he had any advice for prospective future astronauts, the grey haired Slayton replied: 'Decide what you really want to do, and never give up till you've done it.' The speeches over, it was time for exchanges of gifts. Each crew carried ten national flags; in addition, a UN flag carried in Soyuz was handed over for the return leg in Apollo. Commemorative medals and plaques were exchanged, and certificates jointly signed. As a final gesture of goodwill, the cosmonauts were presented with a packet of fast-growing pine tree seeds.

Meal time resembled a Christmas party without any alcohol, though Leonov did his best to provide a substitute by producing tubes of borscht soup covered with vodka labels for his guests. Their main meal was the traditional festive dish of turkey and cranberry sauce, while the cosmonauts tucked into lamb soup followed by chicken. They paused only to take occasional photographs or pose for Leonov's sketched portraits. Several hours after the initial handshake, it was all over. The hatches were sealed behind the retreating astronauts, and, after a false alarm over a slight gas pressure drop, the crews settled down for their rest period.

The second day in docked configuration continued the public relations exercise begun the previous day. There was a succession of crew transfers between craft. Accompanied by Slayton, Brand gained his first close-up view of Soyuz while Leonov joined Stafford in Apollo. 'Howdy partner,' came the breezy greeting as the cosmonaut crossed the threshold. Diplomacy was the order of the day as the crews gave their counterparts and TV viewers on Earth the obligatory guided tours of their craft. Brand could be heard praising the Soyuz as Kubasov showed him around, though Stafford could not help pointing out to Soviet viewers the role of Apollo in the moon landings and record-breaking Skylab missions. Kubasov, in turn, showed his patriotism by reminding the American audience: 'Our country occupies one-sixth of the world's surface. Its population is over 250 million people . . . At the moment, we are flying over the place where Volgograd city is. It was called Stalingrad before. In winter 1942–43, German fascist troops were defeated by the Soviet army here. 330,000 German soldiers and officers were killed or taken prisoner here.' He then presented Brand with a medal similar to those of his two companions, before the two men swapped craft.

Following his steak meal on board Apollo, Leonov accompanied Stafford back to the Soyuz for the afternoon press conference. Many of the spacemen's replies to questions about the future of superpower cooperation were optimistic, though Stafford showed how well he appreciated the realities of detente: 'How this new era will go depends on the determination, commitment and faith of the people of both our countries and the world.' Kubasov reassured his audience: 'We have found out that we can work together in space and co-operate . . . We have complete mutual understanding and everything we have planned we have accomplished.'

Vance Brand had been allotted the task of describing the Florida fly-over, but the highlight of the session was Leonov's astronaut portrait exhibition. More tree seeds were presented, this time to Stafford, and the two halves of a special flight medal were ceremonially joined. The broadcast closed with unscheduled speeches of friendship in each other's languages by the two commanders. 'I'm sure our friendship will continue after this flight,' predicted Stafford. 'We love each other very, very much,' concurred Leonov.

Yet again, the gremlins in Apollo refused to allow a full night's sleep after the hectic diplomatic activity. Only two hours after settling down, they were rudely awakened by an alarm. The spacecraft's atmosphere was slowly venting into space through the waste management valve which had inadvertently been left open. The second interruption half an hour later involved transmissions from air traffic control at Atlanta Airport being picked up on the Apollo communication circuit!

After some five hours of sleep, the astronauts were up and preparing for a busy day of docking experiments. With the Soyuz crew once more dressed in their pressure suits, Deke Slayton unlatched the two craft and fired the thrusters to back slowly away. By delicate manoeuvring, he was able to place Apollo between Soyuz and the sun, enabling the cosmonauts to photograph an artificial solar eclipse. The Soyuz docking unit was then extended while Apollo's equipment was fully retracted so that the Soviet craft could become the 'active' partner as Slayton closed in for the second docking. With the brilliant sun reflecting from the green and white Soyuz and obscuring the docking target, and Slayton complaining that the Soviet craft was lying at the wrong attitude, the approach and capture took much longer than anticipated. Apollo hit the Soyuz slightly off centre, resulting in a rough docking and hasty requests for damage checks by Moscow.

After a hasty meal, it was time for the last separation. The cosmonauts once more settled down in the Soyuz descent module prior to undocking, then Slayton pulled away for the final joint experiment. The American craft was to aim light beams at reflectors located on the Soyuz and pick up the returned beams without contaminating the intervening space with gas from the thrusters. The intention was to measure the amounts of oxygen and nitrogen in space by means of a spectrometer on board Apollo. It was, however, a highly complex manoeuvre which gave Slayton and his companions many headaches before reaching a successful conclusion.

It was time for final farewells. As Apollo moved up and behind Soyuz, allowing the Soviet craft to steadily drift ahead, Leonov radioed, 'Thank you very much for your big job.' The cosmonauts continued their fungi experiments and carried out a springclean of their cabin, then retired early in order to adjust their internal clocks to Moscow time once more. Their American comrades had a longer day of experiments ahead, followed by another four days with up to 17 hours each day scheduled for their scientific programme. Apart from work with the furnace and the separation of living cells, the crew made the first observations of the universe at extreme ultraviolet wavelengths, and made numerous photographic and observational studies of the Earth's surface, ranging from deserts to icebergs, from oceanographic surveys to spotting oil slicks and water pollution.

On awakening, the cosmonauts continued their programme of Earth and solar observations. Their craft was checked and the main engine fired to lower the orbit as a preliminary for re-entry the next day. They enjoyed another chat with their less celebrated comrades in the Salyut 4 space station, but mission control was becoming concerned over tiredness. Medical sensor readings also suggested a slightly irregular heartbeat from Leonov. The commander protested: 'I don't feel any fatigue, I don't have a headache. Everything is normal.' Mission control was not convinced and ordered him to take a dose of calcium to steady his metabolism.

The morning of 21 July saw the cosmonauts stow all loose equipment in the descent module and don their pressure suits prior to sealing the hatch. On the ninety-sixth orbit, with final instructions from ground control completed, the automatic re-entry sequence began with a three minute engine burn aimed at braking the craft. Soon after, the orbital and service modules were jettisoned, leaving the crew in the spherical module designed to withstand the tremendous frictional heating of re-entry. With millions of Russians watching the first ever live broadcast of a spacecraft returning to Earth, the communications with the crew were lost as the craft was enveloped in a ball of flame. Right on schedule, the main drogue opened, followed by the single red and white striped main parachute. Television cameras carried by a fleet of helicopters zoomed in on the slowly descending capsule. The heatshield was jettisoned to expose the retro-rockets, the only defence against a shuddering crash-landing for the two occupants. Viewers' pulses momentarily raced as the capsule was obscured by a cloud of dust on contact with the dry, barren steppe. The module reappeared moments later, now resting on its side, but safe and sound. The retro-rockets had fired as planned, only a few feet above the surface, to cushion the touchdown. As wild applause and backslapping broke out in both Moscow and Houston, the hitherto empty landscape was suddenly inundated by crowds of people who appeared from nowhere in helicopters, trucks and cars. The hatch opened to reveal the two returning heroes. Kubasov clambered out first, pushing back his helmet visor as the rescue crew embraced him. Leonov emerged a few minutes later, waving aside offers of assistance but staggering slightly in the unfamiliar gravity. The men completed the ritual by signing their names in chalk on the side of the capsule; Leonov added the word 'Thanks'. They were then flown back to Baikonur for a news conference.

At the moment of impact, the Apollo crew were still asleep: 15 minutes later, they were awakened with the good news. Stafford commented: 'Give them my best. Sure glad to hear everything's OK.' Moscow controllers reciprocated by sending best wishes for the successful completion of the American flight, while Academician Boris Petrov pointed out the significance of seven men being in space at the same time, the largest number up to that date.

The next two days passed relatively peacefully as the American crew continued their experiments. One of the highlights was a TV news conference in which the participants at both ends could speak directly to each other. The crew admitted to problems of overwork and overcrowd-ing in the cramped spacecraft, despite the bonus of the extra room in the docking module. 'We were always bumping into each other,' commented Brand. 'We needed a traffic cop.' Nevertheless, they had succeeded in retaining their sense of humour. When asked at what age a man would be too old to fly in space, Deke Slayton replied that his 91-year-old aunt would have no problem. Some six hours later, the unique docking module was jettisoned ready for orbital tracking in an experiment to refine existing knowledge about the Earth's gravitational field. The day ended in a lighter vein as they attempted to measure each other's inside leg for analysis by space doctors in Houston.

Homecoming day was 24 July. The morning was spent completing the experimental programme and tidying up the command module. During their chores, they discovered some film cassettes inadvertently left behind by the cosmonauts, so the crew promised priority treatment for their Soviet friends' property after splashdown. On orbit 138 the main engine fired perfectly over the Southern Ocean, slowing the craft for its fiery descent. The service module was jettisoned seven minutes later, exposing the blunt heat shield for the coming ordeal. The shield functioned exactly as intended, but as the craft reappeared after radio blackout, all was not well with the crew. As Stafford read out the final checklist during re-entry, Brand missed an instruction to activate the automatic landing system switches designed to deploy the parachutes at the correct altitude. When the parachutes did not appear on schedule, Slayton told Brand to deploy them manually. However, the thrusters were still programmed to fire automatically and stabilize the craft, so Stafford decided to cut off their fuel. Unfortunately, the nitrogen tetroxide oxidizer used in the thrusters continued to escape from the jets and through a pressure equalizing valve into the spacecraft. As they descended the last few miles towards the Pacific, the astronauts began inhaling this highly poisonous gas. Unknown to the watching millions, the men were coughing violently as they struggled to stay alive. Slayton turned the oxygen flow rate on full in a desperate attempt to dispel the gas, but the crew were in a bad way when Apollo hit the waves 'like a ton of bricks' and turned upside down. Stafford, seated in the centre couch, undid his straps and promptly fell forwards into the docking tunnel, but succeeded in grasping the emergency oxygen masks. There was utter chaos in the cramped cabin as Stafford and Slayton clasped the life-giving oxygen to their faces and tried to revive Vance Brand who was already unconscious. It seemed an interminable 40 seconds before Brand at last revived. An angry Stafford shouted to the frogmen splashing around outside: 'Get this f...ing hatch open!' Finally, he did the job himself, enabling the three of them to gulp in the fresh sea air. The first Apollo had killed three men and the final mission had almost achieved the same dreadful notoriety.

The ceremonies went ahead as planned. Forty-five minutes after splashdown the module was winched aboard the USS *New Orleans* with the crew still inside. Unaware of the flirtation with death, NASA officials told the world of another flawless Apollo splashdown. After the brass band welcome, the crew, still rubbing their eyes, accepted the traditional congratulatory telephone call from the President, but, once below decks, they wasted no time in familiarizing the doctors with their ordeal. All celebrations were immediately cancelled as they were confined to bed and underwent intensive medical examination. On reaching Honolulu, the crew were installed in an Army hospital under intensive care, and only released for the triumphant return to Houston after two weeks recuperation. The doctors reported that all three had suffered from blistered lungs, but further bad news awaited Slayton. X-rays showed a small shadow on his left lung; apparently it had been there before the flight, but no one had noticed, or at least no one had said so. An exploratory operation lifted the gloom, however; it was a benign lesion, with no signs of cancer. Slayton was able to accompany his fellows to Russia in September for the reunion with comrades Leonov and Kubasov and a meeting with Soviet Communist Party Chairman Brezhnev. The cosmonauts returned the compliment the following month.

Not surprisingly, Deke Slayton never flew again in space. He was appointed manager for the approach and landing test programme of the Space Shuttle, later going on to oversee the first Shuttle operational missions. Approaching his fifty-eighth birthday, he resigned from NASA to become a vice-chairman of Space Services Inc., a private company aiming to send up satellites, including an imaginative scheme to launch urns filled with people's cremated ashes into outer space. Tom Stafford, promoted to major general, resigned from NASA to pursue his Air Force career. He took command of the Air Force Flight Test Center, going on to become a Deputy Chief of Staff before resigning in 1979. He is currently vice chairman of Gibraltar Exploration Ltd in his home state of Oklahoma. Vance Brand remained with NASA during the lean years, eventually flying in space more than seven years later as commander of the fifth Shuttle mission.

Alexei Leonov never flew in space again, having left his substantial imprint on space history. He is now deputy head of the Yuri Gagarin Cosmonaut Training Centre. Valeri Kubasov commanded Soyuz 36 on a visit to Salyut 6 in 1980 with a Hungarian co-pilot as guest. All five men remain good friends and met for a reunion in Washington exactly ten years after their link-up, later meeting Soviet President Gromyko to discuss Soviet-American cooperation in space.

As for ASTP, it has so far proved a dead end. The space cooperation treaty expired in May 1977, but its successor fell foul of the freeze in relations between the two superpowers. Only in the last couple of years has there been any suggestion of the possibility of another joint mission. The docking module, successful though it was, proved a one-off construction since the Apollo spacecraft never flew again. In this sense, the Soviets did not gain any significant intelligence concerning American technology; all of the knowledge they gained could have been acquired from open literature anyway. The Americans gained some useful insights into the hitherto secret Soviet space programme, and suffered quite a shock when they realized how unsophisticated were the methods employed by the country which had sent up Sputnik and Gagarin and had seemed poised to dominate the space race in the sixties. Perhaps the greatest achievement of ASTP was, however, to show people on both sides of the Iron Curtain that human fears, emotions and behaviour are similar no matter what the ideological background. For a brief span of time, there was a bond of friendship and cooperation which transcended any arbitrary political boundaries.

14

The Longest Journeys
The Salyut Space Stations, 1974–85

Since the early days of manned spaceflight, travellers returning from orbit had explained to sceptical Earthbound audiences that the human eye could pick out minute detail on the surface of the planet. To imaginative military intelligence staff, however, this was a piece of information of more than passing interest. Since most American manned missions flew orbits of low inclination to the Equator, the potential spies in the sky passed over the less strategically important tropical regions, but for the Soviets it was a different proposition. With a typical orbital inclination of more than 51 degrees, it was relatively easy to disguise a photographic reconnaissance mission as a routine Earth resources survey while the craft drifted innocently over the USA and its NATO allies in Europe. The Salyut 3 space station, launched on 25 June 1974 into a low 150 mile (240 km) orbit, aroused suspicions over the nature of its experiments among Western observers when two military officers were launched aboard Soyuz 14 to rendezvous with it.

Commanding officer was Colonel Pavel Popovich, making his second flight after a 12 year interval, and alongside was 43-year-old air force lieutenant colonel Yuri Artyukhin, a cosmonaut since 1963 but making his first venture into space. A little over a day after launch on 3 July 1974, the two officers successfully docked with their new home, only the fourth time this had been achieved by the Soviets. It nearly proved to be a short-lived stay as the sudden appearance of solar flares sent lethal radiation streaming out into space, leading some experts to recommend a quick return. Close monitoring of onboard radiation meters, however, showed the men were not in danger, partly due to their low orbit, so the crew were able to settle in and take stock of their new surroundings.

Viewed from the outside, the new Salyut differed from its predecessor in having three rotatable solar panels in T formation so that there was no longer any need to manoeuvre the station in order to point the panels at the sun. The 18 ton Salyut was much smaller than Skylab, with two modules subdivided into work and rest areas totalling only one-third the internal volume of the American station. The living area included the 'bedroom', a kitchen equipped with a stove for heating their soups and drinks and a small table fitted with clamps for holding down crockery, a tape recorder for listening to music, a library and a chess set. To make life more comfortable and make adjustment easier, the station also contained a shower and a new scheme of interior decoration – the 'floor' was painted a dark colour while the 'ceiling' was lighter.

Exercise was considered a vital part of the flight plan, for the Soviets had much less experience of long term weightlessness than the Americans. Four daily periods of exercise totalling two hours were scheduled, using a moving-belt running track and a special sweatshirt with elastic cords designed to place loads on the mens' bones and muscles as they ran on the spot. Another innovation were the two large spherical air-locks positioned beside the reconnaissance camera and suitable for ejection of body wastes or of film capsules.

Soviet news releases concentrated on the economic value of the onboard experiments as the flight progressed. Long photographic sessions were intended to examine geological structures in a search for new mineral deposits, soil resources and glaciers. Western observers were not convinced, especially when it became clear that a so-called solar telescope seemed to spend most of the time pointed towards Earth, and that some ground to space communications were being conducted in code. There was, however, some attempt at serious scientific observation, notably with a spectrograph intended to study changes in the heat balance of the atmosphere due to aerosol particles, smoke and dust. Other studies included a 'microbiological cultivator' full of bacteria and a system for recycling water. Much of the 15 day mission was necessarily taken up with checking out the new space station, helped by discussions with engineers on the ground equipped with a full-scale replica of Salyut. As the flight drew to an end, the crew stored their experimental results, film and log books, then shut down some systems and left the station in an automatic mode ready for the next occupants. They soft-landed safely on the Soviet steppes on the afternoon of 19 July, completing a highly successful mission which the authorities hoped would be the forerunner of a much-needed string of Soviet triumphs.

The expected launch of a Salyut relief crew duly materialized on 26 August, when Soyuz 15 carried another all-military duo towards the space station. In command was 32-year-old air force lieutenant colonel Gennadi Sarafanov, a cosmonaut since his recruitment in 1965 at the age of 23. His flight engineer was 48-year-old air force colonel Lev Demin, the oldest space rookie up to that time and the first grandfather to fly in space. He trained as a pilot but was told to give up hopes of a flying career due to shortsightedness, so he switched to engineering and concentrated on research.

Soyuz 15 reached orbit on schedule, but over the next 24 hours everything went wrong. Observations showed that, instead of closing on the space station from behind, the ferry craft approached too quickly and had to pull away to a safe distance. Apparently, the automatic docking system had malfunctioned, causing the craft to veer out of control and preventing a manual docking. At 8 am Moscow time on 28 August, Tass announced that, 'Having approached Salyut 3 many times, the space craft is being prepared to return.' The re-entry capsule completed its emergency landing at night.

With the joint Apollo–Soyuz flight approaching, the Soviets hastened to reassure the Americans that it was a minor hiccup with a battery-powered version of the Soyuz and it would not affect the forthcoming flight. As usual, the authorities were reluctant to go into details, but it was eventually explained by Vladimir Shatalov that the craft had been testing a new automatic remote-controlled approach and docking system which would one day be installed in an unmanned supply craft. As for Salyut 3, it had completed its useful life. A data capsule was ejected from an air-lock to be recovered in the Soviet Union on 23 September, further evidence that a spying mission had been continuing under ground control since the Soyuz 14 crew had departed. After further tests of its propulsion and navigation systems, the space station was commanded to re-enter over the Pacific Ocean on 24 January 1975 and burnt up in the atmosphere. The Soviets had still not succeeded in sending more than one crew to an orbital station, but they intended to continue trying. Even before the destruction of Salyut 3, the next space station in line had been placed in orbit.

Salyut 4 was launched from Baikonur into a low orbit on 26 December 1974 but its altitude was soon raised to about 220 miles (350 km), indicating that its primary purpose was rather different from that of its predecessor. Sixteen days after the space station entered orbit, its first occupants were sent up aboard Soyuz 17 in another night launch. The crew's composition also pointed towards a more scientific flight programme: although the commander was an air force colonel, 43-year-old Alexei Gubarev, his flight engineer was a leading civilian engineer and spacecraft designer, Georgi Grechko. The bushy-haired Grechko had learned his trade under chief designer Korolev and had gone on to make a name for himself in his work on the Luna soft landers.

The two cosmonauts docked with Salyut 4 on 12 January. On entering the station they found a note left 'on the doormat' by Soviet engineers: 'Wipe your feet.' Their new home was of a similar size and outward appearance to its predecessor, but was decked out in soft green, yellow and blue shades, and fitted with equipment similar to that used in Skylab. Communication with the ground could be achieved through a teleprinter, including transmission of personal messages; the station was automatically guided by an 'autonomous navigation system'; and there was a water-regeneration system which extracted water vapour from the cabin atmosphere and provided water for drinking and washing. Adaptation to weightlessness proved more of a problem than expected, however, despite the provision of a bicycle ergometer in addition to the treadmill and muscle-loading suits. Nearly a whole week passed before the cosmonauts were said to be fully adapted – Gubarev's complaints that he was suffering from a common cold were put down to zero gravity by ground controllers. Cosmonaut Nikolayev explained:

> Doctors believe that this [adaptation] could have happened earlier if the cosmonauts had not assumed voluntarily certain additions to their normal work load and if they had followed more rigidly the recommended sleep periods which they shortened during their first few days in space because of a great amount of work to do on board.

The crew were ordered to hold in check their 'irresistible and avid urge for work'.

The busy schedule was brought about by the wide range of experiments carried on the station and requiring attention in addition to the two and a half hours of daily exercise and the numerous household chores. One 'Oasis' experiment included attempts to grow peas in a special cultivator, a necessary forerunner of the space greenhouses which would be needed on long interplanetary missions. Physics and astronomy were covered by experiments with a solar telescope, an infrared telescope–spectrometer for studying the Earth's atmosphere and two X-ray telescopes designed to detect celestial radiation sources. These were carried out with a minimum of fuss or publicity, causing the mission to be virtually ignored by the Western media. At one stage, Grechko gently reprimanded ground controllers who had switched on a TV camera: 'Publicity isn't the best thing during our work.' Four weeks passed, enabling the crew to exceed the flight time of the Skylab 2 mission and become the third most experienced spacemen in history; then observers detected preparations for a return to Earth. With their precious cargo of scientific results and film carefully stowed in the Soyuz, the men returned to a bumpy landing in the secondary recovery zone north east of Tselinograd. Wind speed exceeded 40 mph, cloud cover was low and visibility was poor, but the landing was sufficiently accurate to enable a rapid retrieval of the crew. The two cosmonauts had been told to wear their gravity suits during re-entry to guard against blackouts, and they emerged tired but triumphant. Gubarev had lost about 6 lb in weight and Grechko about 11 lb, deficits which it took about a week to make up. It took considerably longer for them to recover normal vitality.

The unspectacular, trouble-free success of Soyuz 17 led to Soviet hopes that the next mission would lay to rest the bogy which seemed to dog all relief crews. Soyuz 18 was launched on 5 April with two experienced cosmonauts on board, air force colonel Vasily Lazarev and civilian flight engineer Oleg Makarov. Nine minutes later, it was all over. The first stage with its strap-on boosters had fired normally and shut down on schedule, but, unknown to the crew, three of the six latches which join the lower and upper stages fired prematurely, severing a vital electrical connection. It was now impossible for the remaining three latches to be fired, so the lower stage was still dragging along beneath the second stage when the latter's engines fired. The whole craft began to veer rapidly off course, a deviation detected by the booster's gyroscope. After only four seconds of operation, the second stage engines were automatically cut off and the first recorded launch abort began. The spacecraft shroud and descent module of the Soyuz were separated from the booster, leaving the surprised crew at an altitude of 90 miles with insufficient velocity to attain orbit.

The crew were now 1,000 miles downrange and condemned to a low velocity, high stress re-entry with a potentially disastrous landing at the end of it. One of their main worries was the possibility of a touchdown in China, causing them to seek repeated assurances from Moscow that they would avoid this misfortune. Despite the inbuilt lift provided by their capsule, the men were forced back into their sets with a numbing 14 to 15G, unable to do anything but trust to luck and the design of their automatic re-entry system. Their craft plummeted to Earth near the Siberian town of Gorno-Altaisk in the foothills of the Altai mountains, some 200 miles from the Chinese border, but the parachute and braking system worked perfectly. The shaken crew staggered from the descent cylinder into a chilly Siberian wilderness, and prepared for a long wait. They lit a fire to keep warm and help location by search parties. Once found, they were whisked away to Baikonur and then to Moscow for medical treatment and debriefing. Soviet authorities declared the men were both unhurt, though it seems hard to believe there were no adverse effects of their ordeal: one unofficial account told of how their capsule rolled down a steep slope on hitting the ground, and came to rest teetering on the edge of a precipice held only by its parachute lines which had caught in some pine trees. Whatever the truth of this, it was the end of the line for 47-year-old Lazarev. He retired from active duty to help train future cosmonauts, including the foreign 'guests' selected in the late 1970s. Makarov was able to return to the flight list, and eventually succeeded in boarding a Soviet space station in 1978.

Meanwhile, the Soviets had to explain yet another embarrassing failure to their American partners only three months before commencement of the joint project. They confided that the faulty booster had been of an old, outmoded design and that it could not adversely affect the forthcoming mission. Assurance was also given that no problems would arise over Soviet capability of controlling two simultaneous flights, since a new attempt to rendezvous with Salyut would now have to overlap with the Apollo–Soyuz Test Project (ASTP). They would simply use their newly commissioned control centre in Moscow in conjunction with the old centre in the Crimea.

As if to wipe the slate clean and ignore the ignominious failure in April, the next launch was also designated as Soyuz 18 – only success counted in the minds of the Soviet authorities. Another experienced crew, the back-up on the recently aborted flight, was launched towards Salyut 4 on the afternoon of 24 May. The new commander was Pyotr Klimuk, now promoted to lieutenant colonel, while his civilian flight engineer was Soyuz 9 veteran Vitali Sevestyanov. To everyone's relief, the launch phase went according to plan, enabling the crew to carry out a routine rendezvous and docking on 26 May. When the crew entered the mothballed Salyut, they found a message left by the previous occupants and intended for Lazarev and Makarov: 'Welcome to our common home'.

After a few days checking systems and reactivating the experiments, the crew settled down to continue the work begun by the first occupants. Blocks of time, often lasting several days, were allocated to specific tasks, such as Earth resources photography, solar activity observation, and medical tests. This type of scheduling was less flexible than that applied to the American Skylab and regarded as only half as efficient by Western experts. Nevertheless, when interesting solar activity was reported by the Crimean Astrophysical Observatory, the crew were able to switch their attention to the onboard solar telescope.

Soviet determination to extend their space endurance record and widen their experience combined with a desire to gain further prestige from the ASTP mission. The crew were instructed to wear their special load suits at all times while they were awake, and to increase their exercise periods. The Soviet 30 day record was passed at the end of June but was ignored by the Soviet press in a highly unusual show of restraint. On 8 July Sevestyanov celebrated his fortieth birthday with a meal that included spring onions grown on board. Less successful were the experiments with peas in the Oasis cultivator: most of them sprouted but died within four weeks. Sevestyanov blamed the hot TV lights during a live broadcast, but it was later revealed that all was not well with the Salyut air conditioning. The resultant rise in humidity seems to have caused windows to mist over and walls to run with condensation. Some reports suggest that the crew were driven to repeated requests for an early termination of their discomfort, only to be turned down. Whatever the truth concerning their apparent distress, the two men were able to view the launch of Soyuz 19 (the ASTP craft) at the end of their fifty-second day in orbit, and later exchanged greetings with their comrades prior to the docking of Soyuz 19 with Apollo 18. For the next few days, space was more densely populated than ever before as four Soviets and three Americans travelled around planet Earth. Academician Boris Petrov dubbed them 'The Magnificent Seven'. Soyuz 19 returned to Earth on 21 July followed by Apollo on 24 July. As the Americans bade farewell to manned spaceflight for the next six years, the two cosmonauts on Salyut 4 began to pack their bags and close down the space station. Undocking and re-entry were successfully completed two days later.

After 63 days of weightlessness, no one was sure what their condition would be, so doctors ordered the cosmonauts to wear vacuum suits during re-entry in order to prevent pooling of the blood in their legs. In an unprecedented gesture, the second live broadcast of a Soyuz landing filled national TV screens to show the Soviet people and the world the triumphant culmination of another technological and scientific break-through. Doctors equipped with stretchers rushed to the landing site, but the cosmonauts insisted on walking out to greet the media. Nevertheless, both men were to suffer feelings of dizziness, nausea and general weakness over the next few days and several days passed before they could take a ten minute walk. Doctors were generally pleased with their rate of recovery, which was said to be faster than that of the Soyuz 17 crew. One cause of concern was the major loss of haemoglobin from their blood, a 25 per cent decrease in the case of Sevestyanov. The engineer and broadcaster retired from the active list to work on improving Soyuz systems, while his younger partner eventually returned to active duty.

Meanwhile, the ageing Salyut 4 continued to orbit and carry out a variety of manoeuvres and observations under automatic control from the ground. More significant was the docking of an unmanned craft, designated Soyuz 20, with the space station on 19 November. Although not the first time the Soviets had demonstrated an automatic rendezvous and docking capability, it was, however, a major step towards the perfection of a robot supply ship which could be used to support long endurance crews on future Salyuts. On this occasion, there was no transfer of fuel, though it did carry a biological cargo which included tortoises and bulbs. The two craft remained locked together for 90 days before the Soyuz returned to Earth on 16 February 1976. Salyut lingered in orbit for another year, its altitude slowly decaying through atmospheric drag. Shortly before natural re-entry was due, ground controllers fired the main engine for the last time to give her the coup de grace. After more than two years of service, Salyut 4 broke up harmlessly over the northern Pacific on 3 February 1977.

Salyut 4's successor had been in place for more than seven months by the time it fell from orbit. Few details of Salyut 5 were released by the Soviet authorities, leading to incorrect speculation by Western experts that it had two docking ports to accommodate supply ships like Soyuz 20. Two weeks later, on 6 July 1976, an all-military crew of colonel Boris Volynov and lieutenant-colonel Vitali Zholobov was sent up in Soyuz 21 in order to activate the new station, confirming suspicions that Salyut 5 might be another reconnaissance platform for the Soviet armed forces. Additional clues were the low 150 mile (240 km) orbit and the initial training of Zholobov as back-up for a mission to Salyut 3. The crew were confidently expected to break the 63-day Soviet record and stay on until they passed the Skylab all-time record. To help them achieve these targets, Salyut 5 carried a moving treadmill instead of a bicycle ergometer, as well as dumb-bells and the usual load suits.

Apart from a couple of hours' exercise each day, the crew's time was spent carrying out an extensive programme of military and scientific experiments. One early task was to monitor Soviet air and sea man-oeuvres off eastern Siberia in an evaluation of space spying tech-niques. Other activities were less controversial. Extensive photo-graphy of the Soviet Union was carried out with the aim of discovering new mineral deposits and geological structures, including a survey of the proposed route for a new Baikal–Amur railway. Using a hand-held spectrograph, the cosmonauts studied atmospheric pollution, while more sophisticated equipment was used to study solar ultraviolet radiation and the upper layers of Earth's atmosphere. Keeping them company during their long, lonely sojourn was a variety of biological specimens, including a tank of pregnant guppy fish, fertilized frog-spawn, green peas, bacteria and, as usual, fruit flies. Their materials processing experiments began on 13 July with operation of a furnace to smelt the metals lead, bismuth, tin and cadmium. They used an open cooling area where the molten materials could solidify into spheres as they floated. Another special processor was used to produce semi-conductor crystals in a pilot trial of future space manufacturing. One further innovation was a simulation of a fuel transfer without pumps to overcome weightlessness, relying instead on surface tension.

Although the Soviet media regularly gave out the routine comment that the crew were 'feeling well', some concern began to show as they entered their second month. Dr Igor Pestov admitted that the cos-monauts were tending to overwork, they were losing weight and suffering from raised blood pressure. Further signs of stress were revealed as they reached their seventh week of weightlessness, over halfway to the Skylab record. Despite the use of a slide projector, a portable teleprinter for messages, a library and other entertainment for their leisure hours, the crew began to complain of 'sensory deprivation' and made frequent requests for news from Earth. Arrangements were made for them to speak to children from the USA and many Communist countries as they toured the Crimean control centre on 14 August. After this welcome break, the crew returned to their routine flight plan, apparently on their way to the coveted record. Then, suddenly, after 49 days, it was all over.

The Soviets usually announced well in advance when a mission was drawing to a close. On this occasion, the announcement came only ten hours before re-entry. This sudden decision, combined with an unusual landing in darkness, suggested an unexpected spacecraft malfunction or a rapid deterioration in the crew's health. Certainly, there was evidence that the crew had been suffering both physically and psychologically during their confinement, a burden which doctors had tried to relieve by broadcasting music. However, American reports suggested a fault in the station's environmental control system which had spread an 'acrid odour' through the craft. Unable to correct the situation before it became unbearable, the crew were forced hurriedly to pack their bags and head for home. It looked like another banana skin on the Soviet road towards permanent space stations. Both men were said to be well and to have recovered quickly from the rigours of their flight, but neither of them entered space again.

While the Soviets considered their next move with the Salyut, they sent up a solar-winged Soyuz on an eight-day independent mission. Flight commander was colonel Valeri Bykovsky, making his second space flight 13 years after his first trip on board Vostok 5. Alongside him was a former air force pilot turned spacecraft engineer, 41-year-old Vladimir Aksyonov, a cosmonaut since 1973. Soyuz 22 was launched into a low orbit at the unusually high inclination of 65 degrees. This flight path carried the craft over the main areas of NATO exercises which stretched from Norway to Turkey. Soviet authorities revealed that the docking mechanism had been replaced by a multi-spectral camera designed by the East Germany company of Karl Zeiss Jena, for the ostensible purpose of photographing geological and geographical features in both East Germany and the Soviet Union. The official press proclaimed the camera as the first foreign instrument installed in a Soviet spaceship, a claim that was not strictly true but complemented nicely the pre-flight announcement that 'guest' cosmonauts from socialist countries would be trained for future space station visits. Apart from the spying aspect of the mission, detailed stereo-imaging showing details as small as 90 feet across was used to survey a wide range of Earth resources, including forests, mineral deposits, tidal flats and the proposed route of the new Baikal–Amur railway. The crew landed safely on the barren steppes on 23 September.

Only three weeks later, Soyuz 23 was launched on another type of reconnaissance mission – this time to the dubious hospitality of Salyut 5. The crew were both rookies, though they had been in training for such a mission since 1965, and, as befitted the nature of their mission, both men were military officers. Alongside spacecraft commander Vyacheslav Zudov was another lieutenant colonel, Valeri Rozhdestvensky, but there the similarity ended. The flight engineer was a naval pilot, the first representative of his service arm to board a Soviet spacecraft, and was a former diving unit commander. During a pre-flight press conference, he told reporters: 'I think my pre-cosmic profession may yet come in handy.'

The drama began on their second day in orbit after initiation of the automatic rendezvous system. The cheerful, chatty crew suddenly fell silent as they became aware of a malfunction in the approach control system. Unable to use the manual override, and restricted to battery power only in the ferry version of the Soyuz, they had no choice but to head home as soon as was practicable. They were aware of the atrocious weather conditions and the fall of darkness over their landing zone, so it was two very nervous cosmonauts who prepared for an emergency re-entry, the fourth such trauma in the past nine missions. The cosmonauts needed a special brand of courage during those days of constant uncertainty and danger.

All went according to plan until their parachute was caught up in a raging blizzard, carrying the helpless crew towards Lake Tengiz, a 20 mile wide salt lake and a treacherous watery trap. The descent capsule hit the water over a mile (2 km) from the shoreline. Although it remained floating on the choppy lake, the internal temperature rapidly plummeted towards the outside temperature of minus 15°C, forcing the cosmonauts to change into their thermal survival suits and pray for rescue. No Soviet manned spaceflight had ever splashed down on water, although the cosmonauts did practice emergency procedures. Drawn to the site by the Soyuz radio beacon, the rescue crews still had to overcome the half frozen quagmire which surrounded the lake and the almost zero visibility. Rafts launched on to the lake failed to reach their target when they encountered ice floes, so frogmen desperately struggled to attach a line for the hovering helicopters. At last they succeeded, only to find that the capsule was too heavy to be lifted by helicopter, so the recovery team decided on the next best alternative. The craft was dragged across the lake and the soggy lake shore in one of the most undignified ends to any manned space mission. Once on terra firma, a fleet of helicopters filled with doctors, technicians and space officials descended on the capsule. It was the end of six terrifying hours for the crew, who were grateful to be alive. Chief cosmonaut Vladimir Shatalov praised the 'high courage and heroism' of the rescue crews in the 'very difficult' recovery conditions. Neither of the cosmonauts flew in space again.

Another four months passed before the Soyuz 23 back-up crew attempted another link-up with the ageing Salyut 5. Commander was the experienced air force colonel Viktor Gorbatko, and flight engineer was another air force officer, lieutenant colonel Yuri Glazkov. Both men were trained in space walking techniques, leading to speculation among Western observers that they would carry out the first Soviet EVA in eight years. This did not take place. Other surprises were in store, notably the mission duration of only 18 days. Rendezvous and docking were successfully completed on the evening of 8 February 1977, with the final approach carried out under manual operation. Mission control then gave the unusual command for the crew to spend their sleep period in the Soyuz before entering the space station. This may have been a precaution in case the cabin atmosphere in Salyut 5 was abnormal, but there was no need for concern, as Glazkov reported all was 'comfortable' when he entered the next day.

During the next two weeks, the crew repaired an onboard computer and replaced other units, but their most unusual task was the total replacement of the Salyut's atmosphere using bottled compressed air. An official statement declared: 'The air in the station was quite satisfactory to the crew and the doctors, but it was decided to test the system which is important for prolonged expeditions.' The more sceptical observers saw the operation as justification for their belief in Salyut's air conditioning problems. Other work carried out by the crew involved similar experiments to those previously completed, including materials processing, biological, medical and Earth observations. Apparently, the main objective of the mission once more was military reconnaissance, however, for the early return of the crew was obviously planned well in advance. Results from the mission were brought back by the cosmonauts on 25 February and by a special capsule ejected from the space station the following day. This seemed to signal the end of the space station's useful life, though it remained in orbit until commanded to burn up during re-entry on 8 August. As for the cosmonauts, they were spared the agonies of their predecessors, though Soyuz 24 still made a hairy landing in conditions of high wind, poor visibility and driving snow. After their relatively short stay, the crew adjusted to normal gravity after only one day, though they did comment that a 'back expander' would have been useful to combat the tendency to bend forwards in zero gravity. Glazkov never flew in space again, though Gorbatko continued to train for space station missions.

The objective of Soviet space activity was the operation of permanently manned space stations, and a major step towards achieving that goal was taken on 29 September 1977 when Salyut 6, described as a 'second generation space station' was sent into orbit. In outward appearance, the new Salyut differed little from its predecessors, consisting of three main cylindrical modules to which were attached two solar 'wings' and a third solar panel projecting out like an upright fin. The main alteration was the addition of a second docking port at the rear of the station. In order to achieve this modification, the main engine assembly had been redesigned, with the engine compartment widened to cater for the relocated engines and the new docking port. This relatively simple modification meant that a manned Soyuz ferry could dock at the forward end while the aft port could be used simultaneously by a Soyuz or a robot supply ship. It was clear that the Soviets were aiming to keep their crews in orbit for long periods with the help of regular supply shipments and relief crews.

The active life of this latest monument to Soviet technology was planned to begin with a showpiece opening session to be carried out by the crew of Soyuz 25. The men chosen for this initiation ceremony were two of the new breed of cosmonauts who had been trained specially for space station activities. Commander was 35-year-old lieutenant colonel Vladimir Kovalyonok, an air force officer and former paratroop instructor. His senior by three years was the flight engineer, civilian electronics expert and spacecraft designer Valeri Ryumin.

Soyuz 25 was launched in flawless fashion from Baikonur on the morning of 9 October 1977. Film shown on Soviet TV showed the crew walking to their rocket on the same pad used to launch Sputnik almost exactly 20 years earlier. Viewers watched the rocket until it disappeared through the cloud cover, carrying the cosmonauts and a copy of the new Soviet constitution to their historic rendezvous. To add extra meaning to the event, it was seen as a way of celebrating the sixtieth anniversary of the glorious Bolshevik Revolution. Unfortunately, the celebrations turned sour.

After an apparently normal rendezvous and approach, the airwaves went silent. There was a lengthy delay, then the media admitted: 'Due to deviations from the planned procedure for docking, the link up was called off'. Further reports were no more helpful, simply adding that there had been an 'unexpected malfunction'. Tracking information and radio monitoring from Kettering, England, suggested that the two craft had spent up to eight hours and five orbits in close proximity before the mission was called off. Apparently no attempt was made to use the second docking port at the rear of the space station. With the limited power supply on the Soyuz, there was no alternative but to make a humiliating retreat. The propaganda spectacle had backfired, providing yet another Soviet embarrassment, the seventh failure in 11 missions to space stations since 1971. Soyuz 25 landed safely in the normal recovery zone on the morning of 11 October. The orbital module to which the Soyuz docking mechanism was attached had been ejected and had burned up during re-entry, so there was no evidence with which technicians could assess the hardware. The crew would have to wait a while longer for their occupation of Salyut 6.

The failure of the initial occupation attempt meant a drastic rethink of the scheduled programme. If the Salyut docking mechanism really had malfunctioned, this could only be checked by a qualified engineer carrying out a first-hand inspection during a space walk. Although the Soviets had recently begun training for EVA once more, and had

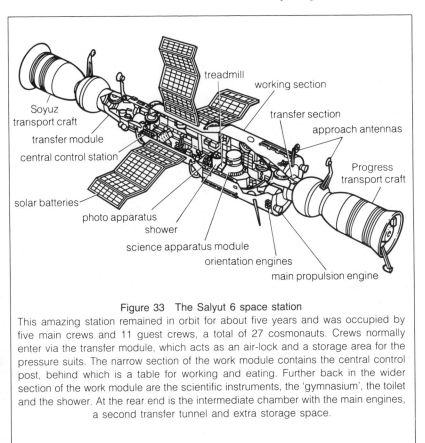

treadmill

working section

Soyuz transport craft

transfer section

approach antennas

transfer module

central control station

Progress transport craft

solar batteries

photo apparatus

shower

science apparatus module

orientation engines

main propulsion engine

Figure 33   The Salyut 6 space station
This amazing station remained in orbit for about five years and was occupied by five main crews and 11 guest crews, a total of 27 cosmonauts. Crews normally enter via the transfer module, which acts as an air-lock and a storage area for the pressure suits. The narrow section of the work module contains the central control post, behind which is a table for working and eating. Further back in the wider section of the work module are the scientific instruments, the 'gymnasium', the toilet and the shower. At the rear end is the intermediate chamber with the main engines, a second transfer tunnel and extra storage space.

developed an improved pressure suit equipped with its own portable life support system, the task ahead would need a cool head on the shoulders of an experienced engineer. Such a man was available in the form of the Soyuz 17 veteran and brilliant spacecraft designer, Georgi Grechko. Now 46 years old, Grechko was pushed into a hurried training programme, displacing the less experienced Alexander Ivanchenkov. His nominal superior as spacecraft commander was a 33-year-old air force officer, Yuri Romanenko, who was making his first flight into space. The two men had very different backgrounds and very different temperaments – Grechko was a mild, easy-going character while even Romanenko admitted to having a 'volatile' temperament – and with only two months to get to know each other, the crew's interaction promised to be interesting.

Soyuz 26 was launched towards the waiting space station on

10 December 1977 with these words from Vladimir Shatalov ringing in their ears: 'The working programme this time will be rather crowded.' Avoiding the suspect forward docking port, the craft was successfully docked at the rear end of Salyut: already the provision of a second port was paying off. After extensive checks, the men moved into their new home and began the long process of activation. Over a week went by before they were ready to check out the forward docking mechanism. Both cosmonauts moved into the transfer tunnel and donned their new EVA suits and backpacks on 20 December. Using a precautionary lifeline, Grechko floated out of the air-lock, then worked his way along the hand rails so thoughtfully provided by the Salyut's designers. His partner remained in the air-lock, just poking his head out far enough to hand over a TV camera, tools and lighting equipment. It didn't take long for Grechko to reach his target and complete a close examination. 'The butt end is brand new as though just taken off a machine tool,' he reported. 'There are no scratches, traces or dents on it. All of the docking equipment – lamps, electric sockets, latches – all is in fine order. The receiving cone is also clean, without a single scratch.' It seemed as if Soyuz 25 had never made contact with the space station, but Grechko's clean bill of health was a welcome relief to all concerned.

EVA was a rarity on Soviet missions – this was the first such excursion since 1969 – so Grechko decided to linger a while and admire the vast expanse of black emptiness which blanketed the blue and white Earth. He might never have such an opportunity again. As the craft drifted out of radio contact, the discipline of the military officer cracked; Romanenko decided he would also take an unauthorized look at the wonders of the universe. It was almost the last decision he ever made. A startled Grechko watched as his companion began to float away from the air-lock hatch and drift into the void which surrounded them. Romanenko suddenly realized he had forgotten to attach his lifeline. A desperate cry for help alerted Grechko to his commander's plight. The only way he could prevent Romanenko from dying a lingering death in perpetual orbit was to catch him before it was too late. In a great, despairing lunge, Grechko grabbed hold of the trailing tether and managed to reel in a highly relieved cosmonaut. Back inside the transfer tunnel after a 20 minute space walk, both men decided not to mention the incident to ground control. But they were not yet out of the woods.

Before the crew could remove their pressure suits and return to the main living quarters of Salyut, they had to repressurize the air-lock. The gauge showed that an escape valve was still open, and while this remained the case, there was no possibility of raising the air pressure to normal. The men were trapped. Back in contact with ground controllers, they discussed the chances of space walking to the Soyuz ferry at the other end of Salyut and attempting to break in, but both sides admitted this was probably doomed to failure. Their only chance was reckoned to be a false reading on the gauge, so they offered up a silent prayer and allowed some air to enter the tunnel: to their delight, the pressure held firm. Ignoring the escape valve, the relieved crew raised cabin pressure to normal, took off their suits and floated back into the work module. It had been one of the longest 90 minute sessions anyone could remember.

Although Salyut was nowhere near as large as the American Skylab, with a maximum width of only 13½ feet and an internal volume comparable to that of an average caravan, the Soviets had used their past experience to make life as tolerable as possible. Automatic navigation and manoeuvring systems meant the crew could largely ignore the orbital path and its minor variations, and so concentrate more on their experimental work. The space station interior was decorated with bright coloured cloth and given extra insulation as a means of reducing noise from onboard equipment. A toilet and a collapsible shower were provided, though the crew only used the shower about once a month, preferring daily sponging instead. The water condensation unit which recycled moisture in the cabin atmosphere meant they had plenty of water for drinking and reconstituting dehydrated food. Other food was packaged in tubes or canned, and there was a limited supply of fresh items which was replenished by visiting ferries. The cosmonauts' individual preferences were taken into account when designing the menu, though meals were cycled so that a minimum of 6 days passed before they were repeated. Four meals a day were scheduled, adding up to 3,200 calories a day. This high calorie diet and the one to three hours of daily exercise were intended to help maintain peak physical condition under the stress of long-term weightlessness. Someone even remembered to load a living Christmas tree and presents on to Salyut for the crew's pleasure: unfortunately, they didn't have anything stronger than orange juice with which to drink the toast.

On 10 January 1978 a two-man Soyuz crewed by newcomer air force lieutenant colonel Vladimir Dzhanibekov, and wiry old veteran Oleg Makarov, was launched to attempt the first-ever dual link-up with a space station. It was a critical test of the entire Salyut concept – failure would be an unmitigated disaster for the whole programme. Designers were uncertain about the stability of three craft attached in a row like a string of sausages, so Grechko and Romanenko prudently retired to their Soyuz in case an emergency evacuation was required. The forward docking port operated perfectly, however, as Soyuz 27 nosed into a secure dock as the craft passed over the Soviet Union nearly 26 hours after launch. Within three hours of this momentous event, Salyut 6 became the first four-man space station. The resident crew returned to their semi-permanent home to welcome their first visitors. Soviet TV audiences were shown Makarov and Dzhanibekov being almost dragged through the hatch to be greeted with delighted bearhugs and a toast in fruit juice drunk from plastic tubes. 'Come in and show yourselves,' called the smiling crew. 'Oh, they've brought piles of letters for us. I hope you've paid the postage. We've got no money up here.'

Apart from letters, the visiting crew brought a supply of books, *Pravda* newspaper and research equipment, so there was little time for much work other than the obligatory medical tests. With the help of the resident crew, the two newcomers adapted more quickly to the zero gravity environment. Dzhanibekov checked out the Salyut electrical system, then both men transferred their individually contoured couches to the old Soyuz 26, the craft in which they would return to Earth. Loaded with film and research results, Soyuz 26 departed from the space station on 16 January freeing the aft docking port for future ferry craft and leaving the fresh Soyuz for the use of Grechko and Romanenko. The crew landed safely on the snow-covered steppes and walked unaided to the waiting helicopters. They had proved the viability of the three-craft assembly, with an overall length of 120 feet and a weight of nearly 32 tons.

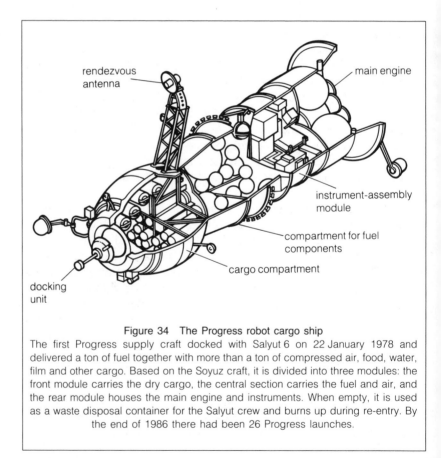

Figure 34   The Progress robot cargo ship
The first Progress supply craft docked with Salyut 6 on 22 January 1978 and delivered a ton of fuel together with more than a ton of compressed air, food, water, film and other cargo. Based on the Soyuz craft, it is divided into three modules: the front module carries the dry cargo, the central section carries the fuel and air, and the rear module houses the main engine and instruments. When empty, it is used as a waste disposal container for the Salyut crew and burns up during re-entry. By the end of 1986 there had been 26 Progress launches.

The long-term crew did not have long to wait for their next visitor. This time it was a robot supply ship called Progress 1, a derivative of the manned Soyuz which had been carefully developed and tested over the past decade. To a fanfare from the Soviet media, the new craft completed automatic docking at the recently vacated rear docking port on 22 January. In another first for the Soviets, Progress demonstrated a capability to resupply indefinitely an orbiting space station and its crew. No longer would a mission's duration depend on the consumption period of the original supply of consumables on a spacecraft. Progress looked like a standard Soyuz, but the interior had been completely designed to carry the vital supplies: dry cargo was stored in the front module, with four propellant tanks and several more spherical tanks containing compressed air and nitrogen located in the central section. It took more than a week for the crew to complete the transfer of all of this cargo. Starting in the front section, they first unloaded the mail and newspapers sent up by friends and family, then, taking care not to upset the stability of the combined craft, they turned to the replacement parts. Among the welcome additions were food and water, new bed linen and clean clothes, including underwear. Altogether, the crew moved more than a ton of cargo before the compartment was emptied and available for loading with refuse.

On 2 February the vital transfer of fuel and oxidizer began through the lines which connected the two craft. Using high-pressure nitrogen gas to push the volatile propellant into Salyut's tanks meant that most of the available electrical power was consumed by the compressor pumps, so all unnecessary equipment had to be powered down for the next two days. Carefully monitored by ground control, the six sessions of pumping proved dramatically how well the Soviet designers and engineers had done their homework. Having prolonged Salyut's active life through the resupplying operation, Progress further extended the station's operational activity by acting as a tug and firing it into a higher orbit. Then, its useful life over, the remarkable craft separated from Salyut and was ordered to brake for a fiery path to destruction over the Pacific Ocean.

By 12 February the crew were breaking more new ground as they passed the previous Soviet space endurance record: the 84-day Skylab record was in their sights. The Soviets had done everything possible to make the prolonged flight endurable, from sending up fresh fruit and bread, and entertainment in the form of a cassette recorder complete with music cassettes, to providing a 'freshness' device to ionize the cabin atmosphere for a more bracing sea air quality. Apart from the scheduled exercise periods every day, the crew had to wear 'Chibis' low pressure trousers for their entire working day in order to pull more blood from their upper bodies and make their hearts work harder. They tended to be rather lazy when it came to exercise, however, pleading pressure of work as their excuse. There were weekly chats with loved ones as well as specially arranged discussions with a variety of sportsmen and well-known personalities to relieve the boredom. Meanwhile, the two men had to learn to tolerate each other's company and try to prevent their idiosyncracies from getting on each other's nerves. Despite their different backgrounds and temperaments, they made great efforts to be considerate and cooperative, thereby avoiding the minor irritations which could fester and grow into a major conflict. In this they seem to have been remarkably successful, though Tass did report that Grechko tended to spend his spare time 'with a camera and sketch pad near the portholes in the transfer tunnel' while Romanenko continually checked over the station's systems and equipment.

February passed with the crew continuing a varied programme of experiments. The Splav furnace delivered by Progress 1 was installed in the forward air-lock for materials processing; the majority of the remaining time was taken up with Earth resources photography using the East German MKF-6M camera to survey selected areas of the Soviet Union. An infrared telescope was used to examine Earth's ozone layer, with additional observations of the planets, stars and galaxies, but the crew seem to have gained most satisfaction from looking after their biological specimens, even though the results were not always successful. Despite loving care, the peas and wheat stubbornly refused to grow beyond their early stages in the Oasis container, while the Earth-born tadpoles swam in a completely disorientated manner, unlike their space-born relatives. Most frustrating of all were the fast-breeding fruit flies which were used for genetics studies: some managed to escape and proved highly adept at avoiding the swatting efforts of their floating adversaries.

With the overall space endurance record about to be passed, the crew's morale was boosted once more by the arrival on 2 March of the first truly international space crew on board Soyuz 28. Alongside the Soviet commander, Alexei Gubarev, was a Czech army air force captain, 29-year-old Vladimir Remek. The Soviets revealed that the young 'guest' cosmonaut had been in training, along with another of his countrymen, two Poles and two East Germans, for more than a year, during which time he had lived with Gubarev and his family. It was noted that the young jet pilot was only slightly older than Gubarev's son. More important than his age or his ability, however, was his suitability as the official representative of his country – as a member of the Communist Party and the son of a deputy defence minister, his credentials were impeccable. No matter that he had been asked to lose 40 lb in weight to qualify for the programme, he was about to become a national hero in another Soviet propaganda victory which beat the American 'guest' astronaut' programme by some four years.

TV pictures showed more bearhugs and kisses as the four cosmonauts greeted each other in true Russian style. Tradition continued with the cherry juice toast and the snack of bread and salt around the table. It took some time for the celebrations to die down, but the rest of the week was taken up with a series of experiments for which Remek had been specially trained. Further use of the Splav furnace was made to produce pure crystals and new conducting alloys; observations of brightness of stars as they set below Earth's horizon were conducted; algae were cultivated using special nutrients; and oxygen concentrations in human tissue during zero gravity were measured. After nearly seven days of intense activity, the two visitors loaded their craft with their results and headed home for a soft landing in a snow-covered field. The 'Interkosmos' programme had got off to a very promising start.

Almost as soon as they had left, Grechko and Romanenko began preparing for their own departure, stowing film and other research materials on board Soyuz 27. Worried about their ability to withstand the stresses of re-entry after more than three months of weightlessness, doctors ordered them to step up their exercise routine during the final week. The space station was finally evacuated on 16 March after 96 days of continuous occupation. As might be expected, the cosmonauts found considerable difficulty in adapting to normal gravity. They were placed on stretchers and carefully lifted from the capsule into the waiting helicopters. Every movement was an effort, even breathing, and they were unable to stand unaided for several days. Doctors reported reduced heart volume and shrunken calf muscles, as well as more unexpected responses. They 'tried to swim out of their beds' when they awoke in the morning, and had difficulty picking up a cup of tea or even turning a radio dial. One specialist added: 'They are both still up there in space, not only physically, but in their thoughts.' With the help of special trousers to assist in walking, the crew gradually made a full recovery, and went on to make further contributions to the Salyut effort.

Having broken the three-month barrier, the Soviets were in no mood to sit back and mark time. Their American rivals had no plans for further long duration missions in the foreseeable future: all of their eggs were in the Shuttle basket, but the re-usable orbiter was falling further and further behind schedule, leaving near-Earth space free for Soviet domination. After absorbing the lessons of the Soyuz 26 mission, Soviet doctors modified the exercise regime and training of the next long term crew to occuy Salyut 6. The 96-day record of Grechko and Romanenko was already under threat.

On 15 June 1978 Soyuz 29 blasted off from Baikonur to rendezvous with the space station, carrying air force colonel Vladimir Kovalyonok, commander of the ill-fated Soyuz 25, with civilian engineer and spacecraft designer Alexander Ivanchenkov. This was 37-year-old Ivanvhenkov's chance to prove himself after being dropped in favour of the more experienced Grechko the previous year. Docking took place at the forward port two days after launch. On entering Salyut, they found a greetings note from its former residents. All seemed well, so they wasted no time in reactivating the station and commencing their programme of experiments with the Splav furnace.

Only ten days after the crew entered Salyut, they received their first visitors – Pyotr Klimuk, a veteran of two space missions yet still only 35 years old, and a 36-year-old Polish air force officer, Miroslaw Hermaszewski. One of the main factors in favour of his selection seems to have been that his brother was a leading air force general. Apart from a useful propaganda exercise, it seems difficult to appreciate any advantages to be gained from the timing of this mission. Among the mementoes brought by the visiting crew was a container of soil from a Russian village and from Warsaw, 'earth made sacred by the blood of Soviet and Polish soldiers'. The week passed with a series of joint experiments, including use of the furnace for production of exotic semi-conductor materials and photography of southern Poland. By 5 July Kovalyonok and Ivanchenkov were once more left to their own devices as their visitors returned to Earth in Soyuz 30.

For the orbiting cosmonauts, the respite was brief. A robot supply ship, Progress 2, arrived at the back door on 9 July, loaded with fresh food, water, air, fuel, film and replacements including a new instrument panel which showed the position of the station above the Earth, and a new Kristall furnace which was better insulated than the Splav and so could be installed in the work module.

The extra air supplies were particularly useful, since the cosmonauts carried out a routine space walk while the Progress was still attached and air was inevitably lost each time the air-lock was used. Wearing the same suits as the Soyuz 26 crew, but with the straps adjusted to make a comfortable fit, the crew spent most of the pass over Soviet territory chatting to ground controllers. By the time they stepped out of the air-lock, they were passing over Korea and moving out of radio contact. While Kovalyonok edged out towards the externally attached experiments, using the handholds provided, his partner remained in the hatch, only his head and shoulders projecting out into space. The men took the opportunity of watching a spectacular sunset over the south Pacific, then switched off the spotlights to enjoy the 35 minute night pass. To their surprise and delight, the darkness was suddenly illuminated by the bright streak of a nearby shooting star. It was a good omen. Back in sunlight over South America, the crew set to with a will, retrieving and replacing scientific equipment for measuring the effects of cosmic rays and micro-meteorites, and the Medusa experiment of cassettes containing biopolymers which had been installed by Georgi Grechko. They were so efficient that they finished the work ahead of schedule, causing ground control to suggest that they should get back inside. Kovalyonok was in no hurry to comply: 'We would just like to take our time. This is the first time in 45 days that we've got out in the street to take a walk.' Their space walk finally ended after two hours. Once again, it had been beamed back to Earth through a portable colour TV camera, to the delight of Soviet viewers.

Two more Progress craft docked with Salyut during the prolonged stay by Kovalyonok and Ivanchenkov, each time bringing vital supplies which enabled continuation of the record-breaking flight. Progress 3 was sent up on 7 August, only three days after its predecessor burned up in the atmosphere. No fuel was carried since Salyut's tanks had only recently been topped up, but there were a number of unusual items apart from the usual water, oxygen and samples for the furnaces. The food rations included a number of delicacies intended to sharpen the cosmonauts' appetites and revive their flagging taste buds – strawberries, fresh milk, onions and garlic. It was now an accepted peculiarity of long duration spaceflights that the crews' appreciation of food diminished, even their favourite dishes tending to resemble sawdust in the mouth. Another item of cargo was particularly pleasing to Ivanchenkov – his request had been noted, and his guitar was carefully packed among the more traditional items.

After the departure of Progress 3, the rear port was left free for the next visitors, the Vostok veteran Valeri Bykovsky and his 41-year-old East German companion, air force officer Sigmund Jähn. They duly docked in Soyuz 31 on 27 August, to be greeted with the now traditional bearhugs, toast, bread and salt, and long welcoming speeches for the benefit of the watching millions. Among the gifts brought by the new crew were a Russian doll, a Mishya bear as a symbol of the forthcoming Moscow Olympics, and souvenir watches. More significant from the old hands' point of view were the extra delicacies brought to supplement their menu: onions, garlic, lemons, apples, milk, soup, honey, pork, Bulgarian peppers and gingerbread. The following week passed with long photographic sessions using the East German MFK-6M and KATE cameras, and further materials processing. The results of these experiments were brought back in the ageing Soyuz 29, leaving the fresh Soyuz 31 for the use of the long term crew. It was clear that there was a long haul still ahead for Kovalyonok and Ivanchenkov.

These two intrepid voyagers were now faced with an unprecedented practical problem: their ferry craft was docked at the aft end, the only docking port equipped for use by the robot Progress ships. If any further supplies were to reach them in the Progress, they would have to park the Soyuz at the front end of their home. Accordingly, on 7 September the two cosmonauts loaded their ferry with vital research data in case they had to beat a hasty retreat back to Earth, and backed away several hundred feet from the giant Salyut. With the space station's orientation system switched off so that it would swing gently towards them under the influence of Earth's gravity, the crew were able to pitch the Soyuz over using the minimum amount of fuel. The nose of the Soyuz was effectively swung through 180 degrees until the docking probe was aligned with the forward end of the station and driven home to complete a pleasant 'drive in the country'.

Their third Progress arrived at the vacated port on 6 October, loaded with fuel to replace the large amounts used during manoeuvres of Salyut related to the observational programme., By now, the crew were into their third month in space and every day that passed was breaking new ground. Doctors pronounced themselves happy with the cosmonauts' condition, the only minor irritations being occasional headaches, fatigue and cold feet. Despite the Chibis trousers which pulled blood to the legs, the men were suffering from poor blood circulation, so they were delighted to receive two pairs of fur boots among the Progress 4 cargo. As for the headaches, they seemed to be related to a build-up of carbon dioxide in the atmosphere, a problem easily remedied by more frequent changes of the air purifiers. Other noteworthy items on the robot ship were the food supplements, including a variety of fresh, freeze-dried and canned meats, curds with nuts, and two boxes of sweets. There were also two new 'Penguin' suits to replace the worn out muscle-toning suits which were such a valuable part of their exercise programme, two pairs of slippers, dividers to section off the cabin and give more privacy, and a selection of music tapes. Despite their slowly declining physical condition, the crew now had the job of unloading the cargo down to a fine art and completed the task in half the scheduled time. Refuelling then followed on 12 October, and the Progress engines were fired to raise Salyut's orbit a week later. The space station was now ready for a long vacation prior to the next record breaking attempt by future occupants. Progress 4, having completed its work, separated from the station to prepare for its fiery death on 26 October.

As the flight drew to a close, more attention was paid to fitness by the doctors. They ordered the crew to wear their Chibis trousers throughout the day and to step up their exercises in the Penguin suits and on the treadmill. Intensive monitoring of their physical condition showed they were in a much better state to withstand re-entry and normal Earth gravity than their Soyuz 26 predecessors, an evaluation which was born out after their return on the afternoon of 2 November. The record-breaking crew had completed almost 140 days of zero gravity, some one and a half months longer than any previous crew. Yet the Soviet people watched spellbound as the two heroes waved off the proffered helping hands and clambered unaided from their capsule. Kovalyonok was sufficiently fit to indulge in a small, moving celebration of their homecoming, though he later admitted: 'It was not easy . . . to bend and pick up a little lump of earth because of the Earth's gravitation, but I forced myself to do so. And this smell of the sun-drenched steppe will remain with me for ever.' The euphoria also spread to his companion.

'We were literally intoxicated by the fresh air,' admitted Ivanchenkov. Both men exhibited the tell-tale signs of prolonged weightlessness: a weight loss of 6–9 lb, reduction of calcium and other minerals from the bones, wasted muscles, including a reduction in heart volume, and general fatigue. Nevertheless, they astounded the doctors with the rapidity of their recovery. By the second morning, they were able to take a short walk in the park, an unprecedented feat for returning long term space travellers. Their pulse fluctuations and inability to judge the weight of their bodies or other objects gradually returned to normal over the next week, and by the second week they were participating in short jogs or games of tennis. Indeed, the recovery was so swift that they were allowed to return to Star City and a triumphant homecoming ceremony on 14 November.

As 1979 dawned, the Soviet space programme was on the crest of a wave after escaping from the trough of despair occasioned by the disasters and disappointments characteristic of most of the decade. Cosmonauts had at last overtaken the cumulative man-hours total in space achieved by the USA, and there was no intention of relinquishing this domination. The long term goals of permanent orbiting space stations and interplanetary travel meant that the zero gravity barrier had to be pushed back still further. The ageing Salyut 6 still had a role to play.

On 25 February 1979 the next crew to occupy the mothballed space station was launched on board Soyuz 32. It was the start of the most remarkable – and lonely – voyage up to that time. Neither of the participants was very experienced: commander Vladimir Lyakhov, a 37-year-old air force lieutenant colonel, had been a cosmonaut since 1967 but was only now making his first venture into space. Hardly more familiar with space was the 39-year-old civilian flight engineer, Valeri Ryumin, whose flight time amounted to a mere two days aboard the aborted Soyuz 25.

The new crew could have been forgiven for thinking they had been mistaken for maintenance men instead of highly trained pilots and engineers during the first month in space. As if the usual adaptation to zero gravity was not enough, they were expected to repair or replace components that had never been intended for replacement while at the same time initiating the various scientific experiments. Using a new tool kit which included a special soldering iron, the cosmonauts found work to do in almost every nook and cranny of the space station – the treadmill and exercise cycle, the communications and TV systems, the air conditioning and life support systems. They even managed to replace the head on their video recorder. The hectic pace continued after Progress 5 docked with Salyut on 14 March, bringing a ton of fuel and nearly 1½ tons of other cargo. Among the items the crew unloaded were six new carbon dioxide detectors, a new tape recorder, a walkie-talkie system, a linen dryer, bath shampoo, clothes and various foods, including five kinds of bread. Particularly important to the crew was the first black-and-white TV system which allowed two-way link-ups with scientists, families or famous personalities on the ground. It was probably this system more than any other which helped the men retain their sanity in the long months ahead when there was no other company but their own.

With the Progress attached, the crew initiated the most difficult and time-consuming maintenance operation of all, one for which they had trained prior to launch. Kovalyonok and Ivanchenkov had reported that there appeared to be a leak in the membrane which separated liquid fuel from nitrogen gas in one of the three fuel tanks. Fortunately, the space station could still operate on only two tanks, but the damaged tank could not be repaired and had to be closed down. With Salyut set slowly spinning to help separate the fuel and gas, the men transferred as much fuel as possible to another tank on Salyut or to an empty container on Progress, then vented the fuel remnants into space with the help of compressed nitrogen. The entire process took a full week, and though ground control took over during periods of radio contact, much of the work had to be entrusted to the two cosmonauts. It was not to be the last demonstration of human repair capabilities in orbit on this mission.

Having boosted Salyut into a higher orbit, Progress 5 was ordered to separate and self-destruct, leaving the aft port free for the first human visitors of this mission. On 10 April Soyuz 33 was launched amidst some of the worst weather known for a manned lift-off, with winds gusting up to 40 mph. Should anything go wrong, there was a risk of a diplomatic incident, for, alongside the balding two-time veteran Nikolai Rukavishnikov, the first civilian commander of a space mission, was 38-year-old Bulgarian air force major Georgi Ivanov. However, the decision to go ahead seemed to have paid off as the craft safely entered orbit and began its automatic approach to the space station.

By the afternoon of 11 April Soyuz 33 was less than 2 miles (3 km) from Salyut and clearly visible against the black sky. Rukavishnikov described the final main propulsion burn of the rendezvous: 'When the actual time of approach arrived, the motor was fired to slow lateral velocity. Then we noticed that something was wrong.' The Soyuz main engine cut out after only three seconds, its uneven thrust causing the ship to shake. The crew held on until the vibration ceased, uncertain as to exactly what had happened. To complicate matters, Ryumin reported seeing an unusual lateral glow from the engine. A second attempt to fire the engine failed, and analysis of telemetry by ground controllers indicated a serious malfunction. There was no choice but to abort the mission, using the single-thrust back-up engine. Georgi Ivanov tried to lift the gloom by suggesting they had a meal and opened a celebration gift originally intended for the Salyut crew. 'We opened the handsome red box tied with a multicoloured ribbon and fortified ourselves. I had very little, Georgi took a good drink.' The contents of the box were not revealed!

Re-entry took place on 12 April. The hurried return was necessitated by the lack of power on the Soyuz and the need to aim for the prime landing zone during a night landing. Fortunately, the back-up system worked, though it failed to cut out automatically and had to be shut down by Rukavishnikov. In only the second ballistic re-entry by a Soyuz craft, the crew were subjected to deceleration forces of 8G, twice normal loads. Recovery teams reported sighting the glowing spacecraft in the night sky like some alien invader from outer space, but this enabled them to pinpoint the touchdown site quickly. Although the capsule rolled on to its side after hitting the ground, the crew were able to scramble out unharmed, though rather shaken. Rukavishnikov summed their views: 'It may have been only two days, but it seemed like a month.'

Until the engineers sorted out exactly what had gone wrong, there was no chance of any more visitors to Salyut 6, so Ryumin and Lyakhov would have to soldier on alone. The days passed in a fairly rigid routine intended to maximize their efficiency while maintaining their health: they had an eight hour working day, five days a week, with an extra two hours set aside for exercise, two hours for meals and another two hours for leisure. On weekends the team of psychologists arranged a varied programme of two-way conversations with family, friends, favourite actors, scientists and sportsmen. Although there were the inevitable minor disagreements, the men seemed to get along reasonably well in their claustrophobic environment, and Ryumin, in particular, appeared to thrive in zero gravity – he became the first cosmonaut to put on weight during a mission.

Apart from the usual experiments with the two topographic cameras and the 'submillimeter' telescope, the crew worked for long periods with a new Kristall furnace delivered by Progress 5. It was another experiment, named 'Biogravistat', also brought by Progress 5, which gave them most pleasure, however, along with their space garden. Seeds planted on the rotating centrifuge part of Biogravistat developed roots twice the length of those on stationary trays, but plants still stubbornly refused to grow and develop properly. Among the extensive garden planted by the crew were items such as garlic, onions, cucumber, wheat, lettuce, peas, parsley and dill which could supplement their normal diet. Unfortunately, most of these failed to grow, and none produced any seeds, though the men did appreciate eating herbs which they had personally tended. Orchids fared no better, but the Soviets refused to give up on flowers in space, and sent up a tulip on board Progress 6, followed by others on the unmanned Soyuz 34. Apart from their scientific interest, it was thought they would bring the crew a touch of spring.

As the flight continued into summer, the Soviet authorities were faced with several problems. Resupplying the space station was relatively straightforward – robot Progress cargo ships docked with Salyut on 15 May and 30 June, bringing the routine supplies along with replacement parts and new equipment. However, uncertain of the cause of the Soyuz 33 failure, they were obliged to cancel the scheduled visits by Hungarian and Cuban guests. Apart from the psychological problems which might affect the two lone voyagers, there was a more practical difficulty. Their Soyuz 32 craft would deteriorate, perhaps to a dangerously unsafe condition, during a six month sojourn in space. Without a visiting crew to swap ships, how could it be replaced with a fresh craft? The only answer was to send up a Soyuz ferry under automatic control. Thus it was that Soyuz 34 replaced Progress 6 at the aft end of the station on 8 June after passing exhaustive tests of its modified engine. The old Soyuz 32 was loaded with samples from the furnaces, film and other items of scientific interest, including the contents of a vacuum cleaner and a cast of a possible meteorite impact on a docking unit. The removal of the ship for an automatically controlled soft-landing left the front port free for the transfer of Soyuz 34, and demonstrated once again the value of the robot systems so painstakingly developed by the Soviets. A faultless switch of the Soyuz on 15 June left the aft port available for Progress 7 two weeks later.

The major item of cargo on this next supply ship was a specially designed radiotelescope, which arrived like a furled umbrella but measured 33 ft (10 m) in diameter when deployed from the aft docking port. It took some time for the crew to attach the antenna to the edge of the docking unit, link up its controls and power supply, but by 18 July all was ready. Progress 7 pushed gently away, using springs instead of thrusters in case the antenna was damaged. Once the ferry was at a safe distance, the umbrella opened out, dwarfing the space station. Over the next three weeks, the crew carried out wide-ranging studies of distant objects in space as well as Earth observations, often in conjunction with an observatory in the Crimea. It was a time-consuming occupation involving teamwork from both men, and the crew must have been relieved when it came to an end. They were familiar with the mechanism, having carried out training prior to the mission, so they did not foresee any difficulties in stowing the antenna. However, to their surprise and annoyance, the antenna tangled on a mooring bar after the explosive charges fired, and, no matter how much they swung Salyut from side to side, the dish refused to budge. Unless the situation could be remedied, there would be no more spacecraft docking at the station's back door.

Ground controllers were in a cleft stick. They desperately wanted to free the antenna and clear the docking port, but they realized that this could only be done on a difficult, unrehearsed space walk by a tired crew who were less than 100 per cent fit. In the end, the cosmonauts made the decision for them by insisting on the EVA. Ryumin struggled to open the hatch in the forward air-lock, then pushed himself out into space. By the time they had installed hand rails on the station's outer skin, the sun was setting and radio contact was fading, so the men lingered anxiously in the darkness, trying to appreciate the view. Ryumin later described the stars as 'huge diamond pins on black velvet'. Back in sunlight once more, the engineer moved gingerly along the entire length of the Salyut, his lifeline playing out behind him. No cosmonaut had ever worked at the rear of a space station, but Ryumin managed to find a firm handhold as he peered at the tangled mesh which hung limply before him. Snipping away at the nearest wires with pliers from his tool kit, he managed to loosen the dish sufficiently to push it free with his boot. It was a perspiring but relieved cosmonaut who informed mission control: 'The antenna's gone.' On the way back, Ryumin found enough energy to collect samples left outside the station by earlier space walkers, but he had had enough exercise for one day. The exhausted cosmonaut hardly had the strength to release his backpack when it snagged as he was moving back into the air-lock, but the impromptu excursion had been a triumph.

With Salyut back in operational status, the crew could prepare for the return home with a clear conscience. During nearly half a year in orbit, they had taken some 9,000 photographs, completed over 50 smeltings, and carried out innumerable observations with the infrared and radio telescopes, the Yelena gamma ray instrument, the Biogravistat and other scientific equipment. Now came the ultimate test; how would they fare on the return to normal gravity?

Salyut was finally mothballed on 19 August as the cosmonauts pulled away in their fully loaded Soyuz 34. Immediately after touchdown, the capsule was surrounded by medical teams anxious to discover the cosmonauts' condition. Weighed down by their pressure suits, they were almost dragged from the descent module to nearby reclining chairs where they were greeted with flowers and a mug of tea. Ryumin remarked that the flowers felt heavy on his chest, while a doctor commented that 'even a cup of tea feels like a full meal in their stomachs'. Ryumin also had some difficulty in speaking, and both men found they wanted to push off from the ground and float when the doctors allowed them to stand up and take a brief stroll. Recovery was rapid, especially for Ryumin, whose weight had not changed during his 175 day incarceration. Lyakhov, two years his junior, had suffered a weight loss of 12 lb and took longer to adapt, but doctors expressed delight with their overall physical condition. Material proof of Soviet progress in acclimatization came only eight months later when Ryumin was blasted into space for his second long term acquaintance with the resilient Salyut 6.

Valeri Ryumin had very little warning of his new adventure. On returning to duty after a well-deserved period of recuperation, he was assigned to training the next long term crew to visit Salyut, Leonid Popov and Valentin Lebedev. Then, only a month before the scheduled launch, Lebedev sustained a badly damaged knee during a trampoline accident. The unfortunate cosmonaut was rushed to hospital, but was clearly out of the reckoning for the forthcoming flight. Normally there would be no cause for concern – the back-up engineer would step into the vacancy – but on this occasion the understudy was considered too inexperienced, and after long deliberation it was decided to ask Ryumin to step into the breach. The logic was infallible: as the Soviet media explained, he 'unarguably knows the station better than anyone else'. Ryumin's feelings about the sudden switch are not recorded.

The station which he and Popov reoccupied on 10 April 1980 had not been entirely redundant during its eight month vacation. An unmanned version of a new craft, Soyuz T-1, had been launched in December, and spent three months docked with Salyut in a test programme for its improved systems. No sooner had this craft undocked than Progress 8 arrived at the aft port loaded with fuel, spares and supplies for the next occupants. Ryumin and Popov hardly had time to settle into their new home before they were obliged to unload its cargo hold and replenish the fuel tanks. An intensive repair session followed as the crew worked to make Salyut shipshape once more. It must have been a strange sense of *déjà vu* for Ryumin, returning so soon after his previous visit. On opening the letter he had written for Lebedev and Popov, he wryly commented: 'I am not in the habit of writing letters to myself.'

As might be expected, Ryumin found little trouble in readapting to the weightless environment, and he was able to give valuable advice to his rookie companion. Apart from the usual load suits and gymnasium equipment, the crew were given special wrist cuffs to help redistribute their blood, but they were allowed one day in four without any exercises in a significant concession aimed at improving their motivation.

Once the station was activated and re-stocked, the crew were able to begin the experimental programme, with particular emphasis at first on materials processing and plants. One innovation was a miniature greenhouse intended for growing orchids, but sadly the petals fell off the flowers as soon as they were on board. Other plant experiments aimed at improving the crew's diet continued with the space garden, and the Biogravistat was reactivated after a new motor was delivered by Progress 9 on 29 April. The most exciting breakthrough came in September when the crew reported that an Arabidopsis plant had succeeded in flowering, though it failed to produce any seeds.

Other experiments, in particular the detailed photographic surveys of land and sea resources, were often related to the needs of the socialist allies whose guests visited Ryumin and Popov at regular intervals. The first of these was a Hungarian air force officer, Bertalan Farkas, who teamed up with ASTP veteran Valeri Kubasov on Soyuz 36. Among the unusual items carried by Farkas were his daughter's favourite toy rabbit, a typical Hungarian meal of goulash, paté de foie gras, fried pork and jellied tongue, and a range of scientific equipment jointly designed by Hungarian and Soviet experts. Among the 21 experiments carried out during the last week of May were the manufacture of crystals and new metallic alloys using the two Salyut furnaces, and attempts at producing the new 'wonder drug' interferon in space. Although Farkas's home country was often in darkness during flyovers, the crew did successfully conduct a programme of geological and ecological mapping, tied in with simultaneous observations from aircraft. Important information was collected on potential mineral deposits as well as soil and vegetation conditions within Hungary. Farkas returned to Earth in Soyuz 35 on 3 June, some 7 lb lighter but none the worse for wear after a minor bout of dizziness early in the mission.

Once Soyuz 36 was shunted around to the front of the space station, the rear port was available for the first manned Soyuz T ferry. In a period of intense activity for the Salyut crew, their next visitors were launched from Baikonur on 5 June in what the Soviets described as an 'improved' spaceship. Although outwardly similar to the long duration Soyuz with its two solar 'wings', the Soyuz T-2 had also been provided with an onboard computer, a new orientation and manoeuvring system, and a single centralized fuel system for the main engine and the attitude control thruster. Assigned to be the guinea-pigs for this test flight were newcomer Yuri Malyshev, a 38-year-old air force officer, and civilian engineer Vladimir Aksyonov, a 45-year-old spacecraft designer who had been involved in the space programme since 1957. Approach and docking were intended to be computer-controlled, but things did not quite work out as planned. Closing in on Salyut from an unusual angle, the computer selected an approach sequence which the crew had not yet practised in simulators, so they felt obliged to make a more traditional manual docking.

After three days of joint flight tests with Salyut, the new craft returned to Earth carrying samples and results from the station. Re-entry also included some innovations: for the first time the orbital module separated before retro-fire, thus saving some ten per cent of the vital fuel, and the landing engines were given extra thrust to save the returning cosmonauts from an undignified jolt when they hit the welcome Soviet soil.

While Soviet engineers evaluated the performance of their new baby, the guest cosmonauts and their space chauffeurs had to make do with the old fashioned version. Soyuz 37 and Soyuz 38 each delivered major propaganda coups at a time of heightened tension between the Soviet Union and the Western allies. The first of these was launched on 23 July,

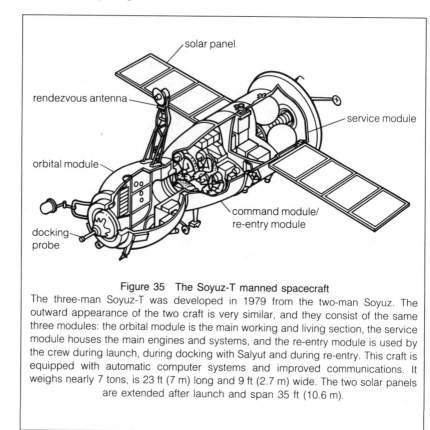

Figure 35   The Soyuz-T manned spacecraft
The three-man Soyuz-T was developed in 1979 from the two-man Soyuz. The outward appearance of the two craft is very similar, and they consist of the same three modules: the orbital module is the main working and living section, the service module houses the main engines and systems, and the re-entry module is used by the crew during launch, during docking with Salyut and during re-entry. This craft is equipped with automatic computer systems and improved communications. It weighs nearly 7 tons, is 23 ft (7 m) long and 9 ft (2.7 m) wide. The two solar panels are extended after launch and span 35 ft (10.6 m).

midway through the controversial Moscow Olympics which had been boycotted by the USA. Adding insult to injury, the Soviets announced that the research pilot alongside veteran Viktor Gorbatko was a former Vietnamese fighter pilot, Pham Tuan, who had shot down an American B-52 bomber during the Vietnam War. Among the tasks which Tuan was to carry out, the Soviets mentioned Earth photography and ecological observations to assess the damage caused by American defoliants and fire bombs during the war. Other elements of the week-long programme included atmospheric observations related to the destructive typhoons which hit Vietnam each year, coastal surveys focusing on tidal erosion and deposition, and an attempt to grow an oriental aquatic fern. The mission ended on 31 July in Soyuz 36, leaving the fresh craft behind. As usual, the remaining ferry was transferred to the front end the next day, the 115th of the current flight.

There was a long quiet spell for Popov and Ryumin over the next six weeks, during which the flight engineer enjoyed the dubious privilege of celebrating his second birthday in succession in orbit. With Salyut's stock cupboard and fuel tanks overflowing after three Progress visits between April and early July, the crew remained untroubled by space traffic, manned or otherwise. Among the cargo delivered by Progress 10 were a colour TV monitor, to replace the black and white unit, and a Polaroid instant camera (a rare piece of American equipment) for their souvenir snapshots. By mid-September the station still held a staggering 2½ tons of dry stores, more than had been on board when the crew first arrived.

Corrections to Salyut's orbit, made through use of Soyuz 37's engines and a test firing of a back-up propulsion system on Salyut which had lain redundant for the past two years, indicated that a new ferry was about to be launched. Soyuz 38, sent up on 18 September, carried the first black spaceman (once again the Soviets had beaten the Americans to the punch), a 38-year-old Cuban air force pilot named Arnaldo Mendez, alongside his Soviet host, Yuri Romanenko. This seventh international mission once again showed the world the advantages of growing up in a socialist society: Mendez had overcome the early loss of his parents and abject poverty as a shoeshine boy to become the first Latin American citizen in space and a credit to the Cuban Revolution. Among the 15 experiments allocated to the following seven days were several bio-medical sessions, and observations of sugar crystallization linked to Cuba's main agricultural product. Photography of the Caribbean was restricted by lack of daylight, though Ryumin and Popov did survey the island of Cuba at other, more favourable, times. Mendez was also obliged to wear special shoes which supported his arches in an imaginative experiment to determine whether they aided his adaptation to weight-lessness. Meanwhile, Yuri Romanenko commented how much the station's equipment had changed since his last visit nearly three years earlier. The two visitors returned to Earth in a pinpoint landing on 26 September, to be greeted with the usual avalanche of awards and medals.

Omission of the Soyuz changeover so frequently practised in the past gave advance notice that the main mission was drawing to a close. Progress 11 docked with Salyut at the end of September, but its cargo was not intended for Ryumin and Popov, although they were obliged to unload the ship. The Soviets clearly intended to reoccupy the station, even though it had long ago passed its designated lifespan; the decision would depend on the status report delivered by the returning crew. Using the Progress engines to compensate for atmospheric drag, Salyut was placed in a safe orbit on 8 October. Three days later, yet another remarkable feat of endurance came to an end when Ryumin and Popov bade farewell to their venerable home and came down to Earth in Soyuz 37.

Although the cosmonauts' time in orbit was only ten days longer than that of their predecessors, and, as such, could not be ratified as a record,* it was further proof of growing Soviet experience and expertise in conquering the problems of long duration space travel. Valeri Ryumin, admittedly by accident, had clocked up almost a whole year in space, sufficient for a journey to the planet Mars, yet suffered no lasting ill effects. Indeed, the returning cosmonauts were in remarkably good shape: both men had actually put on weight, about 10 lb in the case of Ryumin and 7 lb for Popov. Doctors also revealed that they had grown more than an inch in height as their muscles and spinal tissue had stretched slightly, but this was only temporary. They were able to go for a 30 minute walk the day after their return and indulged in a game of tennis before the week was out. Clearly, the limit to human endurance in space had not yet been reached, an opinion seconded by the cosmonauts. 'If it was needed to prove we could go to Mars, then Leonid and I would volunteer right now,' commented Ryumin, the first man to travel 150 million miles.

During their prolonged mission, the crew had carried through a monumental workload: 3,500 photographs of Earth with the MKF-6 camera, another 1,000 with the full-scan KT-140 camera and thousands more with hand-held cameras; 40,000 spectra of the atmosphere and surface; nearly 100 samples of semi-conductors and other new materials; and a continuous programme of repairs which took up about one-quarter of all the available time. This rigorous maintenance routine had proved highly worthwhile for the cosmonauts were able to report that Salyut 6 was still in operative condition more than three years after launch. This success was all the more galling for the American onlookers who had been forced to sit back and watch Skylab debris scatter all over Australia as the first, and so far only American space station disintegrated in the atmosphere. A European-built Spacelab was intended for the Shuttle, but the first test flight of this reusable orbiter was still six months away, and the new craft would have an endurance limit of only ten days.

Salyut 6 still had a useful part to play in the Soviet plans for the future. Its orbit was modified after the first completely automatic refuelling from Progress 11 in mid-November, preparing the elderly station for further visitors a fortnight later. The launch of Soyuz T-3 on 27 November 1980 marked another important step on the Soviet road to recovery – for the first time since the disastrous flight of Soyuz 11 in June 1971, there was a three-man crew on board. This had been made possible by rearrangement of the interior so that it could accommodate three fully suited

---

* Official endurance records were only recognized if they exceeded the previous record by at least 10 per cent.

cosmonauts: the commander sat in the centre couch with the flight engineer to his left and the research engineer on his right. For this test flight the position of commander was held by Leonid Kizim, a 39-year-old air force officer who was making his first venture into space. His two engineers were both civilians: Gennadi Strekalov had spent most of his life in the space design bureau, though only now was he being given the opportunity to fly in space; Oleg Makarov made up for the inexperience of his two companions as he set off on his fourth mission, the first cosmonaut to achieve this distinction.

After the failure of the automatic docking system on Soyuz T-2, there were no hitches this time. Since Progress 11 was still attached, the ferry pulled into the forward port the day after launch. For the next two weeks the three-man crew threw themselves into the most rigorous repair session yet. Using the special tool kit, they succeeded in replacing parts which the station's designers had never expected, or intended, to be removed. Their major task was the installation of a new hydraulic pump unit for the thermal control system, a job which entailed sawing through one of the metal supports and collecting the dangerous metal filings, as well as making sure no liquid coolant escaped. With the help of a direct communications link to Popov and Ryumin, who were relaxing in the Northern Caucasus, the crew also succeeded in replacing parts of the major control and telemetry systems, and in repairing the fuel transfer system. Surprisingly, the labourers found time to conduct a few experiments with the Splav furnace, some plants and a holographic camera based on a laser system developed in conjunction with Cuban scientists but completed too late for inclusion in the September flight by Arnaldo Mendez. Inevitably, something had to give, and that something was the exercise programme. As a result, Kizim and Strekalov apparently suffered a certain amount of 'stress' during the return to Earth on 10 December.

The replacement of Progress 11 by the next in the series during January once more indicated that there was still some life in the old station. Sure enough, on 12 March 1981, a two-man crew was launched aboard Soyuz T-4 to occupy the ageing Salyut. Mission commander was stocky air force colonel Vladimir Kovalyonok, making his third trip to Salyut 6 (the first was aborted after a docking failure). His flight engineer was a 41-year-old rookie Viktor Savinykh, who had begun his engineering career with the State railways before joining the spacecraft design office. Savinykh had only joined the cosmonaut detachment in 1978 but had succeeded in jumping the queue to become the one hundredth human in space and the fiftieth cosmonaut.

From the start it became obvious that this new crew were going to continue at the hectic pace set by their predecessors: on their second day aboard the station, they spent 15 hours in unloading their Soyuz and transferring water from Progress 12 docked at the rear. A busy experimental schedule began with planting seeds in the greenhouse, and further repairs were carried out to Salyut, including the release of a jammed solar panel which had reduced power to the station. This repair was given high priority since the cabin temperature of the laboratory had dropped sufficiently to cause the walls to run with condensation.

Progress 12, loaded with space rubbish, separated from Salyut on 19 March to make room for Soyuz 39. This visiting craft carried another of the Intercosmos guests, a Mongolian with the most unpronounceable name of any space traveller to date. He was 33-year-old Jugderdemidyin Gurragcha, the son of a livestock herder who was called up in 1968 and served as a wireless operator before studying aviation in the Soviet Union and becoming a radio mechanic on helicopters. More significantly, he graduated from the Zhukovsky Air Force Academy in Moscow the year before he began training at Star City. Accompanying Army captain Gurragcha was commander Vladimir Dzhanibekov, on his second call to Salyut. The last week in March was spent with the joint international crew surveying Mongolian territory for possible mineral and water resources, and evaluating the state of pasture land. Most additional experiments were extensions of previous activity, though the holographic camera was used for the first time to transmit images to Earth. Most important was the propaganda aspect of the mission, giving the Soviets a rare opportunity to broadcast to the world the advances made in Mongolia during the 60 years since its 'people's revolution'.

A fortnight after Gurragcha returned through the mist and drizzle of the Soviet steppes, the American Shuttle at last escaped from Earth's gravity and blasted into orbit. As one era began for the USA, another came to an end for the Soviet Union. Soyuz 40, the last of the old-style manned ferries, carried the final Warsaw Pact guest into space. Romanian army air force lieutenant Dumitru Prunariu had been waiting for his chance to fly since 1978, but, as a sign of their disapproval over Romanian independence in foreign affairs, the Soviets pushed his mission back to the end of the line. At the age of 28, Prunariu was taken under the wing of Leonid Popov, joint endurance record holder from the previous year. A week-long session of Earth resources photography, biomedical tests and studies of the influence of space conditions on structural materials ended on 22 May. It was the sign for Kovalyonok and Savinykh to prepare for the final abandonment of the Salyut.

The last crew to occupy the faithful space station came home on 26 May 1981 after 75 days in orbit as the Soviets announced there would be no more manned flights in the foreseeable future. The statistics told a story of endurance and innovation adding up to a string of successes unparalleled in the history of manned spaceflight. Salyut 6 had been visited by five long-term crews and 11 guest crews involving 27 cosmonauts. Of these, six had been on board the station twice, and eight had benefited from the international Intercosmos programme. Seven crewmen had spent more than 100 days on Salyut, and two had endured more than six months. In addition, the station had been supplied by 12 robot Progress craft and one automatic Soyuz ferry, an unprecedented convoy which totally vindicated the Soviet policy of developing automatic systems whenever practicable. There remained one final job for Salyut. On 19 June, an unmanned module designated Cosmos 1267 docked with the station. With a launch weight of 15 tons, it was almost as large as the station itself. Its significance became apparent when the Soviets announced that it was 'designed to test systems and elements of the design of future spacecraft and for training in the methods of assembly of orbital complexes of a big size and weight'. Over the next six weeks, the experimental craft's propulsion system was used to man-oeuvre the station and raise its orbit like an orbital tug, while ground engineers monitored the stability of the complex. Their work completed, the two craft were commanded to re-enter for destruction on 29 July 1982. On its final journey Salyut 6 had helped pave the way for the modular space station of the future.

Even before the death of Salyut 6, its successor had been placed in orbit. Salyut 7, launched on 19 April 1982, differed very little from the earlier second generation space station, the main differences being an improved navigation system, a strengthened forward docking port and modified solar panels. It was a logical step to consolidate the knowledge gained from the tremendous success of Salyut 6, and an example of the methodical way in which the Soviet space programme had been developing over the past decade. The days of space spectaculars which could be ordered at the whim of political masters had mercifully disappeared (apart from the irresistible urge to send up international crews, women and cosmonauts from various racial backgrounds ahead of their American counterparts).

Tests of the new station gave it a clean bill of health, allowing the first long term crew to be sent up in the early afternoon of 13 May 1982 on board Soyuz T-5. According to the usual Soviet custom of placing military membership over experience or engineering ability, the commander was a 40-year-old air force lieutenant colonel, Anatoli Berezovoi, who was making his first flight after 12 years as a cosmonaut. His flight engineer was spacecraft design expert Valentin Lebedev, now fully recovered from the knee injury which had put an end to his mission aboard Soyuz 35. There was some suggestion of a less than perfect rendezvous and docking using the automatic system. Ground control was quoted as saying: 'We understand it was a hard job for you. We lived through every moment with you. But you are in fine form now after such a business.' The cosmonauts shouted 'We've made it' as their ferry pulled into position at the forward port more than 25 hours after launch.

As a home for two or three people, Salyut 7 was not particularly spacious, but the designers had endeavoured to make life as comfortable and relaxing as possible. The station's living and work areas were well lit, decorated with bright colours and kept at a constant temperature and humidity. A table in the living quarters was fitted with a 'range' for heating the food and devices for holding the dishes and utensils in place, though it could also double as a small work bench. Hot water was available at all times, and there was now a refrigerator for storing fresh food ferried from Earth. A showerstall and toilet–washroom unit were fitted in the work module, along with the gymnasium – the training 'bicycle', the 'running track', the vacuum suit and apparatus to stimulate muscle activity. There was even an effort to reduce that most annoying characteristic of spacecraft for men trying to sleep, the humming sound of the equipment and life support systems.

Experimental work soon got under way. Although much of the apparatus was similar to that carried on Salyut 6 – the cameras and furnaces in particular – the sub-millimetre telescope on Salyut 6 had been replaced by a spectrometer and an X-ray telescope. An improved Oasis greenhouse was also carried in an attempt to provide the vital breakthrough in plant growth. A new video recording system reduced the crew's workload by recording and playing back to Earth the results of visual observations, medical and biological experiments. Life was also made considerably easier by perfection of the automatic orientation and navigation system first tested on Salyut 6. Closer medical monitoring was carried out through a new unit called Aelita.

The first departure from routine came on 20 May when the crew pushed a small satellite called Iskra-2 into space through one of the station's air-locks. It had been designed by students at the Moscow Aviation Institute for use by amateur radio enthusiasts, but the timing of the launch coincided with the opening of the Young Communist League Congress. Iskra's useful lifetime was limited by the low orbit which led to high atmospheric drag.

Progress 13 arrived at the back door on 25 May, loaded with the first supplies for the crew and their future visitors. During transfer of the fuel and water, the Soviets revealed that the water tanks on Salyut 7 were located externally in order to save space inside the station. Among the letters, parcels and fresh food were a number of experiments made in France which were to be used in the near future by French 'spationaut' Jean-Loup Chrétien. After one of the shortest stays on record, Progress 13 separated from Salyut on 4 June, vacating the rear port for the first Western European spaceman, on board Soyuz T-6.

Unlike the earlier international missions, this latest venture attracted a lot of attention from the Western media as well as the usual fanfare from the Soviets. Detailed analysis of the mission profile went as far as to include in-depth discussion of the French obsession with haute cuisine:

dishes to be carried aloft included jugged hare, crab soup, country paté, lobster pilaf rice with sauce à l'armoricaine and Cantal cheese. Not everyone was delighted with the forthcoming mission – some 4,000 scientists signed a petition protesting at the French connection with a Soviet regime which had invaded Afghanistan and forced the suppression of the Solidarity trade union in Poland; yet it was the culmination of many years of Franco–Soviet cooperation in space.

Jean-Loup Chrétien was a French air force colonel who had become one of his country's leading test pilots before being appointed as deputy head of Southern Air Defence Division. During his two years of training for the mission, he had learned to speak passable Russian while his Soviet companions took French lessons. Commander was Vladimir Dzhanibekov, now a silver-haired veteran of 40 participating in his third space trip. Flight engineer was another space station veteran, Alexander Ivanchenkov. This was not the original make-up of the crew, however, for Dzhanibekov was obliged to step in when the cosmonaut originally selected, Yuri Malyshev, was taken ill in January. Chretien, himself, had begun training on crutches after a parachute mishap.

In the full glare of worldwide publicity, the mission did not start out in quite the way the Soviets would have wished – the Soyuz computer malfunctioned, forcing the crew to complete the docking under manual control. After the traditional bread and salt, followed by the presentation of a bouquet of flowers, the crew celebrated with their French dinner. Chretien was able to peer down at a brightly lit Paris, then selected his sleeping position on the 'ceiling'. There was no time to be homesick, and little room for privacy with five men on board a Salyut for the first time. During the next week, the crew carried out a variety of materials processing, medical, biological and astronomical experiments. These involved production of new crystals and alloys for use as conductors; use of ultrasonic detectors to study the redistribution of blood under zero gravity; photography of very weak light sources and infrared studies of Earth's atmosphere; and observations of bacterial infections when treated with antibiotics during weightlessness. Some of these experiments continued after the visitors left on 2 July, but many of the results were loaded into the returning Soyuz. The cosmonauts were greeted on Earth with two bottles of champagne – one French and one Soviet.

Advance warning of the next Soviet space spectacular was given by former cosmonaut Georgi Beregovoi even before the Franco-Soviet mission ended. The chief of the Yuri Gagarin Training Centre announced that two women were undergoing training for a visit to Salyut 7. Sure enough, on 19 August 1982 Soyuz T-7 was launched with only the second woman space traveller ever on board, almost a year ahead of the first American spacewoman. Thirty-four year-old Svetlana Savitskaya was a very different example of Soviet womanhood from her predecessor, Valentina Tereshkova. She had grown up with flying in her blood: her father was a noted pilot who went on to become an air marshal. Unable to join a flying club at the age of 16 because she was too young, Svetlana took up skydiving and within a year broke three world records. Admitted to the flying club once she reached 18, she went on to break 18 world flying records, including a speed record exceeding twice the speed of sound. In 1970 she became the women's world aerobatics champion. Later she married a pilot and graduated from the Moscow Aviation Institute as an instructor and test pilot, prior to cosmonaut selection in 1980. As befitted such an independent character, there was little sentimentality in her attitude. 'Emotions interfere with flying and I switch them off,' said the cool cosmonaut. As for her opinions on men? 'I don't divide human qualities into male and female ones,' was the terse reply.

There were some suggestions in the Western media that Savitskaya's preparation had been rushed in the desire to beat the Americans. The space experience that was lacking in the space siren and flight engineer Alexander Serebrov, also a newcomer, was more than made up by mission commander Leonid Popov, still the endurance record holder and now on his third visit to a space station. Inevitably, the media spotlight focused on their female companion, however: it was even admitted that seasoned journalists were laying bets on Svetlana's first words after launch. (In fact, they were: 'G-loads growing slowly . . .') Georgi Beregovoi suggested that a woman's touch would prove of enormous benefit in space, exerting 'an ennobling effect on the group'. There was little sign of this among the four men with whom she was to share the cramped quarters: they presented her with an apron on her arrival. Berezovoi teased her: 'Wouldn't you like to take up the role of "Lady of the House?" ' Savitskaya retaliated: 'If you insist, I'll try it on. But let's specify the work rules first. Housekeeping details are the responsibility of the host cosmonauts.'

Most of the week-long mission was taken up with medico-biological studies. Savitskaya was the guinea-pig whose adaptation to weightlessness would be intensively monitored, but other tests were also conducted in relation to motion sickness and eye co-ordination. Further observations with the onboard scientific instruments were undertaken, and the results loaded on to Soyuz T-5 for the return trip on 27 August. Soviet doctors reported that there were no substantial differences in male and female reactions to weightlessness, and added that Svetlana was fit and well. A tired Svetlana, slumped in a chair, answered reporters in her customary gritty manner: 'I'd be happy to do it again.'

The cosmonauts had maintained their busy schedule between visits, including a two and a half hour EVA on 30 July during which Lebedev removed and partially replaced scientific equipment mounted on the station's exterior. More significant activity during the space walk involved use of special tools to assess the effectiveness of different materials and mechanical joints for future space assembly operations. An entry from Lebedev's diary, published after his return, provides a fascinating first-hand account of the cosmonaut's inner feelings and observations:

> Once the exit hatch was opened, I turned the lock handle and bright rays of sunlight burst through it. I opened the hatch and dust from the station flew in like little sparklets, looking like tiny snowflakes on a frosty day. Space, like a giant vacuum cleaner, began to suck everything out. Flying out together with the dust were some little washers and nuts that had got stuck somewhere; a pencil flew by.
>
> I had no fear, nor any excitement at the moment. I installed TV cameras and a headlight, and got down to work. My first impression when I opened the hatch was of a huge Earth and of the sense of unreality concerning everything that was going on. Space is very beautiful. There was the dark velvet of the sky, the blue halo of the Earth and fast-running lakes, rivers, fields and cloud clusters. It was dead silence all around, nothing whatever to indicate the velocity of the flight . . . No wind whistling in your ears, no pressure on you. The panorama was very serene and majestic.

Three further Progress craft docked with Salyut after the departure of Soyuz T-6 in June, bringing welcome mail and supplies as well as providing a back-up propulsion system which could be used for orbital manoeuvres. Among the cargo were cassettes of the cosmonauts' families and of birds singing. Lebedev admitted to reporters that he had never been a gardener but found it very rewarding looking after their kitchen garden of cucumbers, radish and cucumber grass, as well as the specimens of wheat and peas. 'It's nice touching the sprouts with the palm of your hand – they tickle you,' wrote the normally unemotional engineer. Further evidence of the sensory deprivation which crept up on the crew was their tinkering with artificial fragrances. Berezovoi explained: 'By the time of the second expedition, we were able to create the smell of roses and Svetlana was very, very surprised, asking where we had roses in bloom. We missed the smell of flowers, the city noises and city smells.' He went on to explain the importance of listening to taped birdsong, falling rain and rustling leaves. 'You just close your eyes, and suddenly you're back on Earth,' said the commander wistfully.

The psychological problems to be overcome loomed ever larger on the seven-month mission. As Lebedev confided in his diary:

> The most difficult thing in flight is not to lose temper in communication with the Earth and within the crew, because accumulating fatigue makes for frequent slips in contacts and in the

work with Earth. In the crew, too, these explosive moments happen, but no eruption must be allowed, for a crack, if it appears, will grow wider.

As the days passed, Lebedev's mind continually dwelled on home and family, wondering how they were and when he would be able to rejoin them. One lighthearted interlude broke the monotony on 14 July, the birthday of Berezovoi's daughter, Tatiana.

We cooked a pie from bread packets. Felt-tipped pens did duty for candles, and bits of foil acted as imitation flames. There were also electric candles – in the form of four hand torches. To make them eight, the number of years, we set a mirror behind them and the reflections added to eight. We suspended coloured balloons everywhere, and rode upon a vacuum cleaner and a balloon. We entertained the girl over television the best we could.

Towards the end of October, the crew told ground controllers they were happy to continue with their flight. Accordingly, Progress 16 was sent up in mid-November, just as the crew passed the previous endurance record. A few days later, another Iskra satellite for use by amateur radio enthusiasts was ejected from Salyut's air-lock, but, inevitably, the cosmonauts' work rate was slowing after more than six months in orbit. As the mission crept towards its finale, the Soviets announced that the crew had carried out more than 300 major experiments and taken more than 20,000 photographs with the various cameras on board. The latter included numerous requests from engineers involved with major construction work in Siberia, as well as surveys of water resources, geology, mineral resources and forests. Discoveries included a new gas field in the Volga basin, another possible oil and gas source between the Caspian and Aral Seas, and a potential metallic ore site in Siberia. Foresters were given early warning of forest fires and of pest attacks, farmers were informed of potential problems with their harvest, and fishermen were directed to the more productive areas of ocean in an operation estimated to have saved the Soviet fleet more than 20 million roubles. The Soviets also reported the first pilot production of monocrystals in the new Korund furnace, and the successful growth on Earth of seeds obtained in space from the Arabidopsis plant.

With the mission at last drawing to a close, Lebedev found himself thinking of the welcome which awaited them, and, quite unnecessarily, wondered whether their mission would be rated a success. Having stowed their gear and shut down the space station, the two voyagers bade farewell to the giant cylinder which had been their home for the past 211 days and 80 million miles.

Their night landing on 10 December was just a little too rough for comfort, as the wintry weather once more made its mark on a returning manned mission. In almost zero visibility due to a raging blizzard and low cloud, the capsule was blown off course then rolled over several times as it landed on a snow-covered slope. It took 20 minutes before search helicopters spotted the flashing beacon on the Soyuz, but landing proved far from easy as the loose snow aroused by the rotors obscured the ground. One helicopter crushed its landing gear as it set down in a dry stream bed. A second attempt was more successful, enabling rescue crews to escort the freezing cosmonauts to a waiting vehicle. With the helicopters grounded until conditions improved, the weary heroes, weakened by their long spell in zero gravity, were forced to spend one of the most uncomfortable nights of their lives huddled in a tractor cab. Fortunately, there was no permanent effect on the cosmonauts from the ordeal. Doctors reported that the men had healthy appetites and were as fit and well as could be expected, though it took a week of recuperation before they could walk more than a few steps unassisted.

While the Soviets absorbed the lessons of the record-breaking mission. Salyut 7 continued alone in orbit. Not until March 1983 was it joined by another craft, a remote-controlled experimental module labelled Cosmos 1443. The Soviet media described the newcomer as an improved version of Cosmos 1267 which had docked with Salyut 6 in June 1981. The multi-purpose craft could be used as a cargo ship capable of carrying two and a half times as much payload as the Progress ships, or as an inter-orbital tug. Unlike the Progress series, however, the Cosmos could return half a ton of film and equipment back to Earth in a re-entry module. Attached to Salyut 7, the combined assembly was nearly double the length and weight of the station alone, and provided a 50 per cent increase in work space for any future occupants. As the Soviets freely acknowledged, satellite ships of this class would eventually form the basis of permanent space stations in the not-too-distant future. On this occasion, Cosmos 1443 delivered some 4 tons of supplies and equipment to the station, including extra solar panels to be installed by the next crew. Its engines were used in early April to lower Salyut's orbit to around 190 miles (300 km), a sure sign that another manned launch was imminent.

The three-man Soyuz T-8 was duly launched towards the orbital complex on 20 April 1983. Commander was 36-year-old Vladimir Titov, a Soviet air force officer and relative of Vostok cosmonaut Gherman Titov. He and civilian spacecraft designer Gennadi Strekalov had served as back-ups to the previous record-breaking crew, so the evidence seemed to point towards another long flight. Third crewman was another civilian engineer, Alexander Serebrov, one of Svetlana Savitskaya's escorts in 1982.

Their craft was successfully placed in a low chase orbit, then raised after four orbits in the normal manner for the gradual rendezvous sequence. Then came the media blackout. Something serious had obviously gone wrong. Eventually, the Soviets released a brief statement explaining that 'because of deviations from the planned approach regime, the docking was cancelled'. Behind this terse announcement lay hidden the dramatic struggle of the crew to salvage their mission. Some months later, an unusually frank article by Titov described how the Soyuz rendezvous radar failed to deploy early in the flight. Attempts to shake it free by 'wiggling' the craft also failed, so Titov applied for permission to try the first 'seat-of-the-pants' rendezvous and docking with a space station. Estimating angle and distance through the periscope, the crew succeeded in closing the gap so that Salyut was in view by orbit 19 and closing fast. Unfortunately, at this stage the craft moved into orbital night and out of contact with ground control. Illuminated only by a Soyuz searchlight, Salyut's front docking port was not clearly visible to the approaching cosmonauts. Titov closed in to less than 200 yards (160 m) until discretion became the better part of valour. Fearful of a disastrous collision, the pilot hastily reversed thrust and sheered away. Once radio contact was regained, they were advised that the fuel situation was too critical for another attempt. Preparations were made for a hasty re-entry, and the unlucky cosmonauts touched down safely in the normal recovery zone at 5.30 pm Moscow time on 22 April.

The next occasion when the light conditions would be suitable for a docking attempt with Salyut–Cosmos 1443 was in June, an opportunity which the Soviets took by launching Soyuz T-9 with two men aboard on 27 June 1983. Assigned as commander was Colonel Vladimir Lyakhov, companion to Valeri Ryumin on the 175-day mission to Salyut 6 four years earlier. This was clear evidence that a long duration flight was planned. Accompanying him was 40-year-old civilian engineer Alexander Alexandrov, making his first trip into space. A specialist in spacecraft control systems, the rookie cosmonaut had applied for selection in 1965 but had been turned down, only to see his perseverance rewarded in 1978. The new crew docked successfully at Salyut's rear port, their Soyuz craft bringing the total weight of the three craft complex to about 47 tons. Settling into the new environment proved easier second time around for Lyakhov, though Alexandrov did admit to a rush of blood to his head and a feeling of thirst.

After activating the station, one of their first jobs was to begin unloading Cosmos. Since this was a new task they took their time, clearing the descent module first while all the time filling the empty racks with household rubbish. There was little opportunity to relax, however, during their 12 hour day. Apart from unloading the cargo ship and general housekeeping, there were over 80 experiments to conduct as well as the strict exercise regime. In order to prevent exhaustion, a new work routine was introduced whereby the men would work for half a day on

Saturdays instead of trying to cram everything into extended work days. Nevertheless, they were often caught putting in extra hours during off-duty time in the first month. Despite the grinding work schedule – they shot more film in one week than the previous crew had done in 211 days – they succeeded in remaining healthy and cheerful, actually managing to put on weight.

A vivid illustration of the dangers of spaceflight was given on 25 July when the cosmonauts reported a loud crack against a window. They discovered a small pit in the toughened window created by a micro-meteorite impact. Although this projectile was too small to cause a disastrous decompression, it was a reminder of the space debris and radiation that pose a potential threat to any orbital craft. The Soviet authorities added that it happened at an appropriate time since the crew were practising for emergency evacuation of the station. The standard time for such a procedure was given as 90 minutes, equivalent to one complete orbit, though it could be cut to 15 minutes if the station was in dire straits.

Much of their time during July was taken up with photographing territory of socialist countries, in response to requests from their governments, as well as regions of their vast homeland. During these sessions Lyakhov often shut himself away in the forward transfer module with the lights out to reduce background interference.

As August began, the crew's activities switched more towards biological and technological studies. These included experiments to produce semi-conducting alloys in the furnaces, production of pure biological materials and film records of changes in various materials exposed to space conditions in the air-lock for different periods. The initial results were stored in Cosmos, along with redundant equipment which included faulty regenerators and a malfunctioning memory unit from the Delta navigation computer. Although everything they moved was weightless, the crew complained that manoeuvring large, bulky items made their arms tired and brought on more general fatigue. They were glad when their removals were completed so that Cosmos 1443 could undock for a return to Earth, leaving the front port free for fresh craft. The Cosmos pulled away on 14 August and conducted its own tests before separating into two sections. By the time the descent module soft-landed on 23 August, its place on the nose of Salyut had been taken by Soyuz T-9 and Progress 17 had pulled in at the rear entrance with welcome supplies and mail. One change which the cosmonauts appreciated was the loading of the cargo ship with fresh food only hours before launch instead of several days before.

Refuelling took place during the second week in September. It marked the start of one of the most difficult periods of the Soviet space programme in recent years. Specialist publications in the West began to tell a story of unmitigated disaster in early October as rumours spread about a fuel leak on 9 September which was said to have 'left Salyut 7 with minimal manoeuvring propellant and flying mostly in a free drift mode with 16 of 32 attitude control thrusters not usable'. In addition, the station's back-up main engine was also said to be out of action, and an unnamed American analyst was quoted as saying, 'Salyut 7 is essentially dead in the water'. As if this was not bad enough, news began to filter through from unofficial sources of an explosion on the launch pad at Baikonur during countdown for Soyuz T-10. Initial reports suggested that three crew members had been injured after a launch abort, one of them a woman, but this was later modified when it became clear that only two men had been on board. The unfortunate cosmonauts turned out to be Vladimir Titov and Gennadi Strekalov, fresh from the failed docking of Soyuz T-8. Speculation abounded that the destruction of the Soyuz and its launch pad had condemned the two orbiting spacemen to a lingering death: one newspaper printed the banner headline 'Marooned In Space'. According to this inaccurate reasoning, the old Soyuz T-9 was almost at the end of its safe operating life, and may have been damaged anyway by the fuel loss earlier in the month. Even if another relief craft could be sent up, it was argued, there was no way in which four men could fit inside the three-man Soyuz descent module. The successful docking of Progress 18 on 22 October was simply seen as a slight respite from the threat of starvation or suffocation, though it was further suggested that a rescue might be possible using an American Shuttle.

Soviet officials vehemently denied that there had been a fuel leak or that their cosmonauts were in danger. No mention was made of any launch explosion. Under this blanket of silence, the atmosphere was ripe for wild speculation and rumour-mongering. Once again, the Soviets had brought the bad publicity on themselves through poor public relations and lack of honesty. Details of the launch abort only came out after the International Astronautical Federation (IAF) Congress in Budapest in mid-October, though Soviet citizens were not informed until the completion of the mission, and then only through the external services of Radio Moscow.

Western experts had been expecting a manned launch in early October during the normal Soyuz launch window, so the preparations for a 27 September launch suggested a mystery reason for such a hurried mission. Perhaps this rush was the basic cause of the drama which followed. Apparently, there was a fuel leak some 90 seconds before completion of countdown when a valve failed to close. A fire spread so rapidly that the wiring to the automatic abort system on the Soyuz craft was severed, so ground controllers rushed to send the required signal before the men disappeared in a fireball. By the time the small escape rockets fired, the rocket beneath them was rapidly turning into a tower of orange flame. The shroud on the nose of the booster separated from the rest of the rocket and sent it into the night sky above the cosmodrome. Perhaps ten seconds later, the descent module was cut free of the orbital module, enabling the back-up parachute to deploy and the heat shield to fall away, exposing the three soft touch-down braking rockets on the bottom of the descent module. Touchdown occurred about 2 miles away from the blazing launch pad. Rescue teams found Titov and Strekalov shaken but unharmed: the only medicine they required was a large glass of vodka.

The Soviets announced 'substantial changes' in the Salyut flight programme after this setback, supposedly to offset depression which often beset long-term crews when they felt lonely and irritable. There was certainly some excuse for Lyakhov and Alexandrov to feel depressed, for, unknown to Western analysts, the station's cabin temperature was down to 18°C with high humidity leading to miserable, damp conditions on top of all their other worries. Furthermore, the rumours concerning the fuel leak on board Salyut, though exaggerated, had been based on fact. Salyut's power supply had been substantially reduced, possibly because of the leak of corrosive fuel, causing the unpleasant environmental effects and drastically restricting the experimental work possible. In this context, the hurried launch of Titov and Strekalov (now known as Soyuz T-10A) began to make sense: the men were intended to act as a repair crew, assembling the extra solar panels which had been delivered by Cosmos 1443. If Salyut was to remain a viable space laboratory, it was imperative that repairs be initiated as soon as possible. With the failure of Soyuz T-10A, Lyakhov and Alexandrov were now obliged to make the attempt themselves, despite their lack of training for this specialist task. So, instead of going home in September, the cosmonauts found themselves preparing for two tricky space walks at the beginning of November.

The first excursion began on 1 November during a period of maximum sunlight for each orbit. Alexandrov exited first from the hatch, pulling himself some 20 ft (6 m) along the shell of the station with the help of hand rails before anchoring himself in position by the giant solar panel. Lyakhov followed with the assembly tools and the container which held the folded additional panel. The men then connected the plug and socket units before winding the panel into position with a winch, just like raising a flag. Most of the deployment sequence was carried out in TV and radio contact with ground control so that advice could be passed on from two cosmonauts conducting a simulation in a water tank at Star City. Installation was completed after 2 hours 50 minutes, five minutes less than the second EVA which took place on 3 November. On the second occasion, the additional panel was raised on the opposite side of the main panel to the first auxiliary. Doctors monitored the physical performance of the men very carefully, especially since they had earlier been complaining of fatigue and muscle

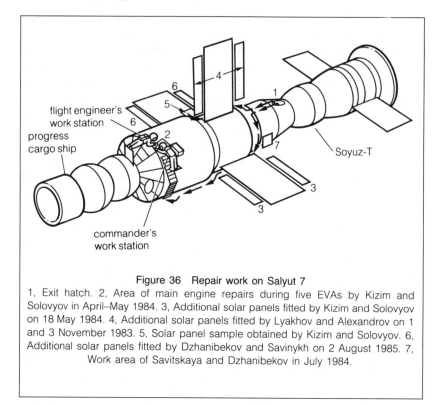

Figure 36   Repair work on Salyut 7
1, Exit hatch. 2, Area of main engine repairs during five EVAs by Kizim and Solovyov in April–May 1984. 3, Additional solar panels fitted by Kizim and Solovyov on 18 May 1984. 4, Additional solar panels fitted by Lyakhov and Alexandrov on 1 and 3 November 1983. 5, Solar panel sample obtained by Kizim and Solovyov. 6, Additional solar panels fitted by Dzhanibekov and Savinykh on 2 August 1985. 7, Work area of Savitskaya and Dzhanibekov in July 1984.

contractions, but all went well in an operation described by the Soviet media as 'a landmark in Soviet space exploration'. Once again, the value of men in space had been admirably demonstrated.

Concern over the men's health continued during the final weeks of the mission: doctors ordered them to add more salt to their diet in order to retain more body water and stabilize blood volume. Increased exercise was also prescribed in the usual end-of-flight effort to tone up lazy muscles. Progress 18 unloading was completed by mid-November, allowing the cargo ship to be undocked for destruction. A test firing of the Soyuz T-9 propulsion system proved how wide of the mark the scaremongers had been in the Western media, and on 23 November the two cosmonauts floated into the craft for a safe return to Soviet soil.

Fog in the landing zone prevented their being airlifted to Baikonur until the following day, so they spent most of the night talking to rescue personnel. After 150 days in space the men were weak and easily tired, but their recovery was swift. At the post-flight press conference, Lyakhov finally admitted that there had been 'a slight leak' in the propulsion unit. Training chief Vladimir Shatalov added: 'We switched off that part of the subsystem and the station continued to function . . . The failure had no effect on the craft's ability to manoeuvre, and other systems worked normally. The incident didn't affect either the flight or the research programme.' Western observers understandably greeted this announcement with a fair degree of scepticism.

As 1984 dawned, the Soviets continued to pour scorn on the reports in Western media; they described the Salyut 7 as in 'excellent working order'. Even more pointed was the comment that the station would soon receive further crews and was 'expected to continue in operation for a very long time'. The two additional solar panels had been installed simply to provide the extra power needed by the larger furnaces that would be used to produce semi-conductor materials, and to compensate for the deterioration of solar panel efficiency under the micro-meteorite bombardment.

On this occasion, the Soviets were as good as their word. A three-man crew was blasted into orbit on 8 February 1984 in a craft labelled Soyuz T-10 – in the official Soviet mind, the previous aborted mission had never existed. Commander of the latest crew to occupy Salyut 7 was Colonel Leonid Kizim, now on his second flight to a space station during a 19-year stint in the cosmonaut corps. His flight engineer, 37-year-old Vladimir Solovyov, had the usual qualification of civilian spacecraft designer. The nearest he had been to flying in space was when he and Kizim accompanied Frenchman Patrick Baudry on the back-up crew of Soyuz T-6. Also a space newcomer, and out of the usual cosmonaut mould, was the third crewman, 34-year-old physician Oleg Atkov, a specialist in ultrasonic investigation of heart action. The young doctor had helped develop a portable ultrasonic cardiograph used on the 211-day mission, a clear sign that this was to be another long-duration attempt which would be closely monitored by the space doctor and his colleagues on the ground.

Manual docking with Salyut was successfully completed on 9 February. Soon afterwards, the cosmonauts crossed the threshold of their new home to be greeted by a letter and bunch of flowers from the previous crew. They wasted no time in activating the station and carrying out minor repairs during a longer than usual working day of six hours. Atkov eagerly began his studies of body adaptation, reporting some flow of blood to their heads but no major discomfort. An automatic Progress craft docked at the rear door on 23 February, carrying more than 2 tons of fuel, food and scientific equipment. In order to enable them to continue their research, the cargo ship unloading was spread over several weeks. Meanwhile, Atkov began operation of a new cardiograph brought by the Progress and helped his colleagues activate the Yelena gamma ray detector. Among the Earth photography sessions were the first observations of India, a programme which would continue during the imminent visit of the first Indian cosmonaut.

The man in question was Indian squadron leader Rakesh Sharma, a 35-year-old test pilot who had been in training at Star City since September 1982. One of his first tasks was to learn the Russian language sufficiently to understand technical instructions. He was then able to practise on the Soyuz and Salyut simulators, study space navigation, spacecraft design and space medicine and undergo sea survival training. In October 1983 he was assigned to the prime crew along with Yuri Malyshev and Nikolai Rukavishnikov, but the unlucky Rukavishnikov (he had yet to set foot on a space station after three flights in a 17 year career) was taken ill and back-up Gennadi Strekalov stepped into the breach, now fully recovered from the launch abort five months earlier.

Amid a blaze of publicity, Sharma and his companions blasted off on board Soyuz T-11 on 3 April 1984 in a rare live broadcast from Soviet TV which was also transmitted to Indian viewers. VIP guests at the cosmodrome included the Indian Defence Minister and the ambassador to the Soviet Union, but there had earlier been something of a minor diplomatic incident when Soviet officials refused to allow Indian journalists access to the launch site. Western media attention mainly focused on Sharma's food and avowed intention to practise yoga instead of more traditional exercises. His traditional menu included curry, pineapple and mango juice, crisp bananas and mango fruit bars. Also in his personal luggage were a number of video cassettes of Indian films and a cassette of Ravi Shankar's music. Sharma's week-long visit would clearly be quite a culture shock for his Soviet companions.

Another live TV broadcast showed the Soyuz docking with Salyut during its eighteenth orbit. Shortly afterwards, the crews met in a series of warm embraces amid solemn ceremonial. Sharma brought portraits of Indian leaders to be placed on the walls of the work module, a national flag, emblems of participating Indian agencies, and some soil from the Mahatma Gandhi memorial. It was the first time that six people had occupied a Soviet space station, hardly the most spacious of vehicles at the best of times. Two days later, there were 11 humans in space simultaneously as the cosmonauts were joined by five Shuttle astronauts, but there was no direct contact between the craft as the East–West divide was carried into space.

Much of the experimental time was taken up with photographic surveys of India and the surrounding ocean intended to identify sources of oil and other minerals, to chart remote mountain areas, land utilization and coastal erosion. Results from the mission were later said to have added to knowledge of glaciers and potential freshwater supplies in the northern mountains of India, and to have improved knowledge of plankton distribution as well as pollution in the main fishing zones. Another practical observation was sent back on the fourth day when the men reported a big forest fire in Burma. Additional work with the furnaces led to an alloy of silver and germanium being produced. But it was not all hard work. Time was allotted to interviews with Mrs Gandhi, the Indian Prime Minister, with the press and with members of the crews' families. Results of the mission were loaded into Soyuz T-10, the craft to be used for the return trip. Sharma and his companions touched down safely on 11 April in a frost-covered field, to be greeted by eager reporters. The Soviets commented that the crew had been so keen they had almost exhausted Salyut's film stocks. A beaming Sharma autographed the descent module and claimed that his yoga had helped alleviate backache during the mission.

Once the ballyhoo was over, the three experimenters in orbit were able to settle down again into their normal routine. They shifted the fresh Soyuz to Salyut's nose so that Progress 20 could pull alongside on 17 April, but the new ship's cargo differed in one important respect from that of previous supply ships – equipment for repair of the propulsion system was included. It was the signal for the initiation of the most intensive space walk programme yet. As Kizim and Solovyov prepared for their first excursion, the Soviets once again demonstrated their newfound flexibility and capacity for innovation when a folded framework with a work platform attached was deployed from the Progress at the station's rear end. This was to be the site of intense repair activity over the next few days.

Leaving Atkov inside the cabin to monitor their progress and physical condition, the crew exited from the transfer tunnel on the morning of 23 April. Solovyov took the lead, swinging along the hand rails to the equipment bay, then slowly fitted the folding ladder which Kizim passed to him. Conditions for such activity were at their optimum, with long periods of sunlight and long communications sessions possible during the passes over Soviet territory, but it was tricky work in the bulky pressure suits and gloves. One earlier cosmonaut had likened it to threading a needle in boxing gloves. Once Solovyov and Kizim had tested their work stations – one on the ladder and one on the Progress platform – and fixed the tool boxes in place, it was time to head back to the hatch. All they had achieved in four and a quarter hours, a Soviet space walking record, was the preparatory stage of what was obviously going to be a long and difficult series of repair sessions.

The second EVA followed three days later, at last revealing what had been suspected long before by Western observers. Official reports explained that the crew 'opened the protective cover of the switched-off part of the back-up feed line of the unified propulsion system and installed a valve'. That completed, the line was purged and its sealing checked before the crew could return inside for a well-deserved rest. They had spent another five hours in space, a session extended for about an hour by that most frustrating of obstacles, a nut reluctant to turn. Despite this, the eager crew asked for the next stage to be brought forward, so EVA 3 took place on 29 April. This time a new conduit was added to the reserve system and its sealing tested, an operation which took nearly three hours. A delighted mission control advised the cosmonauts to have a good rest. The next few days were taken up with light duties and greetings to May Day revellers on Earth. A similar operation was repeated on 3 May as a second extra conduit was installed during almost three hours outside the hatch.

Soviet reports suggested that the station's back-up fuel system was now successfully repaired, but the space walking cosmonauts were not finished yet. Progress 21 replaced its sister ship on 10 May, bringing more equipment for installation as well as the usual mail, food, water and air. On this occasion the record-breaking space walkers were to affix two small solar panels to one of the main array in a repeat of the November EVAs. This fifth excursion took place on 18 May, Kizim and Solovyov hoisting the two new panels on either side of the large solar 'wing' after a struggle with the portable winch. On this occasion, however, they had the benefit of intensive training and so were able to complete the operation in one three hour session, though their task was made somewhat easier by Atkov rotating the main panel through 180 degrees so that the men hardly had to shift position. Everyone was delighted with the standard of efficiency and 'high workmanship' demonstrated by the crew. Kizim's pleasure increased still further a week later when ground control informed him that he was a father again – his wife had just presented him with a daughter.

Normality settled in at last with the arrival of Progress 22 on 30 May, marking the return to the less spectacular side of a cosmonaut's life – Earth photography, medical and biological studies, and observations with the X-ray telescopes. Targets for photography included the remote areas of Siberia, the harvests of the Ukraine and the Atlantic and Pacific Oceans. Refuelling was completed during the first week in July, enabling Progress 22 to separate for destructive re-entry and make way for the next human visitors. As expected, Soyuz T-12 blasted off from Baikonur on 17 July amid another fanfare of publicity. Commander of the mission was that experienced space chauffeur, Vladimir Dzhanibekov, participating in his fourth short-term flight to a space station. Alongside the silver-haired veteran was newcomer Igor Volk, an experienced test pilot and engineer who had joined the cosmonaut group in 1978 at the age of 41 and now had civilian status. However, it was neither of these crew members who attracted the publicity, but the redoubtable Svetlana Savitskaya. Svetlana, now aged 35, had been promoted to flight engineer for this mission, and was undoubtedly a courageous and capable pilot, but the Soviets, as always, made the most of the propaganda value from her presence by pointing out that she was the first woman to fly twice in space. A few days later she notched up another first by walking and working outside the station. The Americans had been beaten yet again.

The visiting crew spent a busy week carrying out experiments on the space station. These included production of vaccines and antibiotics, photography of materials exposed to space, Earth resources photography and studies of Earth's atmosphere. Most media attention was predictably directed to the pioneering space walk by Savitskaya and Dzhanibekov on 25 June designed to test 'a new general-purpose hand-operated tool'. Carrying the new universal tool, Dzhanibekov moved out first to the work position close to the hatch. Svetlana followed close behind, anxious to begin work. Flight Director Valeri Ryumin told her: 'Well, if you're so impatient, you'd better start.' Using the electron beam gun, she set about cutting through a sample of titanium, then welded two metal sheets together, tried out soldering with lead and tin, and finally sprayed a silver coating on to two black discs. The first lady of space found the work tiring and complained that the bright sunlight got in her eyes, but she professed herself pleased with her efforts. Dzhanibekov, who had been recording the moment for posterity with a TV camera, then swapped places for a repeat performance. Finally, the duo collected samples left on Salyut's exterior and headed back inside after 3 hours 35 minutes of EVA. Dzhanibekov later commented: 'You work much better with a woman beside you.' The woman's touch also had its impact on the other crew members: Solovyov, who normally hated potatoes, was persuaded to eat some after she had cooked them, and it was also reported that some crewmen began shaving twice a day! The historic mission ended on 29 July with the soft-landing of Soyuz T-12 and the inevitable siege by reporters.

During their visit, the T-12 crew had taken the opportunity of briefing Kizim and Solovyov on the details of another repair operation. So, for the sixth time in three and a half months, the two cosmonauts drifted out of the air-lock in the afternoon of 8 August 1984. Following the procedure they had learned from a videotape and discussions with their comrades, they used a new pincer-like tool to seal off the fuel pipe which had leaked the previous year. At last the repair sequence was finally completed. On their way back to the hatch, they cut off a sample from a solar panel so that it could be analysed on Earth. The five hour space walk brought their total EVA time to about 24 hours, a remarkable feat for a programme which had clocked up a mere eight hours outside the air-lock before November 1983.

The remainder of the mission was something of an anticlimax. Yet another Progress supply ship arrived in mid August with fuel and equipment, including the Soviet–French X-ray spectrometer called Sirene. This instrument was set up in the vacuum of the adapter module and used during September to make astronomical observations of exotic structures such as black holes, pulsars and the Crab Nebula. Another block of work involved joint studies with unmanned satellites, ships and aircraft to evaluate observations of large water bodies such as the Caspian Sea. Atkov's intensive medical check-ups continued, while an experiment to separate DNA samples, essential for genetic engineering, was salvaged by an in-orbit repair job. By the end of September, the crew had passed the previous endurance record and were well into uncharted territory. They began to wind down their experiments ready for the long-awaited return to mother Earth. Departure from the mothballed Salyut took place on 2 October after a routine meal and donning of pressure suits.

After almost 237 days in weightless conditions, it was not surprising to see the three cosmonauts lifted gently from their capsule on to special lounge chairs prior to preliminary medical tests in a tent set up near the landing site. They looked very tired, though pleased to be home and enjoying the 'dense and tasty air'. Re-adaptation was a gradual process as they learned once more, like children, to walk unaided, swim and play sports. Doctors likened their condition to that of people who had been bedridden for a long time. Solovyov commented: 'I wake up in the morning and my first thought is, "Why didn't I break the bed?" That's how much I feel the weight.' Another strange sensation was the increase in height of several inches: none of their clothes fitted properly, though this was purely temporary. The decrease in bone calcium and blood volume took some time to reverse, but by 25 October the men were fit enough to withstand a full press conference, followed a few days later by an award ceremony led by President Chernenko. For once, the ailing President could not be accused of exaggeration when he described the mission as 'an unforgettable page in the chronicles of space exploration'.

As the space station approached its third birthday, the perennial speculation arose once more concerning the next step along the road for the Soviets. Further visits to Salyut 7 were expected – one report based on comments by Nikolai Rukavishnikov stated that a flight by an all-woman crew was imminent – but as the time period without a crew stretched to six months, there was no sign of such visits materializing. Meanwhile, the tone of Soviet statements underwent a subtle change: instead of the station being in a state of preparation for a new crew, a Tass update commented that 'the planned programme of work . . . has been completely fulfilled'. Soon after this negative statement, it became clear that all telemetry from the station had ceased. With a lifeless Salyut in a decaying orbit, there seemed little doubt that there had been a major malfunction on board which had condemned the craft to a premature death. Three more months passed with the cold, empty craft tumbling out of control, assigned to the scrap heap by most observers, until the announcement came through on 6 June 1985 that a two-man crew had been launched on board Soyuz T-13 in a last ditch effort to breathe life into the patient.

The launch of Vladimir Dzhanibekov and Viktor Savinykh was a calculated gamble by the Soviets, though the difficulty of their task was not immediately clear. Apparently, a long term crew of Vladimir Vasyutin, Viktor Savinykh and Alexander Volkov had been in training for a routine Salyut occupation, but a radical revision of the plan was required after the ground controllers lost contact with the station. Savinykh remained a first choice for a rescue mission since he was an experienced cosmonaut and an expert on Salyut design, but there remained the thorny problem of who would accompany him. Dzhanibekov was chosen since he had all the relevant experience, having already completed four space missions, a 'seat-of-the-pants' approach to Salyut after a radar failure, and a long space walk as well as a repair briefing for the previous record-breaking crew. For the veteran cosmonaut, wrenched from the peaceful contemplation of a weekend with his family, the visit of Alexei Leonov resulted in mixed emotions, though it meant the fulfilment of his 'most cherished dream' – a prolonged orbital expedition.

It was evident that the approach to an uncooperative target such as Salyut would be a major headache, but the designers of Soyuz did their best to help by installing controls alongside the portholes which gave the commander a wider view of the target. As Dzhanibekov concentrated on the view from his windows, Savinykh called out distance and speed, and kept an eye on the spacecraft systems. 'Its crimson colour gradually grew lighter, and finally became white, the shade of ivory', recalled the five-time veteran. 'As we neared the craft the separate elements of its construction became distinguishable, and we saw the wings of the solar batteries. At first they seemed to be correctly orientated towards the sun – a glimmer of hope. But within a few minutes it was obvious that this was an optical illusion.' It was clear that resuscitation of the ailing station would be a great problem, even if docking could be achieved.

With consummate skill, Dzhanibekov steadied the Soyuz so that it hovered less than 600 feet (180 m) from the space station, performed a somersault and pulled alongside the slowly tumbling Salyut. Docking was finally executed at 12.50 pm Moscow time on 8 June. After the usual checks, the crew opened the hatches and entered the transfer tunnel. While they tested the atmosphere for fumes – there was a possibility that a fire had disabled the station – the cosmonauts noticed the unusual 'oppressive silence and stillness of the air'. They moved on into the large work section, wearing gas masks as a precaution and guided only by their torches, since the metal window shades were pulled down. On removing their masks, they noted a 'stagnant machine shop smell'. At the table they discovered a packet of rusks, tied up with adhesive tape, and some salt tablets, together with a letter, a souvenir doll and some eau de cologne left behind by the previous crew. It was a touching moment, but, as they looked despairingly around the darkened cabin, it looked more and more like a hollow welcome. Further investigation did nothing to change their minds: the windows were covered in hoar frost, the cabin temperature was below zero, there was no power in the batteries and the drinking water was frozen. As Dzhanibekov later admitted, 'We should have simply returned home, but, with help from Earth, we began to look for a way out of the apparently hopeless situation.'

The search began for a way of bringing Salyut back to life. Without power, the station could not be revived, so the first priority was to link each chemical battery in turn to the solar panels. After severing the cable and reconnecting the first battery so that they bypassed the entire circuit, the cosmonauts sat back and hoped that the idle panels would receive sufficient sunlight during each orbit to create a current to the battery. To their delight, the battery began to charge. As each battery reached full charge, so the crew switched over to the next during a night pass. There was no time for sleep until their work was completed, but the conditions were far from ideal for efficient workmanship. In the absence of ventilation the carbon dioxide they exhaled tended to accumulate around them, resulting in tiredness and headaches. Despite their fur boots, hats and gloves, the low temperature also took its toll, eventually forcing them to work only in daylight and retire to the Soyuz cabin every 40 minutes for warmth. When the time came for sleep, they closed the hatch and settled down in the Soyuz cabin. There was still the problem of drinking water. Until they could defrost the tanks, they were forced to cut their daily ration, utilizing every drop available from the EVA suits and condensation from various pipes and hoses. As for hot food and drink, they hit upon the idea of improvising a stove from a metal container and a powerful photographic lamp: after 30 minutes or so, they were able to heat up packets of food, tea and coffee. Both men agreed that this was their salvation.

It was not until 14 June that Tass was able to announce that the station was now working normally, though tests of systems continued and the water supply had still not been reactivated. The first experimental work began on 19 June with a series of Earth photography sessions tied in with satellite observations and ground studies, but of more significance was the arrival of Progress 24 at the back door carrying much-needed water, food and spare parts. These included new chemical batteries and ventilation fans which they fitted as soon as possible. At last the two overworked cosmonauts were able to settle into the normal work routine with weekends off. Progress 24 was loaded with rubbish and separated from Salyut on 15 July, having successfully secured the station and its salvage crew. It was replaced almost immediately by another ship, apparently of the Progress type though designated Cosmos 1669 (at one stage, however, the Soviet media referred to it as Progress 25).

On board the new freighter was a variety of experimental equipment, and a set of small solar panels. These were added to the third main panel during a five hour EVA on 2 August, completing the work begun in

November 1983. They were helped in their installation by more mobile pressure suits and improved helmets with larger visors and small lamps fitted for use during night passes. The space walk ended with retrieval of biological samples and emplacement of new samples, including a micrometeorite detector. It was the final proof, if further verification was needed, of the value of man in space for the repair and modification of all-too-vulnerable hardware.

The remainder of August passed peacefully as the crew at last found time to conduct Earth observations, analysis of coatings exposed to space, and growth of crystals. They were also delighted to raise cotton shoots successfully from seed in the Oasis greenhouse, despite the sub-zero temperatures which had prevailed a few weeks earlier. At the end of the month, Cosmos 1669 was undocked and de-orbited, once more freeing the rear port for a fresh visitor. With the station now back in full working order, it was time for the original crew to take up their allotted place. Soyuz T-14 was launched on 17 September with Vasyutin and Volkov on board. Seated in the third couch was rotund, veteran flight engineer, Georgi Grechko, brought in for a brief visit to Salyut before returning to Earth with Dzhanibekov. Before the flight he explained the rationale behind the change of policy which entailed Savinykh remaining on the station. 'With his help the newcomers will quickly get into their stride,' he said. 'We expect good results from their joint work because none of the previous experience will be lost. This is the way we think future crews of large stations will work.' So for the first time, a space station did not have to be mothballed on completion of a mission.

Dzhanibekov felt a certain sadness as he moved his sleeping bag from the 'commander's post' on the 'ceiling' to the transfer tunnel to make way for the incoming commander. After a week of intensive atmospheric studies, he and Grechko chatted late into their sleep period, discussing their work and the activities of the three men on the other side of the hatch. Finally, on 25 September it was time to complete their packing and bid their comrades farewell. It was an emotional parting, especially for Dzhanibekov and Savinykh, the two men who had endured such privations and overcome such obstacles a few months earlier. In an unusual departure from tradition, Soyuz T-13 did not immediately re-enter the atmosphere, but spent a day station-keeping with the space station and practising approach techniques. When the cosmonauts finally came to Earth, they were greeted by warm sunshine, flowers and congratulations. Both men were granted the highest awards, and Dzhanibekov was promoted to major-general in recognition of his outstanding services to his country.

While the returning hero was informing the nation and the world of the hair-raising problems he and Savinykh had faced, the new crew settled down to receive a giant 20-ton cargo module, Cosmos 1686, which had been launched the day after Soyuz T-13 landed. The robot

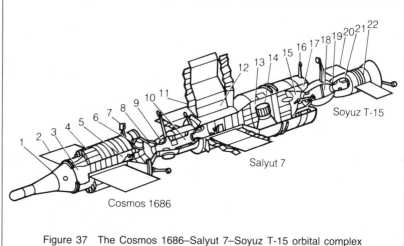

Figure 37   The Cosmos 1686–Salyut 7–Soyuz T-15 orbital complex
Cosmos 1686: 1, Re-entry vehicle. 2, Solar battery. 3, Transfer tunnel. 4, Fuel tank.
5, Operational-service unit. 6, Inner transfer hatch. 7, Antenna of the radio-technical
docking system. 8, Docking unit. **Salyut 7 station**: 9, Hatch for going out to space.
10, Transfer compartment. 11, Additional solar battery. 12, Main solar battery. 13,
Working compartment. 14, Research instruments compartment. 15, Transfer
module. 16, Antenna of radio-technical docking. 17, Engine compartment. **Soyuz
transport ship**: 18, Living compartment. 19, Hatch for transfer to living compartment.
20, Solar battery. 21, Descent capsule. 22, Instrument compartment.

craft docked at Salyut's front port on 2 October, delivering research
equipment and general supplies. It differed, however, from its
predecessors in that the descent module contained an array of telescopes,
a clear indication that the crew would be kept busy with experiments for
a long time to come.

October passed with the three men collecting samples of dust from the
near passage of comet Giacobini–Zinner, studying high energy particles
in space with an improved gamma spectrometer, tending their plant
seedlings and observing the behaviour of different materials under zero
gravity. They were so busy that unloading of Cosmos 1686 was relegated
to a spare time occupation, though time was allocated for more intensive
physical exercise than normal. On 25 October Tass stated that the
cosmonauts were in 'good health and feeling well', the last such reference
to be made. Routine announcements after that date indicated that the
normal programme was continuing, but the usual reassurance of fitness
and good health was noticeably missing. Then, on 13 November
Western eavesdroppers picked up scrambled signals between the station
and the ground, employed only when secret information was being
transmitted. Was something wrong?

The next (and final) Tass statement on 15 November revealed that the crew were wearing the Chibis suit while continuing their range of experiments. Then came the bombshell. On the afternoon of 21 November, it was suddenly announced that the crew had returned to Earth earlier that day. Great play was made of the scientific return from the mission, which included photography of 16 million square kilometres of Soviet territory and 400 sessions of experiments, but the key to the mystery was left till the final paragraph: the mission was terminated early 'due to Vladimir Vasyutin's sickness and the need for hospital treatment for him'. After a preliminary check-up on the landing site, doctors confirmed that he needed a thorough examination and whisked him away to Moscow. His companions were described as not feeling too bad considering the time they had spent in space – more than two months for Volkov and nearly half a year for Savinykh. In a press conference held less than 24 hours after touchdown, an unusually hurried occasion, they admitted that Vasyutin had not been able to complete his work programme, but paid tribute to his devotion to duty during the descent. No further light was shed on the nature of the illness which had resulted in the first known termination of a mission due to ill health.

Unconfirmed reports circulated during the next month, suggesting that Vasyutin suffered 'some inner inflammation', a possible reference to appendicitis, but this seemed to be contradicted when extracts from Savinykh's diary were published on 29 December. In it, the engineer referred to 'a slight uneasiness' in Vasyutin's behaviour, a loss of appetite and sleep, eventually culminating in the cosmonaut deteriorating into 'a bundle of nerves'. This was interpreted widely as indicating a nervous breakdown by the mission commander. The next day, the three men were re-united at the Kremlin for an award ceremony led by President Gromyko. Apparently recovered from his 'illness', Vasyutin thanked his comrades for doing 'their utmost to support and help me'.

There must remain a giant questionmark over Vasyutin's future as an active cosmonaut and the contribution he will be able to make as the Soviets take ever more daring steps towards their occupation of near-Earth space. It was undoubtedly a setback for Soviet prestige and for the Salyut 7 scientific programme, though hardly a body blow. Salyut 7 continued to orbit with Cosmos 1686 still attached as 1986 dawned, but the Soviets were looking ahead. They announced that two Syrian cosmonauts were in training, and continued to mention the possibility of a manned mission to Mars by the turn of the century. The immediate step, however, was to establish the long-promised permanent space station.

15

The Space Glider

The Shuttle Missions, 1981–5

On 12 April 1981, the twentieth anniversary of Gagarin's momentous space voyage, as the clocks registered 7 am Cape Canaveral time, a fireball brighter than the Florida sun illuminated Pad 39A. More than a million spectators held their breath as giant billowing white clouds spread rapidly across the barren flat coastal strip. The blast flattened wire fencing, demolished public address systems and scorched grass up to a mile away. Then came the deafening roar, the sound shaking the foundations of buildings three miles from the epicentre. Within a matter of seconds, the source of all this confusion had cleared the tower and was heading into the early morning sky, a white delta-winged dart clutching vertically on to three pointed columns like some strange eastern monument. This was the product of nearly a decade of stretching American technology and ingenuity to the limit, and of investing American taxpayers' money to the tune of more than nine billion dollars. This occasion, coming on the date of one major space breakthrough, would now also be remembered for the inauguration of the world's first re-usable spacecraft. After the Space Shuttle, travel into space would never be the same again.

Seated on the Shuttle's flight deck were the two men whose task it was to put the new craft through its paces. For the commander, chief astronaut John Young, and his pilot, Robert Crippen, it was the moment of truth after so many years of waiting. To them had fallen the unenviable job of testing a craft that had never before been flown, the first time in history that a spacecraft had been launched without prior unmanned rehearsals. At the time of their selection in March 1978 the mission had seemed less than a year away, but a series of technological problems had pushed the date back until another two years had passed. Young, now aged 50 and needing to wear glasses to correct long-sightedness, trained for up to 18 hours a day for his fifth journey into space. His companion, Bob Crippen, aged 43, had been waiting more than 14 years for his first space voyage since his selection for the USAF Manned Orbiting Laboratory programme in October 1966. When that project was cancelled, Crippen transferred to NASA in September 1969, but was too late to be involved in the final Apollo missions or Skylab other than as a support crew member. Now the years of waiting and training were over as the two former Navy flyers and test pilots became the first Americans to fly in space for six years.

Not surprisingly, Crippen's pulse rate jumped to 130 while the old hand alongside him only registered about 85. Young later admitted that even he was nervous during the final seconds of the countdown, but 'I was just so old my heart wouldn't go any faster'. The two men had good reason to be nervous. As Young said, 'Anyone who's not apprehensive about climbing on top of the first-time flight of a liquid hydrogen–oxygen rocket ship really doesn't understand the problem.' Only two days earlier, he and Crippen had lain on their backs for six hours awaiting launch only for it to be postponed due to a computer failure. Even more disquieting were the development problems associated with the Shuttle's rocket motors: there had been five engine fires and on one test run an engine had actually exploded. Their launch proved much smoother than expected: 'as smooth as glass', said Young; like 'riding a fast elevator' added Crippen. Only a later review by engineers revealed that the shock wave caused by the wall of flame from the Shuttle Columbia and the two solid fuel rocket boosters was four times greater than had been estimated, and exceeded the Shuttle's safety limit.

The first milestone successfully achieved, the Shuttle rode its smoke trail to an altitude of 29 nautical miles before jettisoning the two boosters over the Atlantic. They parachuted gently into the ocean ready for recovery and refurbishment for a later Shuttle mission. Another six minutes of flight using the fuel from the giant external tank saw Columbia reach orbital velocity of 17,000 mph. As the main engines cut off and the tank was jettisoned to burn up in the atmosphere, the crew knew that they were 'go' for the next two days in space. 'What a view!' was Crippen's cry of delight as he gazed upon the breathtaking Earth panorama which filled his window. 'It hasn't changed any,' came John Young's laconic Southern drawl.

Columbia had entered an elliptical orbit of 152 × 65 miles (245 × 106 km), but this was circularized by firing the Orbital Manoeuvring System (OMS) engines. These and the 44 small thrusters proved very efficient propulsion systems. With routine orbital adjustments controlled by the onboard computers, the astronauts found Columbia a delight to fly, though Crippen later remarked that the thrusters sounded like 'muffled howitzers' and really shook the craft each time they fired. Free of the seat belts and brown pressure suits, the two men began to adjust

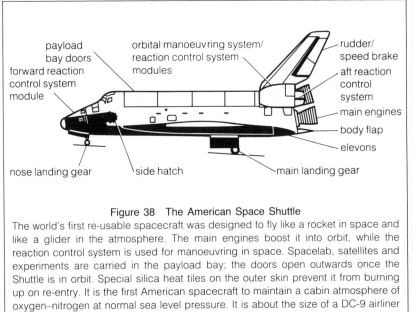

payload bay doors
forward reaction control system module
orbital manoeuvring system/ reaction control system modules
rudder/ speed brake
aft reaction control system
main engines
body flap
elevons
nose landing gear
side hatch
main landing gear

Figure 38 The American Space Shuttle
The world's first re-usable spacecraft was designed to fly like a rocket in space and like a glider in the atmosphere. The main engines boost it into orbit, while the reaction control system is used for manoeuvring in space. Spacelab, satellites and experiments are carried in the payload bay; the doors open outwards once the Shuttle is in orbit. Special silica heat tiles on the outer skin prevent it from burning up on re-entry. It is the first American spacecraft to maintain a cabin atmosphere of oxygen–nitrogen at normal sea level pressure. It is about the size of a DC-9 airliner and can carry up to nine crew as well as a maximum of 18 tons of cargo.

to the unfamiliar weightlessness. Crippen took no chances: he had already swallowed a motion sickness pill, and moved gingerly around the relatively spacious flight deck, swivelling his whole body rather than risk spacesickness by turning his head.

The astronauts' enjoyment was somewhat suppressed, however, when they opened the payload bay doors, exposing the cargo bay and enabling excess heat to radiate into space. Crippen noticed some unfamiliar dark patches on the OMS pods either side of the tail fin: 'Hey, John, we've got some tiles missing.' Of the 30,000 silica tiles which gave Columbia her heat-resistant skin, the astronauts could see one which had been completely removed and 15 others which were damaged. This was not the first occasion such damage had occurred, and led to considerable press speculation on Earth. One major fear was the so-called 'zipper effect' whereby a whole series of tiles, one after the other, might be pulled free during re-entry. Examination by the crew showed that the wings and tail were secure, but they were unable to see the underside, the area most susceptible to the fiery heat of re-entry. NASA officials assured the press that there was no need for concern, and ruled out a space walk by Crippen to inspect the damage. Nevertheless, they did make the surprising revelation that top secret cameras in Florida and Hawaii would photograph Columbia's underbelly as it passed overhead.

The crew did not have time to worry over their fate, with a long list of checks to see how the new craft was operating. As was the habit of the flight planners, hardly a minute went by without some new test or operation on the checklist. There was little time for sightseeing or snapshots of planet Earth until the rest period began at the end of the day. Bob Crippen had the privilege of acting as the first chef on board the Shuttle. Neither man slept well that first night, but after a second busy day it was a different story.

Mission control roused the men with a bugle call and loud music; Young ventured to prepare their breakfast. A final test of the flight control system, then they stowed all loose gear ready for re-entry, donned their pressure suits, programmed the computers and closed the payload bay doors. All was set for the historic return to Earth, the first time an entire spacecraft would carry its crew back to a runway landing.

The OMS engines fired to slow the craft, then Columbia's nose pitched up to expose the underbelly to the 1,600°C furnace of frictional heating. Hitting the top of the atmosphere at more than 24 times the speed of sound, Young and Crippen watched Columbia's outer skin turn a pinkish-red. With their visors down and suits pressurized, they sat out the 15 minute radio blackout as the Shuttle plunged steeply towards California. 'Hello, Houston, Columbia here,' came the routine call as the contact was re-established, much to the relief of all concerned on the ground. With speed down to mach 7, the crew could see the Californian coast approaching. 'What a way to come to California!' exclaimed a delighted Crippen. Although the computers had controlled the craft perfectly, Young decided to take manual control for the last two roll reversals. Speed continued to drop as he swung Columbia eastward over the landing strip at Edwards Air Force Base, home of so many test pilots who had graduated to astronaut status. A double sonic boom echoed around the dry lake basin as the world's largest glider dropped below the speed of sound. More than a quarter of a million people had flocked to the Mojave Desert to witness this historic occasion. The world's press had been scattered over the landing field just in case, as one NASA official phrased it, 'Columbia failed to make the base limits'. He need not have worried. As Columbia straightened after completing the final gliding turn, Young was right on line. Joined by a small escort fighter, Crippen talked down the airspeeds until touchdown at 185 knots, twice the speed of a normal jetliner landing. The tyres bit into the hard desert surface, kicking up small amounts of dust, and, oh so slowly, the nose wheel came down for a perfect landing. 'Welcome home, Columbia. Beautiful, beautiful,' said a delighted Houston Capcom. 'Do you want us to take it up to the hangar?' enquired an equally delighted John Young.

Columbia was immediately surrounded by a fleet of safety vehicles. The crew had to sit tight while they checked for noxious gases and linked up external power and cooling systems. Fifty-five minutes passed before

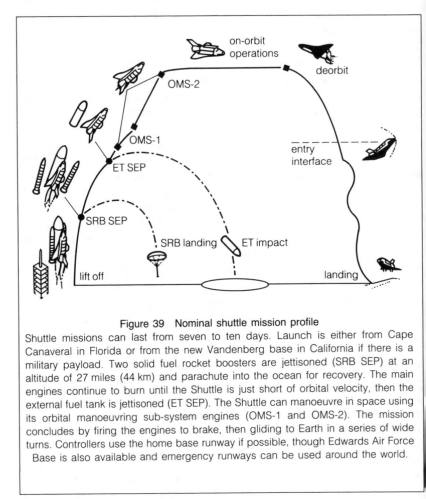

Figure 39   Nominal shuttle mission profile

Shuttle missions can last from seven to ten days. Launch is either from Cape Canaveral in Florida or from the new Vandenberg base in California if there is a military payload. Two solid fuel rocket boosters are jettisoned (SRB SEP) at an altitude of 27 miles (44 km) and parachute into the ocean for recovery. The main engines continue to burn until the Shuttle is just short of orbital velocity, then the external fuel tank is jettisoned (ET SEP). The Shuttle can manoeuvre in space using its orbital manoeuvring sub-system engines (OMS-1 and OMS-2). The mission concludes by firing the engines to brake, then gliding to Earth in a series of wide turns. Controllers use the home base runway if possible, though Edwards Air Force Base is also available and emergency runways can be used around the world.

they were allowed to leave through the side hatch. Young exited first, wasting no time in stalking around the orbiter's exterior and shaking hands with the technicians. The normally unemotional astronaut could be seen punching the air and whooping with delight. Crippen stayed on board a few minutes longer until a relief crew took over. They were late attending the end-of-mission ceremony where their wives and some 2,000 VIPs were waiting, but no one really cared. The wave of enthusiasm which spread over all those who had witnessed the historic event, whether at the Cape, at Edwards or simply on TV, came across loud and clear from the returning crew. Young gave his verdict: 'A really fantastic mission from start to finish. The human race is not too far from the stars.' Crippen confirmed that the long wait was 'well worth it'. 'We are really in the space business to stay,' he added. President Reagan, recovering from the attempt on his life the previous month, telephoned his congratulations to Houston and in a written statement to the astronauts declared, 'Your brave adventure has opened a new era in space travel'.

NASA had named six more men for the initial Shuttle test flights in its March 1978 announcement: Joe Engle, Richard Truly, Fred Haise, Jack Lousma, Vance Brand and Gordon Fullerton. When the prototype Shuttle, named Enterprise, began a series of preliminary tests in 1977, it was Haise and Fullerton who first rode piggyback on the specially converted Jumbo Jet, then separated for the first glide test from an altitude of more than 25,000 ft (8,000 m). The crew, quite reasonably, expected to take Columbia into space on her maiden flight, but when it became clear that Young and Crippen had claimed that mission, Haise resigned from NASA in June 1979 to join Grumman Aerospace as vice president. In the reshuffle that followed, Brand dropped back to command the fifth Shuttle mission, and Lousma teamed up with Fullerton. The second flight of Columbia was given to the back-up crew for the maiden flight, Joe Engle and Richard Truly. Both men were on their first trip into space although they had been in training as astronauts for more than 15 years, and Engle had three times flown the rocket powered X 15 to the threshold of space in the mid-1960s. For both men, it was the climax of years of hard work and frustration. Engle had been allotted to the final Apollo moon landing, then been replaced by geologist Harrison Schmitt; Truly had been involved in the USAF Manned Orbiting Laboratory project until it was cancelled in 1969. Now all their hard work during the Enterprise approach and landing test flights was paying off.

The second flight by Columbia involved its own frustrations, however, as a series of technical hitches pushed the launch date remorselessly back from September to October and then to 4 November. On that occasion the countdown stopped with 31 seconds to go as a second computer hitch automatically stopped the clock. While technicians tried to clear the problem, a build-up of pressure in the Shuttle's auxiliary power units forced officials to scrub the launch. It was later discovered that clogged oil filters were the culprits. Only on 12 November, Truly's forty-fourth birthday, did the gremlins relent. After enjoying a celebratory cake with his breakfast, Truly and Engle were willed into space by the watching thousands.

A few hours later, mission control told the astronauts, 'We think you are looking at a minimum mission.' One of Columbia's three fuel cells, the source of electrical power and drinking water for the craft, began to malfunction. Controllers decided to follow the rule book, and ordered the crew to return after only two days instead of the scheduled five. The crew worked overtime to carry out as much as possible of the flight plan, notably the first tests of the Canadian-built remote manipulator arm, a 50 ft long jointed mechanism equipped to carry payloads and satellites. One shot from the arm's camera showed the crew holding a sign saying 'Hi, Mom'. The Shuttle also carried its first scientific cargo, a pallet in the cargo bay which included a new Imaging Radar and instruments to measure Earth pollution. These instruments provided valuable results as Columbia flew upside down with its payload bay doors open. Analysis of radar images of the Sahara showed what appeared to be ancient river beds and human settlements previously hidden beneath the sands.

After 36 orbits, Columbia braked over the Indian Ocean, then swooped over the Pacific in a re-run of the April landing. This time, Engle swung the craft through a series of rolls and manoeuvres designed to give stress data, but most of the landing phase was automatic, using a new microwave scanning beam landing system. Only when Columbia levelled off for the final approach to Edwards runway did Engle take over manual control. Plunging through extensive cloud cover, he brought Columbia to a perfect landing on the dry lake bed. Checks showed that no tiles were lost. The Shuttle had become the first craft to enter space more than once, and had achieved '90 to 95 per cent' of the mission objectives.

Two more test flights were scheduled before the Shuttle could be recognized as fully operational. The third flight of Columbia began exactly as planned on 22 March 1982, with 46-year-old Skylab veteran Jack Lousma commanding and former MOL recruit Charles Fullerton alongside. Local hoteliers and tourist officials offered up their thanks as the Shuttle lifted off only one hour late into an overcast sky. Despite some problems with the auxiliary power units only seven minutes into the flight, Columbia reached orbit safely. Another recurrent problem was spotted the next day when 25 heat tiles were seen to be missing from the nose and 12 more had been shaken loose from the rear. Ground controllers reassured the crew: 'We think there is no concern.' Lousma replied, 'If you found out otherwise, I don't think we'd want to know.' The first two days proved exhausting as the crew battled to overcome minor faults while completing their busy work schedule. A thermostat problem kept the cabin temperature either too warm or too cold, while loud static in their headsets kept them awake each time they passed over Asia. Already affected by spacesickness, the two men then reported an even more embarrassing problem: 'The Shuttle commode is out of action.' Dosed up with pills, mission control told them to reduce their workload and take a longer sleep period in order to 'get the crew back on the straight and narrow'.

Further technical hitches continued to dog the mission, most seriously a loss of three radio channels on 26 March. Columbia's attitude was altered according to plan in order to test the craft under different heating conditions: first the tail was pointed towards the sun for 28 hours, then the nose turned to the sun for 80 hours and finally the open cargo bay was exposed to maximum heat for 28 hours. The crew also operated the remote manipulator arm for its first lifting task, despite the failure of the arm's TV wrist camera. Fullerton called it 'a fantastic piece of machinery'. Meanwhile, a science package in the cargo bay collected data on conditions for future experimental missions, as well as various solar and micro-meteorite experiments. In the Shuttle's cabin were two canisters containing moths and honeybees, provided by a Minnesota high school student, Todd Nelson, as a study on the effects of weightlessness.

Throughout the mission, ground controllers were glancing over their shoulders at the weather. Even before the launch, Edwards Air Force Base had been ruled out by a rainstorm which turned the desert into a mudbath. The best alternative landing site was Northrup Air Strip at White Sands, New Mexico, but strong winds and sandstorms were a continual threat. Optimistic forecasts were dashed on 29 March as astronaut John Young flew over the strip and reported winds gusting up to 55 mph with visibility down to 50 yards. 'I think we ought to knock this off', was Young's advice. Mission control agreed, so Columbia was waved off for another 24 hours. Already, equipment and personnel had had to be transported to White Sands by two special trains. Now it looked as though the Shuttle would have to attempt its first landing on the runway at Cape Canaveral. Then, overnight, the capricious weather changed again. After its longest mission so far, 8 days and 129 orbits, Columbia landed at White Sands in perfect conditions. Some 50,000 spectators had braved the desert to welcome the astronauts home. There were a few nervous flutters as Lousma left his landing gear up until almost the last minute, then lifted the craft's nose as if preparing to take off once more. Despite the hitches, NASA officials were delighted with the outcome. Fullerton summarized the astronauts' viewpoint: 'It's an unbelievably beautiful flying machine.'

For Jack Lousma, it was his last spaceflight. He resigned in 1983 to become an aerospace consultant in Michigan. 'I left because I felt I had done everything I wanted at NASA,' explained Lousma. 'There is a lot to do and a great deal of emotional stress. It's not as glamorous as it looks. It is plain hard work.' His wife added: 'Jack felt that there was more to Jack Lousma than flying in space.'

The fourth and final test flight of Columbia was commanded by one of Lousma's group 5 comrades, Thomas Mattingly, making his first venture into space since orbiting the Moon with Apollo 16 ten years earlier. Alongside him was Henry 'Hank' Hartsfield, a former Air Force test pilot and MOL astronaut before joining NASA. Although Hartsfield was now a civilian, the fourth Shuttle flight had the distinction of carrying the first military payload, infrared and ultraviolet scanners being tested for future military surveillance satellites. TV coverage of the mission, already reduced as the novelty factor wore off, was further cut by the restricted views of the cargo bay allowed by the Defense Department. Other cargoes were less controversial: the electrophoresis experiment to test new methods of manufacturing drugs was the first to be provided by a commercial firm; two medical experiments using the astronauts as guinea-pigs were contributed by High School students; and the first 'Getaway Special', a $10,000 package of nine self-contained experiments, was paid for by Gilbert Moore of North Ogden, Utah, and presented to Utah State University. The 'Special' got off to an inauspicious start when it refused to operate, but a 'hot-wiring' bypass by the crew succeeded, to the delight of the students. 'One small switch for NASA, a giant turn-on for us,' they radioed.

The mission was also affected by those ever-vulnerable tiles. While on the launch pad, Columbia had been subjected to a battering from torrential rain and hail. Despite this, the Shuttle was launched right on time, only three months after the touchdown at White Sands, but engineers insisted on turning the Shuttle's underside to the sun in order to dry the tiles and prevent damage by ice.

There were two other mission highlights: the crew were able to speak to crowds at the Knoxville World's Fair on 1 July, describing the view across the Mississippi valley, and the next day Mattingly tried out one of the new 'off-the-hook' EVA pressure suits while standing in the air-lock. The only problem was, he said afterwards, 'that I didn't get to open the door'.

Columbia returned on 4 July, Independence Day, to a red, white and blue welcome at Edwards Air Force Base. Among the estimated half a million spectators was President Reagan, delighted to take the opportunity of shaking the returning heroes by the hand and making a rousing patriotic speech. As the President spoke, the second Shuttle, named Challenger, took off on the back of a Boeing 747 for the Cape. Not without reason did he liken the Shuttle's progress to the driving of the golden spike which completed the first transcontinental railroad, 'an entrance to a new era'.

This new era began with the first operational mission of the Shuttle in November 1982. On board was a four man crew, commanded by the 51-year-old ASTP veteran Vance Brand, with former MOL astronaut Robert Overmyer as pilot. Marine colonel Overmyer was the only military crewman, for the other two occupants were scientists Dr William Lenoir and Dr Joseph Allen. Recruited by NASA in 1967 Lenoir was an assistant professor in electrical engineering at Massachusetts Institute of Technology while Allen had been a nuclear physicist prior to selection. They were now the first of a new astronaut category, the mission specialists, who were responsible for the commercial and scientific aspects of the mission while the overall spacecraft command and control remained with the two pilots.

In the payload bay during a perfect launch on 11 November were two communications satellites, the first real money-earners for NASA's 'space truck', though at the bargain basement fee of $17 million, they repaid a mere fraction of the $250 million bill for the entire mission. On the flight deck was another unfamiliar sight – four helmeted men in light blue coveralls instead of bulky pressure suits. This time there were no ejection seats: it was either slide down an escape wire if a disaster threatened on the launch pad or ride the Shuttle high enough to glide back to Earth. If the Shuttle ran into trouble in orbit, the pilots were provided with the new $2 million dollar EVA suits while the rest of the crew had to zip themselves into insulated and pressurized 'rescue balls'. These suits, scheduled for their first practical test during the flight, proved to be the biggest let-down of the five-day mission. First, Joe Allen's backpack failed to circulate oxygen through his suit properly, then Lenoir's suit failed to pressurize fully. Instead of entering the payload bay to practise at a model work station and try out satellite repair techniques, the two astronauts spent most of Monday attempting suit repairs. It was particularly frustrating after the spacesickness which had dogged the early part of the mission and brought Overmyer and Lenoir uncomplimentary publicity. Press speculation suggested red-hot Jalapeno peppers might be the culprit.

This setback contrasted with the success of the first two days when both satellites were released from the cargo bay and fired into geostationary orbits more than 22,200 miles (35,700 km) above the Earth. Once the Shuttle's orbit was accurately fixed, Overmyer orientated Columbia with its right wing pointing towards Earth and the open payload bay facing back the way the craft had just passed. Allen and Lenoir fed commands into the computer to open the clam-like thermal shield inside which the satellite sheltered, then rotated a turntable to spin the satellite and its attached rocket motor at around 50 revolutions per minute. Finally, a powerful spring was released to push the satellite gently free of the cargo bay. Retreating to a safe distance and turning Columbia's hardened belly to the satellite, the crew then waited for the satellite's control centre to fire the rocket motor for ascent to the desired orbit. After the first day's launch, mission control told the astronauts: 'It's going to be awfully hard for you guys to match yesterday.' When the second launch went equally smoothly, the delighted crew displayed a sign to TV viewers which read 'Satellite deployment by Ace Moving Co. – we deliver'. The mission ended just as perfectly on the concrete runway at Edwards soon after sunrise on 16 November.

After the accelerating turnaround of Shuttle flights during 1982, 1983 was rather a let-down for NASA and led to potential customers expressing doubts over the Shuttle's ability to meet the original deadlines. The maiden flight of the second Shuttle, Challenger, was put back from January to March when potentially disastrous leaks of hydrogen gas resulted from hairline cracks in the orbiter's three main engines. Then, as launch date approached, the Shuttle and its precious satellite cargo were coated with dust during a storm at the Cape, despite the weather seal around the craft. Thorough cleaning and inspection were necessary, so it was not until 1.30 pm Cape time on 4 April 1983 that Challenger finally got off the ground. Meanwhile, the publicity given to the spacesickness on the fifth Shuttle flight had penetrated the normally tough hides of some astronauts. In the relatively spacious cabin of the Shuttle, half of the astronauts had experienced 'space adaptation syndrome', beginning 12 hours or so after launch and continuing for the next two to three days. Skylab veteran Paul Weitz, assigned to command Challenger, summarized his comrades' feelings: 'I think it's between me and my doctor and is nobody else's business.'

Weitz, now aged 50, was the only member of the four-man crew to have spaceflight experience. Also on the flight deck were USAF colonel Karol Bobko, 45, and mission specialists Dr Story Musgrave, 47, and Donald Peterson, 49. They were the oldest American crew ever to fly, further evidence that the low acceleration forces – maximum 3G on launch and 1½G on re-entry – and shirtsleeve environment of the Shuttle did not require the peak physical fitness and sharp reflexes of the first pioneers. Dr Musgrave, a qualified physician and specialist in aerospace medicine had been given the specific task of trying to learn more about the dreaded spacesickness.

Challenger proved to be a lighter yet more powerful sister to Columbia. Installed in the cargo bay was a payload of 21 tons, nearly 50 per cent heavier than the payload on the previous flight. Among the weight savings was the replacement of the ejection seats by lighter, operational seats, but there was no cause for alarm as Challenger soared into a brilliant blue sky.

Apart from the usual prototype industrial experiments and three small 'Getaway Specials', the main cargo was a Tracking and Data Relay Satellite (TDRS), the first of three such satellites intended to relay communications between Earth stations and satellites or Shuttles. The largest and most advanced relay satellite yet developed, TDRS-A should have been placed in geostationary orbit above Brazil on the morning of 5 April, but space launches remain a high risk affair. The crew watched the $100 million satellite and its attached booster push safely away from the cargo bay, then backed off for the first stage burn. Everything seemed fine, but then they learned that the upper stage had cut out early, placing the satellite in the wrong orbit. For several hours, ground controllers struggled feverishly to separate the satellite from the booster and deploy its solar panels, eventually regaining control. However, the fact remained that the main purpose of the mission was a miserable failure.

The rest of the flight was redeemed by the first American space walks in nine years. Wearing the EVA suits which had caused problems on the previous mission, Musgrave and Peterson moved into the air-lock for three hours to breathe oxygen and so purge the nitrogen from their blood while hanging from racks. Encouraged by a phone call from the President, the two men moved into the cargo bay and cavorted around in a series of space gymnastics, restrained only by 50 feet long tethers. They then got down to the serious business of trying out their special tools and work techniques, watched by their two partners through the flight deck windows and by millions of TV viewers in the USA. Their activity lasted the best part of three orbits and more than four hours, and successfully tested the new, flexible pressure suits. Even ground controllers entered into the spirit of the occasion. 'While you're under the hood, Story, why don't you check the oil?'

The new Shuttle handled beautifully during re-entry on 9 April, swooping across the Pacific to land on the main runway at Edwards. For the first time, the pilots had the benefit of a 'heads up display' of spacecraft data on a transparent screen before their eyes. Immediately *Challenger* landed, NASA announced that ground crews would work around the clock seven days a week to get the craft back into space as quickly as possible.

To everyone's delight, the hard work paid off, as *Challenger* returned to orbit in another flawless launch on 18 June 1983. The seventh mission of the Shuttle saw a revival of the media attention and ballyhoo which had once attended every flight but was now a rarity. But it was not the flight programme which excited the flocks of reporters, but the presence on the flight deck of America's first spacewoman, 32-year-old physicist Dr Sally Ride. One of six women selected by NASA as mission specialists in 1978, she tried in vain to remain outside the press jamboree and maintain a dignified stance in the face of the inevitable banal interrogation. 'I didn't come into this programme to be the first woman in space. I came in to get a chance to fly in space,' she protested. But it was no use. The rise of Sally Ride was subjected to microscopic analysis: how she won a tennis scholarship to an exclusive girls' school in Bel Air at the age of 15, went on to win a major college tennis championship and was urged to turn professional by Billie Jean King, but turned instead to science and gained a doctorate in X-ray physics in 1978, the year she applied to NASA. Out of the thousands who applied, only 35 were chosen. One of the others was a young astronomer, Dr Steven Hawley. Love blossomed in the simulators and lecture rooms, resulting in a marriage ceremony in 1982 conducted by the bridegroom's father in his back garden. The space age bride flew in solo for the wedding, then whisked her husband away for their honeymoon in the back seat of the plane. Now the public were deluged with trivia as Ms Ride accompanied four men into space: no, she wasn't wearing a bra ('No G-forces up there'), she had not taken any lipstick or perfume; her first words were, 'It's an E ticket,' a reference to the ticket that gets the best rides at Disneyland. Even commander Bob Crippen, who had specially wanted Ride because of her skill with the remote controlled arm, could not hold back a wisecrack: 'Nice Ride!' The other, almost unnoticed, members of the crew were pilot Frederick Hauck and mission specialists John Fabian and Norman Thagard. A qualified physician, Dr Thagard had been added to the already training crew in December 1982 specifically to conduct medical tests related to spacesickness in orbit.

The seventh Shuttle, mission STS-7, proved an unqualified success in a technological as well as public relations sense. The first day ended on a high note with the spring ejection of a Canadian communications satellite and was followed the next day by the almost identical launch of an Indonesian satellite, both the work of Ride and Fabian. With these out of the way, the crew were able to concentrate on the remaining experiments in the cargo bay, a Shuttle Pallet Satellite (SPAS), a terrestrial applications payload loaded with materials processing experiments, and seven 'Getaway Specials' including a colony of 150 Carpenter ants. The German-built SPAS was the focus of most attention on the third day as Ride and Fabian used the arm to pick up the satellite and dump it overboard. Challenger then waltzed around the SPAS before returning to grapple it with the arm in a vital test of the remote manipulator's capabilities. Another series of manoeuvres followed in the afternoon before SPAS was reberthed. Apart from providing power for seven microgravity experiments, SPAS also carried cameras which provided the first pictures of the entire Shuttle in orbit. The satellite was held aloft during the fourth day to test further the stability of the arm and the craft while the Shuttle's thrusters were firing.

Challenger returned to Earth on the morning of 24 June, but had to divert from the Cape to Edwards because of rain in Florida. Sally Ride climbed down the steps ahead of her companions to face a barrage of reporters and offers from Hollywood agents. The cool spacewoman firmly rejected the advances: 'I didn't go into the space programme to make money or be famous.'

A celebrity of a different kind was among the crew of the next Shuttle mission. Forty-year-old Lieutenant-Colonel Guion Bluford had the distinction of being the first black to fly on an American craft. One of three mission specialists on the Challenger STS-8 mission, Bluford had flown 144 combat missions over Vietnam in the mid-1960s before becoming a pilot instructor, completing further studies which culminated in a doctorate in aerospace engineering while serving as a development engineer at Wright–Patterson Flight Dynamics Laboratory. His fellow mission specialists were 34-year-old Dale Gardner, a Navy fighter pilot and engineer selected at the same time as Bluford, and Dr William Thornton, a physician delegated to study spacesickness and, at 54, the oldest person ever to enter space. Commander was the pilot of the second Shuttle, Richard Truly, with another Vietnam combat veteran and test pilot alongside in the form of 39-year-old Navy commander Daniel Brandenstein.

Spectators and photographers flocked to the Cape for the first night launch of a Shuttle, necessitated by the orbital requirements of the Indian satellite installed in the cargo bay. At 2.32 am Cape time on 30 August 1983, in between a series of spectacular thunderstorms, the flat Florida landscape lit up as an awesome ball of flame erupted from the Shuttle and its solid fuel boosters. The crew watched in fascination as a fiery red glow surrounded the craft. Truly described the view: 'It got brighter and brighter. When the boosters separated it was 500 times brighter than I remember.' Gardner tried to turn his head for a better view of the magnificent spectacle. 'I damn near blinded myself,' he commented.

Once in orbit, Bluford began operation of one of the main experiments, an electrophoresis system designed to separate living cells in a pilot scheme aimed at one day producing new medical advances such as live insulin-generating cells which could be used to treat diabetics. Among other experiments were a self-contained cage with six rats as passengers and 12 'Getaway Special' canisters. Inside eight of these were 260,000 postal covers commemorating the twenty-fifth anniversary of NASA and intended for sale to collectors.

On the second day, Bluford ejected the Indian communication and weather satellite from the cargo bay over the Indian Ocean. Ground controllers later reported that it had reached orbit, but its solar panels had not deployed properly, and more than a week passed before it became fully operational. The third day saw Gardner use the remote arm to lift its heaviest load yet, a dumb-bell shaped test article weighing over three tons. Using automatic and manual control, sometimes with only video cameras as aids, Gardner grappled the unusual load from all angles like a one-armed weightlifter. Once more, the arm passed with flying colours.

Meanwhile, an equally important communications test was taking place between the Shuttle, the ground and the TDRS-A satellite which had finally been edged into the correct orbit by ground controllers after its lopsided launch from Challenger four months earlier. Exhaustive testing showed that TDRS-A was fit and ready to transmit the large volume of data due to be returned by the first Spacelab mission. Challenger re-entered the atmosphere on the morning of 5 September for the first Shuttle night landing. For safety reasons, Edwards was chosen as the venue, and, with dozens of practice night landings behind them, Truly and Brandenstein brought Challenger back to a perfect touchdown amidst the glare of high-intensity arc lights. The only sour note emerged later, when NASA revealed that the crew had been within 14 seconds of disaster. The lining of a solid fuel booster's nozzle had almost burnt through during the launch; had this occurred, the Shuttle would have gyrated wildly off course as the thrust was directed off alignment. The first flight of Spacelab would have to be delayed for an inquiry.

Frenzied efforts by the space agency meant that Spacelab 1 was only put back for a month, eventually lifting into orbit aboard Columbia on 28 November 1983. Some of the scientists were unhappy that their experiments would be adversely affected by the poorer lighting conditions for observations of the Earth's surface and atmosphere, but they were outvoted, so the European-built Spacelab went ahead. With more than 70 experiments carried in the pressurized laboratory on an instrument pallet in the cargo bay, it was scheduled to be the longest and busiest Shuttle mission yet. Because of the purely scientific nature of the mission, the six-man crew was also composed differently from normal. The commander was John Young, now aged 53, and taking part in a record-breaking sixth spaceflight. His co-pilot was a former Vietnam combat veteran and test pilot, 38-year-old Major Brewster Shaw. To these men would fall the difficult task of completing some 200 manoeuvres in orbit in order to orientate the instruments correctly. The two mission specialists were both scientist–astronauts, 53-year-old Skylab veteran Owen Garriott and 46-year-old astronomer Robert Parker, making his first flight since NASA selection 16 years previously. Their main task was to look after flight engineering and onboard systems, and maintain communications between the laboratory, the ground and the pilots.

The final two seats were occupied by a new category of astronaut, the payload specialists. Their job was to oversee the wide range of experiments on board, so they had no training in flying the Shuttle but had travelled the world for training by the scientific investigators in their specific experiments. It was clearly impossible for two men to be qualified in all scientific areas, so NASA and the European Space Agency (ESA) selected and trained scientists who were proficient in at least one discipline and then broadened their horizons.

At least they had direct communication with the investigators on Earth during the mission, though this inevitably led to considerable stress on the payload specialists. In order to minimize the stress and enable maximum efficiency, the crew was split into two shifts: the 'Red Team' comprised Young, Parker and West German physicist Ulf Merbold; the 'Blue Team' was made up of Shaw, Garriott and biomedical specialist Byron Lichtenberg. Forty-year-old Merbold, a specialist metallurgist, had the distinction of being the first non-American passenger on a US spacecraft. He was one of four candidate payload specialists selected by ESA in 1977, and had spent the past five

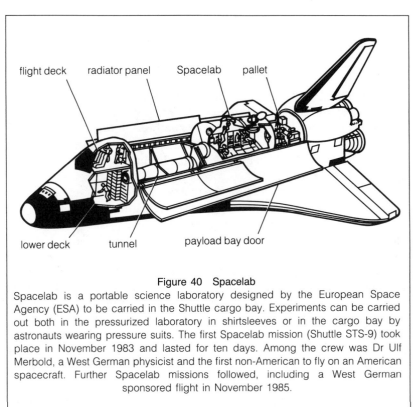

flight deck   radiator panel   Spacelab   pallet

lower deck   tunnel   payload bay door

Figure 40   Spacelab

Spacelab is a portable science laboratory designed by the European Space Agency (ESA) to be carried in the Shuttle cargo bay. Experiments can be carried out both in the pressurized laboratory in shirtsleeves or in the cargo bay by astronauts wearing pressure suits. The first Spacelab mission (Shuttle STS-9) took place in November 1983 and lasted for ten days. Among the crew was Dr Ulf Merbold, a West German physicist and the first non-American to fly on an American spacecraft. Further Spacelab missions followed, including a West German sponsored flight in November 1985.

years in preparation for this mission. Lichtenberg was a curiosity, a Vietnam war hero with two DFCs to his credit, who had gone on to a doctorate in science while continuing to serve as a fighter pilot for the Massachusetts Air National Guard.

The Shuttle Columbia, modified to carry the large payload, 9½ tons heavier than anything previously carried, placed Spacelab into a perfect 150 mile (250 km) high orbit. The payload bay doors were opened and Spacelab's systems were powered up, enabling Garriott, Merbold, and Lichtenberg to break new ground by floating through a connecting tunnel to check out the unique laboratory.. It was then left to the 'Blue Team' to kick off the scientific research while the 'Red Team' settled down for their first sleep period.

Spacelab carried experiments in five major areas – life sciences, Earth studies, space physics, astronomy and materials science – and the special requirements for some of these investigations led to Columbia's flying an unusual 57 degree inclined orbit which carried it as far north as the Soviet Union and as far south as the tip of Chile.

Tests in biology and physiology occupied most of the time on the first two days, with considerable emphasis on investigations into the adaptation of the human body to weightlessness. A camera intended to photograph astronauts' eye movements failed when the flashgun refused to fire, but ground controllers told the crew simply to describe their reactions in words. Potentially more serious was the breakdown of a data link between the craft's computer and four experiments in the cargo bay. The crew had to work hard to bypass the breakdown, causing mission controller Harry Craft to comment: 'We're pressing you pretty hard, but you're doing a heck of a job to keep up.' Feelings of nausea among the crew did little to help.

This was to be the pattern for the remainder of the mission, a complex research programme punctuated by irritating equipment failures which necessitated repair attempts by the troubleshooting astronauts. Fortunately, the feelings of sickness had retreated by the end of the third day, though vigorous head movements still caused the symptoms to return. A long-standing theory was disproved in an experiment which involved hot air being blown into one ear while cold air was blown into the other; standard theory predicted that there would be no effect in zero gravity, but the crew reacted as they would on Earth. They also had to endure other biological experiments from the regular measurement of blood pressure and taking of blood samples to being accelerated, spun or 'dropped' with elastic cords, and placing the head in a dome with a pattern of rotating dots to give the sensation of spinning. It was during one of these experiments that the pressure on the crew became apparent. Ground control interrupted to ask them to start an entirely different experiment. 'You guys need to recognize there are two people up here trying to get all your stuff done,' snapped Parker. 'I think you might be quiet until we get one or all of them done.'

The materials processing began on the third day, with Merbold switching on three high temperature furnaces. The results were encouraging, though two of the furnaces had to be repaired by the crew. Lichtenberg also saved a fluid physics test by in-orbit repairs, while a moveable camera experiment to photograph the upper atmosphere was only saved by Young and Shaw manoeuvring the Shuttle to compensate for a mechanical failure which prevented the camera from pointing properly. A unique mapping experiment nearly bit the dust when the new metric camera's film magazine jammed, but it was successfully freed by Parker. In gratitude, ground scientists donated 80 frames of film for the crew to use for their own Earth photography. The crew took full advantage during what was their penultimate day in space; the mission had already been extended by one day to allow additional time for research, including several fluid physics demonstrations improvised by the crew and the ground investigators.

On their ninth day in space, the crew stowed the last of the equipment and prepared for re-entry. Then, just under four hours before the scheduled de-orbit engine burn, first one and then a second main computer failed. Commander Young blamed two strong jolts from Columbia's attitude control thrusters. Although there were three more back-up computers available, mission controllers decided to delay the landing while the problem was evaluated. Although one of the computers was re-started, a failure also occurred in one of the units used for navigation. Columbia was showing her age.

She eventually landed seven and a half hours late at Edwards Air Force Base. One of the computers failed again on landing, and several small fires caused by fuel leaks were discovered around two of the auxiliary power units after touchdown. NASA officials denied there was any danger to the crew, but initiated inquiries into the mission difficulties. Columbia was retired for an extensive two-year refit. Meanwhile, the delighted scientists began analysis of the fruits of the ten-day mission: it was estimated that Spacelab had yielded 50 times the information gathered during 24 weeks of work aboard Skylab in the early 1970s.

1984 began with a dress rehearsal for a satellite rescue mission. The Shuttle Challenger carried a five-man crew and two communications satellites into orbit on 3 February, but the mission was dogged by hardware problems. Most serious was the loss of both satellites due to misfires by their booster rockets, setbacks which caused major palpitations in the insurance offices of New York and London. Each satellite was insured for around $100 million and was declared a write-off by its owner – Westar should have provided the USA with telephone links while Palapa was the second in a series of Indonesian communications satellites. Between these financial disasters, another experiment had to be prematurely curtailed as a 6½ ft diameter mylar balloon, intended as a target for practising tracking and rendezvous techniques, exploded during inflation after being ejected from the cargo bay. The crew reported they were able to sight and track a large fragment of the balloon, and even achieved a radar lock on it out to a distance of 40,000 ft (12 km), but it was hardly an auspicious start to the mission.

Fortunately, the second part of the eight-day flight proved a triumph. On the morning of 7 February, mission specialists Bruce McCandless and Robert Stewart left Challenger's air-lock for an historic six hour period of space walking. McCandless, now aged 46, had been an astronaut since 1966 but had never before flown in space; Stewart, five years his junior, was the first US Army officer to fly on board an American spacecraft, and had joined NASA in 1978. Much of McCandless' time and energy over the years had gone into developing an astronaut manoeuvring unit which could imitate the rocket propelled backpacks beloved of science fiction writers and enable an astronaut to float free of all tethers for the first time. Now his hard work was about to pay off. He floated to the manned manoeuvring unit (MMU), which was mounted on the side of the cargo bay like some strange armchair, attached it to his suit and checked it out. Then mission commander Vance Brand informed the Hawaii ground station, 'Bruce is in free flight.' As a TV camera on the remote arm showed him flying around the cargo bay, McCandless radioed: 'That may have been one small step for Neil, but it was a heck of a big leap for me.'

While Stewart practised with a work station in the cargo bay and waited for his turn, McCandless completed a series of gymnastic fly-pasts, then received clearance from Brand to head into the black beyond. Moving backwards at a steady 2 mph, a live audience around the world was able to gaze in wonder at the Buck Rogers type figure as he flew to a distance of about 150 ft (45 m) then back to the orbiter. A camera mounted on his helmet sent back pictures of the Shuttle as it passed over the southern USA. 'There are some jealous folks down there. Looks like you're having fun,' commented Mission Control. 'When you put in for a long translation, the thing shudders and rattles and shakes,' said McCandless as he prepared for a second voyage. This time, he moved to a distance of 320 ft (96 m), a small object framed against the black sky and the blue curvature of the Earth. 'The view you get out here is like the difference between the view you get flying a heavy aircraft and looking out little windows compared to flying a helicopter,' he said.

Back in the cargo bay, Stewart helped him mount a special docking adapter unit which would be used on the next mission to rescue a satellite. Several practice docking runs were completed with a pin mounted on the payload bay wall, then Stewart donned his NMU. For the next 65 minutes he repeated the incredible manoeuvres of his partner, though mostly out of TV range. 'It's a piece of cake,' radioed the happy astronaut as he floated away from Challenger, but a more cautious Vance Brand advised him to slow down. The last part of the EVA saw McCandless place his feet in a restraint on the end of the remote arm and practise satellite repairs, then be lifted around the cargo bay by mission specialist Ronald McNair, who was controlling the arm.

The sixth day of the mission was devoted almost entirely to scientific experiments attached to the SPAS platform while it was anchored in the cargo bay. The SPAS had first flown on the seventh Shuttle mission in June 1983, and so was the first satellite ever to be refurbished and flown again. However, plans to lift the SPAS with the remote arm so that the astronauts could practise docking while it was held aloft had to be abandoned when an electrical problem arose in the arm's 'wrist joint'. So the second EVA on 9 February was roughly a repeat of the first, but without the long distance trips and with the emphasis on evaluating the MMU and its docking capabilities. One unscheduled piece of activity

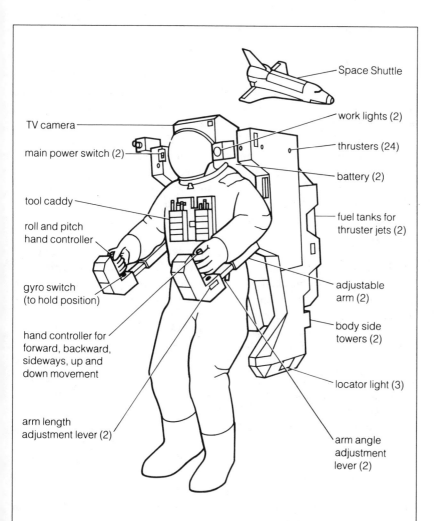

Space Shuttle

work lights (2)

TV camera

main power switch (2)

thrusters (24)

battery (2)

tool caddy

roll and pitch
hand controller

fuel tanks for
thruster jets (2)

gyro switch
(to hold position)

adjustable
arm (2)

hand controller for
forward, backward,
sideways, up and
down movement

body side
towers (2)

locator light (3)

arm length
adjustment lever (2)

arm angle
adjustment
lever (2)

Figure 41   The manned manoeuvring unit (MMU)
This backpack enables an astronaut to work untethered more than 100 ft (30 m)
from the spacecraft. It weighs 300 lb and contains enough oxygen and electric
power to supply the astronaut for seven hours. Movement is made possible by 24
tiny jets which propel nitrogen gas. Top speed is around 66 ft (20 m) a second. It
was first used by Bruce McCandless, mission specialist on Shuttle 41-B, on
7 February 1984. It was later tested on the same mission by Robert Stewart, and in
subsequent satellite rescue missions.

involved a chase after a broken foot restraint: Brand manoeuvred Challenger with such skill that McCandless was able to reach up from the cargo bay and grab the stray. The final chore was a test to transfer fuel to a satellite, a possible future role for the space 'garage': the successful dummy run brought more than six hours of EVA to a close.

Brand brought Challenger back to her home base at the Cape on 11 February, the first time a landing at the launch site had been achieved. It meant that the space agency would save both time and money during preparations for the next launch, a vital step in speeding up the launch programme. Yet the weather nearly let them down: ground fog at the Cape was an issue until almost the last moment, but test approaches by John Young eventually led to the opinion that it would clear in the early morning sun. As the Shuttle flew across the Florida coast for a perfect landing, a ground team patrolled the runway to ensure that no alligators, wild pigs or other wildlife strayed across the Shuttle's path.

The time had now come to put into practice all the experience gained by the dress rehearsals for a satellite rescue. The object in question was the Solar Maximum Mission (Solar Max) satellite, disabled since November 1980 by three blown fuses which affected its attitude control system. Scientists were keen to resume their studies of solar activity, even though the peak number of sunspots and solar flares had now passed, so plans for a Shuttle repair attempt were formulated in early 1981, although the Shuttle had yet to fly. One factor in favour of the attempt was the design of Solar Max, the first scientific satellite built with possible retrieval and repair in mind. Over the years, the satellite's orbit had decayed, but to reach it still required the first use of a direct ascent insertion, whereby Challenger's main engines boosted the Shuttle directly to an altitude of 250 miles (460 km), a Shuttle record.

The historic mission began from Pad 39A on the morning of 6 April 1984. On board was a five-man crew, headed by Robert Crippen who was making his third Shuttle flight in three years. His four companions were space rookies: pilot Francis (Dick) Scobee, a 44-year-old Vietnam veteran and former test pilot; mission specialists George Nelson, Terry Hart and James van Hoften, all astronauts from the class of 1978. Nelson was an astronomer and expert in solar physics, Hart was an experienced fighter pilot and electrical engineer, while van Hoften had seen combat duty in south east Asia before continuing advanced studies in civil engineering and biomedical flows related to the cardiovascular system – a crew of many talents whose skills and ingenuity would be severely tested in the days to come. Van Hoften nearly did not make the flight: the day before the launch, he was forced to bring his T 38 training jet to an emergency stop when one of its engines flamed out on the runway.

With the successful entry into orbit of Challenger, space became more populated than ever before: apart from the five Americans, there were five Soviets and an Indian on Salyut 7 and 3,300 honey bees travelling around the Earth. The honeybees were the subject of a student experiment to study construction of honeycombs in zero gravity. There were 57 other experiments loaded on to a 10 ton, 12 sided cylinder called the Long Duration Exposure Facility (LDEF), a passive satellite designed for release from the Shuttle and recovery by another mission some months or even years later. Loaded in trays for exposure to the space environment on this first trial of the LDEF was a variety of material, including some 31 million tomato seeds. In another interesting innovation, Challenger carried special movie cameras intended to produce spectacular films for projection on planetarium domes, and high resolution 70 mm film cameras.

The LDEF was gently launched overboard for its 10 months journey through space on the second day, leaving the crew free to concentrate on Solar Max. Challenger gradually closed on her target, commander Crippen moving to within 200 ft of the gold and silver satellite. The rescue team of Nelson and van Hoften donned their spacesuits and entered the cargo bay. Nelson attached himself to the MMU, then van Hoften helped him mount the docking pin to the front, ready for his excursion to the slowly spinning Solar Max as soon as the sun appeared above the curved horizon. TV views transmitted via Hawaii showed Nelson coolly propel himself across the gulf, timing his arrival to miss the rotating solar panels. Firing the MMU thrusters to move to the right in time with the satellite, 'Pinky' Nelson skillfully positioned himself for capture. As the suspense built up among the watching crew and the audience on the ground, Nelson tried three times to close the docking attachment's jaws around the pin on the satellite, but each time they failed to hold. Crippen was becoming concerned: 'Do you want to reattempt it or do you want to come on back? Watch out for the solar array there.' It was clear to both men that the contact between Nelson and the satellite had started Solar Max wobbling out of control. 'I've really got some rates going on the satellite there,' reported Nelson. 'One more try, I've got 1,500 psi of fuel.'

Worried that Hart would be unable to use the remote arm to attempt a grab at Solar Max, Crippen made an extraordinary suggestion: 'Pinky, you've started that rotation to your left now. We really need to stop that to do a rotating grapple. Is there any way you think you can do it with your hands – if you can grab hold of it?' In Mission Control there was suddenly consternation on engineers' faces, but it was too late to stop Nelson. Viewers on Earth saw the amazing sight of an astronaut using both hands to hold on to a giant solar panel in a vain attempt to stop the cartwheeling of a 2½ ton satellite. His efforts only made matters worse, and with his propellant running low, Houston instructed him to return to Challenger. Attempts to capture the errant satellite with the remote arm proved just as fruitless. 'I can't believe it,' said a frustrated Nelson.

Challenger moved off to a safe distance while an alternative strategy was considered.

During the night, the action moved to the Goddard Space Flight Center where engineers used all their ingenuity and skill to bring Solar Max back under control before its batteries were drained. Helped by a fortunate alignment with the sun, they succeeded in stabilizing Solar Max and bringing it back to full power. With the Shuttle short on propellant for its forward reaction control system, the crew were instructed to use the remote arm instead of an EVA to snare their quarry. As Challenger passed out of radio range over the Indian Ocean, telemetry showed the manoeuvring arm was beginning to operate. A tense six minutes crept by as controllers awaited the outcome. Then came a message from Crippen as they tuned in to the Australian ground station: 'OK, we got it – and we are in the process of putting it on the flight support station.' A chorus of cheering echoed around Mission Control. 'Outstanding!' commented Capcom Jerry Ross.

By 9.50 am Cape time, Hart had gingerly placed his precious load on its support cradle in the cargo bay. If in-flight repairs failed, the crippled satellite could now be returned to Earth for closer inspection, but Nelson and van Hoften were confident that this would not be necessary. The next morning, 11 April, saw the two astronauts hard at work, unbolting the faulty attitude control box on the side of Solar Max. While Nelson floated freely around, van Hoften worked perched on the end of the remote arm, instructing Hart where to carry him. 'The tools are working real great. They're a pleasure to work with,' was van Hoften's cheerful comment. With most of the experiments now linked up, the men turned to a trickier task, opening the protective skin and unscrewing the main electronics box. Handling the tiny screws with pressure gloves was likened by the men to 'surgery in boxing gloves', but they maintained a sense of humour as they worked. 'I lost my first screw and I caught it,' said van Hoften. 'Ox has his screws loose,' quipped Nelson. So successful were the world's first satellite repair crew that they finished the installation of the new electronics box in less than an hour, well ahead of schedule. 'We lost two screws – one disappeared over the tail but I don't know where the other is,' reported Nelson.

Their main tasks completed, the two men hitched a ride on the remote arm to deposit their tools in a storage box, then Nelson rode back to Solar Max to inspect the docking pin. 'The only [unusual] thing I see is this little button that sticks out about a quarter of an inch.'

Sufficient time remained for van Hoften to don his MMU for some 50 minutes of flying practice around the cargo bay. By the time they clambered back into the air-lock, the astronauts had spent 7 hours 18 minutes outside, longer than any other American spacemen other than the Apollo 17 moon walkers. Meanwhile, Solar Max was lifted aloft by the remote arm, and, early on 12 April, was released overboard.

Delighted ground controllers reported that the satellite was behaving perfectly. Not surprisingly, the crew appeared on TV wearing signs which read: 'Ace Satellite Repair Co.'. As Crippen told viewers, 'We proved that repairing satellites is a do-able thing. Satellite servicing is something that's here to stay.'

Challenger was due to land at the Cape, but was waved off for one orbit as thick cloud closed in on the runway. The seven-day mission ended in sunny California after the weather refused to improve in Florida.

The twelfth Shuttle mission, scheduled for June 1984, was also to be the maiden voyage of the third Shuttle, Discovery. On board would be a crew of six, commanded by Henry Hartsfield, the only space veteran among them. Of the three mission specialists, two drew particular attention: Dr Steven Hawley, husband to America's first spacewoman, Sally Ride, and Dr Judith Resnik, a 35-year-old electrical engineer who was about to become the second American woman to experience space travel. Most media attention, however, was focused upon the sixth crewman, 36-year-old payload specialist Charles Walker. He had the distinction of being the first fare-paying passenger on the Shuttle: his employer, McDonnell Douglas, had negotiated a bargain fare of $80,000, taking advantage of NASA's desire to attract commercial customers. Apart from some knowledge of emergency procedures and Shuttle household chores, Walker's training exclusively dealt with operation of a commercial electrophoresis experiment sponsored by his employer and Johnson & Johnson. Walker had helped to design the experiment and had trained astronauts on earlier missions in the operation of small-scale tests of the equipment. Now there was to be a prolonged large-scale manufacturing test with Walker as the obvious choice to supervise it.

Unfortunately, Discovery's gremlins had other ideas. A computer fault on 25 June caused the launch to be scrubbed just eight minutes from lift-off. Worse was to follow. Launch control commentator Mark Hess told the story: 'Ten . . . we have a go for main engine start . . . seven, six, five . . . we have main engine start . . . we have a cut-off . . . we have an abort by the on-board computers of the Orbiter Discovery!' Casual observers noted only a small puff of smoke from Discovery's tail-end, with the Shuttle obstinately rooted to the launch pad. Behind the scenes, a real drama was unfolding. One engine of the Shuttle had ignited and a second was about to start when the computers noticed that a main fuel valve in the third engine was not working, and so the whole system was shut down with only four seconds to launch. The solid fuel boosters had fortunately not ignited – there was no way to shut them off once they started firing. Then, just as everything seemed under control, launch officials noticed a fire on the underside of the Shuttle's tail, close to the fully loaded external fuel tank. Inside the giant torpedo-shaped tank were hundreds of tons of potentially explosive liquid hydrogen and oxygen. As technicians turned high pressure hoses on to the fire, controllers spoke to Hartsfield about an emergency evacuation of the cabin, but they eventually decided to ride the storm rather than risk breathing in toxic fumes by heading for the slide-wire to ground level. More than 40 minutes passed before a shaken crew was able to emerge and be led away to a safe refuge. As a former test pilot, Hartsfield was predictably philosophical, though his less experienced crewmates probably felt more emotional about the close shave with death. 'We have been training for this for the last 16 months and we had all reached an emotional pitch where we really had to do this thing. It was very disappointing to have to pull out at the last second,' said Discovery's commander.

Discovery's maiden flight had to be put back until late August, thereby causing the cancellation of its second mission. NASA put a brave face on it and altered the payload to include three communications satellites due to be carried on the mission designated 41-F. Another computer problem caused a two-day postponement on 28 August, but Discovery finally made it off the launch pad on 30 August, despite a six minute delay when a light aircraft strayed into restricted airspace. The new Shuttle was the lightest yet, though its cargo of more than 21 tons was the heaviest of any Shuttle mission. Once in orbit, Discovery behaved well, successfully deploying the three satellites on each of the first three days of the flight. Second out of the cargo bay was a military Syncom satellite, the first designed specifically for the Shuttle. Mission specialists Hawley and Richard Mullane controlled a 'Frisbee' launch of the horizontally mounted satellite, the first time a payload had been rolled overboard in such an undignified manner. Another unique experiment on this mission was the deployment of an extendable solar panel, an ultra-light series of panels which could reach up to 100 ft above the cargo bay yet fold up into a container only 7 inches deep. With Resnik at the controls, the panel was extended successfully and shaken by thruster firings to test its stability. Meanwhile, Walker was fully occupied with his sample processing, proving his worth by solving various mechanical problems and unscheduled shut-downs.

One problem arose late in the mission which affected all crew members. A large icicle developed on the waste water dump outlet, effectively preventing use of the toilet if urine built up in the storage tanks and could not be unloaded overboard. Discovery was turned into the sun in an effort to melt the ice, a potential hazard if it broke off and hit the Shuttle's skin. Meanwhile, the crew were obliged to resort to the plastic bags used during Apollo flights, a procedure which they did not appreciate. 'We decided that those Apollo astronauts must have been real men,' commented one crewman. 'You don't want to hear what Judy has to say,' added another. Eventually, it was decided to risk using the remote arm to dislodge the culprit, despite the risk of damaging some heat tiles on the Shuttle's wing. Hartsfield took responsibility for the operation, and, guided mainly by TV pictures, succeeded in knocking off most of the icicle. Mission Control began to designate the crew as the 'Icebusters', but another problem threatened the crew during their last sleep period. Oxygen used for cabin pressure and fuel cells was found to be leaking, causing Mission Control to wake the crew early and contemplate a quick return to Earth, although a back-up set of tanks was available. When the situation stabilized, the original flight plan was retained, so Discovery returned to Edwards soon after sunrise on 5 September. The press were particularly interested in Ms Resnik's hair: pictures from orbit had shown her long hair standing upright in zero gravity, and one official had unkindly referred to her as looking 'like a Brillo pad'. The astronaut's hair stylist promised a more functional hair-do for her next mission.

Two women were among the record crew of seven to fly aboard Challenger a month later: Sally Ride was making her second flight, and Dr Kathryn Sullivan, a 32-year-old geologist and specialist in remote sensing techniques, was to work on an imaging radar experiment as well as become the first American woman to walk in space – the Soviets had been first in the propaganda stakes, as usual, a few months earlier. Commanding the mainly scientific mission was Robert Crippen, piloting a Shuttle for the fourth time. Other crew members were pilot Jon McBride, a Vietnam combat hero and former test pilot; mission specialist David Leestma, a brilliant aeronautical engineer who had only been an astronaut since 1980; payload specialist Paul Scully-Power, an Australian oceanographer employed by the US Navy; and payload specialist Marc Garneau, a leading Canadian authority on communications and electronic warfare. NASA officials emphasized the need for 'good housekeeping' on such a crowded craft in order to minimize problems and friction between crew members, although the presence of women on board was expected to modify the behaviour of the males, just as the presence of Sally Ride had influenced her companions the previous year.

Despite being the thirteenth Shuttle flight, mission 41-G went off without any major hitches. The only satellite carried, designed to measure the solar radiation balance of the Earth, was launched by the remote arm on the first day after brief exposure to the sun in order to unfreeze its solar panels. The 57 degree inclination of the orbit carried Challenger over a variety of targets for the high resolution imaging radar and large-format camera, and the camera designed for measuring air pollution. Failure of a dish antenna on the orbiter meant a lot of extra work for the pilots and crew as they had to store data from the imaging radar on tape, then manoeuvre the Shuttle until the antenna pointed at the TDRS satellite so that it could be transmitted to Earth. This and other minor problems, notably a hiccup in the air conditioning, caused the space walk to be postponed for two days. At last, on 11 October,

Leestma and Sullivan entered the cargo bay for a successful test of a new satellite refuelling technique. As Sullivan took on the role of 'plumber's mate' and enjoyed a final session of sightseeing and acrobatics, she was obviously having a great time. The flight, earlier dubbed 'Murphy's Mission' by the sceptics, ended in a blaze of success; even hurricane Josephine, once a threat to a Florida landing, held off so that Challenger could return to its launch site soon after midday on 13 October.

The year ended with a spectacular satellite rescue mission. Although the insurance underwriters had paid out $180 million to the owners of the two satellites stranded in the wrong orbits after launch from the Shuttle in February, they were prepared to take a bold gamble by asking NASA to attempt a salvage operation. The mission specialists entrusted with the tricky task were Joseph Allen and Dale Gardner, both experienced spacemen. Also on board were two rookies, pilot David Walker, another combat veteran of south east Asia and a former test pilot, and the first mother to go into space, 35-year-old Anna Fisher, the wife of astronaut William Fisher and mother of a baby daughter. Throughout the mission press photographers delighted the nation by showing pictures of year-old Kirstin proudly watching her mother mount the Shuttle and perform on the TV screen.

Before the rescue could be attempted, the new communications satellites had to be deployed successfully, a job which was completed with no difficulty by the third day. The next day was occupied with catching up manoeuvres by the orbiter on the first errant satellite, Palapa B-2, spinning uselessly 210 miles above the Earth. Commander Frederick Hauck eventually brought Discovery to within 35 ft of Palapa, easily within range of the two astronauts in their rocket-powered backpacks. As Allen moved into the cargo bay and spotted the satellite hovering close by, he exclaimed: 'Holy smoke, Palapa is right there, my friend! Let's go get it.' TV viewers in the States were treated to spectacular live pictures of the activity as Allen drifted across to the satellite, bearing a capture device called a 'stinger', like a skeleton umbrella, in front of him. Once the stinger was inserted into the engine nozzle of the satellite, he fired the backpack thrusters to stabilize Palapa so that Fisher could grab it with the remote arm. So far so good. By grasping the stinger, Fisher brought Palapa and Allen into the cargo bay so that Gardner could cut off a protruding antenna and install a bridging device. Half an hour later, the two space walkers were still struggling to attach this bridge, causing Hauck to inform Houston: 'We've got a problem here. We are proposing to go to the no A-frame procedure.'

Following this back-up procedure, Allen removed his MMU, then, imitating the Greek hero Atlas, he grabbed hold of the great drum, which weighed 1,200 lb on Earth, and held it above his head. For the next 77 minutes, he remained in this position while Gardner beavered away below, removing the stinger and replacing it with an adapter so that the precious load could be secured in the special cradle in the cargo bay. Then, with commander Hauck continually warning them to be careful in case they damaged either the Shuttle or the satellite, the two salvagemen gently pushed Palapa into position. After a six hour EVA, the men were granted a rest day on 13 November prior to their next adventure. Meanwhile, Mission Control decided they had done such a good job using the hands-on method that it would be repeated for rescuing the Westar satellite.

On the morning of 14 November it was Gardner's turn to float across to the satellite and stabilize it. Allen then joined him, standing on the end of the remote arm, and began his Superman stint all over again. At one stage, Walker, a spectator in the cabin, spotted a torque wrench drifting free from Gardner, causing the astronaut to chase after it hurriedly, but the base adapter was safely installed with more than an hour to spare. As the second satellite was stowed in its cradle, the Lutine bell rang at Lloyd's in London to signal a successful salvage operation. A delighted crew were asked by Capcom if they would like to pick up any more strays: 'We've got enough propellant for some more, so if you guys see any more satellites let us know and we'll get a good fix on them.' Gardner told Houston: 'You know, we've all been looking out the window at those two satellites in the bay and none of us can quite believe we've really got them in there.' When asked in a press conference about what they planned for an encore, Gardner drily replied that he had to pay some bills when he got home, while Allen admitted he would have to cut the grass. Such is the life of space heroes! Discovery returned safely to Florida on 16 November.

In contrast to the high profile given to the satellite rescue mission, the first Shuttle flight of 1985 was subjected to an unprecedented news blackout. Discovery's five-man military crew, including Major Gary Payton, the first Department of Defense astronaut, were sent on a top secret three-day mission to launch a giant spy satellite over the Soviet Union. In contrast to the usual open policy of NASA, the space agency was obliged to meet USAF requests for silence concerning the mission details and its purpose. After a delay on 23 January due to freezing weather at the Cape, the Shuttle blasted off the next day. NASA refused to announce the exact launch time, though a nine-minute warning was given beforehand. All communications between the crew and the ground were scrambled for security reasons. Considerable criticism of the role of the military in space came to the surface, including revelations that the military space budget was now double that of the civilian space programme. Nevertheless, NASA announced after the mission that more top secret Department of Defense missions were planned in the next few years, beginning with the launch of two more military satellites by the Shuttle Atlantis in October 1985. Further controversy surrounded press attempts to probe into and reveal details of such secret missions, a policy which Defense Secretary Weinberger described as 'the height of journalistic irresponsibility'.

Another politician stepped into the limelight in January when it was announced that Senator Jake Garn, 52-year-old chairman of the sub-committee which oversees NASA's budget, would fly on the Shuttle Challenger in February. Grateful for the chance to gain support for further expenditure on the Shuttle and the proposed Space Station, NASA officials were, nevertheless, uncertain what to do with the Mormon politician from Utah. Garn simply commented that they could do anything they wanted with him, so it was eventually decided to use him in the undignified role of human guinea-pig. While Garn became the butt of cartoonists and humourists in the media – one wit declared that his main duty would be 'to throw up' – NASA also went through its own series of embarrassing moments as a long run of mishaps hit the mission.

Most of the crew assigned to this mission, originally designated 41-F, had expected to fly in August 1984, but their flight had been cancelled after the delay to Discovery's maiden voyage. Reassigned to fly with Garn and Frenchman Patrick Baudry, their flight was again delayed through tile problems, then cancelled altogether when a fault appeared in the main payload, the TDRS satellite. The crew were shuffled round to replace the next crew in line, except that Baudry had to drop out in favour of engineer Charles Walker, and the Canadian Anik satellite was placed aboard Discovery instead of Challenger. After all this confusion, mission 51-D finally managed to get off the ground on 12 April 1985, though even then there was a 55 minute delay when a ship strayed into the drop zone of the solid rocket boosters. Discovery just blasted off ahead of the rapidly deteriorating weather.

Further embarrassment followed after the successful deployment of Anik on the first day. A second communications satellite, spun out of the cargo bay on 13 April, refused to fire its booster rocket and remained languishing in full view of the frustrated crew. This mishap did little to improve the mood of the flying Senator – he was already suffering from spacesickness, much to the delight of the doctors (and perhaps a few other observers). With an important American defence satellite out of commission, NASA teams went into a huddle to rustle up a way out of the predicament. Unwilling to risk a dangerous space walk, they came up with the idea of a 'space flyswatter' which could be attached to the end of the robot arm then used to flick on the booster activation switch. Discovery's cabin looked like a craft workshop as the astronauts used the plastic cover from a flight plan book, part of an aluminium window shade, a metal rod and some vacuum cleaner hose to fashion the flyswatter and a 'lacrosse stick' under instructions from Houston.

On 16 April, mission specialists David Griggs and Jeffrey Hoffman floated out into the cargo bay on the first totally unplanned space walk of the American space programme. Neither man had been in space before, though Griggs was a highly experienced fighter plane and test pilot, working on NASA research programmes even before he became an astronaut in 1978. Hoffman was also a 1978 recruit, though of a very different background, having gained a doctorate in astrophysics. Although it took some time to adjust to their environment, they were more than happy when they returned inside after some three hours. 'Is that not beautiful or is that not beautiful?' asked Griggs as they admired their handiwork. With the flyswatter in place, the way was open for the rescue attempt.

On 17 April commander Bobko closed in again on the errant satellite while Rhea Seddon, another space first-timer, prepared to manoeuvre the robot arm. Her first try at snagging the switch on the slowly spinning satellite failed, but her aim was better next time around. Four solid tugs on the lever succeeded only in ripping the plastic flyswatter; the switch refused to move. Syncom would have to remain a derelict for the time being at least. Re-entry on 19 April also proved far from straightforward. Discovery had to be waved-off for one orbit while rain cleared from the Cape, then the Shuttle burst a tyre when Bobko applied differential braking to bring it back on line when a gust of wind blew it to one side of the runway. Despite the setbacks, Garn confirmed he was an even more committed supporter of the Shuttle programme than before the flight.

Only ten days after Discovery's return, the next mission blasted off using Challenger, a record for the beleaguered programme. On board this scientific flight, designated Spacelab 3 even though it was the second time the European-built laboratory had flown, were seven crew and various animals. Three of the astronauts were aged over 50: Don Lind had waited 19 years for his first flight since joining NASA (he had come nearest when he served on the Skylab back-up crews); William Thornton was a medical specialist on his second flight; and Dutch chemical engineer Lodewijk van den Berg was a payload specialist tending his own crystal growth experiment. The other payload specialist was Chinese-born physicist Taylor Wang, assigned to look after a fluid mechanics experiment. The menagerie included 24 rats and two caged squirrel monkeys, resulting in the predictable headlines of 'Monkey Business In Space' when things started to go wrong.

Among the minor irritations which faced the astronauts within the first 24 hours were a broken air-lock, a blocked tap and a small satellite which refused to leave the cargo hold. Dr Thornton also ran into a number of unexpected and unpleasant problems with his equipment: first, a vacuum device to analyse astronauts' urine ran riot and began spraying the cabin, necessitating an 'extensive clean-up'; then a combination of monkey faeces and dried rat food escaped from the cages.

Thornton described it as 'some sort of puffed cereal broken up into tiny pieces', but the comments of his companions were unprintable. When mission control reminded commander Robert Overmyer that they were on the air the exasperated Overmyer apologized: 'I guess we'll have to clean up our act considerably.' Potentially more serious was the breakdown of Wang's fluid experiment, but the frustrated scientist stubbornly stated: 'I'm not coming home until I've fixed it,' and after several days' work, fix it he did. His experiment, linked with colleagues on the ground, proved one of the main successes of the mission. Another noteworthy success was the various experiments designed to produce pure crystals that would be used in X-ray, gamma ray and infrared detectors. Some spectacular photographs of aurorae were also obtained as the Shuttle flew on its highly inclined path around the Earth. There was some concern that the cargo bay doors had not closed properly, but Shuttle landed safely at Edwards Air Force Base on 6 May.

The wait of Dan Brandenstein's crew, displaced by Bobko's early in the year, came to an end with the next flight of Discovery on 17 June. Among the crew of seven were another woman first-timer, Shannon Lucid, and French air force pilot Patrick Baudry, back-up for the 1982 Franco-Soviet mission to the Salyut 7 space station. But most media attention was reserved for the first royal spaceman, Prince Sultan Salman Abdul Aziz Al-Saud. The 28-year-old nephew of the Saudi Arabian King Fahd was chosen through his position as Acting Director of the Saudi Arabian TV Commercial Department, for one of Discovery's main payloads was the Arabsat telecommunications satellite. The Shuttle cargo bay was crammed full for this mission, with two more communications satellites (one American, one Mexican) and a small X-ray detecting satellite called Spartan which spent nearly two days in independent flight before retrieval and return to Earth.

One of the most controversial aspects of the mission, and one which re-emphasized the increasing military role of the Shuttle in the eyes of its critics, was a 'Star Wars' test of a laser beam directed at the Shuttle from Hawaii. This low-powered laser was part of a new missile tracking system intended to destroy incoming Soviet missiles during a nuclear attack. Its first trial on 19 June failed when the beam missed a small mirror mounted on the Shuttle after Discovery's computers were fed the wrong information and pointed the craft in the wrong direction. A second attempt on 21 June went faultlessly, however, and the crew sent back TV pictures of the flashing blue light accompanied by music from the 1812 Overture.

All four satellites were successfully launched during the first part of the mission, much to the relief of the Sultan, enabling the specialists to get down to other work. Al-Saud happily went about mixing oil and water to observe the effects of zero gravity, assisted Baudry in his medical experiments and snapped away during five daylight passes over his homeland. Arab journalists also noted how he studied the Koran in orbit and observed the crescent moon through Discovery's window at the end of the religious feast of Ramadan. The mission ended on 24 June at Edwards Base in California amid general praise from all concerned. The only sour note came when one of the wheels sank into the bed of the runway which had been softened by rain: a towing job was required on the multi-million dollar craft.

After one of the most successful missions of the entire programme, confidence was high among NASA officials when Challenger blasted off on 29 July 1985 with Spacelab 2 and a crew of seven on board. Among the specialists were two men who had become astronauts in August 1967 yet were only now venturing into space. At the age of 58, astronomer Karl Henize was the oldest person ever in orbit; geophysicist Anthony England had stayed with NASA through the excitement of the Apollo moon landings, then left to join the US Geological Survey. Returning to Houston in 1979, he was looking forward to studying the Earth from a height of 240 miles (390 km) instead of ground level.

Challenger was initially set for launch on 12 July, but as the clock reached T minus three seconds, the computers detected a malfunctioning valve and the engines shut down, leaving the Shuttle rooted to the pad with the crew swaying in the breeze and wondering what had happened. Recycled to 29 July, Challenger's eighth flight began perfectly, but then sensors on one of the Shuttle's main engines detected a rapid temperature rise. Once again, the safety system cut in, shutting off the faulty engine. Flight controllers were confident that the crew could attain orbit with only two engines operational, so they were told to 'Abort to orbit'. Then another crisis arose: a temperature sensor on another engine indicated overheating once again. With Challenger more than eight minutes into the mission but still short of orbital velocity, the controllers were placed in an almighty quandary. Should they order the crew to attempt an emergency landing, possibly in the sea if they missed the island of Crete, or should they override the sensor? A quick check of engine performance suggested all was well, so the latter course was chosen. The decision proved correct, and Challenger struggled into a low orbit of only 75 miles (120 km). Although the manoeuvring system thrusters were able to raise this to around 170 miles (275 km), the orbit was still well below the altitude originally planned for the experiments. Still, the men in space and the team on the ground were merely grateful the craft was still in one piece.

For the first time a Spacelab mission was being flown using only a pallet in the open cargo bay rather than a pressurized workshop. Crew operations were therefore conducted from the mid-deck operations centres. As on previous Spacelabs, the crew were beset by technical problems, most notably with an experimental aiming device which refused to point the onboard telescopes at the sun, and with a solar magnetic field measurement system which shut down. However, to the delight of the crew and their colleagues on the ground, the solar telescope suddenly returned to life of its own accord, and the pointing system's teething troubles were successfully ironed out. By 5 August the mission was going so well that the crew accepted the offer of an extra day in orbit. Challenger landed in California on the following day, carrying 'enough data to keep scientists busy for years'. An immediate examination of the Shuttle's engines confirmed that the launch problem lay with the sensors.

Having tried out various satellite rescue techniques on previous flights, NASA planners were determined not to leave the abandoned Syncom satellite wallowing like a stranded whale. The $80 million satellite, weighing some 7½ tons on Earth, had been waiting since April for someone to switch on its rocket motor. This was clearly a delicate operation – if the motor fired while an astronaut or Shuttle was floating nearby, the consequences would surely be disastrous. Extra training was laid on for the five-man crew who had originally expected a relatively straightforward mission involving the launch of three communications satellites. Discovery's launch was delayed for three days due to bad weather and a failure in the back-up computer, so the twentieth flight of the Shuttle began on the morning of 27 August 1985.

Once in orbit, the crew noticed a problem as soon as they opened the payload bay doors. The sunshield around the Australian communications satellite, Aussat, had jammed after hitting the remote arm: not only was the arm damaged, but the sunshield was unable to protect its valuable cargo from the damaging solar radiation. There was no choice but to launch Aussat as soon as possible, so the crew achieved another Shuttle first by deploying two satellites on the first day. The third satellite, a sister to the stranded Syncom, was successfully fired into orbit a couple of days later, leaving the cargo bay empty and the crew free to concentrate on reviving the sleeping giant.

The men charged with the unenviable task were both well qualified in their own way. William Fisher was a surgeon who specialized in emergency medicine before he qualified as an engineer and joined NASA in 1980. Accompanying him would be James van Hoften, nicknamed 'Ox' because of his 6 ft 4 in, 196 lb physique. A former Navy pilot and combat veteran over Vietnam, the burly van Hoften went on to advanced studies in hydraulic engineering and fluid mechanics before joining NASA in 1978. As commander Engle closed in on their target, eventually station-keeping only 35 ft (11 m) from it, the astronauts were relieved to confirm that it was hardly spinning. The first stage began on 31 August when van Hoften stood on the remote arm so that John Lounge, another space newcomer, could gingerly lift him towards Syncom's activation lever. For the next three hours, he and Fisher took it in turns to wrestle with the massive cylinder until they had it under control and firmly in the grip of the robot arm. Now Fisher's surgical skills came to the fore as he successively shorted out the activation sequencer, unscrewed two panels to install hardware for ground command of the satellite, and by-passed the inoperative sequencer. His final job was to install a battery powered unit which enabled the main antenna on Syncom to fold outwards. After a record 7 hours 8 minutes outside the cabin, the satisfied crew called it a day.

The morning of 1 September saw the space repairmen venture into the cargo bay to complete their job. Fisher fitted a delaying mechanism which would prevent Syncom processing ground commands until Discovery was safely out of the vicinity, then grabbed hold of a handling bar he had fixed on the previous day. Lounge then released the grip of the remote arm in order to pick up van Hoften. With both men holding on, Engle manoeuvred the Shuttle into the best position for the satellite to be thrown overboard, but then they hit a snag. Automatic thruster firings threatened to throw the satellite out of alignment, so that it was all the men could do to prevent the massive cylinder from striking the ship. As Fisher let go, van Hoften was raised on the end of the arm looking like some mythological hero carrying the world on his shoulders. 'If something happens and I'm about to lose it, I'm going to give it a heck of a push and bail out,' warned the unhappy Ox. 'It all looks good. Take what you've got,' advised a closely watching Engle. So van Hoften released his grip and gave the satellite its initial spin with five hefty pushes on a small 'spin-up bar'. After more than four hours of manhandling the awkward Syncom, the men deserved the congratulations of all concerned. Ground controllers were later able to confirm reactivation of the disabled satellite, confirming the opinion of Steve Dorfman, President of Hughes Communications, that it was 'the most remarkable salvage mission in the history of the space programme'.

Discovery landed in California on 3 September. The only casualty was a $120 screwdriver lost overboard during one of the space walks.

After this magnificent boost to its prestige, NASA would have preferred to continue the high profile publicity for the maiden flight of the fourth Shuttle, Atlantis. However, the debut mission had been dedicated to a secret defence payload, revealed in the media as two military communications satellites, so a blanket of silence came down for the second time on a Shuttle mission. To confound Soviet intelligence agencies, the exact time of launch was not revealed until just before lift-off, and no details were given concerning the welfare of the crew or their classified payload. The name of military payload specialist William Pailes was also kept secret until a few days before the flight, and, of course, all interviews with the crew were definitely out. It was a far cry from the open (some said too open) information policy which had characterized NASA since its inception in 1958.

The low key debut of Atlantis was followed by another mission that received minimal exposure in the popular press, except in West Germany. Fitted aboard Challenger for the 30 October launch were 76 experiments entirely paid for by West European countries, and the mission had been sponsored to the tune of $64 million dollars by the West German government. Not surprisingly, there were three European astronauts on board the Spacelab D-1 flight, and the payload was supervised from a ground control centre near Munich, the first ever such delegation of authority. The two West German space rookies were physicists Ernst Messerschmid and Reinhard Furrer, both in their forties and plucked from their academic work by a newspaper advertisement in 1978 which read: 'astronauts wanted'. Third payload specialist was Dutchman Wubbo Ockels, another physicist who had been a back-up crewman for the first Spacelab mission two years earlier. Altogether, Challenger carried a record crew of eight, ensuring that both the craft and the mission schedule were packed tight. There was one woman on board, first-timer Bonnie Dunbar.

Apart from some problems with a fuel cell and communications links which delayed entry into the pressurized laboratory for a few hours, the crew found few obstacles to the successful accomplishment of their objectives. Split into two teams, just as on previous Spacelab missions, the Blue Team of Nagel, Dunbar and Furrer stayed on duty for the first 12 hours, then gave way to the Red Team of Bluford, Buchli and Messerschmid: Ockels and commander Hartsfield alternated between shifts as required. Only during the brief handover periods did the entire crew get together for activities such as meal preparation. One oddity was Wubbo Ockels' 'bedroom': using a new sleeping bag which he had helped to develop, the Dutchman slept in the laboratory instead of on the mid-deck like his companions.

Much of the working space in the module was taken up by a space-sled which ran along its entire length. It consisted of a chair mounted on rails which could be accelerated backwards or forwards at different speeds. All five Spacelab specialists took their turns in wearing the heavily instrumented helmet during these nausea-inducing rides. Among the other experiments were the Biorack, loaded with seedlings of water cress and maize, materials processing aimed at producing pure crystals and a test of a new orbital navigation system. The NAVEX project was saved, like so many experiments in the past, by onboard repairs necessitated by a wiring problem.

Challenger returned to Edwards Base with the mass of valuable data on 6 November. As the orbiter rolled along the runway, Hartsfield tried out an improved system of nosewheel steering designed to overcome the problem of crosswinds which had kept recent Shuttle landings away from the confined runway at Cape Canaveral.

When President Reagan had informed Congress on 25 January 1984 that he was 'directing NASA to develop a permanently manned space station – and to do it within a decade', the idea of construction in orbit was still an untried dream. All previous space stations had been assembled on Earth then blasted into orbit on top of giant booster rockets, but the concept of multi-modular stations built in space and capable of sustaining six or more people on a permanent basis was still only a glint in the eyes of the theoreticians. The first practical demonstration of construction techniques had to wait until the flight of Atlantis at the end of 1985, when the seven-member 'Ace Construction Co.' set off to make history.

The men assigned to try out various structures during two EVAs were both space rookies who had joined NASA in 1980. Army lieutenant colonel Sherwood Spring was a highly decorated Vietnam war hero, having completed 375 combat missions in fighter aircraft and helicopters during the 1960s before going on to test fly both types of aircraft as Head of the Ordnance Systems Branch at Patuxent River, Maryland. His companion was Air Force major Jerry Ross, four years his junior but a leading flight test engineer who had gained a master's degree in mechanical engineering and worked on ramjet design before becoming an astronaut. Also on board was the first Mexican spaceman, electronic engineer Rudolfo Neri Vela, who was going along mainly to monitor the Morelos communications satellite: the mission launch was altered so that this vital link could be established as quickly as possible after the disastrous Mexican earthquake. Other newcomers were the only female crew member, biologist Mary Cleave, and former fighter and test pilot Bryan O'Connor of the Marine Corps.

There were a few racing heartbeats during launch when a couple of sensors in the Shuttle's engines performed erratically, but otherwise the mission began smoothly. Three communications satellites were flawlessly launched, including Morelos, during the first three days while McDonnell Douglas engineer Charles Walker calmly carried on with his latest experiments to purify pharmaceutical products. Then on 29 November Spring and Ross donned their pressure suits for the first trial of the space assembly techniques they had tested in water tank training. Using a triangular jig as the basis for attachment of tubular beams, they erected a 45 foot tower in about the same time it had taken them on Earth and in about half the time they had been allocated. Tethered to either side of the cargo bay by slide wires, the men were so engrossed in their work that they merely switched on their helmet lights and kept building instead of pausing for breath during the night passes.

Task one completed, they dismantled their new creation, stowed the beams, and began the first of eight exercises using larger beams to construct a pyramid. This time they took it in turns to leave the cargo bay and float freely while installing the sections passed up by the partner down below. In case of a mishap, each section of beam structure was attached to a tether which could be quickly released once it was firmly in position. By the end of the changeover after four pyramids had been erected, Spring was beginning to feel the strain of such intensive manipulation while wearing the stiff EVA gloves. 'On that last one I had no idea which way was up or down,' commented the tired astronaut. Before concluding the five and a half hour excursion, Spring deployed a small visual satellite target for tests with the Shuttle's autopilot software. Ground observers were full of praise for their efforts, detecting 'clear evidence of a learning curve' for both crewmen.

The second EVA took place two days later as Atlantis once again passed over Central America. During another trial of the 45 foot tower, the men completed nine of the ten triangular bays while in their fixed work positions, then Ross added the top bay while standing on the robot arm controlled by Mary Cleave, and clipped on a rope to simulate an electrical cable. He next held the entire structure above his head as Cleave slowly raised the arm, demonstrating the ability to manoeuvre such large assemblies by hand. 'Very little force is required to start or stop motion', he reported confidently. With the tower safely returned to its assembly jig, Spring hopped on to the arm to practise removal and repair of the structure, then took his chance to perform an orbital Charles Atlas display. Finally, after stowing the beams once more, they returned to pyramid assembly, this time using the robot arm with Spring on board. TV viewers received more spectacular pictures of Ross silhouetted against the blue and white Earth with the massive structure above his head. The highly successful space construction crew headed back inside for a well-deserved rest after six and a half hours full of significance for the future of the American space programme.

Atlantis returned to California two days later after a prolonged search for underground water sources in drought-stricken Ethiopia. NASA officials glowingly spoke of one of the most successful missions ever, and looked forward to an even more promising New Year. But a sign of things to come appeared in the form of the refitted Shuttle Columbia. Planned for a 18 December launch, the flight was called off after final systems checks fell behind schedule, then postponed after a solid booster rocket overheated and shut down the countdown. It was a bad sign for a space agency under pressure to turn the Shuttle into the long-promised space commuter which could pay its way.

16

The Shuttle is Grounded
The Challenger Disaster, 1986

Nineteen eighty-six was to be a red letter year for the two space superpowers. In the Soviet Union a new civilian space agency had been established to oversee the peaceful exploitation of outer space. With the faithful old Salyut 7 space station no longer occupied, it seemed only a matter of time before a third generation replacement was sent up to join it. Meanwhile, under the impetus of the Presidential directive, the American space agency was pressing ahead at full steam to meet the demands for a defensive 'Star Wars' system and an operational space station by the early 1990s, as well as the demands of its civilian and scientific customers. As NASA's critics so often pointed out, the Americans had put all their eggs in one basket with the Shuttle, a gamble that had so far not paid off. The nine billion dollar ship was not paying its own way, neither was it the reliable space taxi service which had originally been promised. So 1986 would be the year when NASA launched 15 Shuttle missions, the largest number so far, and go back on line for a target of 30 or more per year by the end of the decade.

Vital to achievement of this target was the continued operational status of all four Shuttles. To this end, NASA officials were relieved when the first ship, Columbia, returned to the fold in December 1985 after a two-year refit, but almost straight away the Shuttle jinx caused the twenty-fourth mission to be postponed, leaving the seven-member crew strapped forlornly in their seats with nowhere to go. After another delay early in January to enable technicians to take a holiday, the fourth attempt ground to a halt with only 31 seconds of the countdown remaining when a malfunction cropped up in Columbia's main engines. Postponement number five came when thick cloud blanketed the Cape and the emergency landing sites; number six was due to another faulty engine valve; number seven was forced by heavy rain in Florida. By now the frustrated crew had endured many hours in the Shuttle's cabin without moving an inch, and the sceptics were dubbing it 'mission impossible'. Columbia had gained the unfortunate distinction of more launch delays than any previous Shuttle. Small wonder that lucky charms were strewn around the cabin entrance when the crew arrived for yet another attempt on 12 January.

Leading the parade of astronauts to the launch pad that day was US Congressman William Nelson, representative for the area of Florida that included Cape Canaveral. It had been a particularly galling time for Nelson, since he headed the House Space Science and Applications subcommittee and was one of NASA's leading political supporters on Capitol Hill. Like his predecessor, Senator Jake Garn, the second American politician in space had volunteered to act as a guinea-pig for motion sickness studies and to help out with other experiments studying proteins that are linked to cancer. Some confusion arose with the crew list since it also included George Nelson, an astronomer who was looking forward to studying Halley's Comet from the ideal vantage point above Earth's interfering atmosphere. Other first-timers were pilot Charles Bolden, a Marine fighter and test pilot; Robert Cenker, an engineer with RCA's Astro-Electronics Division; and Franklin Chang-Diaz, a Costa Rican-born nuclear physicist who was the first Hispanic-American to fly in space.

Columbia made a spectacular sight as it finally blasted off in the pre-dawn darkness, followed on its way by the relieved cheers of the ground crews. The flight's nickname was changed to 'mission accomplished' after the RCA communications satellite, the only major payload in the cargo bay, was successfully released later in the day, but the celebrations proved premature. When George Nelson tried to photograph Halley's Comet using a special telescope-camera system, he found that the batteries on the light intensifier had gone dead. Even after these were replaced, the equipment still refused to function, so he was reduced to snapping exposures of the comet without its aid. Mission controllers decided to cut the flight by a day, but even the weather refused to play its part. The crew were obliged to remain in orbit for two more days until the clouds and rain abated at the Cape. A frustrated agency spokesman muttered: 'Here we go again. Everybody's pretty disappointed that these things keep happening to us.' Columbia finally touched down at Edwards Air Force Base in California instead of Cape Canaveral on 18 January. Already the ambitious launch schedule for the year was threatened, though NASA engineers plunged into an all-out effort to return Columbia to Florida and complete preparations for a 6 March blast-off to study Halley's Comet before it disappeared for another 76 years.

Apart from the Astro mission to observe the famous comet, the Shuttle fleet was allocated to a varied payload inventory over the next 11 months. The twenty-fifth mission, scheduled for 22 January, would also study Halley and launch a sophisticated TDRS communications relay satellite. Other time-critical launches of scientific payloads were due in the summer: two satellites were to be launched towards the planet Jupiter, the European-built Ulysses solar probe and the Galileo craft designed to survey the Jovian system. In August, the Hubble Space Telescope the most important astronomical project of the century and predicted to revolutionize our understanding of the universe, was to be placed in Earth orbit. Furthermore, in addition to the commercial payloads, there were four missions sponsored by the Pentagon, at least one of which would lift-off from the new spaceport at Vandenberg in California. The reputation of the Shuttle, and NASA in general, was on the line. Would they be able to deliver?

First in the queue was mission 51-L, loaded with a scientific satellite to make ultraviolet studies of Halley's Comet and a second tracking and data relay satellite for worldwide coverage of Earth-orbiting spacecraft. Commanding the seven-member crew was 46-year-old Francis Scobee, a Vietnam veteran and test pilot during 22 years of service with the USAF before he retired in 1980. The 6 ft 1 in father of two was on his second Shuttle flight, having previously served as pilot on the Solar Max rescue mission in April 1984. His pilot on this mission was 40-year-old Navy commander Michael Smith, also a Vietnam war hero, an instructor at Patuxent River Test Pilot School, and holder of a master's degree in aeronautical engineering. The greying 6 ft 1 in Smith was married with three children, and making his first trip into space.

The mission specialists all had experience of space travel. There was 39-year-old Air Force major Ellison Onizuka, who was a distinguished flight test engineer at Edwards Air Force Base before astronaut selection in 1978, and participated in the top secret defence mission of Discovery in January 1985. Japanese-American Onizuka grew up in Hawaii and dreamed of going to the moon. A master's graduate in aerospace engineering, 5 ft 9 in tall Onizuka had two children. Ronald McNair also had humble beginnings; a black who grew up in South Carolina, he went on to obtain a doctorate in physics from the Massachusetts Institute of Technology and became a specialist in laser development and their space applications. The moustachioed McNair was 36 years old, and had two young children. He joined NASA in 1978 and first flew on mission 41-B in February 1984. The third mission specialist was 36-year-old Judith Resnik, who had become the second American woman in space in August 1984. A classical pianist with a doctorate in electrical engineering, she was divorced and had no children. It was expected she would need all her skill with the robot arm on this mission to pick up the Spartan satellite.

There were two members of the crew who had received minimal training for a Shuttle flight other than learning the basics of living and working in such a complex craft. Gregory Jarvis, age 41, was the payload specialist in charge of the TDRS satellite. As an electrical engineer, he worked on designing missile systems and advanced tactical communications satellites before joining Hughes Aircraft Company as a subsystem engineer. He was married, but had no children. The least experienced person to board Challenger was also the darling of the American media and the most famous school teacher in the country. Thirty-seven-year-old Christa McAuliffe had taught a variety of subjects, including English, American History and Economics, in the 15 years prior to selection as the first private citizen to go into space. She had been presented to the world by Vice President Bush after defeating more than 11,000 other applicants in the NASA 'Teacher In Space' project. She had swayed the judges with her bubbling, extrovert personality, her enthusiasm for the Shuttle programme, and her ability to communicate her experiences to a young audience. Describing her role as a 'space participant' rather than an astronaut, she declared her intention of keeping a journal which would hopefully convey 'the ordinary person's perspective'. Her lawyer husband, two children and parents joined a group of her pupils from Concorde High School, New Hampshire, in the VIP viewing area for the launch on the crisp, bright morning of 28 January 1986.

The lead up to the launch had not been without its share of problems. Put back three days to 25 January, it seemed plagued by bad weather: although a bad forecast for the 26th was not fulfilled, the NASA staff decided there was insufficient time to load the fuel and round up the crew. The next day, a handle on the crew entry/exit hatch jammed and took so long to be replaced that the weather deteriorated, necessitating a third launch scrub. Things did not look much better as morning broke on 28 January: temperature overnight had been well below freezing, resulting in the growth of large icicles on the launch pad. Engineers expressed concern that these could break off and damage Challenger's heat tiles; others feared a reduction in efficiency of the O-ring seals on the solid fuel rocket boosters. No Shuttle had ever been launched with temperatures only a few degrees above freezing, but, after a launch pad inspection and consultations with senior advisers, Shuttle director Jesse Moore gave the green light. Only later was it revealed that the deep doubts of some engineers had not been passed on by their superiors to Mr Moore.

At 11.38 am Cape time on a clear Florida day, the Cape reverberated to the familiar roar of main engine ignition followed by clouds of smoke and flame from the solid fuel boosters. In a matter of seconds, Challenger cleared the tower and began the roll which would carry the craft out to the east across the ocean. Spectators gazed into the brilliant blue sky, tracing the progress of the twenty-fifth Shuttle flight by its billowing trail. A calm, authoritative voice from the public address system echoed around the site: '4.3 nautical miles, down range distance 3 nautical miles'. As Challenger passed through the period of maximum dynamic pressure from the atmosphere, her engines were boosted to 104 per cent thrust.

Unknown to anyone in the cabin or on the ground, there was already a jet of flame hungrily licking around the giant orange fuel tank from the right-hand booster rocket. 'Challenger, go with throttle up,' reported mission control. 'Roger, go with throttle up,' came the reply from Dick Scobee. They were the last words received from the seven astronauts. Seconds later, the Shuttle suddenly disappeared amid a cataclysmic explosion which ripped the fuel tank from nose to tail. The delta-winged craft which had been perched on the side of this tank was torn free by the blast and disintegrated in mid air. Commander Scobee just had time to open up his radio channel, but was cut off before he could speak. Co-pilot Michael Smith, suddenly aware that something was terribly wrong, exclaimed 'Uh, Oh.' Some of the crew activated emergency oxygen supplies, but with no effect. Although the crew's cabin seems to have remained largely intact, the aerodynamic pressure exerted on the human passengers killed any who survived the explosion. The shattered remnants of the craft plummeted nine miles into the ocean, spread over hundreds of square miles. Relatives, friends and pupils in the viewing area could only watch spellbound, eyes staring, mouths wide open, at the white cloud and its two horns which grew in the otherwise cloudless sky. To add to the sense of unreality, the calm voice on the PA continued reading off the altitude and velocity. Then the worst fears of everyone were confirmed:

Flight controllers here looking very carefully at the situation. Obviously a major malfunction. We have no downlink. We have a report from the flight dynamics officer that the vehicle has exploded. The flight director confirms that. We are looking at checking with the recovery forces to see what can be done at this point.

There was nothing more to say.

It was the worst accident in nearly 25 years of manned spaceflight, and the first time any American astronauts had been lost during a mission. A shock wave spread throughout the nation. President Reagan postponed his State of the Union address, and appeared on TV to praise the dead pioneers: 'The crew of the Space Shuttle Challenger honoured us by the manner in which they lived their lives. We will never forget them, nor the last time we saw them, this morning, as they prepared for their journey and waved goodbye, and "slipped the surly bonds of Earth to touch the face of God".' He later appointed a Presidential Commission to investigate the accident, chaired by former Secretary of State William Rogers and including former astronaut Neil Armstrong, first US woman in space Sally Ride and former test pilot General Chuck Yeager.

It signalled the most traumatic period in NASA's 27-year history, with the repercussions rebounding like shock waves throughout the entire organization. NASA administrator James Beggs was already suspended pending a fraud investigation related to his time as an executive with General Dynamics, and he was soon replaced by former NASA administrator James Fletcher. Jesse Moore, Shuttle programme director, was shifted to a position as head of Johnson Space Center and was replaced by former astronaut Rear-Admiral Richard Truly. In another major reshuffle, veteran Philip Culbertson was relieved from his duties as NASA general manager, and the director of Marshall Spaceflight Center, William Lucas, took early retirement. While the commission of inquiry revealed ever more disquieting evidence of a 'flawed' procedure for launching, the agency was further shocked by the uncompromising attitude of the people who put their necks on the line each time the Shuttle blasted off, the astronauts. Chief astronaut John Young went on the record with a blasting condemnation of existing policy: 'The space programme will only succeed in the future if competent and highly qualified men and women who fly the Shuttle have confidence in the system.' Many of his colleagues made it abundantly clear that they did not have such confidence following the revelations of the inquiry, causing Richard Truly to confirm that the next launch would have 'a robust margin for error'. He promised that it would be a daylight launch in warm weather, crewed by career astronauts and targeted to land in the wide open spaces of Edwards Air Force Base. Nevertheless, a number of experienced astronauts voted with their feet and resigned from NASA, including Garriott, van Hoften and Stewart.

The 256-page report of the Rogers Commission appeared at the beginning of June 1986. It was a damning indictment of NASA management, revealing a litany of engineering faults and human errors which amounted to a scandalous disregard for the astronauts' lives. It was shown that engineers working for the SRB manufacturers, Morton Thiokol, and at the NASA Marshall Spaceflight Center in Alabama had repeatedly expressed concern over the poor performance of the O-rings which sealed the joints between the rocket sections, and the sealing system had been rated as 'Criticality 1' – failure would almost certainly lead to loss of the craft. Despite evidence of damage to the seals on 14 out of the 24 successful Shuttle missions, and reservations expressed by engineering when it was revealed that sub-zero temperatures prevailed at the Cape, NASA and Morton Thiokol management gave the go-ahead for launch of mission 51-L. Most disturbing for the astronaut corps was the revelation that these flaws and doubts had existed since the early design stage of the re-usable craft, yet they had never been consulted and were totally unaware of the dangers. As commission member Richard Feynman commented, NASA was playing 'a kind of Russian roulette' every time the Shuttle was launched. Not surprisingly, one of the major recommendations of the Commission was a greater role for astronauts and engineers in approving launches.

The other recommendations included a complete redesign of the SRB joints, the study of astronaut escape systems and greater safety margins for Shuttle landings, and a more realistic flight rate with parallel development of expendable boosters so that there was no longer over-reliance on one launch system. It was accepted that NASA had suffered from trying to meet too many demands from all sides so that its resources were stretched to the limit. In the end, the pressures to meet a certain number of launches per year led to vital engineering concerns being pushed to one side or ignored. The Commission concluded that 'NASA must establish a flight rate that is consistent with its resources' so that safety becomes the top priority and the risks to astronauts' lives are reduced as much as is humanly possible.

The Rogers inquiry also detailed the chapter of incidents which led to the destruction of Challenger only 74 seconds after lift-off. Analysis of photographs taken immediately after ignition showed a puff of black smoke emerging from the aft section of the right solid rocket booster. The cold weather had made the rubber O-rings so brittle that they no longer sealed the joint properly. As Challenger cleared the tower, gas was already blowing by the rings, though the gap was then temporarily plugged by burned rubber and putty. Unfortunately for the crew, Challenger experienced the worst vibrations of any flight to date as it was buffeted by gusts of wind for almost 30 seconds. Building up speed towards the point of maximum aerodynamic pressure, the seals fractured once more with devastating effect. At 58.7 seconds, a small flame like a blowtorch appeared in the side of the SRB, unnoticed by anyone on the ground or in the Shuttle. It began to burn through the skin of the main fuel tank as well as one of the struts which held the rockets to the tank. Less than 14 seconds later, the strut gave way, enabling the pointed nose of the SRB to swivel inwards and pierce the giant fuel tank. Liquid oxygen poured out through the gash in its side to ignite almost instantaneously in a huge ball of flame. There was no escape for the seven crew on board Challenger – the craft broke up under the sudden aerodynamic stress and fell into the calm blue ocean.

While NASA was cancelling all Shuttle launches until at least February 1988, the Soviet Union was making its long-awaited move to set up a permanent orbital base. In another demonstration of their new-found confidence, the Soviets made the most of American discomfiture by announcing in advance the launch of a two-man Soyuz towards their third generation space station called Mir (World). Soviet TV viewers were given the rare treat of watching a live broadcast as cosmonauts

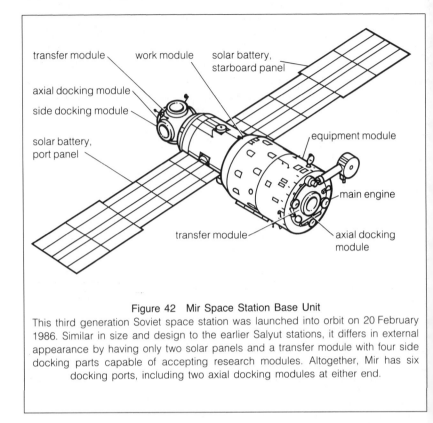

Figure 42   Mir Space Station Base Unit
This third generation Soviet space station was launched into orbit on 20 February 1986. Similar in size and design to the earlier Salyut stations, it differs in external appearance by having only two solar panels and a transfer module with four side docking parts capable of accepting research modules. Altogether, Mir has six docking ports, including two axial docking modules at either end.

Vladimir Solovyov and Leonid Kizim blasted off in their Soyuz T-15 craft on 13 March 1986 to dock with the new station.

It soon became clear that the three-week-old Mir was much more spacious and comfortable than the Salyut predecessors. This had been achieved by removing most of the specialist experimental equipment so that the 20-ton Mir could be divided into small crew cabins, each with its own desk, sleeping bag and armchair. Other comforts included a gymnasium equipped with rowing and cycling machines, and improved life support and ventilation systems. Life for the cosmonauts was also made easier by seven onboard computers intended to take much of the drudgery out of controlling the work of the station. The cylindrical station also differed from the Salyuts in having two central solar wings and six docking ports, four of them arranged at 90° angles around the front end. These are intended to take additional modules like those tested with Salyuts 6 and 7 which can be given a variety of scientific and experimental equipment.

The selection of the world endurance record holders as the first crew for Mir suggested another long mission, but after extensive tests of onboard systems, including use of tracking satellites to communicate with ground control, and unloading Progress 25 and 26, Kizim and Solovyov set their sights on the Salyut 7 station which was orbiting some 2200 miles ahead of Mir. On 5 May, Soyuz T-15 pulled away from Mir, and dropped into a faster, lower orbit. Docking with Salyut 7 was successfully completed the next day. Supplies and equipment ferried across from Mir were unloaded and installed by the busy crew before they settled down to a routine session of experiments. Their historic contributions to exploitation of near-Earth space had not ended, however, for on 28 and 31 May they stepped out of Salyut for the first live broadcasts of Soviet spacewalks. Such was Soviet confidence that advance warning was given. Viewers saw the men erect a 40 foot (12 m) high frame, fix instruments to it, then dismantle it in a rehearsal for future large-scale space assembly. More records were shattered as the men chalked up their eighth spacewalk and a total of 31 hours 40 minutes of EVA. They transferred back to Mir on 26 June, leaving Salyut 7 to be boosted into a higher orbit after ending its active life. Kizim and Solovyov finally mothballed Mir and returned to Earth on 16 July after 125 days in space. Between them they have now notched up 737 days in space, more than any other cosmonauts.

As further evidence that the balance of power swung back towards the Soviet Union in the middle of 1986, a modernized craft known as Soyuz TM was successfully tested at the end of May. After many years in the wilderness following its pioneering successes in the early 1960s, the Soviet Union is now pursuing a rational development programme which may leave it well ahead of its American rivals in decades to come. If they can succeed in building their own Shuttle and the long-predicted giant booster rocket, there is no reason why Soviet cosmonauts should not set foot on the moon, or even Mars before the end of the century. Meanwhile, the Americans must overcome their inward soul-searching, as they did in 1967 after the Apollo 1 disaster, and come out of their trial with a stronger, more rational space policy. There is no doubt they have the technology and the will to succeed, but progress must never be bought at any cost. Is it too much to ask that the superpowers may one day get together to share these costs and the associated risks, just as they did for a few days back in 1975?

Appendix

Soviet and American Manned Spaceflights

Soviet Manned Spaceflights

Launch date	Mission	Crew
12.4.61	Vostok 1	Gagarin
6.8.61	Vostok 2	Titov (G.)
11.8.62	Vostok 3	Nikolayev
12.8.62	Vostok 4	Popovich
14.6.63	Vostok 5	Bykovsky
16.6.63	Vostok 6	Tereshkova
12.10.64	Voskhod 1	Komarov, Feoktistov, Yegorov
18.3.65	Voskhod 2	Belyayev, Leonov
23.4.67	Soyuz 1	Komarov
26.10.68	Soyuz 3	Beregovoi
14.1.69	Soyuz 4	Shatalov
15.1.69	Soyuz 5	Volynov, Khrunov, Yeliseyev
11.10.69	Soyuz 6	Shonin, Kubasov
12.10.69	Soyuz 7	Gorbatko, Filipchenko, Volkov (V.)
13.10.69	Soyuz 8	Shatalov, Yeliseyev
1.6.70	Soyuz 9	Nikolayev, Sevestyanov
22.4.71	Soyuz 10	Shatalov, Yeliseyev, Rukavishnikov
6.6.71	Soyuz 11	Dobrovolsky, Volkov (V.), Patsayev
27.9.73	Soyuz 12	Lazarev, Makarov
18.12.73	Soyuz 13	Klimuk, Lebedev
3.7.74	Soyuz 14	Popovich, Artyukhin
26.8.74	Soyuz 15	Sarafanov, Demin
2.12.74	Soyuz 16	Filipchenko, Rukavishnikov
11.1.75	Soyuz 17	Gubarev, Grechko
5.4.75	Soyuz 18A	Lazarev, Makarov
24.5.75	Soyuz 18	Klimuk, Sevestyanov
15.7.75	Soyuz 19	Leonov, Kubasov

Launch date	Mission	Crew
6.7.76	Soyuz 21	Volynov, Zholobov
15.9.76	Soyuz 22	Bykovsky, Aksyonov
14.10.76	Soyuz 23	Zudov, Rozhdestvensky
7.2.77	Soyuz 24	Gorbatko, Glazkov
9.10.77	Soyuz 25	Kovalyonok, Ryumin
10.12.77	Soyuz 26	Romanenko, Grechko
10.1.78	Soyuz 27	Dzhanibekov, Makarov
2.3.78	Soyuz 28	Gubarev, Remek
15.6.78	Soyuz 29	Kovalyonok, Ivanchenkov
27.6.78	Soyuz 30	Klimuk, Hermaszewski
26.8.78	Soyuz 31	Bykovsky, Jähn
25.2.79	Soyuz 32	Lyakhov, Ryumin
10.4.79	Soyuz 33	Rukavishnikov, Ivanov
9.4.80	Soyuz 35	Popov, Ryumin
26.5.80	Soyuz 36	Kubasov, Farkas
5.6.80	Soyuz T-2	Malyshev, Aksyonov
23.7.80	Soyuz 37	Gorbatko, Tuan
18.9.80	Soyuz 38	Romanenko, Mendez
27.11.80	Soyuz T-3	Kizim, Makarov, Strekalov
12.3.81	Soyuz T-4	Kovalyonok, Savinykh
22.3.81	Soyuz 39	Dzhanibekhov, Gurragcha
15.5.81	Soyuz 40	Popov, Prunariu
13.5.82	Soyuz T-5	Berezovoi, Lebedev
24.6.82	Soyuz T-6	Dzhanibekov, Ivanchenkov, Chretien
19.8.82	Soyuz T-7	Popov, Serebrov, Savitskaya
20.4.83	Soyuz T-8	Titov (V.), Strekalov, Serebrov
27.6.83	Soyuz T-9	Lyakhov, Alexandrov
8.2.84	Soyuz T-10	Kizim, Solovyov, Atkov
3.4.84	Soyuz T-11	Malyshev, Strekalov, Sharma
17.7.84	Soyuz T-12	Dzhanibekov, Savitskaya, Volk
6.6.85	Soyuz T-13	Dzhanibekov, Savinykh
17.9.85	Soyuz T-14	Vasyutin, Grechko, Volkov (A.)
13.3.86	Soyuz T-15	Kizim, Solovyov
6.2.87	Soyuz TM-2	Yuri Romanenko, Alexander Laveikin

American Manned Spaceflights

Launch date	Mission	Crew
5.5.61	Mercury 3	Shepard
21.7.61	Mercury 4	Grissom
20.2.62	Mercury 6	Glenn
24.5.62	Mercury 7	Carpenter
3.10.62	Mercury 8	Schirra
15.5.63	Mercury 9	Cooper
23.3.65	Gemini 3	Grissom, Young
3.6.65	Gemini 4	McDivitt, White
21.8.65	Gemini 5	Cooper, Conrad

Launch date	Mission	Crew
4.12.65	Gemini 7	Borman, Lovell
15.12.65	Gemini 6	Schirra, Stafford
16.3.66	Gemini 8	Armstrong, Scott
3.6.66	Gemini 9	Stafford, Cernan
18.7.66	Gemini 10	Young, Collins
12.9.66	Gemini 11	Conrad, Gordon
11.11.66	Gemini 12	Lovell, Aldrin
11.10.68	Apollo 7	Schirra, Eisele, Cunningham
21.12.68	Apollo 8	Borman, Lovell, Anders
3.3.69	Apollo 9	McDivitt, Scott, Schweickart
18.5.69	Apollo 10	Stafford, Young, Cernan
16.7.69	Apollo 11	Armstrong, Aldrin, Collins
14.11.69	Apollo 12	Conrad, Gordon, Bean
11.4.70	Apollo 13	Lovell, Swigert, Haise
31.1.71	Apollo 14	Shepard, Roosa, Mitchell
26.7.71	Apollo 15	Scott, Worden, Irwin
16.4.72	Apollo 16	Young, Mattingly, Duke
7.12.72	Apollo 17	Cernan, Evans, Schmitt
25.5.73	Skylab 2	Conrad, Kerwin, Weitz
28.7.73	Skylab 3	Bean, Garriott, Lousma
16.11.73	Skylab 4	Carr, Gibson, Pogue
15.7.75	Apollo 18	Stafford, Brand, Slayton
12.4.81	Shuttle STS-1	Young, Crippen
12.11.81	Shuttle STS-2	Engle, Truly
22.3.82	Shuttle STS-3	Lousma, Fullerton
27.6.82	Shuttle STS-4	Mattingly, Hartsfield
11.11.82	Shuttle STS-5	Brand, Overmyer, Lenoir, Allen
4.4.83	Shuttle STS-6	Weitz, Bobko, Musgrave, Peterson
18.6.83	Shuttle STS-7	Crippen, Ride, Fabian, Hauck, Thagard
30.8.83	Shuttle STS-8	Truly, Brandenstein, Bluford, Gardner, Thornton
28.11.83	Shuttle STS-9	Young, Shaw, Garriott, Parker, Lichtenberg, Merbold
3.2.84	Shuttle 41-B	Brand, Gibson, McCandless, McNair, Stewart
6.4.84	Shuttle 41-C	Crippen, Scobee, Nelson (G.), Hart, van Hoften
30.8.84	Shuttle 41-D	Hartsfield, Coats, Resnik, Mullane, Hawley, Walker (C.)
5.10.84	Shuttle 41-G	Crippen, McBride, Sullivan, Ride, Leestma, Scully-Power, Garneau
8.11.84	Shuttle 51-A	Hauck, Gardner, Allen, Fisher (A.), Walker (D.)
24.1.85	Shuttle 51-C	Mattingly, Shriver, Onizuka, Buchli, Payton
12.4.85	Shuttle 51-D	Hoffman, Williams, Seddon, Bobko, Griggs, Garn, Walker (C.)
29.4.85	Shuttle 51-B	Overmyer, Gregory, Lind, Thornton, Wang, Thagard, van den Berg
17.6.85	Shuttle 51-G	Brandenstein, Creighton, Fabian, Lucid, Nagel, Baudry, Al-Saud

Launch date	Mission	Crew
29.7.85	Shuttle 51-F	Fullerton, Musgrave, England, Bartoe, Acton, Bridges, Henize
27.8.85	Shuttle 51-I	Engle, Covey, Lounge, Fisher (W.), van Hoften
3.10.85	Shuttle 51-J	Bobko, Grabe, Stewart, Hilmers, Pailes
30.10.85	Shuttle 61-A	Hartsfield, Nagel, Buchli, Bluford, Dunbar, Furrer, Messerchmid, Ockels
26.11.85	Shuttle 61-B	Shaw, O'Connor, Cleave, Spring, Ross, Vela, Walker (C.)
12.1.86	Shuttle 61-C	Gibson, Bolden, Hawley, Nelson (G.), Cenker, Chang-Diaz, Nelson (W.)
28.1.86	Shuttle 51-L	Scobee, Smith, Resnik, Onizuka, McNair, McAuliffe, Jarvis

Note At the beginning of 1984, NASA decided to use a new system of mission designation, so that STS-10 became mission 41-B. In this case, the first number (4) represented the fiscal year October 1983–October 1984, the second number (1) represented the Cape Canaveral launch centre, and the letter (B) gave the launches for the year in alphabetical order. When the new launch complex at Vandenberg in California eventually came into use, the second number used would be 2, to distinguish it from Cape Canaveral.

Sources

The author wishes to acknowledge the following sources of material quoted in the text.

Chapter 1 Into the Unknown

p. 11 *National Geographic*, vol. 127, no. 1, p. 144.

Chapter 2 Have You Come From Outer Space?

p. 12 *Yuri Gagarin*, Novosti Press Agency Publishing House, 1977.
p. 13 *In Gagarin's Trail*, Novosti Press Agency Publishing House, 1980.
p. 14 *Yuri Gagarin*.
p. 15 David Baker, *The History of Manned Spaceflight*, New Cavendish, 1981, p. 71.
A. Romanov, *Spacecraft Designer*, Novosti Press Agency Publishing House, 1976, p. 64.
p. 17 *Yuri Gagarin*.
p. 18 *Romanov, Spacecraft Designer*, pp. 71, 73–4.
p. 19 Baker, *History of Manned Spaceflight*, p. 100.
p. 20 Romanov, *Spacecraft Designer*, p. 76.
p. 22 Romanov, *Spacecraft Designer*, pp. 80–81.
p. 24 *Daily Mirror*, 17 June 1963.
p. 25 Romanov, *Spacecraft Designer*, p. 80

Chapter 3 From Freedom to Faith

p. 28 *National Geographic*, vol. 120, no. 3, p. 420.
p. 30 Ibid., p. 441.
p. 31 Ibid., p. 443.
p. 32 Ibid., p. 441.

p. 33 Virgil Grissom, *Gemini!*, Macmillan, 1968.
p. 35 Ibid.
p. 37 Tim Furniss, *Manned Spaceflight Log*, Jane's, 1983, p. 16.
p. 38 *National Geographic*, vol. 121, no. 6, p. 794.
p. 39 Ibid., pp. 800, 802, 803.
p. 41 Ibid., pp. 809, 811.
       Baker, *History of Manned Spaceflight*, p. 119.
p. 42 Ibid.
       Tom Wolfe, *The Right Stuff*, Bantam, 1980, p. 272.
p. 43 *National Geographic*, vol. 121, no. 6, pp. 817–18.
p. 44 Wolfe, *The Right Stuff*, p. 280.
       National Geographic, vol. 121, no. 6, p. 822.
p. 45 Ibid., pp. 824, 826.
p. 46 Ibid., p. 827.
p. 47 Wolfe, *The Right Stuff*, p. 157.
p. 51 Baker, *History of Manned Spaceflight*, p. 126.
p. 52 Ibid., p. 127.
p. 53 Wolfe, *The Right Stuff*, p. 313.
p. 55 *National Geographic*, vol. 119, no. 5, p. 730.
p. 57 Wolfe, *The Right Stuff*, pp. 321, 322.
p. 58 Furniss, *Manned Spaceflight Log*, p. 23.
p. 59 *Daily Mirror*, 16 May 1963.
p. 60 Baker, *History of Manned Spaceflight*, p. 162.
p. 61 Ibid., pp. 160–61.
p. 62 Mitchell Sharpe, *Living in Space*, Aldus, 1969, p. 33.
p. 63 Baker, *History of Manned Spaceflight*, p. 163.

Chapter 4 Sunrise, Sunset

p. 65 Romanov, *Spacecraft Designer*, pp. 88, 58–9.
p. 66 Ibid., p. 86.
p. 67 Ibid.
p. 68 Ibid., p. 90.
pp. 69–70 Ibid., pp. 89, 90.
p. 71 *Aviation Week and Space Technology*, 27 December 1965.
p. 72 Romanov, *Spacecraft Designer*, p. 92.
p. 73 Ibid., p. 91.
pp. 74–5 Ibid., p. 93.
p. 77 Baker, *History of Manned Spaceflight*, p. 196.
p. 78 Romanov, *Spacecraft Designer*, p. 99.
p. 79 Ibid.

Chapter 5 The Heavenly Twins

p. 83 *National Geographic*, vol. 125, no. 3, p. 363.
p. 86 Grissom, *Gemini!*
p. 87 Ibid.
p. 88 Furniss, *Manned Spaceflight Log*, p. 33.

p. 90 Baker, *History of Manned Spaceflight*, p. 206.
        National Geographic, vol. 128, no. 3, p. 443.
p. 91 Ibid.
        Daily Mirror, 4 June 1965.
p. 92 NASA Fact Sheet 291–B, 1965.
p. 93 *Daily Mirror*, 5 June 1965.
p. 94 *Daily Mirror*, 8 June 1965.
        National Geographic, vol. 128, no. 3, p. 447.
p. 95 Baker, *History of Manned Spaceflight*, p. 211.
p. 97 *Daily Mirror*, 25 August 1965.
p. 98 Baker, *History of Manned Spaceflight*, p. 214.
        Daily Mirror, 28 August 1965.
p. 100 Grissom, *Gemini!*
p. 103 *Daily Mirror*, 6 December 1965.
p. 104 Baker, *History of Manned Spaceflight*, p. 224.
p. 105 *Daily Sketch*, 13 December 1965.
        Baker, *History of Manned Spaceflight*, p. 225.
p. 107 *National Geographic*, vol. 129, no. 4, pp. 539, 543.
p. 108 NASA Fact Sheet 291–D, 1966.
p. 109 Ibid.
        National Geographic, vol. 129, no. 4, p. 547.
p. 110 *National Geographic*, vol. 129, no. 4, pp. 547, 549.
        NASA Fact Sheet 291–D, 1966.
p. 112 *Daily Mirror*, 17 March 1966.
p. 113 *Daily Mirror*, 18 March 1966.
p. 114 NASA Fact Sheet 291–E, 1966.
p. 116 Baker, *History of Manned Spaceflight*, p. 234.
p. 117 NASA Fact Sheet 291–F, 1966.
p. 119 Ibid.
p. 120 *Daily Mirror*, 7 June 1966.
p. 121 Michael Collins, *Carrying the Fire*, W. H. Allen, 1975.
p. 122 Ibid.
p. 123 Ibid.
p. 124 Ibid.
p. 125 NASA Fact Sheet 291–H, 1966.
p. 126 Ibid.
p. 127 Ibid.
p. 128 Furniss, *Manned Spaceflight Log*, p. 46.
p. 130 NASA Fact Sheet 291–I, 1966.
p. 131 *News of the World*, 13 November 1966.
        NASA Fact Sheet 291–I, 1966.
p. 132 *Daily Mirror*, 14 November 1966.
        NASA Fact Sheet 291–I, 1966.
p. 133 Ibid.

Chapter 6 Too Far, Too Fast

p. 138 Grissom, *Gemini!*
p. 139 *Weekend*, 14 June 1967.

p. 140 Ibid.
p. 141 H. Young, *Journey to Tranquillity*, Cape, 1969.
p. 142 Grissom, *Gemini!*
p. 143 John Mansfield, *Man on the Moon*, Constable, 1969.
p. 144 *Spaceflight*, August 1967, p. 294.
p. 145 Ibid.
p. 146 *Daily Mirror*, 25 April 1967.
p. 147 *Daily Telegraph*, July 1975.
p. 148 James Oberg, *Red Star in Orbit*, Harrap, 1981, p. 92.

Chapter 7   Fly Me to the Moon

p. 150 Young, *Journey to Tranquillity*.
p. 151 *Daily Mail*, 13 October 1968.
p. 152 *The Times*, 16 October 1968.
       Baker, *History of Manned Spaceflight*, p. 310.
p. 153 *Sunday Times*, 20 October 1968.
p. 154 *Daily Mail*, 23 October 1968.
p. 156 *Daily Mail*, 22 December 1968.
p. 158 *National Geographic*, vol. 135, no. 11, p. 604.
p. 159 Ibid.
p. 160 *National Geographic*, vol. 135, no. 11, pp. 610–13.
p. 161 Ibid., pp. 613, 616.
p. 162 Ibid., pp. 616, 619.
p. 163 Ibid., pp. 620, 622, 626.
p. 164 Ibid., pp. 627, 629.
p. 165 Ibid., p. 630.
p. 166 *Daily Mirror*, 28 December 1968.
p. 170 *The Times*, 6 March 1969.
p. 172 *The Times*, 7 March 1969.
p. 173 *The Times*, 8 March 1969.
p. 174 NASA Mission Report MR–3, 1969.
p. 175 *The Times*, 19 May 1969.
p. 176 Baker, *History of Manned Spaceflight*, p. 333.
p. 177 NASA Mission Report MR–4, 1969.
       The Times, 22 May 1969.
p. 179 Ibid.
p. 180 Ibid.
       Furniss, *Manned Spaceflight Log*, p. 62.
p. 181 NASA Mission Report MR–4, 1969.
       Baker, *History of Manned Spaceflight*, p. 337.
p. 182 Ibid., p. 338.
p. 183 *The Times*, 27 May 1969.

Chapter 8   One Giant Leap

p. 185 Collins, *Carrying the Fire*.
p. 186 *The Times*, 7 August 1969.
p. 187 Collins, *Carrying the Fire*.

p. 188   Baker, *History of Manned Spaceflight*, p. 343.
p. 189   Ibid., p. 344.
p. 190   Ibid.
p. 191   *Sunday Times*, 20 July 1969.
p. 192   Baker, *History of Manned Spaceflight*, p. 346.
         The Times, 7 August 1969.
p. 193   Baker, *History of Manned Spaceflight*, p. 346.
p. 194   Ibid., p. 347.
p. 195   *National Geographic*, vol. 136, no. 6, pp. 753, 762.
p. 196   Ibid., p. 738.
p. 197   Ibid., pp. 738–9.
p. 198   Mansfield, *Man on the Moon*.
         Young, *Journey to Tranquillity*.
         National Geographic, vol. 136, no. 6, p. 739.
p. 199   Ibid., pp. 739, 746.
p. 200   Ibid.
p. 201   *National Geographic*, vol. 136, no. 6, p. 747.
p. 202   Young, *Journey to Tranquillity*.
p. 203   *The Times*, 22 July 1969.
         Baker, *History of Manned Spaceflight*, p. 353.
p. 204   Ibid., p. 354.
p. 205   Ibid.
         Collins, *Carrying the Fire*.
p. 206   *The Times*, 23 July 1969.
p. 207   *The Times*, 7 August 1969.
p. 207–8   Collins, *Carrying the Fire*.
p. 209   Ibid.
p. 210   *National Geographic*, vol. 136, no. 6, p. 789.
p. 211   Collins, *Carrying the Fire*.
p. 212   Ibid.
p. 215   Baker, *History of Manned Spaceflight*, p. 365.
         The Times, 15 November 1969.
p. 216   Ibid.
         Baker, *History of Manned Spaceflight*, p. 367.
p. 217   *The Times*, 19 November 1969.
p. 218   *The Times*, 20 November 1969.
p. 219   NASA Mission Report MR–8, 1969.
         Newsweek, 1 December 1969.
p. 220   NASA Mission Report MR–8, 1969.
p. 221   Ibid.
         Newsweek, 1 December 1969.
p. 222   Ibid.
p. 223   Ibid.
p. 224   *The Times*, 25 November, 1969.

Chapter 9   We've Had a Problem

p. 227   *Sunday Times*, 12 April 1970.
p. 228   *The Times*, 13 April 1970.

p. 229 Baker, *History of Manned Spaceflight*, p. 376.
        NASA Mission Report MR–7, 1970.
p. 230 Baker, *History of Manned Spaceflight*, p. 377.
p. 231 NASA Mission Report MR–7, 1970.
        Baker, *History of Manned Spaceflight*, p. 377.
p. 232 Henry Cooper, *13: The Flight that Failed*, Angus & Robertson, 1973.
p. 233 Baker, *History of Manned Spaceflight*, p. 378.
        Cooper, *13*.
p. 235 Ibid.
p. 236 Ibid.
        Baker, *History of Manned Spaceflight*, p. 381.
p. 237 NASA Mission Report MR–7, 1970.
p. 238 *The Times*, 17 April 1970.
p. 239 Baker, *History of Manned Spaceflight*, p. 384.
p. 240 Ibid.
        NASA Mission Report MR–7, 1970.
        The Times, 18 April 1970.
p. 241 NASA Mission Report MR–7, 1970.
        Baker, *History of Manned Spaceflight*, p. 386.
        The Times, 18 April 1970.
p. 242 Furniss, *Manned Spaceflight Log*, p. 72.

Chapter 10   The Mountains of the Moon

p. 245 *Daily Telegraph*, 2 February 1971.
p. 246 Ibid.
        Daily Telegraph, 3 February 1971.
        Baker, *History of Manned Spaceflight*, p. 403.
p. 247 *National Geographic*, vol. 140, no. 1, pp. 136–7.
        Furniss, *Manned Spaceflight Log*, p. 75.
p. 248 *Daily Telegraph*, 6 February 1971.
p. 249 *National Geographic*, vol. 140, no. 1, p. 143.
        Baker, *History of Manned Spaceflight*, p. 405.
p. 250 *National Geographic*, vol. 140, no. 1, p. 143.
        Baker, *History of Manned Spaceflight*, p. 406.
p. 251 *Daily Telegraph*, 9 February 1971.
p. 252 Ibid.
        Daily Telegraph, 10 February 1971.
p. 253 *Daily Telegraph*, 27 July 1971.
p. 254 *National Geographic*, vol. 141, no. 2, p. 260.
p. 256 Baker, *History of Manned Spaceflight*, p. 415.
p. 257 Ibid.
        National Geographic, vol. 141, no. 2, p. 235.
        James Irwin, *To Rule the Night*, Hodder and Stoughton, 1974, p. 55.
p. 258 *National Geographic*, vol. 41, no. 2, p. 237.
        NASA Mission Report MR–10, 1971.
p. 259 *Sunday Times*, 1 August 1971.
        National Geographic, vol. 141, no. 2, p. 241.

p. 260   Baker, *History of Manned Spaceflight*, p. 418–19.
         Sunday Times, 1 August 1971.
p. 261   *National Geographic*, vol. 141, no. 2, pp. 245–6, 247, 249.
p. 262   Baker, *History of Manned Spaceflight*, p. 421.
         National Geographic, vol. 141, no. 2, p. 258.
p. 263   Ibid.
         Daily Telegraph, 3 August 1971.
p. 264   Baker, *History of Manned Spaceflight*, p. 422.
         Daily Telegraph, 3 August 1971.
p. 265   *National Geographic*, vol. 141, no. 2, p. 260.
p. 266   Ibid., p. 265.
p. 269   Baker, *History of Manned Spaceflight*, pp. 432, 433.
p. 270   Ibid., p. 433.
         Furniss, *Manned Spaceflight Log*, p. 82.
p. 271   *Daily Telegraph*, 22 April 1972.
p. 272   Ibid.
         Sunday Times, 23 April 1972.
p. 273   NASA Fact Sheet, Apollo 16, 1972.
         Daily Telegraph, 24 April 1972.
p. 274   Ibid.
p. 275   NASA Fact Sheet, Apollo 16, 1972.
         Daily Telegraph, 29 April 1972.
p. 277   Baker, *History of Manned Spaceflight*, p. 438.
p. 278   Ibid., p. 439.
p. 279   Ibid., p. 440.
         Daily Telegraph, 11 December 1972.
         Daily Telegraph, 12 December 1972.
p. 281   NASA Fact Sheet, Apollo 17, 1972.
         Daily Telegraph, 12 December 1972.
         Baker, *History of Manned Spaceflight*, p. 442.
p. 283   *Daily Telegraph*, 13 December 1972.
         Baker, *History of Manned Spaceflight*, pp. 443, 444.
p. 284   *Daily Telegraph*, 14 December 1972.
p. 285   Baker, *History of Manned Spaceflight*, pp. 446, 447.
         Daily Telegraph, 15 December 1972.
p. 286   Baker, *History of Manned Spaceflight*, p. 448.
         NASA Fact Sheet, Apollo 17, 1972.
p. 287   *Daily Telegraph*, 20 December 1972.

Chapter 11 Plunge into Despair

p. 290   Reginald Turnill, *Spaceflight Directory*, Warne, 1978, p. 334.
         The Times, 19 November 1968.
p. 291   *The Times*, 17 January 1969.
p. 292   Ibid.
p. 295   *The Times*, 15 October 1969.
p. 296   Turnill, *Spaceflight Directory*, p. 298.
p. 299   *The Times*, 26 April 1971.
         Turnill, *Spaceflight Directory*, p. 300.

p. 301 *The Times*, 8 June 1971.
p. 302 *The Times*, 1 July 1971.

Chapter 12 Skylab Sets the Record

p. 309 *National Geographic*, vol. 146, no. 4, p. 451.
        Baker, *History of Manned Spaceflight*, p. 484.
p. 311 Baker, *History of Manned Spaceflight*, pp. 485, 486.
p. 312 *National Geographic*, vol. 146, no. 4, p. 452.
p. 313 Ibid., p. 453.
        Daily Telegraph, 21 June 1973.
p. 314 *National Geographic*, vol. 146, no. 4, p. 453.
p. 315 Ibid., p. 460.
        NASA Mission Report MR–14, 1973.
p. 317 *National Geographic*, vol. 146, no. 4, p. 462.
p. 318 Ibid., p. 463.
p. 319 Turnill, *Spaceflight Directory*, p. 216.
        Daily Telegraph, 22 December 1973.
        National Geographic, vol. 146, no. 4, p. 468.
p. 320 NASA Fact Sheet, Skylab, 1974.
        National Geographic, vol. 146, no. 4, p. 464.
p. 321 NASA Fact Sheet, Skylab, 1974.
        NASA Mission Report MR–15, 1974.
p. 322 *National Geographic*, vol. 146, no. 4, p. 468.
        NASA Mission Report MR–15, 1974.

Chapter 13 The $500,000 Handshake

p. 326 Oberg, *Red Star in Orbit*, p. 139.
p. 329 *Daily Telegraph*, 16 July 1975.
p. 330 Furniss, *Manned Spaceflight Log*, p. 101.
        Orberg, *Red Star in Orbit*, p. 141.
p. 331 *Daily Telegraph*, 17 July 1975.
p. 332 Turnill, *Spaceflight Directory*, pp. 71, 72.
p. 333 Ibid., pp. 73, 74, 76.
p. 335 Baker, *History of Manned Spaceflight*, p. 519.
        Daily Telegraph, 20 July 1975.
p. 336 Turnill, *Spaceflight Directory*, p. 79.
p. 337 *Daily Telegraph*, 22 July 1975.
        Turnill, *Spaceflight Directory*, p. 82.
p. 338 Ibid., p. 84.

Chapter 14 The Longest Journeys

p. 343 Kenneth Gatland, *Manned Spacecraft*, Blandford, 1976, p. 241.
p. 344 *Aviation Week and Space Technology*, 27 January 1975.
        Turnill, *Spaceflight Directory*, p. 307.
p. 350 Ibid., p. 314.
        Daily Telegraph, 18 October 1976.

p. 351   *Aviation Week and Space Technology*, 26 February 1877.

p. 352   *Daily Telegraph*, 11 October 1977.

p. 354   *Aviation Week and Space Technology*, 19 December 1977.
Aviation Week and Space Technology, 2 January 1978.

pp. 355–6   *Daily Telegraph*, 12 January 1978.

p. 358   US Senate Report, Soviet Space Programs, 1984, p. 616.

p. 359   *Daily Telegraph*, 20 March 1978.

p. 360   Soviet Space Programs, pp. 622, 620.

pp. 362–3   Ibid., p. 622.

p. 364   *Aviation Week and Space Technology*, 23 April 1979.

p. 365   Soviet Space Programs, p. 629.
Oberg, *Red Star in Orbit*, p. 197.

p. 367   Soviet Space Programs, p. 627.
Daily Telegraph, 21 August 1979.

p. 368   Soviet Space Programs, p. 631.
Furniss, *Manned Spaceflight Log*, p. 117.

p. 372   Ibid.

p. 375   Soviet Space Programs, p. 466.
Sunday Express, 16 May 1982.

p. 378   Novosti Press Agency *Yearbook*, 1983, p. 176.
Daily Telegraph, 23 August 1982.
The Guardian, 29 August 1982.

p. 379   *Soviet Weekly*, 10 September 1982.

p. 380   Ibid.

p. 382   *The Guardian*, 23 April 1983.

p. 384   *Aviation Week and Space Technology*, 10 October 1983.

p. 386   *Soviet Weekly*, 10 November 1983.

p. 387   *New Scientist*, 25 October 1984.

p. 389   Ibid.

p. 391   *The Times*, 11 August 1984.
Spaceflight, vol. 26, no. 12, p. 466.

p. 392   *Daily Telegraph*, 6 October 1984.
Spaceflight, vol. 27, nos. 7/8, p. 327.

p. 393   *Pravda*, 8 October 1985.

p. 394   Ibid.

p. 395   *Soviet Weekly*, 5 October 1985.

Chapter 15   The Space Glider

p. 399   *National Geographic*, vol. 160, no. 4, p. 481.

p. 400   Ibid., p. 492.

p. 401   Ibid., pp. 491, 498.

p. 402   *The Guardian*, 15 April 1981.

p. 403   Ibid.
Daily Telegraph, 13 November 1981.

p. 404   *Daily Telegraph*, 25 March 1982.

p. 405   NASA Mission Report MR–003, 1982.

p. 406   NASA Mission Report MR–004, 1982.
Daily Telegraph, 5 July 1982.

p. 407   *The Guardian*, 13 November 1982.
p. 408   *Daily Telegraph*, 22 January 1983.
p. 409   *The Times*, 9 April 1983.
p. 410   *The Observer*, 19 June 1983.
         Daily Telegraph, 15 August 1983.
p. 411   *The Guardian*, 31 August 1983.
p. 414   *The Guardian*, 30 November 1983.
         Daily Telegraph, 2 December 1983.
p. 416   *Daily Telegraph*, 8 February 1984.
         Aviation Week and Space Technology, 13 February 1984.
p. 419   *Aviation Week and Space Technology*, 16 April 1984.
p. 420   Ibid.
p. 421   *Daily Telegraph*, 13 April 1984.
         Spaceflight, vol. 27, no. 1, p. 38.
p. 422   *Daily Telegraph*, 29 June 1984.
p. 424   *The Guardian*, 13 November 1984.
         Aviation Week and Space Technology, 19 November 1984.
p. 425   Ibid.
         Spaceflight, vol. 27, no. 6, p. 265.
p. 427   *Spaceflight*, vol. 27, no. 11, p. 418.
p. 428   *Daily Telegraph*, 1 May 1985.
         Tim Furniss, *Space Shuttle Log*, Jane's, 1986, p. 70.
p. 431   *Aviation Week and Space Technology*, 9 September 1985.
p. 434   *Aviation Week and Space Technology*, 9 December 1985.

Chapter 16   The Shuttle is Grounded

p. 436   *The Guardian*, 18 January 1986, p. 102.
p. 439   *Spaceflight*, vol. 28, no. 3, p. 102.
p. 440   *The Guardian*, 4 April 1986.

Index